Astrotheology

Astrotheology

Science and Theology Meet Extraterrestrial Life

EDITED BY
Ted Peters

WITH
Martinez Hewlett, Joshua M. Moritz,
and Robert John Russell

FOREWORD BY Paul Davies

CASCADE *Books* · Eugene, Oregon

ASTROTHEOLOGY
Science and Theology Meet Extraterrestrial Life

Copyright © 2018 Wipf and Stock Publishers. All rights reserved. Except for brief quotations in critical publications or reviews, no part of this book may be reproduced in any manner without prior written permission from the publisher. Write: Permissions, Wipf and Stock Publishers, 199 W. 8th Ave., Suite 3, Eugene, OR 97401.

Cascade Books
An Imprint of Wipf and Stock Publishers
199 W. 8th Ave., Suite 3
Eugene, OR 97401

www.wipfandstock.com

PAPERBACK ISBN: 978-1-5326-0639-7
HARDCOVER ISBN: 978-1-5326-0641-0
EBOOK ISBN: 978-1-5326-0640-3

Cataloging-in-Publication data:

Names: Peters, Ted, editor. | Hewlett, Martinez, editor. | Moritz, Joshua M., editor. | Russell, Robert John, editor. | Davies, Paul, foreword.
Title: Astrotheology : science and theology meet extraterrestrial life / edited by Ted Peters, with Martinez Hewlett, Joshua M. Moritz, and Robert John Russell.
Description: Eugene, OR: Cascade Books, 2018. | Includes bibliographical references and index.
Identifiers: ISBN: 978-1-5326-0639-7 (paperback). | ISBN: 978-1-5326-0641-0 (hardcover). | ISBN: 978-1-5326-0640-3 (ebook).
Subjects: LCSH: Religion and theology. | Space theology. | Cosmology. | Unidentified flying objects—Religious aspects. | Exobiology.
Classification: BL65 U54 A75 2018 (print). | BL65 (epub).

Manufactured in the U.S.A.   JUNE 27, 2018

Unless otherwise noted, Scripture quotations are taken from the New Revised Standard Version Bible, © 1989, 1993 National Council of the Churches of Christ in the United States of America. Used by permission. All rights reserved worldwide.

Scripture quotations also come from the New American Standard Bible® (NASB), Copyright © 1960, 1962, 1963, 1968, 1971, 1972, 1973, 1975, 1977, 1995 by the Lockman Foundation. Used by permission. www.Lockman.org/.

Scripture quotations also come from the Holy Bible, NEW INTERNATIONAL VERSION®, NIV® Copyright © 1973, 1978, 1984, 2011 by Biblica, Inc.® Used by permission. All rights reserved worldwide.

The editors dedicate this book to
Charles (1915–2015) and Frances Townes (1916–2018):
scientists, humanitarians, Christians, and friends.

The editors thank Dr. Richard Procunier for his generous financial support for the astrotheology research and publication project at the Francisco J. Ayala Center for Theology and the Natural Sciences. We also thank Melissa and Joshua Moritz for their dedicated diligence in preparing the manuscript and seeing it through to completion.

Contents

Foreword | Paul Davies | ix
Preface | Robert John Russell | xiii
Abbreviations | xvii
Contributors | xxi

PART 1: *The Tasks of the Astrotheologian*

1. Introducing Astrotheology | *Ted Peters* | 3
2. The Tasks of Astrotheology | *Ted Peters* | 27
3. The Road Map to Other Earths: Lessons Learned and Challenges Ahead | *José G. Funes, SJ* | 56
4. Discovering ETI: What Are the Philosophical and Theological Implications? | *Robert John Russell* | 74
5. The Copernican Revolution That Never Was | *Martinez Hewlett* | 90

PART 2: *The Search for Extraterrestrial Life: Will We Meet Them?*

6. Searches for ET Life in the Solar System—Exobiology, Astrobiology, and the Big Picture | *Margaret S. Race* | 109
7. Exoplanets and the Search for Life Beyond Earth | *Jennifer Wiseman* | 124
8. Yes, We Will Meet Them: The Drake Equation Tells Me So | *Heidi Manning* | 133
9. Yes, We'll Meet Them: A Scientific Argument for ETI | *Martinez Hewlett* | 146
10. God's Self-Communication in a Cosmos Bound for Life | *Oliver Putz* | 160

PART 3: *What Will Happen When We Meet Them?*

11. Extraterrestrial Life and Terrestrial Religion: A Crisis? | *Ted Peters* | 183
12. Jewish Theology Meets the Alien | *Norbert M. Samuelson* | 208
13. Islamic Theology Meets ETI | *Muzaffar Iqbal* | 216
14. Toward a Constructive Naturalistic Cosmotheology | *Steven J. Dick* | 228
15. "ET, Call Church!": Astrosemiotics and Shared Spirituality | *Mark Graves* | 245

PART 4: *Jesus Christ on Earth and Elsewhere*

16. One Incarnation or Many? | *Ted Peters* | 271
17. Many Incarnations or One? | *Robert John Russell* 303
18. Multiple Incarnations of the One Christ | *Peter M. J. Hess* 317
19. One *Imago Dei* and the Incarnation of the Eschatological Adam | *Joshua M. Moritz* 330
20. Extraterrestrial Salvation and the ETI Myth | *Ted Peters* 347

PART 5: *Astroethics and Space Policy*

21. Astroethics and the Terraforming of Mars | *Christopher McKay* | 381
22. Astroethics and Microbial Life in the Solar Ghetto | *Ted Peters* | 391
23. Astroethics and Intelligent Life in the Milky Way Metropolis | *Ted Peters* | 416
24. Concluding Scientific Prescript | *Ted Peters* | 447

Name/Subject Index | 449
Scripture Index | 477

Foreword

Whether or not we are alone in the universe is one of the oldest and biggest of the big questions of existence. For most of human history the puzzle was confined to theology and philosophy, but in recent decades science has made contributions too. The subject of astrobiology, loosely defined as the study of the origin, evolution, distribution, and future of life in the universe, is now a well-established discipline, and it constitutes the framework for this volume of essays.

Astrobiology has been given a considerable fillip in the last few years by the discovery of a plethora of extra-solar planets, and from the general expectation that some fair fraction of them will be "earthlike," offering potential abodes for life. The Milky Way alone may contain billions of earthlike planets. At the more speculative end of the astrobiology spectrum lies SETI, the Search for Extraterrestrial Intelligence, in which astronomers sweep the skies with radio telescopes in the hope of picking up some sort of message or signal from an alien civilization. The possibility of extraterrestrial life in general, and sentient life in particular, are obviously of deep philosophical and theological significance.

Although I am myself not religious in any conventional sense, I have long been fascinated by the interplay of science and religion, for example, the importation of concepts from monotheism into the scientific worldview, and the impact of scientific discovery on our conception of humanity's place in the universe. Astrobiology provides a fruitful case study.

Let me start by attacking a commonly articulated assumption about life beyond Earth. Many commentators slide effortlessly from the statement that habitable planets are common to the conclusion that life is therefore common. Even distinguished scientists are wont to proclaim that the cosmos must surely be teeming with life. But this assertion is to fall into the elementary trap of conflating a necessary with a sufficient condition. While it may well be true that there is a lot of habitable real estate out there, a habitable planet becomes an *inhabited* planet only in the event that life gets going on it, that is, only if non-life is transformed into life (leaving aside for the moment the possibility of panspermia, in which life can spread across space). How likely is that?

Darwin once remarked that it is "mere rubbish" to speculate about the origin of life. "One might as well speculate about the origin of matter," he quipped. In spite of a century of scientific investigation, and a variety of experiments to create the chemical building blocks of life, scientists remain largely in the dark about how a mish-mash of chemicals turns into something as complex and as specific as a living organism. If one doesn't know the process that turned non-life into life it is pointless to try to estimate the odds of it happening. *You cannot work out the probability of an unknown process!* Statements about the probability of this or that planet hosting life are completely unscientific—little more than wishful thinking. Nobody knows the probability; nobody even knows *how to estimate* the probability.

In this state of ignorance we can but list the three broad categories of explanation for how life on Earth might have begun:

1. A supernatural event.

2. A chemical fluke of stupendous improbability.

3. The expected and common outcome of intrinsically bio-friendly laws of nature.

Four hundred years ago it was widely believed (at least in Christian Europe) that life on Earth was the result of an act of special creation by God. The belief in a miraculous genesis persists today among some people of faith. For these people the problem of extraterrestrial life is acute: if the universe is indeed teeming with life, it would also be teeming with miracles, which lends an air of frivolity to something that is surely momentous. It suggests a Cosmic Magician hopping from planet to planet, rearranging molecules here and there into biological organisms, and then withdrawing to leave them to evolve naturally. A conservative position on the "miracle" hypothesis is to hope that life is confined to Earth, thus affirming our special place in the great cosmic order. I shall argue that a miraculous origin should be rejected not just on scientific grounds but on theological grounds too.

To explain my reasoning, let me turn to the abovementioned explanation 2. When I was a student in the 1960s the prevailing scientific view was that life on Earth is the product of a chemical fluke so rare it would have happened only once in the observable universe. "Almost a miracle," was the way Francis Crick expressed it. In those days speculating about extraterrestrial life was a taboo; one might as well have professed an interest in looking for fairies. Jacques Monod, like Crick a Nobel prizewinning biologist, was explicit: "Man at last knows that he is alone in the unfeeling immensity of the universe, out of which he emerged by chance alone." Monod (also like Crick) was an active atheist, and his strong adherence to the "freak chemical event" explanation for life provided a stark and, for him, welcome contrast to the miracle hypothesis of traditional religion: "the ancient covenant is in pieces," he wrote.

The identification of a sterile universe (beyond Earth that is) with atheism and the lack of any cosmic purpose or direction carries with it the flip-side that the discovery of a fecund universe would provide ammunition for those arguing just the opposite. And indeed, today the pendulum has swung very far the other way regarding the prospects for extraterrestrial life. The conventional wisdom among scientists is that life is widespread in the universe—abovementioned explanation 3. It is important to note, however, that this new-found optimism is not based on any actual scientific evidence. No life whatsoever has been found beyond Earth, and the key scientific issues concerning the problem of life's origin haven't changed much since the 1960s. Yet opinion has turned through 180 degrees.

Implicit in the now fashionable belief in a fecund universe is that the transition from non-life to life is not down to chance alone (as Monod asserted) but is somehow "fast-tracked" by some form of chemical "self-organization" or via the operation of a pervasive "life principle" at work in complex chemical systems.

Christian de Duve, yet another Nobel biologist, expressed this belief dramatically: "life is a cosmic imperative!" he wrote. And it's undeniably true that the probability of even a primitive organism forming merely by the random shuffling of molecular building blocks is absurdly, inconceivably, small. So, are the "chemical dice" somehow loaded? Is there some form of innate tendency towards life in complex chemical systems? Are the laws of nature somehow rigged in favor of life? I am deliberately expressing the issue this starkly to point up the sweeping nature of the "cosmic imperative" assumption which tacitly runs through the subject of astrobiology. Well, at this time there is no evidence for a cosmic imperative or life principle, no known innate drive to complexification with "life" as its end state. Belief in the fecundity of the universe is simply an act of faith, a feeling that "life will out" wherever suitable conditions prevail.

In spite of the current lack of evidence for a cosmic imperative, support for it could come at any time. It is merely necessary for us to discover a second sample of life—"Life 2"—that is, a form of life derived from an independent genesis. If life has happened twice to our knowledge, it will have happened many times in the universe. The discovery of Life 2 could come in a variety of ways. We might find evidence in the spectra of light from the atmospheres of extra-solar planets. Or perhaps SETI will succeed by detecting traces of non-human technology. There is also a distinct possibility that we will find life on Mars, which is the most earthlike planet in our astronomical neighborhood. However, as I pointed out over twenty years ago, Mars and Earth have for billions of years been exchanging rocks blasted into space by asteroid and comet impacts, and it seems very likely that this has contaminated Mars with terrestrial microbes; Mars is therefore compromised. In my view, the best hope for discovering a second sample of life is to look right here on Earth. If, as astrobiologists today frequently assert, life starts readily in earthlike conditions, then it should have started many times over on our home planet. How do we know it hasn't? Has anybody looked? I had the thought that there might be Life 2 microbes right under our noses (or even in our noses!) back in 2001, and a few years later I organized a workshop to explore the idea. The challenge as we saw it is to find a terrestrial microbe whose biochemical innards are so different from ours that a second genesis is indicated. Just *one* "alien" microbe would suffice to establish the cosmic imperative.

What would be the impact on theology of such a discovery? On the one hand it would reinforce the scientific position that life is a thoroughly natural phenomenon, emerging from complex chemistry via a sequence of purely physical processes. On the other hand, it would elevate the significance of the living state of matter, implying that life is not (as Monod asserted) a local aberration, a freak sideshow in a basically sterile universe, but is rather a fundamental feature of nature, a truly cosmic phenomenon. To be sure, life on Earth in general, and humanity in particular, would not be "the pinnacle of creation," but nor would we be pointless, meaningless extras in the great cosmic drama. A universe that brings forth life, and perhaps mind, as an integral part of the natural outworking of intrinsically life-friendly laws is one in which we can truly feel at home.

—Paul Davies
Kepler's 545th Birthday
December 27, 2016

Preface

I grew up in Los Angeles in the 1950s and I loved to spend time at the Griffith Observatory. There at night you could see above the smog which occasionally smothered the view of the heavens from my home. Instead the glory of some six thousand stars and the flowing Milky Way were visible from the Observatory's roof. Before climbing to the roof, though, I'd spend hours working my way carefully through the amazing science and astronomy exhibits within the Observatory's hallowed halls. I'd stop first at the huge Foucault pendulum hanging in the entrance whose daily apparent rotation is clear proof of the real rotation of the earth. I'd slowly make my way into the western hall where, hanging on a somewhat obscure wall, was a photo taken from the telescope at Mount Palomar. It appeared to show a star field with tiny lights wafting across a dark, inky background.

But here's the secret, and a secret which always sent chills up my spine: each "star" was actually a galaxy, a whirlwind of billions of stars, and there were hundreds of such galaxies in that photo. It always made me think: how could the universe be so unimaginably vast? What kinds of life are out there across the seemingly endless swaths of space? Will we ever meet them and what would that be like for us? And behind all this, a dimly sensed perception and question: why is there this vast universe in the first place, why does it exist? Intuitively I knew that the best answer to this final question is beyond any that science can give—but it nevertheless builds so gracefully on what science does tell us.

That answer is "God." God is the Creator of our universe with its countless galaxies strewn in all directions. Later at night I'd peer through the 12" Zeiss refractor telescope at the evening's crop of sights—my favorites were Jupiter with its Galilean moons, Saturn with its rings, and the awesome Andromeda galaxy—before returning home with my mom. As I'd lay in my bed, half-asleep, the realization came over me time and again: without God nothing would exist—all these galaxies, their photo in the Griffith

xiii

Observatory, even me. And with this realization, stunned once again at the sheer mystery of existence, I could hardly sleep.

It's been many years since those early visits to the Griffith Observatory, and the questions it triggered have only intensified and increased in number. What is our role in the universe? What is the future of our universe? Are we alone in the universe? Whether there are many other forms of intelligent life in our universe or if we are alone, what does this imply about the meaning of life as such in the universe? What will the discovery of extraterrestrial intelligent life tell us about being human that we could not otherwise discover? And do the answers to these questions speak to the question of how we should live our lives in an ethical and faith-filled way? This book is dedicated to questions like these and more, questions which I believe we all ask in our own way. I hope the essays in this book provide a tentative and promising response to them, and to my own encounter with the immensity of space and the panorama of galaxies without number through those old, but cherished, photos in the Griffith Observatory hallways so long ago.

The scientific background of this book is, of course, standard physical cosmology and evolutionary biology. We now know that the universe is immense and old. According to the Big Bang theory the universe began expanding from a "singularity" of infinite temperatures and density some 13.82 billion years ago. At first its expansion was exponential in time during what's called "inflation," a period lasting from 10^{-43} to 10^{-33} seconds. Even in this incredibly short amount of time the size of the universe grew by a factor of 10^{35}. At about three hundred eighty thousand years the universe had cooled enough that protons and electrons could form hydrogen, and photons "decoupled" from them to move freely through space to form the Cosmic Background Radiation (CBR). Today the CBR has cooled to 2.7 degrees Kelvin and its structure preserves anisotropies that give us information about the very early universe. The first stars formed one hundred to two hundred million years after the Big Bang. After some five billion years, these stars underwent enormous explosions (novae) from which all the heavy elements and second generation stars like our Sun and its planets were formed. It was only then that carbon-based life could evolve on a planet like ours.

Life on Earth began some 3.8 billion years ago. Although scientists are still unclear about the transition from non-life to life, the history of life on Earth is fairly well known. By the end of the Precambrian Period about six hundred million years ago (MYA), single-celled creatures appeared in Earth's primordial oceans. Land plants and insects trace back to the Silurian Period (438–408 MYA). Reptiles and spiders can be found 360 MYA and early mammals from 200MYA. Dinosaurs populated the Triassic, Jurrasic, and Cretaceous Periods until a catastrophic event 65 MYA sealed their fate and allowed mammal species to diversify rapidly. While the earliest hominids date back over three million years, anatomically modern humans (*Homo sapiens sapiens*) go only some two hundred thousand years back (with "archaic" humans originating perhaps three hundred thousand years earlier). *Homo habilis* existed from 2.4 to 1.5 MYA. Fossils of *Homo erectus* place this species between 1.8 million and three hundred thousand years ago. The Neanderthals (*Homo sapiens neanderthalenis*) predate humans by about forty thousand years; they disappeared about twenty-five thousand years ago. According to many scientists these and other hominids possessed many, perhaps all, of the characteristics which we once thought were uniquely human, including language, ritual burying of the dead, toolmaking, and the use of fire. Following their migration out of Africa humans probably bred with Neanderthals; humans of non-African descent typically carry some 4 percent Neanderthal genes. We eventually became the last surviving hominid species, spreading from Africa to Europe and eastward throughout the globe.

But what about life elsewhere in the universe? Presumably even microbial extraterrestrial life would not have emerged before the birth of second generation stars since all but the elements hydrogen and helium were produced by the novae of first generation stars. If the same laws of physics apply throughout the universe then similar chemistries and in turn evolutionary histories will most likely occur wherever there are planets with the right physical conditions for life. While microbial life might be plentiful given these requirements, sentient and particularly self-conscious creatures might be extremely rare, as we will see discussed in some of the chapters here. But suppose creatures do evolve which are self-conscious, will they too produce culture in its many forms—religion, aesthetics, science, technology? Will these cultures be similar in form to ours or entirely different? Again, these knotty questions are pondered in this volume.

Addressing crucial questions like these, which lie at the frontiers of the interaction between theology and science, lies at the heart of the mission of the Francisco J. Ayala Center for Theology and the Natural Sciences (CTNS). Instead of outright conflict between these fields, as urged by militant atheists and fundamentalist Christians, and in place of a "two worlds" or "two languages" strategy that entirely separates them, we at CTNS promote the dialogue and the creative mutual interaction between theology and science. In doing so we carry on the legacy of Ian G. Barbour, pioneer in the field and a leading contributor to its many innovations since the mid-1950s. We are not creationists, Intelligent Designers nor narrow theological dogmatists. We treat theology as a progressive discipline similar to science, and we hope to grow in our theological knowledge as we engage in fruitful dialogue and interact with the natural sciences.

Over the past three decades, this dialogue and interaction has brought together Big Bang cosmology and the doctrine of creation, quantum mechanics and non-interventionist divine action, evolutionary biology and the problem of suffering in nature ("natural theodicy"), the mind-brain problem in light of the neurosciences / cognitive sciences and the question of top-down causality, the future of the physical universe and its challenge to Christian eschatology, to name a few key fields of inquiry. It is the goal of our current research in Berkeley to engage the sciences as we explore the possibilities for life in the universe and ask what we can learn about being human only as we think through what extraterrestrial intelligent life might be like. Hence the subject of this volume.

—Robert John Russell
Epiphany
January 6, 2018

Abbreviations

AHS	*Astrobiology, History, and Society*. Edited by Douglas A. Vakoch. Heidelberg: Springer, 2013
CAO	Robert John Russell, *Cosmology from Alpha to Omega: The Creative Mutual Interaction of Theology and Science*. Minneapolis: Fortress, 2008
CD	Karl Barth, *Church Dogmatics*. 4 vols. Edited by G. W. Bromiley and T. F. Torrance. Edinburgh: T. & T. Clark, 1936–1962
Drake	*The Drake Equation: Estimating the Prevalence of Extraterrestrial Life through the Ages*. Edited by Douglas A. Vakoch and Mathew F. Dowd. Cambridge: Cambridge University Press, 2015
Eerie	Paul Davies, *The Eerie Silence: Renewing Our Search for Alien Intelligence*. Boston: Houghton Mifflin, 2010
ELD	Michael J. Crowe, *Extraterrestrial Life Debate from Antiquity to 1915: A Source Book*. Notre Dame: University of Notre Dame Press, 2008
ELU	*Encountering Life in the Universe*. Edited by Chris Impey, Anna H. Spitz, and William R. Stoeger, SJ. Tucson, AZ: University of Arizona Press, 2013
EMB	*Evolutionary and Molecular Biology: Scientific Perspectives on Divine Action*. Edited by Robert John Russell, William R. Stoeger, SJ, and Francisco J. Ayala. Vatican City State and Berkeley, CA: Vatican Observatory and Center for Theology and the Natural Sciences, 1998

ER1	*Encyclopedia of Religion.* 1st ed. 16 vols. Edited by Mircea Eliade. New York: Macmillan, 1987
ER2	*Encyclopedia of Religion.* 2nd ed. 15 vols. Edited by Lindsay Jones. New York: Macmillan Gale, 2005
ESE	*The Ethics of Space Exploration.* Edited by James S. J. Schwartz and Tony Milligan. Heidelberg: Springer, 2016
ESTE	*Encyclopedia of Science, Technology, and Ethics.* 4 vols. Edited by Carl Mitcham. New York: Macmillan/Thomson Gale, 2005
Exp	*Exploring the Origin, Extent, and Future of Life: Philosophical, Ethical, and Theological Perspectives.* Edited by Constance M. Bertka. Cambridge: Cambridge University Press, 2009
GWF	Ted Peters, *God—The World's Future.* 3rd ed. Minneapolis: Fortress, 2015
Icarus	Jacques Arnould. *Icarus' Second Chance: The Basis and Perspectives of Space Ethics.* Vienna: Springer, 2011
Inst.	John Calvin, *Institutes of the Christian Religion* (1559). In *Library of Christian Classics* XX, XXI. Edited by John T. McNeill. Louisville: Westminster John Knox Press, 1960
LBE	*The Impact of Discovering Life Beyond Earth.* Edited by Steven J. Dick. Cambridge: Cambridge University Press, 2015
LOW	Steven J. Dick, *Life on Other Worlds: The 20th Century Extraterrestrial Life Debate.* Cambridge: Cambridge University Press, 1998
LS	Pope Francis, *Laudato Sí.* 2016 https://w2.vatican.va/content/dam/francesco/pdf/encyclicals/documents/papa-francesco_20150524_enciclica-laudato-si_en.pdf
LW	Martin Luther, *Luther's Works.* American Edition. vols. 1–30, edited by Jaroslav Pelikan. St. Louis, MO: Concordia Publishing, 1955–1967; vols. 31–55, edited by Helmut T. Lehmann. Philadelphia: Fortress, 1955–1986

ABBREVIATIONS xix

Met.	Aristotle. *Metaphysics*. Translated by W. D. Ross. http://classics.mit.edu/Aristotle/metaphysics.html
MW	*Many Worlds: The New Universe, Extraterrestrial Life, and the Theological Implications*. Edited by Steven J. Dick. Philadelphia: Templeton Foundation Press, 2000
NASArm	NASA. *Astrobiology Roadmap*. 2008 https://nai.nasa.gov/media/medialibrary/2013/09/AB_roadmap_2008.pdf
NASAstr	NASA. Astrobiology Strategy. https://nai.nasa.gov/roadmap/
OFH	Martin Rees. *Our Final Hour*. New York: Basic Books, 2003. iBooks: https://itun.es/us/X9EXw.l
OHRS	*The Oxford Handbook of Religion and Science*. Edited by Philip Clayton and Zachary Simpson. Oxford: Oxford University Press, 2006
Origin	Charles Darwin. *The Origin of Species by Means of Natural Selection*. 6th ed. London: Murray, 1872
Plato	*The Dialogues of Plato*. 2 vols. Translated by B. Jowett. New York: Random House, 1892
PPT	*Physics, Philosophy, and Theology: A Common Quest for Understanding*. Edited by Robert John Russell, William R. Stoeger, SJ, and George V. Coyne, SJ. Vatican City State: Vatican Observatory, 1988
RCRS	*The Routledge Companion to Religion and Science*. Edited by James W. Haag, Gregory R. Peterson, and Michael L. Spezio. London: Routledge, 2012
RPP	*Religion Past and Present*. 14 vols. English translation of *Religion in Geschichte und Gegenwart*. Edited by Hans Dieter Betz, Don S. Browning, Bernd Janowski, and Eberhard Jüngel. Leiden: Brill, 2007–2014
SETI	*When SETI Succeeds: The Impact of High-Information Contact*. Edited by Allen Tough. Bellevue, WA: Foundation for the Future, 2000

SRSEI	David Wilkinson, *Science, Religion, and the Search for Extraterrestrial Intelligence*. Oxford: Oxford University Press, 2013
ST	Thomas Aquinas, *Summa Theologica*. In Christian Classics Ethereal Library. http://www.ccel.org/ccel/aquinas/summa.toc.html
ST	Paul Tillich, *Systematic Theology*. 3 vols. Chicago: University of Chicago Press, 1951–1963
ST	Wolfhart Pannenberg, *Systematic Theology*. 3 vols. Translated by Geoffrey W. Bromiley. Grand Rapids: Eerdmans, 1991–1998
TFC	*Touching the Face of the Cosmos: On the Intersection of Space Travel and Religion*. Edited by Paul Levinson and Michael Waltemathe. New York: Fordham University Press, 2016
TI	Karl Rahner, *Theological Investigations*. 22 vols. London: Darton, Longman & Todd, 1961–1976; New York: Seabury, 1974–1976; New York: Crossroad, 1976–1988
TI,ST	Karl Rahner, *Schriften zur Theologie*. 16 vols. Einsiedeln: Benziger, 1954–1984
TIE	Robert John Russell, *Time in Eternity*. Notre Dame: University of Notre Dame Press, 2012
VU	Thomas F. O'Meara. *Vast Universe: Extraterrestrials and Christian Revelation*. Collegeville, MN: Liturgical, 2012
WU	Neil deGrasse Tyson, J. Richard Gott, and Michael A. Strauss. *Welcome to the Universe: An Astrophysical Tour*. Princeton: Princeton University Press, 2016

Contributors

Steven J. Dick is the former NASA Chief Historian and served as the 2014 Baruch S. Blumberg NASA/Library of Congress Chair in Astrobiology. His most recent of numerous books is an edited volume entitled *The Impact of Discovering Life Beyond Earth* (Cambridge University Press, 2015). Minor planet 6544 Stevendick is named in his honor.

José G. Funes, SJ, is a professor of astronomy in the Facultad de Filosofía y Humanidades at the Universidad Católica de Córdoba in Argentina. He is former Director of the Vatican Observatory in Rome. Funes co-edited with Chris Impey and Jonathan Lunine, *Frontiers of Astrobiology* (Cambridge, 2012). He and his fellow researchers are currently working on a project: "Other: An Interdisciplinary Laboratory of Ideas," that looks at the likelihood of life elsewhere in the universe in light of the Drake Equation.

Mark Graves has published forty technical and scholarly works in computer science, biology, psychology, and theology, including the books, *Mind, Brain, and the Elusive Soul* (Ashgate, 2008) and *Insight to Heal: Co-Creating Beauty amidst Human Suffering* (Cascade Books, 2013). He previously worked on software for scheduling astronomical satellites.

Peter M. J. Hess is a Roman Catholic theologian specializing in issues at the interface between science and religion, particularly in evolution, sustainability, and climate change. He is co-author of *Catholicism and Science* (Greenwood, 2008) and editor of two books. He is a fellow of the International Society for Science and Religion, the American Scientific Affiliation, and the International Big History Association.

Martinez Hewlett is professor emeritus of molecular and cellular biology at the University of Arizona, Tucson, and adjunct professor at the Dominican

School of Philosophy and Theology, Graduate Theological Union, Berkeley, California. He has co-authored one of the most widely used textbooks in his field, *Basic Virology* (Blackwell, 3rd ed., 2008). Along with Ted Peters, he has co-authored, *Evolution: From Creation to New Creation* (Abingdon, 2003).

Muzaffar Iqbal is the founder-president of Center for Islamic Sciences (www.cis-ca.org), Canada, (previously, Center for Islam and Science); editor of *Islamic Sciences*, a semi-annual journal of Islamic perspectives on science and civilization, and General Editor of the seven-volume *Integrated Encyclopedia of the Qur'an*, the first English-language reference work on the Qur'an based on fourteen centuries of Muslim reflection and scholarship.

Heidi Manning is a professor of physics at Concordia College in Moorhead, Minnesota. With NASA she has worked instrument development for, and data analysis from, the Cassini Mission (Saturn) and the Curiosity Rover on the Mars Science Laboratory (Mars). For seven years, she taught an Astrobiology class in the honors program at Concordia College.

Christopher McKay is a planetary scientist at NASA Ames Research Center, studying planetary atmospheres, astrobiology, and terraforming. McKay has done research on planetary atmospheres, particularly the atmospheres of Titan and Mars, and on the origin and evolution of life. He is a co-investigator on the Huygens probe, the Mars Phoenix lander, and the Mars Science Laboratory.

Joshua M. Moritz teaches philosophy at the University of San Francisco, theology at the Jesuit School of Theology at Santa Clara University, and theology and science at the Graduate Theological Union in Berkeley. He is Managing Editor of the academic journal *Theology and Science* and he has authored numerous books and articles, including *Science and Religion: Beyond Warfare and Toward Understanding* (Anselm Academic, 2016).

Ted Peters co-edits with Robert John Russell the journal *Theology and Science* at the Francisco J. Ayala Center for Theology and the Natural Sciences at the Graduate Theological Union in Berkeley, California. He is author of *God—the World's Future: Systematic Theology for a Postmodern Era* (Fortress, 3rd ed., 2015) and *God in Cosmic History* (Anselm Academic, 2017). Along with Martinez Hewlett, he has co-authored, *Evolution: From Creation to New Creation* (Abingdon, 2003).

CONTRIBUTORS xxiii

Oliver Putz is Senior Fellow at the Institute for Advanced Sustainability Studies (IASS), Potsdam, Germany, where he works on issues related to religion in the dialogue on climate change and the ecological crisis. He holds a PhD in theology from the Graduate Theological Union, Berkeley, USA, as well as a PhD in biology from the Freie Universität Berlin, Germany.

Margaret S. Race is an astrobiologist serving as Senior Scientist for Planetary Protection and Risk Communication at the SETI Institute, Mountain View, California. She is author or co-author of numerous papers dealing with the ethics of space exploration and planetary protection.

Robert John Russell is the Ian G. Barbour Professor of Theology and Science, the Graduate Theological Union (GTU), and Founder and Director of the Francisco J. Ayala Center for Theology and the Natural Sciences. He is the author of *Time in Eternity* (University of Notre Dame Press, 2012) and *Cosmology from Alpha to Omega: The Creative Mutual Interaction of Theology and Science* (Fortress Press, 2009). Along with Ted Peters, Russell is a founding co-editor of the scholarly journal *Theology and Science*.

Norbert M. Samuelson holds the Grossman Chair of Jewish Studies at Arizona State University in Tempe, Arizona, now emeritus. Author of two hundred articles, author of seven books, and editor of three volumes, Samuelson recently published *Jewish Faith and Modern Science* (Roman & Littlefield, 2009).

Jennifer Wiseman is a senior astrophysicist at NASA's Goddard Space Flight Center, where she previously served as Chief of the Laboratory for Exoplanets and Stellar Astrophysics. She also directs the DoSER (Dialogue on Science, Ethics, and Religion) program at AAAS in Washington, DC.

PART 1

The Tasks of the Astrotheologian

1

Introducing Astrotheology

TED PETERS

Science seeks to discern the laws and order of our universe; religion, to understand the universe's purpose and meaning, and how humankind fits into both.

—CHARLES H. TOWNES, NOBEL LAUREATE[1]

Man's true environment is the universe, and every special environment is qualified as a section of the universe.

—PAUL TILLICH[2]

Lie on beach sand on a clear sunny day and look upward at the sky. What do you see? You see distance. You see magnificence. You see the heavenly vault connoting infinity and majesty.

Then lay on your sleeping bag in the wilderness on a clear night. What do you see? You see distance. You see a dark sky lit up like a Christmas tree by the Milky Way. Your mind fills to the brim and overflows with awe. Infinity enters your soul.

1. Bonnie Azab Powell, "'Explore as Much as We Can,'" interview with Charles Townes, UCBerkeleyNews, June 17, 2005, https://sberkeley.edu/news/media/releases/2005/06/17_townes.shtml.

2. Tillich, *ST*, 2:62.

The human psyche is stirred by our experience with the sky above and with the greater sky behind it, cosmic space. Like an angel, cosmic space comes to us bearing a message from the God who transcends it. The Psalmist responds to the sky's message with humility and gratitude. "When I look at your heavens, the work of your fingers, the moon and the stars that you have established; what are human beings that you are mindful of them, mortals that you care for them?" (Ps 8:3–4).

The scientist looks at the same stars the Psalmist saw. The scientist wonders. The beauty and majesty of our universe stirs the soul. "The encounter is largely spiritual," writes astrophysicist Neil deGrasse Tyson, "and cannot be absorbed all at once; it requires persistent reflection on its meaning and on our relationship to it."[3]

Looking from Earth toward the sky is inspiring. So also is looking at Earth from the sky. Astronaut Eugene Cernan gazed at his home planet from the Moon. "When I was the last man to walk on the moon in December 1972, I stood in the blue darkness and looked in awe at the Earth from the lunar surface. What I saw was almost too beautiful to grasp. There was too much logic, too much purpose—it was just too beautiful to have happened by accident. It doesn't matter how you choose to worship God . . . He has to exist to have created what I was privileged to see."[4] The far reaches of space live in the human soul.

This is where the astrotheologian begins to reflect. We reflect on what we see; and we reflect on what the scientist tells us we are seeing. Just how will we go about the task of astrotheological reflection? In this chapter and the next, we will introduce the field of astrotheology; we will identify the four sources for a theology of nature; we will identify the four immediate tasks of the astrotheologian; and we will face squarely the opportunities and difficulties posed by the creative mutual interaction between science and theology.

If we can imagine a bridge over a chasm—perhaps the Golden Gate Bridge over the raging water pouring into San Francisco Bay from the Pacific Ocean—we will place science on one side and theology on the other. In this book, on the science side of the bridge we will give special place to astrobiology along with SETI, the search for extraterrestrial intelligent life, and METI, messaging extraterrestrial intelligence. On the theology side, we will give special place to astrotheology—that is, to theological topics that find the space sciences relevant for our understanding of God's relation to

---

3. Tyson, *Space Chronicles*, 64.
4. Interview with Eugene Cernan in White, *Overview Effect*, 26.

the world. When the traffic goes both directions, the result will be creative mutual interaction between science and theology.

Astrobiology

This book looks at both ends of the telescope: the near unfathomable distances of cosmic space plus the soul's excitement stirred by the human eye. It takes a scientist to explain what we are looking at. It takes a theologian to understand who is doing the looking.

On one end of the telescope we find outer space. Outer space begins at the point where airplanes can no longer fly, between fifty and a hundred kilometers up from Earth's surface. *Space*, or *outer space*, refers to the "region of the Universe located beyond the part of the Earth's atmosphere in which aircraft can maneuver. The expression "outer space" is used in space law without specific definition or delimitation.[5] Perhaps it is fitting that the use of the term *outer space* is without "delimitation," given that it seems almost endless when we look at it. Is outer space limited? Finite? Or, infinite? Until we can answer such questions with precision, perhaps the concept of outer space should remain without delimitation.

On the other end of the telescope we find amateur astronomers, school children, romantic stargazers, and research scientists. Let us start here with the scientists, especially cosmologists and astrobiologists. Physical cosmology is the scientific study of the universe as a whole, complete with history, evolution, and future. Physical cosmology deals with the whole of material reality. The scope and dynamics of our universe only grow in their ability to amaze us. The 13.82 billion year past, beginning with the Big Bang, leads to a hundred billion year future before entropy reduces everything to an equilibrium. Unless, of course, dark energy does unpredictable things. What we know is fragmentary compared to what is out there, and in here for that matter. "The universe of matter, from tables and chairs to stars and clusters of galaxies, may be but the minor portion of creation."[6] In short, deep time and deep space assault the human mind with wonder and excitement.

If cosmology is the mother domain, its child is astrobiology. Astrobiology deals with what is living within the cosmos. According to Chris Impey at the University of Arizona, astrobiology is "the study of the origin, nature, and evolution of life on Earth and beyond."[7] The astrobiological child names

---

5. French Space Agency CNES: International Council for the French Language, *Dictionaire de Spatiologie*, 101.

6. Glanz, "Cosmic Motion Revealed," 2156.

7. Impey, *Living Cosmos*, 4.

its cosmological parent biophilic, as capable of begetting life. "Like all cosmologies, it makes a claim about the large-scale nature of the universe, and its claim is that life is not only a possible implication, but also a basic property of the universe," comments one of the most informed historians of the extraterrestrial life debate, Steven Dick.[8]

Astrobiology replaces *exobiology*. The term *exobiology* appeared first in some 1959 letters of geneticist Joshua Lederberg to describe transferring the search for the origin of life on Earth to off-Earth sites. It was picked up by Cornell's Carl Sagan, who described *exobiology* as extending the 1952 Miller-Urey experiment on the origin of life to astronomy.[9] Because to date no physical evidence of life beyond Earth is in, Neil deGrasse Tyson, director of the Hayden Planetarium at the American Museum of Natural History in New York, quips, "exobiology is one of the few disciplines that attempt to function, at least for now, in the complete absence of firsthand data."[10]

By the mid 1990s, NASA began using the term, *astrobiology*, to refer to its work on life in the universe. NASA's *Astrobiology Roadmap* of 2003 and 2008 orients the field around three fundamental questions: (1) How does life begin and evolve? (2) Does life exist elsewhere in the universe? (3) What is the future of life on Earth and beyond?[11] According to Christopher McKay at NASA Ames Research Center, "Astrobiology has within it three broad questions that have deep philosophical as well as scientific import. These are the origin of life, the search for a second genesis of life, and the expansion of life beyond Earth."[12] Deep philosophical as well as scientific import, indeed!

In 2014 NASA revised this roadmap with a more detailed *Astrobiology Strategy* identifying six major research areas.[13]

- Identifying abiotic sources of organic compounds
- Synthesis and function of macromolecules in the origin of life

---

8. Dick, "Twentieth Century History," 142.
9. Mesler and Cleaves, *Brief History of Creation*, 184.
10. Tyson, *Space Chronicles*, 36.
11. *NASArm*.
12. McKay, "Astrobiology: The Search for Life," 45.
13. *NASAstr*. The China National Space Administration (CNSA) is launching an ambitious spray of space missions at a rate of four per year to, among other things, examine quantum mechanics at the Space Scale (QUESS); explore dark matter (DAMPE); launch the Hard X-ray Modulation Telescope (HXMT); send Change-4 to land on the other side of the Moon; and with the European Space Agency (ESA) to launch the Solar Wind Magnetosphere Ionospher Link Explorer (SMILE). Normile, "Red Star Rising," 342–45.

- Early life and increasing complexity
- Co-evolution of life and the physical environment
- Identifying, exploring, and characterizing environments for habitability and biosignatures
- Constructing habitable worlds

Looking for Life Beyond Earth

Is it reasonable for us to be looking for life beyond Earth? Yes. "There is growing scientific confidence that the discovery of extraterrestrial life in some form is nearly inevitable," say Margaret Race and Richard Randolph.[14] "Almost beyond doubt, life exists elsewhere," writes David Darling.[15] If the cosmos is biophilic, then the astrobiologist is earning an honest living.

Within the encompassing field of astrobiology, it is common to distinguish between unintelligent and intelligent life. The field of exobiology focuses on the discovery of microbial or biologically simple forms of life, non-intelligent life forms. At the risk of insulting Martian microbes, we will refer to them as ETNL, extraterrestrial non-intelligent life, or even stupid life. Actually, stupid life is already intelligent. It is simply less intelligent than *Homo sapiens*, who run civilization on technology. Perhaps it would be wiser to distinguish between simple life and complex or technologically accomplished life. Be that as it may, in what follows, we plan to use the acronym ETIL or simply ETI to refer to extraterrestrial intelligent life who might be able to communicate with us through technology.

Like galaxies, scientific disciplines come in clusters. Astrobiology is clustered with cosmology, as we have said. Also clustered here are astronomy, astrophysics, spectroscopy, evolutionary biology, bioinformatics, and many other fields. "Astrobiology concerns itself with life in the universe—its origins, evolution, and future. Astrobiology is a highly inter- and multi-disciplinary endeavor, which incorporates both the physical and biological sciences," says Grace Wolf-Chase, a University of Chicago astronomer at the Adler Planetarium.[16] Lucas Mix, a scholar in the Center for Theological Inquiry project on astrobiology at Princeton, emphasizes the multi-disciplinarity of the field. "Astrobiology is the scientific study of life in space. It happens when you put together what astronomy, physics, planetary science,

---

14. Race and Randolph, "Need for Operating Guidelines," 1583, http://www.seti.org/pdfs/m_race_guidelines.pdf.

15. Darling, *Life Everywhere*, xi.

16. Wolf-Chase, "Astronomy," 103.

geology, chemistry, biology, and a host of other disciplines have to say about life and try to make a single narrative."[17]

Of particular interest to us as we proceed will be radio astronomy and the research pursued at the SETI (Search for Extraterrestrial Intelligence) Institute. Optical SETI (OSETI) at Harvard uses an optical telescope; but we will give particular attention to the San Francisco Bay area astronomers using radio telescopes. Radio SETI's goal is to detect intelligent life outside Earth. Among the SETI approaches is the use of radio telescopes to listen for narrow-bandwidth radio signals from space. Such signals are not known to occur naturally, so a detection would provide evidence of extraterrestrial technology.[18] SETI researcher Seth Shostak registers the excitement of the SETI ambiance. "Proof of thinking beings beyond Earth would be one of the most profound discoveries ever."[19]

Excitement regarding meeting new space neighbors has been rising since the mid 1990s because of the discovery of exoplanets. These extrasolar planets orbit other stars within our Milky Way.[20] Astronomers are especially enthusiastic about the possibility of life on Gliese 832C, a planet merely sixteen light years from Earth. More exciting is Proxima Centauri b. Why? It's closer to Earth. "Proxima Centauri b" at 4.2 light-years away is an earthlike planet in the habitable zone—the Goldilocks zone where it's not too hot and not too cold; it's just right. To reside in a habitable zone, temperatures on the planet's surface need to allow liquid water. Placement and width of a habitable zone depends on the brightness of its host star; the dimmer the star the closer must be the planet's orbit. With these criteria combined with planet size in mind, as of this writing Kepler-452b[21] is the most Earth-like, while

17. Mix, *Life in Space*, 4.

18. See the SETI homepage: http://setiathome.berkeley.edu/sah_about.php. Passive SETI refers to listening for extraterrestrial signals coming to Earth. Active SETI or METI (Messaging Extraterrestrial Intelligence) refers to sending messages into space. Former SETI researcher Douglas Vakoch now spearheads METI. "Complementing this existing stress on Passive SETI with an additional commitment to Active SETI, in which humankind transmits messages to other civilizations, would have several advantages, including (1) addressing the reality that regardless of whether older civilizations should be transmitting, they may not be transmitting; (2) placing the burden of decoding and interpreting messages on advanced extraterrestrials, which may facilitate mutual comprehension; and (3) signaling a move toward an intergenerational model of science with a long-term vision for benefiting other civilizations as well as future generations of humans." Vakoch, "Asymmetry in Active SETI," 476.

19. Shostak, "Are We Alone?," 41. See Mitchell Waldrop, "SETI Is Dead," 442–44.

20. See: The Extrasolar Planets Encyclopedia, http://exoplanet.eu/. Near the end of 2013 the number of exoplanets identified approached one thousand, with many more candidates requiring further investigation. "Exoplanet Catalogue Nears 1000," 277.

21. Witze and Krzysztofiak, "Exoplanets," 288–89.

at least three of the seven planets in the Trapist-1 system are contenders.[22] Like pirates on a treasure hunt, planet hunters sense they are getting closer and closer to finding Earth's twin.[23]

Two methods for detecting such extrasolar planets are currently employed by researchers: looking for wobbles and looking for transits. First, radial-velocity surveys analyze the motion of a star induced by its orbiting partner—that is, by measuring a star's wobble astrophysics can speculate that it might be caused by the gravitational pull of an orbiting planet. The High Accuracy Radial Velocity Planet Searcher (HARPS) can detect wobbles of less than one meter per second.

Second, star watchers can engage in visual searches for planets that transit in front of their primary star. When locating a black dot (the shadow side of an orbiting planet) in front of a brightly lit star, telescope viewers can take a series of photos over a period of time to see if it moves in a regular pattern. If so, the black dot might be considered a transit—that is, a planet in orbit. Direct imaging is difficult, as one might imagine, because each star is bright whereas each planet only reflects the star's light. High contrast techniques are being developed. At the present time, these two methods can detect only large planets, nicknamed "Jupiters." If the technology improves, we may in the future find ourselves able to detect earth sized and biophilic objects as well. Once an exoplanet is located, its atmosphere must be analyzed spectrascopically to see whether it could support life. "The obvious biosignatures to look for are oxygen, ozone, and nitrous oxide, unique products of life on Earth which a distant civilization could detect in the spectrum of our atmosphere. Another candidate is dimethyl sulfide, which oceanic phytoplankton produce on Earth."[24] The lines are in the water. How long before we catch that extraterrestrial fish?

Perhaps some fish are already nibbling. In early December 2013 five exoplanets with water in the atmosphere were discovered. The U.S. House Science, Space, and Technology Committee held a hearing titled, "Astrobiology." Steven Dick, who held the Blumberg Chair in Astrobiology at the Library of Congress, told the congressmen: it's time to get ready.[25]

22. NASA, "Largest Batch." https://exoplanets.nasa.gov/trappist1/.

23. "Kepler-186f, a 1.11 [to] 0.14 Earth-radius planet that is the outermost of five planets, all roughly Earth-sized, that transit a 0.47 [to] 0.05 solar-radius star. The intensity and spectrum of the star's radiation place Kepler-186f in the stellar habitable zone, implying that if Kepler-186f has an Earth-like atmosphere and water at its surface, then some of this water is likely to be in liquid form." Quintana et al., "An Earth-Sized Planet," 277. See: Bhattiacharjee, "Almost-Earth," 249.

24. Bhattiacharjee, "A Distant Glimpse?," 932.

25 United States of America House Committee on Science, Space, and Technology,

In the meantime, a controversy has broken out over naming rights. Our dwarf planet Pluto was named in 1930 by a little girl, Venetia Burney, the granddaughter of an Oxford University librarian. With thousands of exoplanets readily findable on our computers now with catalog designations such as Kepler 62f, who will give them handles we can remember? Adam named the animals, and Aristotle classified them. Now, we need this naming skill once again. The International Astronomical Union (IAU) in Paris ordinarily takes charge of nomenclature; but, some entrepreneurs want to sell or auction naming rights. The profits could go to support honorable causes, such as more space exploration. Should one be able to buy the right to name an exoplanet?[26] Perhaps we should name life nurturing planets after our favorite nurses or name dead Jupiters after U.S. Congressmen.

Space Rocks

Like a rock star on stage, astrobiology is the field everyone wants to watch. And also like a rock star on stage, we want to do more than merely watch. We want to dance to its rhythms. In both ancient and modern times, the stars light up both the sky and the soul. "Astronomy compels the soul to look upwards and leads us from this world to another," wrote ancient Athenian philosopher, Plato.[27] One of today's scholars, Harvard astronomer Owen Gingerich, reiterates what Plato previously affirmed. "Cosmology is a voyage of the human spirit."[28]

The sheer scope of the material universe connotes infinity; and infinity connotes divinity. The impression scope makes on the soul includes the question: are we alone? Might God have provided other sentient creatures on other planets to share this magnificent universe with us? Carl Sagan wrote, "space exploration leads directly to religious and philosophical questions."[29] Francis Collins, veteran geneticist and Director of the U.S. National Institutes of Health, asks one of these questions rhetorically: "if God exists . . . why would it be beyond His abilities to interact with similar creatures on

---

"Full Committee Hearing—Astrobiology: Search for Biosignatures in our Solar System and Beyond," December 4, 2013, http://science.house.gov/hearing/full-committee-hearing-astrobiology-search-biosignatures-our-solar-system-and-beyond; and Spitzer, "Congress Ponders Life on Other Worlds," http://membercentral.aaas.org/blogs/capitol-connection/congress-ponders-life-other-worlds.

26. Witze, "Moon and Planet Names," 407.
27. Plato, *Republic* VII.529.
28. Gingerich, "Mankind's Place," 29.
29. Sagan, *Cosmic Connection*, 63.

a few other planets or, for that matter, a few million other planets?"[30] Our inner psyche dances to the rhythms of silent celestial music. Outer space is at work in the inner soul.

The presence of outer space in the inner soul warrants taking the step from astrobiology to astrotheology, from science to spirituality. Theological questions gurgle up from within astrobiology like bubbles in a champagne flute. "The discussion of topics teetering between religion and science has broadened with the rise of astrobiology. Scientific questions about the origin, distribution, and future of life in the universe touch on basic issues of human existence,"[31] observes the late Albert Harrison. Natural observations, scientific definitions, and spiritual connotations warrant a theological analysis and perhaps even a theological construction. That is what we hope to offer in the pages that follow.

Astrotheology

This is a book about astrotheology, the place of cosmic space in Christian theological reflection along with the reflection of Jewish and Muslim thinkers. The authors are a mixture of scientists, theologians, and hybrids—that is, individuals with training in both science and theology. We are not creationists. Nor do we subscribe to the Intelligent Design school of thought. We do not belong to the spiritual-but-not-religious camp (SBNR), even though we applaud contemporary notions of spirituality. We are not New Agers. Nor are we materialists seeking to ridicule allegedly outdated religion. What's left? Sometimes the theistic evolutionists invite us to their barbeques. They like us because we are lovers of God and appreciators of the beauty and order of God's creation. We are grateful for the sciences which aid us in knowing and treasuring the wondrous intricacies of the natural world. God has graced civilization by granting rewards for scientific sweat, by granting the prize of knowledge for running the research race. In our prayers we thank God for science.

Once again, this is a book about astrotheology, the place of cosmic realization within the expression of our faith. Now, you might say, "*astrotheology* is a curious word. It looks a lot like 'astrobiology.'" It should. Astrotheology is an interpretation of astrobiology; and, of course, it is much more. Here is the definition we will be working with: *Astrotheology is that branch of theology which provides a critical analysis of the contemporary space*

---

30. Collins, *Language of God*, 71.

31. Harrison, "Russian and American Cosmism," 39, http://www.tandfonline.com/doi/pdf/10.1080/14777622.2013.801719.

sciences combined with an explication of classic doctrines such as creation and Christology for the purpose of constructing a comprehensive and meaningful understanding of our human situation within an astonishingly immense cosmos. Now, this may seem like a comprehensive definition. It includes, among other things, the origin and future of the universe as scientists picture it. Scholars at CTNS have in recent decades studied the theological implications of Big Bang cosmogony along with the expanding universe and published their research in volumes such as *Cosmos as Creation* and a series of books on divine action in the natural world with the Vatican Observatory.[32] In this volume we turn our focus to one concern within astrotheology: theological engagement with the possibility of extraterrestrial life, either microbial or technologically advanced life.

We must admit at the outset that our employment of the term *astrotheology* is not the only one in current usage. We have no patent on it. The term stimulates considerable excitement in the occult, among neo-pagans and New Age enthusiasts. One alternative use may be worth mentioning here. For many on the internet astrotheology is tied to astrology, especially ancient astrology. Allegedly, looking to the skies inspired our ancestors to worship the impressive phenomena of nature, especially the stars and the planets. Today's astrotheologians of this brand study ancient myths and petroglyphs to recover lost wisdom, wisdom allegedly suppressed by organized religions such as Christianity.[33] One contemporary astrotheologian stresses the esoteric heritage of this wisdom: "this knowledge is the basis and origin for all of our Myths, Legends, Fairy Tales, Nursery Rhymes, and Folk Lore. It is also the Pure Science developed by the very enlightened, wise, and ancient priesthood that give us the Holy Books of all religions."[34] Astrologer-theologians are mythicists, re-interpreting the pre-religious myths that led to the rise of the historical religions.[35] A certain anti-establishment tone accompanies this variant of astrotheology, a tone common to the new religious movements of the late nineteenth and twentieth centuries.

---

32. See: Peters, *Cosmos as Creation*; or Russell et al., eds., *Quantum Cosmology*.

33. Because, allegedly, pagan astrology preceded Christianity, and because Christianity incorporated the very astrotheology it rejected, Christianity is *de facto* a form of paganism. "The knowledge about astrotheology would reveal the Christians' own religion to be Pagan in virtually every significant aspect, constituting a remake of the ancient religion." Murdock, "Astrotheology and the Ancients," http://stellarhousepublishing.com/astrotheology.html#.UMjbpHecmIU. See also Tsarion, "Astro-Theology and Siderial Mythology," http://www.astrotheology.com/astrotheology1.html; and Lyons, "Bet Emet Ministries," http://jesusastrotheology.com/.

34. Bonacci, "Universal Truth."

35. See Acharya, http://www.youtube.com/watch?v=YKW9sbJ3v2w&feature=plcp.

Our concept of astrotheology is not anything like this. We in this book are heir to an earlier term *exotheology*, referring to the theological examination of issues pertaining to extraterrestrial intelligence.[36] Like exotheology, astrotheology incorporates the best scientific knowledge into a critical and constructive theology of nature. Therefore, our use of *astrotheology* should be sharply distinguished from the astrological or occult usage described above. Here are three differences. First, for us, knowledge of the skies is not esoteric. Rather, it is scientific. In principle, scientific knowledge is open and available to all. Second, our knowledge derives from astronomy and related sciences, which replaced astrology and rendered astrology a pre-modern form of pseudo-knowledge. NASA could not design a space ship and send it to Mars based upon astrological alignments. Modern science requires empirical data carefully calculated and assessed by scientific theories. Third, we work from within the circle of theological discourse, not outside. Like other scholars, we subject our theological foundations to critical analysis. We re-think our foundations. And in re-thinking our foundations we give deliberate attention to the role myth plays in both our cultural roots and our sprouting branches. Still, our task is to follow the growth of theological explication fertilized by honest and reliable science, not by myth.

Now that we have sharply set our use of the term *astrotheology* apart from astrology and the occult, I will turn to the relationship between theology and science. Is an alliance between theology and science reasonable? Yes. We work with an assumption we share with the Second Vatican Council (1962–1965): "If methodical investigation within every branch of learning is carried out in a genuinely scientific manner and in accord with moral norms, it never truly conflicts with faith. For earthly matters and the concerns of faith derive from the same God."[37] If astrobiology and related space sciences are performed in a "genuinely scientific manner," we can expect to learn much that will enhance our knowledge of the creator God.

On the basis of this assumption, the astrotheologian incorporates science into a *theology of nature*, wherein what we know by faith is complemented and expanded by what we learn from science. In addition, we may press further. In some instances, science and theology engage one another in *creative mutual interaction* (CMI). When CMI takes place, both science and theology are affected. The two fields bring distinctive resources to any interaction; but each field feels the impact of the engagement. With CMI, the bridge between science and theology supports traffic going in both

---

36. See "Exotheology," *Wikipedia*.
37. Pope Paul VI, *Gaudium et Spes*, 36, in Abbott, *The Documents of Vatican II*, 234.

directions. The method of the astrotheologian, as we stipulate below, includes both theology of nature and CMI.

Our own use of the term *astrotheology* relies primarily on the etymology, where *astro* directs our attention to the heavens and *theology* to the study of claims about the divine. The term *astrotheology* comes from the Greek: αστρο, *astro*, "constellation" plus Θέος, *theos*, "God"; and λόγος, *logos*, "knowledge." We prefix *theology* with *astro* to create a multi-disciplinary branch of theology that takes up the relationship between God and the creation, especially the creation of the universe over time. Our picture of God's work over time is informed by the natural sciences, particularly cosmology, astronomy, and evolutionary biology.

In addition, we interpret an intellectual tradition which includes the 1714 publication of the book, *Astro-Theology, or a Demonstration of the Being and Attributes of God from a Survey of the Heavens*. The author, William Derham (1657–1735), was an Anglican clergyman and chaplain to the future King George II. Derham's own version of the history of science is broken into three epochs, the Ptolemaic, the Copernican, and then his third: the post-Copernican system-of-the-universe era. Accordingly, said Derham, each star is itself a sun like ours with a family of orbiting planets, also like ours. These planets orbiting fixed stars, he declared "to be habitable worlds; places . . . accommodated for habitation, so stocked with proper inhabitants."[38] Derham could not prove this. So, he prayed for either a direct divine revelation or better scientific instruments to confirm or disconfirm his speculation. The task of astrotheology in Derham's era was to glorify God by stressing the immensity and magnificence of God's creation. When we turn to the twenty-first century, we cannot simply become disciples of Derham. Astrotheology's task has become a bit more modest by asking: just how should theologians assess and interpret the findings of astrobiology, especially the search for extraterrestrial life?

One feature of Derham's version of astrotheology is not carried forward by those of us contributing to the present book, namely, the argument for God's existence based upon design seen in nature. Because Derham could see design when surveying "the heavens," he deduced that God must be a cosmic designer. Our contemporary colleague, physicist and astrobiologist Paul Davies, similarly appeals to the argument for design: "If life is widespread in the Universe, it gives us more, not less, reason to believe in cosmic design."[39] Contemporary Durham University cosmologist and theologian David Wilkinson is critical of both William Derham and Paul Davies:

38. Cited by Crowe, *ELD*, 125.
39. Davies, "Biological Determinism," 15.

"The whole design argument has fundamental weaknesses . . . the possibility of alternative hypotheses . . . the design argument at most could lead only to a cosmic architect using existing material . . . Davies' God is more of a demiurge—a craftsman god rather than the supreme creator being."[40] Our approach in the chapters to come will be closer to that of Wilkinson, yet with a bow of appreciation for the likes of Derham, Davies, and other pioneers in astrotheology.

Now, admittedly, relatively few twentieth and twenty-first-century theologians have incorporated the universe into their description of creation, let alone the scientific details we now know about the universe. By omitting reference to the cosmos, theologians inadvertently limit the scope of God's creation to planet Earth. Very few religious scholars have speculated about extraterrestrial life, either ETNL or ETIL. The few who have allowed their imaginations to soar have, for the most part, positively embraced the prospect of sharing our creation with alien civilizations.

Most systematic theologians who have given thought to the matter of sharing our cosmos with extraterrestrial neighbors are ready to lay out the welcome mat. Georgetown University theologian John Haught contends that "SETI is a project that Christian faith should have no difficulty supporting. It would be humbling, but entirely healthy, if some day we found out that we are not alone."[41] The colorful Tübingen theologian, Hans Küng, says "we must allow for living beings, intelligent—although quite different—living beings, also on other stars of the immense universe."[42] Notre Dame scholar, Thomas O'Meara, speculates with anticipation. "There might be a number of modes of supernatural life with God, a variety of God's more intimate life shared with intelligent creatures in a billion galaxies."[43] In short, thinking about cosmic space and the possibility of extraterrestrial life has begun in some theological quarters.

Perhaps more significant than the systematic theologians for astrotheology has been the indefatigable scientific work of the *Specola Vaticana*, the Vatican Observatory. The idea for a Vatican Observatory was tendered in Rome already in 1582; and it was formally established by Pope Leo XIII in 1891. Today, the ever curious and diligent Jesuit astronomers scan the heavens looking for scientific jewels and listening for the music of the spheres.

---

40. Wilkinson, *SRSEI*, 117.
41. Haught, *Responses to 101 Questions*, 66.
42. Küng, *Eternal Life?*, 224; see also Küng, *Beginning of All Things*, 131–36.
43. O'Meara, "Christian Theology," 25.

"The universe sings God's praises because it is beautiful," write George Coyne and Alessandro Omizzolo; "it is beautiful because God made it."[44]

Our point here is that the doctrine of creation should be broadened in scope to include the entire universe, including its past and future. Further, our understanding of God's creation must incorporate what we can learn from the exciting data exploding these days within the natural sciences. Still further, we must attend to empirical and speculative research on the question: are we alone? We need to work from within a cosmic vision. Antje Jackelén, Archbishop of Sweden, lays before us the concept of *cosmovision*. "[This] raises the question of a cosmovision—a vision of what it means to inhabit a world vast in space and vast in time, and what the ultimate end of this cosmos might be."[45] The concept of creation now includes all things, all things in the cosmos. Recognizing how outer space makes its home in the inner soul leads to cosmovision.

Varieties of Theologies

Not every spiritual thinker or theologian will put out the welcome mat for extraterrestrial neighbors. The welcoming position we develop in this book may be disturbing to our detractors. Who might our detractors be? I will dub them: The Bible-Against-Aliens school of thought.

The Bible-Against-Aliens school of theology would not put out a welcome mat for extraterrestrial visitors, because its disciples deny the existence of ETI. According to the Christian Answers Network, for example, "the Bible does not teach that intelligent life exists elsewhere in our universe. Although our all-powerful God could have created such life had He desired, it seems rather obvious from Scripture that He did not."[46] Note the tentative or interpretive tone to tendering this judgment. It is significant that the Bible does not forbid belief in the existence of other worlds with creaturely inhabitants. So, the absence of teaching becomes here the teaching of absence.

Some creationists adhere to the Bible-Against-Aliens position. Answers in Genesis, a *biblical creationist* group, flatly pits the Bible against both evolution and ETI. "Extraterrestrial life is an evolutionary concept; it does not comport with the biblical teachings of the uniqueness of the earth and

---

44. Coyne and Omizzolo, *Wayfarers in the Cosmos*, 160.

45. Jackelén, "Cosmology and Theology," 138.

46. Christian Answers Network, http://www.christiananswers.net/q-eden/edn-co12.html.

the distinct spiritual position of human beings."[47] The same seems to apply to the *scientific creationists*, though less dogmatically.

> Creation scientists apply this law to its logical conclusion: Material life has not existed forever, and life only comes from other life. Therefore, the source of material life must be nonmaterial life. The Bible says nothing to indicate that God created life anywhere but Earth. But it does not explicitly deny it. Some have speculated that God's omnipotence and glory might be expressed by many planets with life. However, Scripture strongly implies that no *intelligent* life exists elsewhere . . . The second person of the holy trinity incarnated on Earth alone, took on human nature, died for the sins of those with whom He was the kinsman redeemer relationship, then ascended to the right hand of God the Father. He did not take on Vulcan or Klingon nature, and He will have only one bride—the church—for all eternity. It would therefore seem hard to reconcile intelligent life on other worlds with the doctrine of the incarnation. It would also seem odd for God to create microscopic life on other planets, but we should not be dogmatic on this.[48]

Over against the creationists, the editors of this volume assume that the Bible does not tell us about extraterrestrial life, either microbial or technologically advanced. The Bible neither affirms nor denies life beyond Earth. Hence the Bible-Against-Aliens school must bend its interpretation beyond credulity to render a judgment about what the Bible implies.

The Bible-Against-Alien school of theology is not the only awkward voice we must listen to. So also is the Bible-UFO school or the Ancient Astronaut school. Ancient Astronaut Theorists argue that visits from extraterrestrials account for the miraculous events reported in the Bible. In a later chapter we will look at the ancient astronaut theory, a view adopted by some preachers of the gospel. Perhaps the most respectable of the Bible-UFO theologians is Barry Downing, long time pastor of Northminster Presbyterian Church in Endwell, New York. Armed with advanced degrees from Princeton and Edinburgh, Downing reads his Bible as a pre-modern report of modern flying saucers. Alien spacecraft visited ancient Israel and employed their advanced technology to part the Red Sea and to guide the Hebrews through the wilderness with a pillar of fire. It was a flying saucer, not a cloud, which Jesus rode to heaven on Ascension Day. Still within the orthodox or neo-orthodox Christian camp, Downing believes UFOs are the

---

47. Answers in Genesis, "Taking Back Astronomy."
48. Sarfati, "Bible Leaves no Room," 5.

form which divine providence and redemption take.[49] Our approach here does not follow the Downing method; nor do we embrace the so-called ancient alien theory. We the editors and writers of this book part company from both the Bible-Against-Aliens and the Bible-UFO views.

In stark contrast to either the Bible-Against-Aliens or Bible-UFO views, we hereby establish what I will call the *Bible-Welcomes-Aliens* school of conjecture. The Bible tells us to welcome the sojourner, the stranger, the outsider, the foreigner, the other. St. Paul tells us to (Rom 12:13) "extend hospitality to strangers." Who could be more other or stranger than a space alien? Our disposition to hospitality should be gracious.

The majority of theologians who take up the question of extraterrestrial life fit into this school and anticipate a future engagement with an alien civilization with both caution and excitement. They put out the welcome mat for hypothetical space neighbors. Lutheran evangelical Mark Worthing, a systematic theologian in Adelaide, Australia, would celebrate the prospect of including aliens in our theological worldview. "The verification of the existence of extraterrestrial life would increase our understanding of the scope of God's providence."[50] Welcome to Earth! Our brand of astrotheology joins others in the Bible-Welcomes-Aliens perspective.

The agreements and disagreements among theologians will not mark our point of departure here in this volume. Rather, we will begin with our primal or raw experience with the sky, especially as we experience the sky in the form of outer space. Astrotheology reflects on this experience. "Space constitutes one of the domains on which the human imagination feeds and expresses itself," writes Jacques Arnould, space ethicist at the French Space Agency (CNES) in Paris. "Our fascination with the sky is such that the cultures of all countries and of all eras have not only placed their gods, their paradises and their origins here but also meaning, the destination of their desires and their dreams."[51] More. "Space is not only a place, but also what humans do there and what they do with it. Its limits are not primarily geographical or natural, but mostly those of our knowledge and our scientific and technical ignorance."[52] It is our human experience with the physical cosmos that prompts our reflection and theological speculation. Such theological speculation cannot deal directly with our raw experience of space, to be sure; so we must advance our reflection by engaging the natural sciences which mediate this experience with space. Space comes to us already

---

49. Downing, *Bible and Flying Saucers*.
50. Worthing, "Possibility of Extraterrestrial Intelligence," 71.
51. Arnould, *Icarus*, 36.
52. Ibid., 123–24.

interpreted by science; yet, space still has the power to provoke and inspire the human soul.

Creative Mutual Interaction between Astrobiology and Astrotheology

Before winding up this introductory chapter, let's look a bit more closely at the question: just what role does science play in astrotheology? We must select between options. One option would be to adopt the *two language* model, the view that science speaks one language while theology speaks a different one.[53] Ian Barbour calls this the "independence" model.[54] In the quote at the beginning of this chapter, Nobel physicist Charles Townes seems to rely on this two language model. "Science seeks to discern the laws and order of our universe; religion, to understand the universe's purpose and meaning, and how humankind fits into both." Science speaks the language of facts, whereas religion speaks the language of meaning. Healthy citizens of the cosmos speak both languages. This two language model is both reasonable and reputable, to be sure. However, the astrotheologian just may require a more interactive relationship to astrobiology and related sciences than this independence model affords.

A second and better option would follow the *theology of nature* model, which we introduced above. This is the first model we embrace in this book. According to this model, Barbour observes, "the doctrines of creation, providence, and human nature are affected by the findings of science."[55] The astrotheologian should absorb as much knowledge as possible gained from astrobiology and other space sciences; then, he or she should construct a theological picture of the creation with reliable scientific knowledge as a constituent part. To say it another way, science would provide the food on which the theologian dines and, thereby, grows. As this book proceeds, a theology of nature will dine on delicious science.

However, this is still not enough. In addition, we wish to ask: might the theologian have something to offer the scientist? Might the discipline of

---

53. Peters, "Science and Theology," 171–78. An interesting addition to the traditional list of models is the creative innovation offered by Lisa Stenmark, "disputational friendship." The disputational friendship model of the Science and Religion Discourse (SRD) "expands the relationship between religion and science beyond doctrines and discoveries, and acknowledges that both have a responsibility to the world. A disputational friendship sees religion and science as having a world-building function." Stenmark, *Religion, Science, and Democracy*, 195.

54. Barbour, *Religion and Science*, 84.

55. Ibid., 100–101.

theological reflection suggest a direction of research that could be pursued scientifically? Or, to put it another way: might the astrotheologian offer something to the astrobiologist while taking something in turn? Might the traffic on the bridge between theology and science go both ways? Might we ask for something more than the theology of nature can deliver? Might we ask for a third model?

Yes. Robert John Russell dubs this "more" *Creative Mutual Interaction*, or CMI for short. CMI urges "the development of research programs in theology and science that make novel moves from theology to science as well as standard moves from science to theology."[56] Although the scientist and the theologian may begin speaking separate languages, they engage in dialogue with each other and in time shared understanding grows. A theology of nature incorporating scientific knowledge also grows. Then, in addition, scientific understanding begins to grow, deepen, or expand. We might call this the *fusion of horizons*, a handy term coined by German philosopher Hans-Georg Gadamer.

According to the CMI model, the theologian may lift up an understanding of reality which would prompt a scientific research program. Russell looks forward to the day when the theologian says to the bench scientist: "look here; and here you'll find it." Such a theological prompt will lead to a progressive scientific research program. On that day, traffic will have successfully traveled both directions across the bridge between science and theology. In the meantime, while waiting for that day, we will be satisfied if traffic comes one direction only; we will be gratified if traffic from the space sciences enriches and expands our theology of nature.

Genuine Science versus Scientism

Now, before we offer a conclusion, we must still ask: is a creative mutual interaction between space science and astrotheology at all possible? If the scientific approach to knowledge about the universe is reductionist, atheistic, and materialistic, is there still room for interpreting the cosmos theologically? Yes, there is. The materialism we so frequently associate with science is not itself genuine science; rather, it is an ideology or a metaphysics superimposed upon science. This superimposed metaphysics turns the discoveries and theories of science, which are limited by the empirical data, into an overarching worldview, a totalizing metaphysical system. The word for this add-on to scientific knowledge has a name, *scientism*.

---

56. Russell, *CAO*, 132.

Science + materialist philosophy = scientism

Scientism, in turn, is often used as an ideology in the public arena, an agenda to reshape culture and society. We need to distinguish between healthy science and bellicose scientism. Perhaps the theologian can point this difference out with clarity; and perhaps the theologian can ask the scientist to be honest about what is and is not knowable.

It is crucial to point out that science and scientism can be distinguished. *Scientific American* columnist and skeptic Michael Shermer admits that *scientism* "is a secular religion in the sense of generating loyal commitments (a type of faith) to a method, a body of knowledge, and a hope for a better tomorrow."[57] Scientism is a secular religion that claims science as its scripture and the laboratory as its sanctuary. Transhumanist Simon Young is a missionary for scientism. "Science and technology increasingly offers us the chance to overcome the limitations of the human condition. Therefore, let us believe in science."[58] Our contrary position is this: the astrotheologian wants to pick the ripe fruit of healthy science, not the rotting husks of a fallen scientism.

To change the metaphor, atheism is like a cloak thrown over science. We can no longer see the science itself, only the atheistic cover. Oxford's Richard Dawkins is such an atheistic cloaker. Dawkins would not drive his science across the bridge to meet a theologian, because he denies the very validity of the theologian's enterprise. "The entire thrust of my position is that Christian theology is a non-subject. It is empty. Vacuous. Devoid of coherence or content."[59] In other words, science is the sole provider of what counts as knowledge for Dawkins. This amounts to a belief in science alone, *sola scientifica,* a faith that science will triumph over all non-scientific competitors. In short, this position would preclude dialogue let alone interaction.

Not every research scientist supports scientism, to be sure. The late Indian geochemist and Christian lay person Rustum Roy takes umbrage. "*Scientism must be destroyed.* By *scientism* I mean the absurdly reductionist belief that all truth can be learned and all reality described through science (never defined) and only through science. 'Only' is what distinguishes scientism from what all of us in mainstream science believe about science."[60] The knowledge gained from scientific research does not justify a secular religion; it does not justify an anti-religious religion of scientism. Devoutly

57. Shermer, *How We Believe*, 61.
58. Young, "Introduction," 16.
59. Dawkins, "From the Other Side," 38.
60. Roy, "Scientism and Technology as Religions," 836.

spiritual people of any religion can appreciate science; and they can practice science with all the wonder and delight available to anyone intent on learning about the natural world. There is no good reason to bar the laboratory door to a religiously faithful scientist.

For a creative mutual interaction to take place, both research scientists and research theologians need to ready themselves for dialogue. Dialogue is a two way conversation, a working through of issues that move toward a shared horizon of understanding. The scientist can provide growing knowledge about the natural world in which we live. The theologian can provide a critical analysis of scientific knowledge as well as a hypothetical yet holistic picture of all things oriented toward the God of grace. Theology, like science, can engage in research and search for broader understanding. Truth is the goal. "Theology is a sacred enterprise, to be enacted with awe and probity;" writes Jewish theologian Michael Fishbane, "for it is the ever-new attempt to speak of the reality of God and direct the self toward this truth."[61] Both the scientist and the theologian must place themselves in the chair of the student, ready to learn and grow and expand.

Creative mutual interaction (CMI) starts off with a theology of nature, in which theology appropriates the discoveries and theories of science, gives them a philosophical interpretation, and then reformulates theology in light of this interpretation. But it also goes beyond this. CMI asks whether theology, so reformulated, can shed light on new research directions in science or on choices scientists make between competing research programs. For example, a Christian understanding of special providence might turn to the physics of subatomic processes: quantum mechanics. The theologian could then adopt the Copenhagen interpretation that randomness in these processes reflects real indeterminism in nature; and then the theologian could explore the implications of this interpretation for a theology of God's acts in nature. So far such a move would fall under the rubric of a theology of nature. But CMI might take a step further. CMI might develop a robust theology of God's acts at the quantum level and then lead scientists to expect (predict!) that wherever intelligent life evolves in the universe we will find evidence of moral ambiguity such as we so clearly find in the human condition. The same quantum physical activity on Earth obtains on each exoplanet, therefore, we can predict that the same conditions for life—including intelligent life—obtain. For more on CMI, see the discussion in Russell's chapters below.

---

61. Fishbane, *Sacred Attunement*, 1.

Conclusion

The goal of astrotheology in this book is first that of constructing a *theology of nature*. In this case, the theologian will critically incorporate as much science as is relevant for drawing a picture of reality in which nature and history are oriented toward the one God of grace. In addition, secondly, the theologian will push back. The theologian will challenge the scientific community to pursue the best science, to purify empirical research by shedding excessive claims. Scientists should shed extra-scientific myths about the world we live in. Insofar as we pursue this task, we authors who are both scientists and theologians will engage in *creative mutual interaction*.

Because all things in heaven and earth are oriented toward the one God of grace, the theologian can say a great deal more about reality than what the scientist can say, to be sure. Yet, we observe that most theologians have two ears and one mouth. This suggests that twice as much listening as speaking is the proper ratio. If one of the theologian's ears is cocked toward what the Bible is saying, the other should be listening to what science is saying. When it is time for the astrotheologian to speak, what he or she says will combine scriptural faithfulness with scientific integrity.

Bibliography

Acharya, S. http://www.youtube.com/watch?v=YKW9sbJ3v2w&feature=plcp.
Answers in Genesis. "Taking Back Astronomy." http://www.answersingenesis.org/articles/tba/bible-and-modern-astronomy-2.
Arnould, Jacques. *Icarus' Second Chance: The Basis and Perspectives of Space Ethics*. New York and Vienna: Springer, 2011.
Barbour, Ian G. *Religion and Science*. New York: Harper, 1997.
Bhattacharjee, Yudhijit. "Almost-Earth Tantalizes Astronomers with Promise of Worlds to Come." *Science* 344:6181 (April 18, 2014) 249.
———. "A Distant Glimpse of Alien Life?" *Science* 333:6045 (August 19, 2011) 930–32.
Bonacci, Santos. "Universal Truth." http://universaltruthschool.com/.
Christian Answers Network. http://www.christiananswers.net/q-eden/edn-co12.html.
Collins, Francis S. *The Language of God*. New York: Free Press, 2006.
Coyne, George V., SJ, and Alessandro Omizzolo. *Wayfarers in the Cosmos: The Human Quest for Meaning*. New York: Crossroad, 2002.
Crowe, Michael J. *Extraterrestrial Life Debate from Antiquity to 1915: A Source Book*. Notre Dame, IN: University of Notre Dame Press, 2008.
Darling, David. *Life Everywhere: The Maverick Science of Astrobiology*. New York: Basic, 2002.
Davies, Paul C. W. "Biological Determinism, Information Theory, and the Origin of Life." In *MW* 15–17.
Dawkins, Richard. "From the Other Side: Richard Dawkins Responds." *Science and Theology News* 6.2 (2005) 38.

Dick, Steven J. "The Twentieth Century History of the Extraterrestrial Life Debate: Major Themes and Lessons Learned." In *AHS*, 133–74.
Downing, Barry H. *The Bible and Flying Saucers*. New York: Avon, 1968.
"Exoplanet Catalogue Nears 1000." *Nature* 502:7471 (Oct. 17, 2013) 277.
"Exotheology." Wikipedia. http://en.wikipedia.org/wiki/Exotheology.
The Extrasolar Planets Encyclopedia. http://exoplanet.eu/.
Fishbane, Michael. *Sacred Attunement: A Jewish Theology*. Chicago: University of Chicago Press, 2008.
French Space Agency CNES: International Council for the French Language. *Dictionaire de Spatiologie, Sciences et techniques spatiales*. Vol. 1, Terms and Definitions. Paris: Conseil international de la langue française, 2001.
Gaudium et Spes (Pastoral Constitution on the Church in the Modern World) 36. In *The Documents of Vatican II*, ed. by Walter M. Abbott, SJ. New York: America Press, 1966.
Gingerich, Owen. "Mankind's Place in the Universe." *Nature* 457:7225 (Jan. 1, 2009) 28–29.
Glanz, James. "Cosmic Motion Revealed." *Science* 282:5397 (18 December 1998) 2156–57.
Harrison, Albert. "Russian and American Cosmism: Religion, National Psyche, and Spaceflight." *Astropolitics: The International Journal of Space Politics and Policy* 11.1–2 (2013) 25–44. http://www.tandfonline.com/doi/pdf/10.1080/14777622.2013.801719.
Haught, John F. *Responses to 101 Questions on God and Evolution*. New York: Paulist, 2001.
Impey, Chris. *The Living Cosmos: Our Search for Life in the Universe*. New York: Random House, 2007.
Jackelén, Antje. "Cosmology and Theology." In *RCRS*, 135–44.
Küng, Hans. *The Beginning of All Things: Science and Religion*. Translated by John Bowden. Grand Rapids: Eerdmans, 2007.
———. *Eternal Life? Life after Death as a Medical, Philosophical, and Theological Problem*. Translated by Edward Quinn. 1984. Reprint, Eugene, OR: Wipf & Stock, 2003.
Lyons, Craig. "Bet Emet Ministries." http://jesusastrotheology.com/.
McKay, Christopher P. "Astrobiology: The Search for Life Beyond the Earth." In *MW*, 45–58.
Mesler, Bill, and H. James Cleaves II. *A Brief History of Creation: Science and the Search for the Origin of Life*. New York: Norton, 2016.
Mix, Lucas John. *Life in Space: Astrobiology for Everyone*. Cambridge: Harvard University Press, 2009.
Murdock, D.M. "Astrotheology and the Ancients." http://stellarhousepublishing.com/astrotheology.html#.UMjbpHecmIU.
NASA. *Astrobiology Roadmap*. https://nai.nasa.gov/media/medialibrary/2013/09/AB_roadmap_2008.pdf.
———. "Largest Batch of Earth Size, Habitable Zone Planets." (2017). https://exoplanets.nasa.gov/trappist1/.
———. Astrobiology Strategy. https://nai.nasa.gov/roadmap/.
Normile, Dennis. "Red Star Rising." *Science* 353:6297 (July 22, 2016) 342–45.

O'Meara, Thomas F., OP. "Christian Theology and Extraterrestrial Life." *Theological Studies* 60.1 (1999) 25.
Peters, Ted. "Science and Theology: Toward Consonance." In *Science and Theology: The New Consonance*, edited by Ted Peters, 171–78. Boulder, CO: Westview, 1998.
Peters, Ted, ed. *Cosmos as Creation: Theology and Science in Consonance*. Nashville: Abingdon, 1989.
Quintana, Elisa V., et al. "An Earth-Sized Planet in the Habitable Zone of a Cool Star." *Science* 344:6181 (April 18, 2014) 277–80.
Race, Margaret S., and Richard O. Randolph. "The Need for Operating Guidelines and a Decision Making Framework Applicable to the Discovery of Non-Intelligent Extraterrestrial Life." *Advances in Space Research* 30.6 (2002) 1583–91. http://www.seti.org/pdfs/m_race_guidelines.pdf.
Roy, Rustum. "Scientism and Technology as Religions." *Zygon* 40.4 (2005) 835–44.
Russell, Robert John. *Cosmology from Alpha to Omega: The Creative Mutual Interaction of Theology and Science*. Minneapolis: Fortress Press, 2008.
Russell, Robert John, et al., eds. *Quantum Cosmology and the Laws of Nature: Scientific Perspectives on Divine Action*. Vatican City State and Berkeley, CA: Vatican Observatory / CTNS, 1993.
Sagan, Carl. *The Cosmic Connection: An Extraterrestrial Perspective*. New York: Dell, 1973.
Sarfati, Jonathan D. "Bible Leaves No Room for Extraterrestrial Life." *Science and Theology News* 4:7 (March 2004) 5.
SETI. http://setiathome.berkeley.edu/sah_about.php.
Shermer, Michael. *How We Believe: The Search for God in an Age of Science*. New York: Freeman, 2000.
Shostak, Seth. "Are We Alone? Estimating the Prevalence of Extraterrestrial Intelligence." In *Civilizations Beyond Earth: Extraterrestrial Life and Society*, edited by Douglas A. Vakoch and Albert A. Harrison, 31–42. New York and Oxford: Bergbahn, 2011.
Spitzer, Chris. "Congress Ponders Life on Other Worlds." http://membercentral.aaas.org/blogs/capitol-connection/congress-ponders-life-other-worlds.
Stenmark, Lisa. *Religion, Science, and Democracy: A Disputational Friendship*. Lanham, MD: Lexington, 2013.
Tillich, Paul. Systematic Theology. 3 vols. Chicago: University of Chicago Press, 1951–1963.
Townes, Charles. Interview by Bonnie Azab Powell, June 17, 2005. https://berkeley.edu/news/media/releases/2005/06/17_townes.shtml.
Tsarion, Michael. "Astro-Theology and Siderial Mythology." http://www.astrotheology.com/astrotheology1.html.
Tyson, Neil deGrasse. *Space Chronicles: Facing the Ultimate Frontier*. New York: Norton, 2012.
United States of America House Committee on Science, Space, and Technology, "Full Committee Hearing—Astrobiology: Search for Biosignatures in our Solar System and Beyond; December 4, 2013; http://science.house.gov/hearing/full-committee-hearing-astrobiology-search-biosignatures-our-solar-system-and-beyond.
Vakoch, Douglas A. "Asymmetry in Active SETI: A case for transmissions from Earth." *Acta Astronautica* 68 (2011) 476–88.
Waldrop, Mitchell. "SETI Is Dead: Long Live SETI." *Nature* 475:7357 (July 27, 2011) 442–44.

White, Frank. *The Overview Effect*. Boston: Houghton Mifflin, 1987.
Wilkinson, David. *Science, Religion, and the Search for Extraterrestrial Intelligence*. Oxford: Oxford University Press, 2013.
Witze, Alexandra. "Moon and Planet Names Spark Battle." *Nature* 496:7446 (April 25, 2013) 407.
Witze, Alexandra, and Jasiek Krzysztofiak. "Exoplanets: The Next 20 Years." *Nature* 527:7578 (November 19, 2015) 288–289.
Wolf-Chase, Grace. "Astronomy: From Star Gazing to Astrobiology." In *RCRS* 103–112.
Worthing, Mark W. "The Possibility of Extraterrestrial Intelligence as Theological Thought Experiment." In *God, Life, Intelligence and the Universe*, edited by Terence J. Kelly and Hilary D. Regan, 61–84. Adelaide, Australia: Australian Theological Forum, 2002.
Young, Simon. "Introduction." In *Designer Evolution: A Transhumanist Manifesto*, edited by Simon Young, 15–26. Amherst, NY: Prometheus, 2006.

2

The Tasks of Astrotheology

TED PETERS

Science without religion is lame and religion without science is blind.

—ALBERT EINSTEIN[1]

To be human is to keep faith with the cosmic processes which made [us].

—VICTOR C. FERKISS[2]

In the previous chapter we introduced the undertaking of astrotheology to construct a theology of nature and, at some points, to engage in a creative mutual interaction (CMI) with the space sciences. We took out membership in the *Bible-Welcomes-Aliens* club. In this chapter we will round out methodological considerations, dealing with astrotheology's four sources and four tasks.

---

1. Einstein, *Later Years*, 26.
2. Ferkiss, *Future of Technological Civilization*, 293.

Astrotheology's Four Sources

Let's remind ourselves of what is frequently called the *Wesleyan Quadrilateral*. A word with *quadri* means we've got four things going here. The theologian thinks about matters of faith, and in so doing appeals to four sources: Scripture, history (tradition), reason, and experience. What the *Bible* says will be important in what follows. It always is when thinking theologically. In addition, secondly, we will look at the *history* of astrotheology or, better, the role of the *many worlds* debate within theology. Believe it or not, this concern with extraterrestrial life goes back to three centuries before Jesus; and it has been on the minds of theologians intermittently ever since. Thirdly, with the term *reason* here, we must confess that theologians like philosophical reasoning as much as toddlers like ice cream. After admitting this, we will give additional attention to reason as exercised in the natural sciences, especially those sciences that deal with space and the search for extraterrestrial life. Finally, with the term *experience* we will examine the modern secular experience broadly along with specific claims some people make of contact with extraterrestrials.

The Christian astrotheologian, like other constructive or systematic theologians, seeks to provide a comprehensive contemporary understanding of all of reality oriented toward the God of grace that (1) faithfully reflects on the symbolic articulations of the Bible; (2) traces critically the development of relevant spiritual and non-spiritual ideas in tradition and cultural history; (3) internalizes the canons of reason while analyzing the reasoning taking place in philosophy and in science; and (4) interprets human experience in circular relation to the first three sources. In principle, intellectual thinkers in traditions other than Christianity could engage in a parallel version of astrotheology, altering the list of sources referred to.

Should these four sources be ranked? Should the Bible be granted first position or first privilege? Postmodern theologian Nancey Murphy would say, yes. "First, postmodern conservative theology must maintain some special role for Scripture over against experience as authority for theology; second, it must provide for special acts of God; and third, it must provide for the possibility of making truth claims for Christianity."[3] We agree with Murphy that theology is in the business of making truth claims. We plan to work here with a dynamic interplay of the four sources, trying to avoid playing the Bible as a simple trump card. The Bible hosts the criterial revelation of the God of grace who promises redemption of all creation, to be sure; so it will be given pride of place as we interpret history, reason, and experience.

---

3. Murphy, *Anglo-American Postmodernity*, 118.

Reason and experience will be given special attention throughout this book, because of the creative mutual interaction we wish to foster between theology and science. Emeritus Princeton professor Wentzel van Huyssteen emphasizes that rational reflection upon experience is parallel if not consonant in both theology and science. In fact, experience cannot go uninterpreted; and the interpretation process leads through explanation to better understanding. "In both theology and science, experiential adequacy pivots . . . on the deployment of good reasons: an act of responsible judgment in which we, through believing, doing, and choosing the right thing for the right reasons, become rational persons."[4] Science is reason reflecting on experience (experience in the form of data); and theology reflects on science as reason along with Scripture and tradition.

Experience and reason are universal. Every human person experiences. The Christian interpretation of experience is historically specific. Christians rely on biblical revelation and a single historical tradition. "That which binds all men together . . . cannot be revelation but must be experience. Revelation is the form taken by particular historical creeds, experience is accessible to man as man,"[5] writes philosopher Karl Jaspers. In principle, universal human experience and the reason which tries to explain experience become sources for specifically Christian theology. Experience interpreted by reason in light of Scripture and tradition: that is what the astrotheologian is about.

Reason in the Form of Science

Science is packaged reason. Science is only one brand of reason, to be sure; but it is the preferred brand for the astrotheologian. We have asked the reader to imagine a bridge, perhaps the Golden Gate Bridge. On one side is theology. On the other side is science. If the traffic goes one direction, from science toward theology, the product is a *theology of nature*. To date, very little traffic of note has gone the opposite direction. But when it does, we'll see traffic going both directions. If the traffic goes both directions, the product is *Creative Mutual Interaction*, or CMI.

As we noted in the first chapter, science provides data for theological reflection. "Where theology is reconstructed in light of science," says Robert John Russell, we have a theology of nature.[6] But, when theology proposes assumptions or directions for scientific research, we have creative mutual

4. van Huyssteen, "Postfoundationalism in Theology and Science," 45.
5. Jaspers, *The Origin and Goal of History*, 19.
6. Russell, *TIE*, 72.

interaction. In what follows, the astrotheologian will construct a picture of universal reality oriented toward the God of grace, a picture painted in scientific hues. In addition, the astrotheologian will render a critique of existing scientific assumptions, substitute better assumptions, and attempt to persuade research scientists to see their own work in a clearer light. Our task in this volume is primarily that of a theology of nature; but at certain junctures watch for CMI.

Theology is a field encompassing field, because it attempts to draw the most comprehensive picture of reality that can be conceived. St. Thomas Aquinas describes theology as a "sacred science" with God as the object of this science, along with all other things. "But in sacred science, all things are treated of under the aspect of God: either because they are God himself or because they refer to God as their beginning and end. Hence it follows that God is in very truth the object of this science."[7] We put it this way: a theology of nature tries to comprehend all things in reality in relationship to the one God of grace. In astrotheology, this includes the galaxies, the stars, the planets, and perhaps our future space neighbors. This places the astrotheologian in the classroom of the Bible-Welcomes-Aliens school of conjecture.

Because theology is a field encompassing field, we can see that theology encompasses scientific knowledge without necessarily altering that knowledge. Scientific knowledge is genuine knowledge, to be sure; but its scope is limited to the natural causal nexus. If one were to base an atheistic doctrine strictly on what science can tell us about reality, then atheism would be relying upon a truncated picture of reality. Hybrid physicist and theologian, John Polkinghorne, makes this point forcefully: his "claim is not that atheism is incoherent, but that it explains less than theism can."[8] Or, "theistic belief is more comprehensive and fully explanatory than atheism can be."[9] Science as a source provides reason and experience for a constructive theology of nature.

The Hermeneutic of Experience

With these four classic sources in mind, on occasion we will zero in on the fourth, experience. Experience for the scientist refers to the empirical results of experiments or data. Yet, we can also think of experience much more broadly, as common or daily experience. Experience must be interpreted. Actually, a primitive interpretation of experience belongs inherently

7. Aquinas, *ST* I.Q1.7.
8. Polkinghorne, "Christianity and Science," 62.
9. Ibid., 66.

to experience itself. We will dissect experience, giving considerable attention to the interaction between raw experience and its primitive interpretation, thereby offering our own interpretation of experience at the level of second order discourse, at the level of abstract reflection.

The astrotheologian is concerned about both empirical experiment as well as daily experience. More specifically, the Christian astrotheologian employs a *hermeneutic of secular experience*, interpreting secular and scientific assumptions in light of a scripturally based faith in the transcendent God of Israel. An astrotheologian belonging to a non-Christian tradition may also benefit from this hermeneutic of secular experience; but its theological meaning may differ accordingly. We welcome multi-religious dialogue on such matters.

If by chance the term *hermeneutic* is new to you, let me mention that it has to do with interpretation. A phenomenon is what appears to us; and what appears to us needs to be analyzed. What gets said needs to be interpreted in the context of the unsaid. We interpret new phenomena in light of their history; and we interpret what is said with a tone of suspicion just in case some hidden grab for power or hegemony is present.

The hermeneutic of secular experience was developed by University of Chicago theologian Langdon Gilkey. A *Hermeneutic of Secular Experience* attempts "to see what religious dimensions there may be . . . in ordinary life . . . which will uncover what is normally hidden and forgotten."[10] Further. "What it seeks is to uncover . . . are those aspects of daily experience which the secular mood has overlooked . . . there are levels latent in secular life of which our age is undoubtedly aware but about which it is unable to speak or to think intelligibly. These elements are the dimension of ultimacy presupposed in all our interaction with the relative world, and the presence of ambiguity within our freedom and creativity, of the demonic and the despairing in life as well as the joyful, with both of which secular experience is suffused."[11] Like a fishing boat captain measuring the water's depth with sonar signals, the theologian measures religious and spiritual depths below the secular surface.

The theologian interprets the secular self-understanding of the modern mind. Secular thinking along with its sibling, scientific thinking, require interpretation to understand what is said in light of what remains unsaid. What goes unsaid is that the secular worldview only thinly covers over the preceding religious worldview of Western culture. To think secularly is like trying to paint an old rusty car without first sanding off its previous finish.

10. Gilkey, *Naming the Whirlwind*, 234.
11. Ibid., 260.

To brush on a simple coat of secularity over a two millennia history of spiritual consciousness risks an eventual peeling. When the new paint peels the rusty fenders and rocker panels poke through. Profound spiritual sensibilities cannot be easily covered up by secular rhetoric or even scientific pontifications. University of Chicago theologian David Tracy reminds us, "even in science, we must interpret in order to understand."[12] Such interpretation of secular science is sometimes subtle, sometimes not.

Take the Apollo Program, for example. Even though space exploration is an international and secular program of scientific research, its underlying religious or spiritual dimension has become visible to even secular observers. In a recent article published in *Astropolitics: The International Journal of Space Politics and Policy*, we find this: "Apollo [Space Program] evoked, in a metaphorical and absolutist sense, emotions of awe, devotion, omnipotence, and most importantly redemption for humanity."[13] Theologians need not superimpose religious interpretations on astrobiology or related space sciences; space's spiritual qualities almost stand up and ask for notice. Although spiritual and religious sensibilities are hidden beneath scientific terminology, sometimes the costume is but a thin disguise.

Just as secular and scientific garb clothes and hides an underlying religious sensibility, so also—more generally speaking—culture clothes and hides raw experience. We need to do some undressing to see what is underneath. Part of our interpretive task is to get back to naked human experience. This is not easy, as we have just said; because our basic experience is covered over with layers of cultural interpretation. The renowned University of California sociologist Robert Bellah reminds us, "We cannot disentangle raw experience from cultural form."[14] Even if we cannot disentangle raw experience from cultural form, we can at least point to it.

Our Experience with the Sky

The naked experience relevant here is the human experience with the sky. Our experience with the sky evokes in us a sense of awe. It is no accident that in archaic religions the sky symbolized power and transcendence. To hear thunder and see lightning must have been frightening in ancient times; so the sky gods tended to be thought of as fierce warriors such as Zeus in Greece or Indra in India.

12. Tracy, *Plurality and Ambiguity*, 33.
13. Launius, "Escaping Earth," 49, http://www.tandfonline.com/doi/pdf/10.1080/14777622.2013.801720.
14. Bellah, *Religion in Human Evolution*, 12.

THE TASKS OF ASTROTHEOLOGY   33

What dominates the sky is the sun, of course. The sun is our source of energy, heat, and life. Gods of the sun reigned over ancient Egypt and ancient Mexico. The sky was thought to be beyond, immense, uncontrollable, ultimate. "Even before any religious values have been set upon the sky it reveals its transcendence," writes Mircea Eliade. "The sky symbolizes transcendence, power and changelessness simply by being there. It exists because it is high, infinite, immovable, powerful."[15] French space ethicist Jacques Arnould makes the same point: what stuns us is how the sky's "incommensurable elevation and its terrifying infinity, its cold immutability and its formidable power imposed themselves on the newly-budding human consciousness."[16] In the raw, we experience the sky as the medium of transcendence. However, methodologically, to get to this raw experience with the sky we need to interpret what is said by scientists about cosmology and by theologians about creation.

We in the modern world, curiously enough, think we have conquered the sky. We fly in airplanes above the thunder and lightning. Daily we listen to the weather report, telling us what the sky will do in the next few days. On April 12, 1961, a Russian astronaut, Yuri Gagarin, flew above the sky and the Marxist atheist press reported that he did not find God in heaven. The sky has been demystified, secularized, and scientifically comprehended. It has lost its enchantment. No longer is the sky ultimate. In the modern world, sky gods are now out of a job.

But right behind the old sky lies a new sky: outer space. Outer space confronts human consciousness with a new infinity, a new sense of awe, a new sense of the beyond. The stars seem countless. Even the galaxies seem countless. Of course, the number of each must in principle be finite. Estimates of the number of stars range about 10^{24}, and 10^{12} for galaxies. But, to our eyes and to our minds, the vastness staggers our imagination. The numbers and the distances are astronomical, to take advantage of a pun. Our planet, like a drop of water in the ocean, is swallowed up in an apparently endless sea of immeasurability.

Our religious or spiritual sensibilities get evoked, provoked, triggered. From within us a psychic force wants to explode, to expand the self to embrace or at least comprehend the unfathomable totality. The impenetrable mystery of deep space and deep time dizzy our thought processes, eliciting a sense of reverence, a sense that we are facing ultimacy.

---

15. Eliade, *Patterns in Comparative Religion*, 39.
16. Arnould, "Space Conquest," 45.

Our Thirst for Ultimate Reality

"The night sky is a primal wonder whose infinite nature spurs a longing to understand human existence," writes Space Studies scientist Mark Bullock. "This is a religious impulse and is also the impulse behind cosmology," the science of cosmology.[17]

This observation that with the sky comes a sense of transcendence and wonder renders the very awareness of outer space religious, or spiritual. This raw spiritual experience underlies culture in general, and our scientific sub-culture in particular. Therefore, the hermeneutic of secular experience is in large part a *theology of culture*. This term comes from Langdon Gilkey's mentor, Paul Tillich. "What I like to call 'theology of culture' . . . is the attempt to analyze the theology behind all cultural expressions, to discover the ultimate concern in the ground of a philosophy, a political system, an artistic style, a set of ethical or social principles. This task is analytic rather than synthetic, historical rather than systematic."[18] In short, we subject culture to analysis in order to uncover the more fundamental human experience that is coming to articulation in culture. Science is one of the cultural forms subject to theological analysis.

The word *ultimate* tips us off about human nature. There is at work within us a thirst, a hunger, a craving for ultimate reality. Eliade called this *ontological thirst*, the thirst for ultimate being. "This religious need expresses an unquenchable ontological thirst. Religious man thirsts for *being*."[19] We cannot feel authentic until we feel grounded in what is real, ultimately real. So when we look at human experience with the natural world through the interpretations of science, we ask: just what role does the orientation toward ultimacy play?

This implies that we, the contributors to this book, cannot accept literally what scientists say. Rather, we *interpret* what scientists say, going beyond the literal with our hermeneutic of secular experience. We assume that science does not sufficiently understand itself. Science falls short of self-understanding; therefore, the theologian might be able to reveal something beyond what science itself can say. The titan of twentieth-century Roman Catholic theology, Karl Rahner, saw theology as a science of science.

---

17. Bullock, "Cosmology," 1:137.

18. Tillich, *ST*, 1:39. Of the two forms of theology of culture identified by Robert K. Johnston—separation and engagement—we here follow the *hermeneutic of engagement*. This engagement method presupposes that the divine is present in the wider culture and that the theologian should offer a hermeneutical approach to help make the divine visible. Johnston, "Theology and Culture," 795–805.

19. Eliade, *Sacred and Profane*, 64.

"Theology is a science which is concerned with the interpretation of human existence, a field which is existentially and ontologically prior to man's interpretation of himself at the level of the natural sciences."[20]

As we embark on our science of science, the astrotheologian will ask: what does the presence or absence of extraterrestrial neighbors mean for understanding our cosmos as the creation of a gracious God? Is it reasonable to study in the *Bible-Welcomes-Aliens* school of conjecture?

Fermi's Paradox: Where Are They?

Does planet Earth appear in the skies of aliens living on exoplanets? Might citizens of extraterrestrial civilizations ask questions of ultimacy, questions that ask whether we earthlings belong in their cosmovision? Is it worth our psychic energy to look to the skies for space neighbors? Enrico Fermi would answer "probably not" to such questions.

The renowned physicist was enjoying an outside lunch at Los Alamos one day during the summer of 1950. Fermi and some colleagues were discussing then current news events. One of the topics had to do with a mystery in New York City. Trash can lids were disappearing. An investigation was underway. Another item in the news was a spate of UFO reports. Flying saucer witnesses took up public attention in news papers and radio reports. Could aliens in flying saucers be stealing garbage can lids in New York? Lunch is more enjoyable when the food is accompanied by humor.

Fermi interrupted the chuckles with a serious question: "where *is* everybody?"[21] All at lunch took this question seriously, perhaps because it was being asked by the 1938 Nobel Prize winning physicist, whom many had nicknamed "the Pope" because of his apparent infallibility in rendering scientific judgments. After scribbling some numbers on a random piece of paper, he conjectured: given the time of the universe and what we know about evolution, space should be filled with astronaut traffic. But, to date, no sign of spacefaring alien life. This line of reasoning has become known as the *Fermi paradox*: they should be there but they are not.

Exoplanet hunter Geoff Marcy and others contemporary to us reiterate the continuing significance of the Fermi paradox.[22] Scientists have evidence suggesting that 10 percent of nearby stars support five or more orbiting planets. At least some of these planets could be earth-sized and lodged in the

20. Rahner, "Theological Observations," in *TI*, XI:289.
21. See the account by Impey in *Living Cosmos*, 266–67.
22. See Geoff Marcy's home page on exoplanets: http://www.exoplanets.org/index.html.

habitable zone. But, are they in fact inhabited by an advanced technological civilization? If so, then one would expect that those inhabited planets within a range of twenty light years or so would, by this time, have developed space travel. We should see their craft whizzing by. Yet, we don't. Where are they? Their absence suggests that they don't exist. The Fermi paradox counts as an argument for thinking that life on Earth is unique, or at least rare.

Even with the Fermi paradox in mind, some voice within our soul still asks: are we really alone? We want the answer to be, no, we are not alone. We want to share this wondrous cosmos with space neighbors. How can we introduce ourselves? Just where are they?

The Drake Equation

They simply must be there! But, where? Frank Drake knows where to look. Somewhere in our Milky Way. We will take a moment to examine Drake's proposed resolution of the Fermi Paradox.

Because our method is hermeneutical, we look not just at the stars. We look also at those looking at the stars. How space is known is as important as the space that is known.

We will turn briefly to the beloved and indefatigable Frank Drake and the Drake Equation. Drake along with Carl Sagan and others pioneered the method of employing radio astronomy to listen for intelligible radio signals coming from advanced civilizations within the Milky Way. With Earth ears open, we are ready to hear what might be spoken to us from outer space. This has led to the establishment of SETI, the Search for Extraterrestrial Intelligent Life Institute.

The *Drake Equation*, first formulated by Frank Drake in 1961 (National Radio Astronomy Observatory in Green Bank, West Virginia), looks like this: $N = R^* f_p n_e f_l f_i f_e f_L$.

- N is the number of civilizations in the Milky Way Galaxy whose electromagnetic emissions are detectable
- R^* is the rate of formation of stars suitable for the development of intelligent life
- f_p is the fraction of those stars forming that have planetary systems
- n_e is the number of planets per solar system with an environment suitable for life
- f_l is the fraction of suitable planets on which life actually appears
- f_i is the fraction of life-bearing planets on which intelligent life emerges

THE TASKS OF ASTROTHEOLOGY   37

f_c is the fraction of intelligent civilizations that develop a technology that releases detectable signs of their existence into space

L is the length of time such civilizations send detectable signs.[23]

Critics say, "the Drake Equation is limited as a tool because it's based on guesswork."[24] But, one might respond in Drake's defense: so, what's wrong with a little guesswork?

The value of the Drake equation is not found in knowing the numerical equivalent of N. Rather, the value is that here we have a template for structuring research and filtering incoming data.[25] The Drake Equation is a heuristic device. Nevertheless, playing with numbers is great sport for astrophysicists. Neil deGrasse Tyson plugs in some numbers such as three hundred billion galaxies in the Milky Way. He offers 0.006 for n_e as the number of planets per solar system with an environment suitable for life. Finally, Tyson arrives at "*up to* a hundred civilizations in the galaxy communicating with radio waves now" and "possibly *up to* 5 billion extragalactic, radio-broadcasting civilizations."[26]

As research advances, various numbers can be plugged in. The calculations will change as new information is gathered. As of the present moment, NASA estimates that 10^{21} planets exist in the universe, of which 10^{10} might be earthlike.[27] George Coyne, former director of the Vatican Observatory, estimates that there are 10^{17} earthlike planets in the observable universe.[28] Paul Davies at Arizona State University estimates 10^{23} planets.[29] Robert Strom at the Lunar and Planetary Laboratory at the University of Arizona estimates that the "Milky Way galaxy contains about 8.8 billion Earth-like planets orbiting Sun-like stars in their habitable zones."[30] Astrophysicist Jeffrey Bennett offers an estimate of "*100 billion* habitable planets in our galaxy" supporting at least "100,000 civilizations."[31]

The mere appeal to such big numbers persuades many astrobiologists that contact optimism is justified. Let us reiterate: the value of the Drake

23. *Drake*, xix.

24. Impey, *Living Cosmos*, 281.

25. "The real power of the [Drake] equation is in the assumptions it forces us to make." Miller, *The Aliens Are Coming*, 57.

26. Tyson, "Search for Life," 167–68.

27. *NASArm*, 18.

28. Coyne, "Evolution of Intelligent Life," 180.

29. Davies, "Many Planets," 8.

30. Strom, "We Are Not Alone," http://www.esciencecentral.org/journals/we-are-not-alone-extraterrestrial-technological-life-in-our-galaxy-2332-2519-1000144.pdf.

31. Bennett, *Beyond UFOs*, 181, Bennett's italics.

equation is that it provides a set of assumptions and an inspiration to pursue further research. It is heuristic and hopefully protean. For more than a half century now, radio astronomers have been able to fill their days and nights scanning the skies, listening for a signal from any planet within N.

So far, only silence. By no means does this prove that our listening is in vain. The sky is so big, that, even with fifty plus years of scanning, our amplified ears have listened to only a fraction of what could be there. In addition, this research has yielded extensive new knowledge of what is out there; and it has spawned new developments in the technology for radio reception. SETI gives us a progressive research program, despite the absence to date of a recognizable signal from space neighbors.

Listening for radio communication is not the only method of ETI detection. An alternative is to search for Dyson spheres. An advanced extraterrestrial civilization is likely to invest in harnessing the energy of its respective sun, SETI researchers think. According to Freeman Dyson, for whom the Dyson sphere is named, a highly advanced civilization would erect a technological shell surrounding a host star to catch solar arrays, to harness as much solar energy as possible. Earth's scientists could detect the increased "heat emission as infrared radiation. And they also could see visible light blocked by the huge structures that nearly surround the star."[32] This is to say that there may be more than one way to skin a cat. Whether earthlings listen for radio transmissions or look for Dyson spheres, the assumptions regarding the likely evolution of life in off-Earth locations remain the same.

The chief assumption at work in the Drake equation is the concept of evolution, *cosmic evolution,* a story of evolutionary history that is much bigger than Charles Darwin's story of life. For Darwin, biological evolution on Earth operated according to the dialectical interaction of variation in inheritance with natural selection. Evolution explained speciation. Hence the title of Darwin's major work, *Origin of Species.* In the case of cosmic evolution, however, variation in inheritance and natural selection are not explanatory for the big bang or for star formation or planets and moons orbiting. Nevertheless, our term *evolution* is now upgraded; it now applies to pre-biotic physics as well as biological descent with modification. "The idea of cosmic evolution implies a continuous evolution of the constituent parts of the cosmos from its origins to the present . . . the entire universe is evolving . . . all of its parts are connected and interact, and . . . this evolution applies not only to inert matter but also to life, intelligence, and even

---

32. Chandler, "The New Search," 30.

culture."[33] The reigning concept in the SETI worldview is evolution, especially cosmic evolution.

Imported into SETI's concept of evolution is progress, advance. The assumption at work here is that if a habitable planet has more time to evolve, then it will advance further on the road to intelligence. Intelligence itself advances, allegedly. And this advance leads to the development of science, technology, and communication. Because it appears that this is what has happened with the evolution of life on Earth, we can export this model to other planets. Further, we can extrapolate with a principle: the more time a species has to evolve, the more highly evolved it will become. Life with more time to evolve will progress further. To communicate with a species with more time than we have had on Earth means we will be communicating with a more advanced species, that is, with our own future.

Does this count as empirical science? No. It is an example of speculation. Is this reliable speculation? Its reliability will depend on the accuracy of the assumptions. This leads to a problem in Drake Equation reasoning: can the assumption regarding evolutionary progress toward increased intelligence be confirmed or verified? Not likely. The idea of progress within evolution is doubted and even repudiated by our leading evolutionary biologists, as we will see in later chapters of this book. In the meantime, we only wish to point out the fragility of the assumptions at work in the worldview presupposed in the Drake equation.

Still working with the Drake equation assumptions, SETI scientists project into the future. We can imagine what might happen if we earthlings find ourselves communicating with an extraterrestrial civilization more advanced than ours. Here are the words of Frank Drake. "Everything we know says there are other civilizations out there to be found. The discovery of such civilizations would enrich our civilization with valuable information about science, technology, and sociology. This information could directly improve our abilities to conserve and to deal with sociological problems—poverty for example. Cheap energy is another potential benefit of discovery, as are advancements in medicine."[34] In short, communication with a more advanced alien race will lead to a windfall of benefits to us on Earth: the solution to our energy crisis, the resolving of social conflict, and advances in health and medicine. We have not yet met those who live in the sky; yet we anticipate blessings to fall from the heavens.

This scientized worldview is beginning to look like a revised religious myth of redemption. The ancient yearning for salvation from the gods has

---

33. Dick and Strick, *Living Universe*, 9.
34. Drake, "Interview with Dr. Frank Drake," 5.

become a modern expectation for more highly evolved beings to provide us with a secularized pot of gold. Is this hope already built into the concept of evolution as we know it? Is this a materialist form of religion, an ontological thirst looking for an extraterrestrial nectar? The hermeneutic of secular experience disrobes the worldview assumed by those manning SETI radio telescopes, and beneath the science we find an unmistakably spiritual doctrine of ultimacy. This is not a sin, to be sure; but sometimes we simply want to look for myth behind the scientific mask.

It will be the task of the astrotheologian to explore with unmasking questions. We cannot leave to the scientists to tell us just where the line should be drawn between what we can know empirically and what constitutes scientism. "Scientism," according to Dominican Michael Dodds, "is fundamentally the transformation of the methodology of empirical science into a metaphysics, a move from the quantitative investigation of nature to the assumption that being is always quantitative. While the former is a legitimate methodology, the latter is mere ideology."[35] We need to apply a hermeneutic of secular experience to locate the borders between knowledge and ideology, and to identify the trespassers. The astrotheologian will gladly show appreciation for genuine science, for science that provides us with reliable knowledge about the universe we inhabit. But, when science parades as religion or anti-religion, it will be time to blow the whistle and redirect the parade.

Please avoid getting a wrong impression here. I am by no means dismissing SETI's research agenda. Nor am I throwing a wet blanket on Frank Drake's hopes. From my point of view, I hope Drake's dreams come true. However, I feel it is the theologians' responsibility to point out the limits of empirically gained knowledge. The astrotheologian appreciates genuine science when it remains science and avoids scientism.

Four Tasks for Astrotheologians[36]

We have said: outer space has an effect on the inner soul. The impenetrable vault of the sky conveyed transcendence and stimulated awe among our neolithic ancestors. In our post-Copernican era, scientific instruments have helped us see beyond the sky. Watching galaxies recede through our telescopes enables us to look into our own past, leading us to embed the cosmos in a framework of time and history. Seemingly without end, the distances of outer space communicate once again a sense of infinity, mystery, and

35. Dodds, *Unlocking Divine Action*, 51 n27.
36. This material is revised from Peters, "Astrotheology," 838–53.

transcendence. Grasping the significance and meaning of that which now lies beyond our comprehension is the nectar we tacitly believe could slake our ontological thirst. It is the task of the astrotheologian to grapple with our thirst for ultimate reality and to taste-test the drink offered us by contemporary science.

If the natural sciences deal with *what-questions*, theologians ask *why-questions*. "Theological explanations are composed of groups of statements that are answers to why-questions," observes Philip Clayton.[37] However, we ask *why* only when we know the *what* to ask about. The scientist who tells what is beyond the microscope and the telescope provokes the astrotheologian to ask: why?

As we look ahead, we would like to place four tasks on the astrotheologian's To-Do List, tasks arising from the possibility of engaging extraterrestrial life, either microbial or intelligent life. First, Christian theologians along with intellectual leaders in each religious tradition need to reflect on the scope of creation and settle the pesky issue of geocentrism. Second, the astrotheologian should set the parameters within which the ongoing debates over Christology (Person of Christ) and soteriology (Work of Christ) are carried on. Third, theologians should analyze and critique astrobiology and related sciences from within, exposing extra-scientific assumptions and interpreting the larger value of the scientific enterprise. Fourth, theologians and religious intellectuals should cooperate with leaders of multiple religious traditions and scientists to prepare the public for the eventuality of extraterrestrial contact. Let us look briefly at each in turn.

1. The Scope of Creation

First, Christian theologians along with intellectual leaders in each religious tradition need to reflect on the scope of creation and settle the pesky issue of geocentrism. "The Earth has a philosophical address," says astrobiologist Chris Impey: "Nothing Special."[38] Are we earthlings willing to accept our mediocre status as nothing special?

This raises the question of geocentrism. Geocentrism is a matter of concern in both science and religion. Many in the scientific community are advocating geocentrism due to the challenges of climate change. Moral or ethical geocentrism is on the rise. In an *Astronomy* magazine editorial, David Eicher trumpets: "Earth is a pretty special place, at least for our species,

---

37. Clayton, *Explanation from Physics to Theology*, 148.
38. Impey, *Humble Before the Void*, 53.

and we should take good care of it."[39] In the moral sphere, geocentrism surpasses localism, nationalism, and ethnocentrism.

In scientific epistemology, the term *geocentrism* refers to the presuppositions with which space researchers scan the skies looking for life. Will extraterrestrial life look like it does on Earth? If different, how different? Because on Earth intelligent creatures enjoy a technological civilization, would this apply off-Earth as well? Not likely, contends astrobiologist Nathalie Cabrol. *"To find ET, we must expand our minds beyond a deeply rooted Earth-centric perspective and re-evaluate concepts that are taken for granted."*[40]

That's geocentrism in science. What about geocentrism in the form of Earth-chauvinism in culture and religion? "Just as we don't like to be accused of racism or sexism, so too we should find planetism unacceptable," writes *Zygon* editor and naturalist theologian Willem Drees.[41] We may now add *planetism* to our list of . . . isms.

Here is a specifically spiritual question: if it turns out that we share our universe with extraterrestrial neighbors, would we want to press the claim that planet Earth remains the center of God's activity or that the human race on Earth is the pinnacle of God's creative achievements? Ought the human species on Earth arrogate to itself a central or superior status when compared to off-Earth species? As it turns out, virtually no Christian let alone non-Christian thinker advocates geocentrism, planetism, or anthropocentrism in this spiritual sense. "Human beings are not the centre of the Universe," David Wilkinson reminds us. "In fact, it is the human belief that we are the centre of all things that the Bible calls sin."[42] If we must de-center our terrestrial self-understanding, the most effective first step will be to make a simple observation: our universe is big. Really Big!

Cosmologist Joel Primack and his artist spouse, Nancy Abrams, remind us of the vast scope of today's picture of the universe. "One thing that has become clear from the highly counter-intuitive nature of the scientific universe is that most of the imagery and concepts familiar from biblical and other origin stories cannot describe it . . . [W]e can understand phenomena of size scales that no earlier culture even imagined, let alone understood."[43] Theologian Ronald Cole-Turner recently announced that there may be as many as sixty billion habitable planets in the Milky Way for the theologian

---

39. Eicher, "The *Real* Reality Show," 9.
40. Cabrol, "Alien Mindscapes," 7, doi:10.1089/ast.2016.1536, Cabrol's italics.
41. Drees, "Bethlehem," 69.
42. Wilkinson, *SRSEI* 148.
43. Primack and Abrams, "Cosmology," 101.

to attend to.[44] The universe is immense, and so is the prospect that we may share it with other sentient creatures.

Allegedly, pre-Copernican Europeans had relied upon a belief that the planet Earth was in the center of the universe. This geocentrism allegedly supported their *hubris*, their pride-of-place as earthlings and as human beings, presuming the human race to rank highest among the living creatures. Today, both non-theologians and theologians worry that a geocentric or anthropocentric religion will suffer drastically if a new relationship with extraterrestrials challenges this residual belief system. This is misleading, as we will show in this book. Worse than misleading, it is a myth that needs busting.

The myth that needs busting is what astronomers call the "Copernican Principle." Herman Bondi (1919–2005) coined the term *Copernican Principle* to refer to the de-centering of Planet Earth and the demotion of the human race to marginal status in a giant universe.[45] Since Bondi, this term has become a cipher for a history that didn't happen and a concept that is superfluous. In a contemporary astronomy textbook we find this: "This removal of the Earth from any position of great cosmological significance is generally known, even today, as the *Copernican Principle*. It has become a cornerstone of modern astrophysics."[46]

This belief—the belief that by removing Earth from the center of the solar system Nicolaus Copernicus (1473–1543) removed the human race from a privileged center and, thereby, challenged traditional religious doctrines regarding the importance of Earth—constitutes a myth—called the *demotion myth*, a false belief—that must be busted. According to science historian Michael Keas, "This belief is false."[47] The demotion myth is false because Copernicus himself believed heliocentrism would be consistent with his Roman Catholic religious commitments. The demotion myth is false because principals in the controversy such as Kepler, Galileo, and Cardinal Bellarmine registered no anxiety over the alleged de-centering of Earth or loss of alleged anthropocentrism. The demotion myth is false because the pre-Copernican worldview did not sponsor geocentrism; rather, pre-Copernican Christians believed the heavens above provided the source of goodness in contrast to Earth below as the location of rebellion, filth,

---

44. Cole-Turner, "Planets by the Billions."
45. Bondi, *Cosmology*, 13.
46. Chaisson and McMillan, *Astronomy Today*, 43.
47. Keas, "Copernican Revolution," 23.

and even hell. "Copernicus himself—indeed, most early-modern astronomers—did not actually embrace the idea [of the demotion myth]."[48]

The misleading demotion myth with the alleged need for de-centering must be dealt with by the astrotheologian so he or she can get on to the bigger issue, namely, the scope of God's creation. The scope of creation for the Abrahamic traditions is inclusive of everything, known and unknown, visible and invisible. When biblical Christians speak of creation, it includes all of physical actuality, and more. The immensity of God surpasses the immensity of the universe. After all, since Anselm we have thought of God as that than which nothing greater can be conceived. It is God, not Earth, who is the center of reality. And the circumference of reality as well.

This in itself should settle the problem of alleged geocentrism with its snobby ethical tone, Earth chauvinism. God is the center, figuratively speaking, not Earth. This has always been the case. Yet, this is not widely understood. Unfortunately, even some theologians continue to dub physical centeredness a matter of spiritual concern, making Western religion unnecessarily vulnerable. Theologian Cynthia Crysdale, for example, promotes the demotion myth while worrying about the impact on our self-understanding of contact with ETI. "We have faced this dilemma before: Copernicus and Galileo dethroned the human. Darwin made us mere coincidences of evolution. Slowly the human race is discovering that we're not the center of the universe, but that both space and time are so vast that we are mere blips on the screen. This . . . won't go down lightly."[49] To the contrary, it will go down lightly. It already has. New Testament historian N.T. Wright, to cite a contrary and more accurate example, states as emphatically as euphemistically that "*We are not the center of the universe. God is not circling around us. We are circling around him.*"[50] He adds, "The earth, and we with it, go round the sun of God and his cosmic purpose."[51]

An astrotheologian need not loiter on the question of geocentrism with its Earth chauvinism. What is foremost is the question of scope: does God's creation deal solely with Planet Earth or does it encompass the entire universe with its 13.82 billion year history and perhaps hundred billion year future? Russell argues strenuously for God's providential action at the sub-atomic quantum level and—even though atoms are small they are everywhere!—divine action is the same in Andromeda as it is here. "When we shift to an *in*deterministic world, a new possibility opens up: One can now speak of objective acts of God that do not require God's miraculous

48. Ibid., 29.
49. Crysdale, "God and Astrobiology," 201.
50. Wright, *Justification*, 23, italics in original.
51. Ibid., 24.

intervention but offer, instead, an account of objective divine action that is completely consistent with science."[52] The laws of nature are the same in the most distant galaxy as they are in our own backyard, assumes the astrophysicist. God acts in natural processes everywhere, in distant galaxies and in our gardens. An astrotheologian must become a cosmic theologian.

This first task of astrotheology provides an instance of a more general principle: theology should be open to self-revision when prompted by new knowledge about our world. Theology should not be understood as an attic trunk filled with old and outdated doctrines. Theologians treasure their memorabilia, to be sure; yet, there is more to the ongoing task. "A healthy theology that takes science, and, indeed, all other forms of rational inquiry, seriously must be willing to alter its perspectives when truth so requires," writes Noreen Herzfeld.[53] Widening the scope of the theologian's comprehension of creation is due directly to new knowledge provided by astronomers. Let's thank God for the astronomers!

2. One Incarnation, or Many?

Second, *the astrotheologian should set the parameters within which the ongoing debates over Christology (Person of Christ) and soteriology (Work of Christ) are carried on.* It should be dubbed a mistake to connect the incarnation with geocentrism. The question of multiple incarnations versus a single incarnation is a reasonable one, but not if the option to rely upon a single incarnation appears to justify geocentrism or Earth chauvinism. Let me try to clarify what is at stake here.

The logic of the question looks like this: if Christians claim that God becoming incarnate in Jesus Christ constitutes the decisive divine act of revelation or salvation, then must this incarnate activity be repeated on every planet for every intelligent species? Does Christian theology require a planet-hopping Christ? And, if billions of such habitable planets host life, will this require billions of incarnations? For those who answer negatively to the planet-hopping Christ, God's redemptive act on Earth suffices for the entire cosmos. Might this return us to the abhorred geocentrism?

Philip Melanchthon (1497–1560) illustrates the problem. Despite the fact that the Lutherans at Wittenberg and Nuremberg had been responsible for the publication of Copernicus's *De Revolutionibus*, Reformer Melanchthon argued against the plurality of worlds on christological grounds. "The Son of God is One; our master Jesus Christ was born, died, and resurrected

---

52. Russell, *CAO*, 128.
53. Herzfeld, "'The End of Faith?'" 68.

in this world. Nor does He manifest Himself elsewhere, nor elsewhere has He died or resurrected. Therefore it must not be imagined that there are many worlds, because it must not be imagined that Christ died and was resurrected more often, nor must it be thought that in any other world without the knowledge of the Son of God that men would be restored to eternal life."[54]

Has Melanchthon justified geocentrism on the basis of his Christology? This would be a mistake, in my judgment. The existence or non-existence of other inhabited worlds with intelligent creatures is not a christological question. It is a scientific question. Or, within theology, it is a question about the scope of creation.

The question of multiple incarnations depends in part on whether one thinks of soteriology in terms of revelation or in terms of atonement. If the work of Christ is primarily that of a teacher who reveals the truth about God, then one would tend to embrace multiple incarnations, one for each intelligent species whom God wishes to invite into the divine fellowship. If, on the other hand, one thinks of the work of Christ in terms of atonement—as a work of redemption accomplished on behalf of the entire fallen creation—then a single incarnation would suffice. Thinking this matter through with transparency is one of the services the astrotheologian can render.

Perhaps an impatient reader might exclaim: "why don't those Christians just run off into a corner and hash out their parochial nonsense all on their own? Why contaminate science with religion?" Here is why: because astrotheology is a form of public theology. This commitment to transparency while thinking through a theological conundrum contributes to making astrotheology a valuable form of public discourse. We work with the assumption that a progressive theological understanding of the human condition within the wider creation of God will enhance what can be known scientifically. Theological reflection will actually contribute indirectly to a broader and deeper apprehension of the reality on which our scientists are at work. Fuller Seminary's Hak Joon Lee makes this point about public theology. "Public theologians argue that Christian doctrines of creation, sin, redemption, eschatology, covenant, and ecclesiology are informative for our understanding of the nature, meaning, and destiny of human life."[55] Astrotheology is a branch of public theology, even when theologians are talking with each other.

---

54. Melanchthon, *Initia doctrinae physicae* fol. 43, cited by Dick, *Plurality of Worlds*, 89.

55. Lee, "Public Theology," 50.

3. An Internal Critique of Science

Third, *theologians should analyze and critique astrobiology and related space sciences from within, exposing extra-scientific assumptions and interpreting the larger value of the scientific enterprise.* Although scientists should be respected and honored for what they know and for what they promise, scientific claims should not be given a free pass. Scientific claims should be subjected to critical review by religious thinkers. "Theology can assist in examining some of the assumptions upon which SETI is built," avers Wilkinson.[56]

The theological critique of science by identifying blind assumptions targets two domains: first, mistaken images held within the scientific community about theological matters and, second, assumptions and trajectories that frame the scientific picture itself. Regarding the first, Heidelberg theologian Michael Welker speaks forcefully: "Theology can and must challenge the natural sciences to correct their false perceptions of theological themes and contents."[57] Notre Dame's Michael Crowe proceeds to do just this. "It is sometimes suggested that the discovery of extraterrestrial life would cause great consternation in religious denominations. The reality is that some denominations would view such a discovery not as a disruption of their beliefs, but rather as a confirmation."[58] As we have seen here, correcting mistaken views of religious believers—mistaken by both scientists and theologians in some cases—warrants the theologian's attention.

It is one thing to correct what scientists think about religion. It is another to correct what scientists think about science. We have already distinguished between science and scientism. Because this distinction can on occasion be subtle, the theologian may on such occasions need to enter the internal domain of science with sharpened analytical and critical tools. Quite frequently extra-scientific or even ideological commitments slip into scientific frameworks at the level of assumption. Materialism and ontological reductionism, among other . . . isms, are common. Even atheism in many cases. In the field of astrobiology and its sister, SETI, an over-interpreted variant of Darwinian evolution frames and guides the research program.

The key question has to do with teleology: can we expect evolution over time to progress toward advanced intelligence? No, say evolutionary biologists. "The follies record shows very clearly that there is no central line leading steadily, in a goal-directed way, from a protozoan to man," contends

---

56. Wilkinson, *SRSEI*, 182.
57. Welker, *The Theology and Science Dialogue*, 14.
58. Crowe, *ELD*, 328–29.

George Gaylord Simpson.[59] Harvard's Ernst Mayr thunders, "cosmic teleology must be rejected by science—I do not think there is a modern scientist left who still believes in it."[60]

Despite the fact that leading evolutionary biologists decry the presence of a progressive entelechy or directional purpose in evolution, space researchers frequently work on the assumption that life's genesis is almost inevitable where pre-biotic chemistry is present and, even more suspiciously, that once life gets going it will progress toward increased complexity, toward intelligence, and toward science and technology as we know it. In sum, the presumed purpose of the entire history of our natural cosmos is to produce the very persons studying the cosmos, our scientists. This is a disguised form of geocentrism, now transformed into scientist-centrism. Perhaps even megalomania. Religious intellectuals may wish to point out this hegemonic ideology from time to time.

A constellation of ideas has converged to produce this hegemonic ideology, what we might call the *ETI myth*. Yes, there is myth in the heart of science! "Today, ancient myths are reemerging with a scientific spin and cloaked in space-age garb," writes Albert Harrison, who lumps together SETI science with ufology. "Thus, rather than subject to God's scrutiny, we are watched by naturally evolved entities whose level of intelligence is beyond our ken . . . we are visited by humanoids that drive advanced spacecraft and wear shiny spacesuits."[61] The ETI Myth includes both astrobiology and the UFO phenomenon, unfriendly siblings though astrobiologists and ufologists may be. Harrison now uses the term *cosmism* to refer to a new religion of space worship. "Carl Sagan and Neil deGrasse Tyson are high priests, astronauts are like saints that ascend into heaven, and extraterrestrials are as gods—benevolent, wise, and capable of manipulating space and time."[62] In short, space scientists right along with UFO devotees have dressed themselves in mythical clothing.

Evangelical James Herrick objects to scientized myths that include redemption. He contends that science fiction influences science proper; and this has led to a myth in the heart of science. He uses the term "Myth of the Extraterrestrials" to refer to "the idea that intelligent extraterrestrials exist and that interaction with them will inaugurate a new era in human

---

59. Simpson, "The Nonprevalence of Humanoids," 773.

60. Mayr, "The Ideological Resistance to Darwin's Theory," 131.

61. Harrison, *Starstruck*, 6. In this book, astrotheology will concentrate on SETI, not the UFO phenomenon. "The evidence for UFOs is weak, and the scientific credentials of SETI are strong." Miller, *Aliens Are Coming*, 286.

62. See the interview with Albert Harrison by Ross Anderson, "The Holy Cosmos."

existence."[63] Spiritually deprived modern culture is thirsting for superior entities in space who can save our planet and, according to Herrick, this is a poor substitute for the classic God of theism and its genuine promise of redemption. Herrick fears that the ETI Myth—replete with the alleged evolutionary promise that we can employ science and technology to achieve our own redemption and that our more highly evolved ETI neighbors are already where we are going—will replace the Christian faith, not augment it. "This is the Christian church's challenge today—to reclaim its story and tell it in such a way that it stands out among all the others as authentic, as the Great Story that other stories have often sought to imitate."[64] Or, "The biblical message is that transforming grace rather than an evolving human race is the means of discovering our spiritual destiny. Salvation is the liberating gift, not of benevolent aliens, but of a preexistent, creating and redeeming God."[65]

To avoid possible confusion, let me re-identify the issue at stake. The issue is not whether extraterrestrial neighbors exist. We in this book believe it is worth scientific effort to investigate this possibility. What is at issue is whether scientists ought to engage in extra-scientific speculation exempt from theological criticism. In the case of the space sciences, the theologian must point out the difference between science and scientism. Some of astrobiology's working assumptions regarding biological evolution stretch our imaginations well beyond what current empirical knowledge permits. No warrant exists for astrobiology to prophecy an evolutionary future on this planet or any other planet that leads to utopia. When scientism attempts to perform the tasks of religion—to perform theology without a license—it's time to blow the whistle.

In this book, we come at the version of scientism at work in the space sciences from two directions. From one direction, we note how modern science has methodologically eliminated appeal to final causes; it has rejected purposeful direction in nature. From the other direction, we note how some space scientists have re-introduced purposeful direction by appeal to the doctrine of progress. The former tends toward metaphysical naturalism, while the latter leads to practicing theology without a license. We need not decide who is right on the question of teleology. Rather, we would like to blow the whistle when one position slides over into the other and surreptitiously replaces clerical collars with white lab coats.

63. Herrick, *Scientific Mythologies*, 51.
64. Ibid., 252.
65. Ibid., 261.

Yes, we are critical of establishment science. But, our criticism arises out of a deeper applause, celebration, and near reverence for the best science. By no means should the reader jump to the unwarranted conclusion that the theological authors of this volume are defenders of creationism or sponsors of a public campaign against science. Quite the opposite. We work from within the framework of what we deem the best science, the most fertile science. And this includes the neo-Darwinian model for explaining evolution. We like to attend barbecues with the evolutionary biologists and theistic evolutionists.

Theistic evolutionists, among others, see God's creative and redemptive activity in, with, and under natural processes. Würzburg theologian Hans Schwarz gets it right. "From the theological perspective it could be concluded that God functions with a finite living being as co-creator without exerting any force and without preempting the result which this co-creation causes. For this process God established the external parameters which are now discovered by the natural scientist. God makes it possible through the open process of evolution that ever more complex organisms and systems that react with each other are constructed so that eventually those living beings emerge as we are. God acts invisibly and unnoticeably behind and through the natural processes."[66]

In sum, astrobiology and sister fields should be celebrated for the fertile science that continues to produce new knowledge about our immense and complex universe. However, this celebration is limited to science that remains science. The theologian should offer a critique when science drifts toward disguised ideology or substitute religion. Even more, such a critique calls for a self-analysis on the part of the scientific community. "What is called for, therefore," writes Gilkey, "is a reassessment of science in our cultural life, and one conducted soberly by the scientific community itself and not alone by those on the outside."[67]

4. Space Ethics

Fourth, *theologians and religious intellectuals should cooperate with leaders of multiple religious traditions and scientists to address ethical issues associated with space exploration and to prepare the public for the eventuality of extraterrestrial contact*. No one can predict with precision exactly what is coming. If the day of extraterrestrial contact arrives, re-thinking our terrestrial worldviews should follow. This is likely to be complex, not simple.

66. Schwarz, *Vying for Truth*, 199.
67. Gilkey, *Society and the Sacred*, 84.

Harrison recognizes that "we cannot simply incorporate extraterrestrial ideas without thinking them through, because our systems (supranational, societal, and organismic) have highly interrelated parts, so changes in one arena yield changes in another."[68] Religion is one of those parts, perhaps even foundational for revised worldview construction. Boston University systematic theologian John Hart foresees that "the collaboration of scientists, ethicists, and theologians will enhance both reflection on Contact, and terrestrial-extraterrestrial interaction when Contact occurs."[69] Cooperation and collaboration are the watchwords.

Anticipated contact with intelligent creatures from outer space is the most dramatic of the scenarios. There are many other scenarios regarding space exploration that are less dramatic, yet every one still calls for scrupulous ethical attention. We will divide these ethical issues into two categories: first, space exploration looking for extraterrestrial non-intelligent life (ETNL) or stupid life within our solar ghetto and, second, contact with extraterrestrial intelligent life (ETI) or civilizations elsewhere in the Milky Way metropolis. Pursuing both of these redounds to a new understanding of who we are on planet Earth. We need to think of Earth as a single planetary society where humans deal in concert with one another to address off-Earth concerns. Space is not the private property of one nation, one profession such as science, one religion, one ideology, or any other terrestrial entity in competition with others. The community of moral deliberation on space matters needs to be the entire human community. Might the flip side of addressing outer space be the unification of all humankind on our planet?

With regard to the first category—exploration within our solar ghetto where we expect to discover microbial life or ETNL—a number of issues stand up and demand our attention. Here is our inventory of issues to formulate and address: (1) Does Planetary Protection Apply to Earth Alone? (2) Does Extraterrestrial Life Have Intrinsic Value? (3) Should Space Explorers Invoke the Precautionary Principle? (4) Should We Clean Up Our Space Junk? (5) What Should We Do About Satellite Surveillance? (6) Should Nations Weaponize Space? (7) Who Gets Priority: Scientific Research or Making a Profit? (8) Should Earthlings Terraform Mars? (9) Should Earthlings Colonize Mars? (10) How Should We Protect Earth from Extraterrestrial Threats? (11) Does AstroEthics Require a Single Planetary Community of Moral Deliberation? (12) Should the Common Good Include the *Galactic Commons*? It might be too early to resolve each of these issues, but simply

---

68. Harrison, *After Contact*, 298.
69. Hart, "Cosmic Commons," 390.

formulating them for moral deliberation should provide guidance for public policy.

The above list applies only to non-intelligent life, to microbial life within the solar ghetto. With regard to the second category—contact with intelligent aliens or ETIL who live elsewhere in the Milky Way metropolis—we can imagine three types of creatures who might become our new neighbors. The first would be extraterrestrial beings who are intelligent, but less intelligent than we are. What ethical principles might obtain? Should we treat such aliens as we now treat our animals? The second would be extraterrestrial beings who are approximately as intelligent as we are. Would this require us to treat them with dignity? Would it make a moral difference to us if the aliens are hostile or if they are benevolent? The third would be extraterrestrial beings who are superior to us in intelligence. If these intelligent beings are hostile, might they enslave us? Might we have to adopt a slave ethic and impose it on ourselves? Or, if these super-intelligent aliens turn out to be benevolent, might we have to adopt an ethic of gratitude? Just what scenarios would aid us in preparing mentally and culturally for what might happen?

Planetary readiness informed by wisdom drawn from Earth's historic religious traditions is being called for here. Secular or scientific anticipations are not enough. Religious readiness will be helpful to both spiritual and non-spiritual sectors alike.

What Next?

In this book we would like to examine what appears at both ends of the telescope. At one end is the cosmos as the object: the magnificent and baffling and awe-inspiring universe. At the other end is the subject: the human eye accompanied by a mind that asks: just what am I seeing? Holmes Rolston reminds us that "the most astounding entity in the universe which an astronomer surveys lies just back of the eyes looking into the telescope."[70] The object is in the subject, conditioning and influencing the subject.

The presence of the cosmos within the human soul evokes a sense of ineffability and wonder. It cries out for explanation. Paul Davies dubs the linkage between mind and universe a mystery. "We have cracked part of the cosmic code. Why this should be, just why *Homo sapiens* should carry the spark of rationality that provides the key to the universe, is a deep enigma. We, who are children of the universe—animated stardust—can nevertheless reflect on the nature of that same universe, even to the extent of glimpsing

70. Rolston, *Science and Religion*, 66.

the rules on which it runs. How we have become linked into this cosmic dimension is a mystery. Yet the linkage cannot be denied."[71]

The question of whether we share this universe with other rational creatures is a scientific question, to be sure. But, this question is overloaded with meaning. Arthur C. Clarke is remembered for saying, "Two possibilities exist: either we are alone in the universe, or we are not. Both are equally terrifying." Buckminster Fuller made the same point: "Sometimes I think we're alone. Sometimes I think we're not. In either case, the thought is staggering." The astrotheologian is ready to stagger with this terrifying set of thoughts.

Bibliography

Aquinas, Thomas. *Summa Theologica* in Christian Classics Ethereal Library. http://www.ccel.org/ccel/aquinas/summa.toc.html.
Arnould, Jacques. "Space Conquest and Ritual Practices: Lighting Candles for Ariane." *Theology and Science* 11.1 (February 2013) 44–51.
Bellah, Robert N. *Religion in Human Evolution: From the Paleolithic to the Axial Age.* Cambridge, MA and London: Harvard University Press, 2011.
Bennett, Jeffrey. *Beyond UFOs: The Search for Extraterrestrial Life and Its Astonishing Implications for Our Future.* Princeton: Princeton University Press, 2008.
Bondi, Herman. *Cosmology.* Cambridge: Cambridge University Press, 1952.
Bullock, Mark R. "Cosmology." In *ESTE* 1:437–42.
Cabrol, Nathalie A. "Alien Mindscapes—A Perspective on the Search for Extraterrestrial Intelligence." *Astrobiology* 16:9 (2016) 1–16, doi:10.1089/ast.2016.1536.
Chaisson, Eric, and Steve McMillan. *Astronomy Today.* 8th ed. Boston: Pearson, 2014.
Chandler, David L. "The New Search for Alien Intelligence." *Astronomy* 41:9 (September 2013) 28–33.
Clayton, Philip. *Explanation from Physics to Theology.* New Haven: Yale University Press, 1989.
Cole-Turner, Ronald. "Planets by the Billions." *Metanexus* (April 3, 2012) http://metanexus.net/blog/planets-billions.
Coyne, George V., SJ. "The Evolution of Intelligent Life on Earth and Possibly Elsewhere: Reflections from a Religious Tradition." In *MW* 177–88.
Crowe, Michael J. *Extraterrestrial Life Debate from Antiquity to 1915: A Source Book.* Notre Dame, IN: University of Notre Dame Press, 2008.
Crysdale, Cynthia. "God and Astrobiology." In *Workshop Report: Philosophical Ethical, and Theological Implications of Astrobiology,* 196–207. Washington, DC: AAAS, 2007.
Davies, Paul C. W. "Many Planets, Not Much Life." *Scientific American* 315:3 (September 2016) 8.
———. *The Mind of God: The Scientific Basis for a Rational World.* New York: Simon & Schuster, 1992.
Dick, Steven J., and James E. Strick. *The Living Universe: NASA and the Development of Astrobiology.* New Brunswick, NJ: Rutgers University Press, 2005.

71. Davies, *The Mind of God*, 232.

Dodds, Michael J., OP. *Unlocking Divine Action: Contemporary Science and Thomas Aquinas.* Washington, DC: Catholic University Press of America, 2012.
Drake, Frank. Interviewed by Diane Richards. "Interview with Dr. Frank Drake." *SETI Institute News* 12.1 (2003) 5.
Drees, Willem B. "Bethlehem: The Center of the Universe?" *God for the 21st Century,* edited by Russell Stannard, 67–70. Philadelphia: Templeton Foundation, 2000.
Eicher, David. "The *Real* Reality Show: Let's Cut the UFO Crap." *Astronomy* 43.3 (2015) 9.
Einstein, Albert. *Out of My Later Years.* New York: Philosophical Library, 1950.
Eliade, Mircea. *Patterns in Comparative Religion.* New York: Meridian Book, 1963.
———. *The Sacred and the Profane.* Translated by Willard R. Trask. New York: Harcourt, Brace, and World, 1959.
Ferkiss, Victor C. *The Future of Technological Civilization.* New York: George Braziller, 1974.
Gilkey, Langdon. *Naming the Whirlwind: The Renewal of God-Language.* Indianapolis and New York: Bobbs-Merrill, 1969.
———. *Society and the Sacred.* New York: Crossroad, 1981.
Harrison, Albert A. *After Contact: The Human Response to Extraterrestrial Life.* New York: Plenum, 1997.
———. Interview by Ross Anderson. "The Holy Cosmos: The New Religion of Space Exploration." *The Atlantic* (March 29, 2012) http://www.theatlantic.com/technology/archive/2012/03/when-the-saints-go-blasting-off-space-exploration-as-religion/255136/.
———. *Starstruck: Cosmic Visions in Science, Religion, and Folklore.* New York and Oxford: Bergbahn, 2007.
Hart, John. "Cosmic Commons: Contact and Community." *Theology and Science* 8.4 (November 2010) 371–92.
Herrick, James A. *Scientific Mythologies: How Science and Science Fiction Forge New Religious Beliefs.* Downers Grove, IL: IVP Academic, 2008.
Herzfeld, Noreen. "'The End of Faith?' Theology as Process." In *How Do We Know? Understanding in Science and Theology,* edited by Dirk Evers et al., 67–76. London: T. & T. Clark, 2010.
Impey, Chris. *Humble Before the Void.* West Conshohocken, PA: Templeton, 2014.
———. *The Living Cosmos: Our Search for Life in the Universe.* New York: Random House, 2007.
Jaspers, Karl. *The Origin and Goal of History.* London: Routledge, 1953.
Johnston, Robert K. "Theology and Culture." In *The Routledge Companion to Modern Christian Thought,* edited by Chad Meister and James Beilby, 795–805. London and New York: Routledge, 2013.
Keas, Michael N. "That the Copernican Revolution Demoted the Status of Earth." In *Newton's Apple and Other Myths about Science,* edited by Ronald L. Numbers and Kostas Kampourakis, 23–29. Cambridge: Harvard University Press, 2015.
Launius, Roger D. "Escaping Earth: Human Spaceflight as Religion." *Astropolitics: The International Journal of Space Politics and Policy* 11:1–2 (2013) 45–64. http://www.tandfonline.com/doi/pdf/10.1080/14777622.2013.801720.
Lee, Hak Joon. "Public Theology." In *The Cambridge Companion to Christian Political Theology,* edited by Craig Hovey and Elizabeth Phillips, 44–65. Cambridge: Cambridge University Press, 2015.
Marcy, Geoff. Exoplanets.org. http://www.exoplanets.org/index.html.

Mayr, Ernst. "The Ideological Resistance to Darwin's Theory of Natural Selection." *Proceedings of the American Philosophical Society* 135 (1991) 131.
Melanchthon, Philip. *Initia doctrinae physicae*. Wittenberg, 1550, quoted in Steven J. Dick, *Plurality of Worlds: The Extraterrestrial Life Debate from Democritus to Kant*. Cambridge: Cambridge University Press, 1984.
Miller, Ben. *The Aliens are Coming*. New York: The Experiment, 2016.
Murphy, Nancey. *Anglo-American Postmodernity: Philosophical Perspectives on Science, Religion, and Ethics*. New York: Harper, Westview, 1997.
NASA. *Astrobiology Roadmap*. https://nai.nasa.gov/media/medialibrary/2013/09/AB_roadmap_2008.pdf.
Peters, Ted. "Astrotheology." Chapter 72 in *The Routledge Companion to Modern Christian Thought*, edited by Chad Meister and James Beilby, 838–53. London and New York: Routledge, 2013.
Polkinghorne, John. "Christianity and Science." In *OHRS* 57–70.
"Preface." In *Drake* xix.
Primack, Joel R., and Nancy Ellen Abrams. "Cosmology." In *RCRS* 93–102.
Rahner, Karl. "Theological Observations on the Concept of Time." In *Theological Investigations*. 22 vols. XI:288–308. London: Darton, Longman, and Todd, 1961–1976; New York: Seabury, 1974–1976; New York: Crossroad, 1976–1988.
Rolston, Holmes, III. *Science and Religion*. New York: Random House, 1987.
Russell, Robert John. *Cosmology from Alpha to Omega: The Creative Mutual Interaction of Theology and Science*. Minneapolis: Fortress, 2008.
———. *Time in Eternity*. Notre Dame, IN: University of Notre Dame Press, 2012.
Schwarz, Hans. *Vying for Truth: Theology and the Natural Sciences*. (Göttingen: Vandenhoeck & Ruprecht, 2014.
Simpson, George Gaylord. "The Nonprevalence of Humanoids." *Science* 143 (February 21, 1964) 769–75.
Strom, Robert G. "We Are Not Alone: Extraterrestrial Technological Life in our Galaxy." *Astrobiology and Outreach* 3:5 (2015). http://www.esciencecentral.org/journals/we-are-not-alone-extraterrestrial-technological-life-in-our-galaxy-2332-2519-1000144.pdf.
Tillich, Paul. *Systematic Theology*. 3 vols. Chicago: University of Chicago Press, 1951–1963.
Tracy, David. *Plurality and Ambiguity: Hermeneutics, Religion, Hope*. San Francisco: Harper, 1987.
Tyson, Neil deGrasse. "The Search for Life in the Galaxy." In *WU* 146–69.
van Huyssteen, J. Wentzel. "Postfoundationalism in Theology and Science: Beyond Conflict and Consonance." In *Rethinking Theology and Science: Six Models for the Current Dialogue*, edited by Niels Henrik Gregersen and J. Wentzel van Huyssteen. Grand Rapids: Eerdmans, 1998.
Welker, Michael. *The Theology and Science Dialogue: What Can Theology Contribute?* Göttingen: Neukirchener Theologie, 2012.
Wilkinson, David. *Science, Religion, and the Search for Extraterrestrial Intelligence*. Oxford: Oxford University Press, 2013.
Wright, N. T. *Justification: God's Plan and Paul's Vision*. Downers Grove, IL: IVP Academic, 2009.

3

The Road Map to Other Earths

Lessons Learned and Challenges Ahead

JOSÉ G. FUNES, SJ

> What to think of these stars without any doubt similar to our Sun, destined like the Sun to keep alive an enormous quantity of creatures of every kind? Those immense regions must be inhabited by intelligent beings endowed with reason, capable to know, love, and honor the Creator.
>
> —ANGELO SECCHI, SJ[1]

Are we alone? This question has triggered many scientific projects on the search for life in the universe, especially for extra-terrestrial intelligent (ETIL or ETI) life. It also has aroused a very high level of fascination among the general public and places new challenges and frontiers in the fields of science, philosophy, and religion.

I believe that science fiction movies and literature could be an excellent means through which to introduce this subject, which is barely mentioned in the academic context. Although I acknowledge that this matter is becoming more recognized in the academic world.

1. Secchi, *Le Soleil*, 418.

Here is a brief list of very human and spiritual issues associated with science-fiction movies. The list is incomplete but it renders the idea of the impact that this subject has had in the collective imagination.

- *The Day Earth Stood Still* and *E.T.:* hope for intervention on the part of mediators from distant worlds carrying moral messages to awaken our human conscience
- *Contact* and *Interstellar:* our hope for life beyond death
- *Star Wars:* the battle between light and dark, good and evil
- *Contact* and *Close Encounters of the Third Kind*: divine revelation, angels, people abducted into heavens, etc.
- *Independence Day* and *Resurgence, Mars Attacks:* Earth's humankind might recover original unity and common set of values and goals in the face of a potential extraterrestrial threat
- *War of the Worlds, Avatar,* and *District 9:* conflicts between humanity and an extraterrestrial civilization.

I grew up in Argentina watching *Star Trek,* the original TV series. The very popular TV show opening: "Space: the final frontier. These are the voyages of the starship Enterprise. Its five-year mission: to explore strange new worlds; to seek out new life and new civilizations; to boldly go where no man has gone before,"[2] interprets the very human desire to explore the universe. Our species has a very deep desire to explore the universe that might be rooted in Aristotle's idea that "all human beings by nature desire to know."[3] In other words "human curiosity is the driving force for the scientific development, in which belief systems and philosophy still have their valid place"[4] and certainly for exploring other inhabited worlds.

The quest for ETI life is already part of our daily culture. Even Pope Francis preaching during the daily liturgy in the Vatican, said: "If tomorrow an expedition from Mars—Martians, green, with that long nose and big ears, as painted by children—arrived to Earth and came to us . . . and one said: 'We want the baptism!' what would happen?"[5] In the context of the 2016 US presidential campaign, the *New York Times* reported: "Mrs. Clinton, a cautious candidate . . . has shown surprising ease plunging into the

---

2. Wikipedia, "Where No Man Has Gone Before."

3 Aristotle, *Met.* Book I, Chapter 1, line980a21.

4. Preface of the booklet of the Plenary Session of the Pontifical Academy of Sciences: Arber, "Preface."

5. Radio Vaticana.

discussion of the possibility of extraterrestrial beings."[6] Will we ever learn what Hillary Clinton actually thinks about this subject?

The potential discovery of ETI could be the next giant leap for the human family, at least according to Neil Armstrong who landed on the Moon's Sea of Tranquility in 1969: "One small step for [a] man, one giant leap for mankind."[7]

The cultural, philosophical, and religious implications of the search of inhabited worlds is one of the research projects in which I am involved at the Catholic University of Córdoba. Scientific, philosophical, and religious research is developed in a cultural context. Also, I believe that the cultural framework could drive or modify our ideas if we are open-minded. Scientists are a bit biased in their research. We ought to be realistic and be aware of our own bias. Trying to answer the questions related to ETI from Argentina is a new challenge that I intend to face with critical thinking but also considering the bias that a Latin American culture might have in my research project.

In light of the growing number of discovered Earth-like exoplanets located within the circumstellar habitable zone also known as "the Goldilocks zone," the potential discovery of the existence of an ETI species elsewhere in our galaxy and then our potential contact with them might have a profound impact on our scientific, philosophical, theological, and social comprehension of humanity. Indeed, this question as to the existence of intelligent life elsewhere has been a persistent question asked throughout the history of philosophical and religious thinking as well as in science.[8]

The current scientific search for ETI poses questions that scientists *could* attempt to answer from a purely scientific point of view. However, such an important quest requires complementary reflection from the perspectives of a variety of diverse fields of human knowledge and disciplines. What is life? How and why does life originate? What might be the criteria that we adopt to identify what we might call an ETI or an ET civilization? Which criteria should we use to determine the values on which a civilization is based? What might the impact be of our discovering that we are perhaps the only highly technologically-developed civilization in the entire history of the galaxy so far? What are the implications of our potential discovery that we are only one of many ET highly-technologically-developed civilizations?

---

6. Chozick, "Hillary Clinton."
7. Wikiquote. "Neil Armstrong."
8. See Crowe, *ELD*; and Dick, *LOW*; plus Fantoli, *Extraterrestri*.

When we consider the possibility that extraterrestrial civilizations might be more advanced than we on Earth in science and technology, this reminds us of our own evolution on Earth. Technological progress has been accompanied by developments in consciousness. Might a civilization with a longer evolution have advanced to a higher state of consciousness? If this turns out to be the case, what might we be able to learn from the successes and the failures on the part of such species?[9]

Even if our need to answer these questions of a scientific, philosophical, religious, and social nature is not perceived to be immediate—because we have not yet empirically confirmed the existence of intelligent creatures off-Earth—this quest deserves our attention and reflection because the search itself helps us to understand the origins and future as humanity.

These Big Questions Have Been Asked Before

The question as to whether "other worlds" exist is not a totally new question. Alberto Magno (1193–1280) commented: "One of the most marvelous and noble questions in Nature is whether there is just one world or whether there may be many . . . It's our desire to investigate the matter."[10]

From the time of the Greek philosophers, the debate over "the plurality of worlds" was an intense issue of debate between the Epicureans (who favored of the "plurality of worlds") and Aristotelians (who favor the uniqueness of Earth).[11]

The important historic consequences that this debate generated should merit a special attention in our studies. Consider, for example, the thinking of Giordano Bruno (1548–1600). This thinker adopted the heliocentric ideas of Nicolaus Copernicus (1473–1543), transforming Bruno into a believer in the theory of an infinite number of stars similar to our Sun, with planets orbiting those Suns, all inhabited, in an eternal universe. Bruno criticized Copernicus's idea because Copernicus's thinking was confined exclusively to mathematics, refusing to address the philosophical issues of the new vision of the universe. Therefore, to Bruno, Earth is a planet similar to others that could be designated to be "other Earths."[12] Another point in Bruno's thinking was that he denied the legitimacy of the notion that there exists any "center" of an infinite universe. Due to the limitations of this work, I

---

9. See Frank and Sullivan, "Sustainability and the Astrobiological Perspective," 32–41.
10. Crowe, *ELD*, 6.
11. Ibid., 8.
12. Fantoli, *Extraterrestri*, 44–45.

will simply mention that it is also important to consider the idea of other inhabited worlds as was discussed by Johannes Kepler and Galileo Galilei (1564–1642).[13] A less well-known case is the case of the Jesuit Priest Angelo Secchi (1818–1878), an astronomer and one of the founders of the field of modern Astrophysics who was the Director of the Astronomical Observatory of the Collegio Romano, who was the first to classify stars in spectral classes. Secchi, in the nineteenth century, discussed the existence of other inhabited planets, as to which he was convinced.[14]

In the search for answers to these open questions, I have learned few lessons that I will summarize here. Also I will point out some challenges that lay ahead.

The Road Map to Other Earths Is the Next Copernican Revolution

I would like to start with a biblical scene taken from the book of Genesis. God promised Abraham to be the Father of a multitude of descendants as numerous as the stars in the sky: " He brought him outside and said, 'Look toward heaven and count the stars, if you are able to count them.' Then he said to him, 'So shall your descendants be'" (Gen 15:5).

This is a beautiful image that I would like to reflect upon. God had called Abraham, the Father of three of the world's great traditions—Jewish, Christian, and Muslim—to leave his land and his relatives to go to the Promised Land (Gen 12:1). Afterwards God took him outside to gaze at the stars.

I like to think that looking up at the sky is one of the first human acts. Our most ancient ancestors most likely looked at the unfathomable magnificence of the sky and pondered their own significance in the cosmos they knew. Since earliest times, the beauty of the sky has fascinated humankind. Some of us today miss this raw experience of night sky due to light pollution in our cities; we can't see glories of the Milky Way. Still, it is easy to believe that many ancient people adored the stars, the sun, the moon, and the planets attracted by their magnificence. Vincent Van Gogh said: "When I have a terrible need of—shall I say the word—religion. Then I go out and paint the stars."[15]

---

13. Ibid., and Crowe, *ELD*, 9–13.
14. Secchi, *Le Soleil*, 418; and Secchi, *Le Stelle*, 337.
15. Van Gogh quoted in Adams and Adams, *An Examined Faith*, 259; van Gogh, Wikiquote, "Vincent van Gogh."

Today's astronomers paint the stars, so to speak. The US National Astronomical Observatory with its many telescopes is built in Kitt Peak, in the territory of the reserve of the Tohono O'odham tribe in Southern Arizona. The Tohono O'odham tribe or the "People of the Desert" called astronomers "The People with Long Eyes." Indeed, we are a species with long eyes and Galileo Galilei is the forefather of this "People with Long Eyes." Galileo was the first *Homo sapiens* to point the telescope to the Moon, Jupiter and its satellites, and the stars about four hundred years ago. Today we, astronomers, are fascinated by the universe as much as our forefathers were. We seek answers to the fundamental questions about the universe: Are we alone? Are there other Earths?

The universe is made of a hundred billion galaxies. I find amazing that the number of galaxies is similar to the number of neurons in our brain. Each galaxy contains more than a hundred billion stars. There are as many stars in the universe as sand grains on all the beaches on Earth. And each of those stars might have planets orbiting around. If planets are a common characteristic of stars, we ask ourselves if life is also a common characteristic of the stellar systems. Is Earth a unique case or a common phenomenon? Even the public opinion in the scientific community considers that Earth is not a unique case where life could have developed; there exists some dissident opinion that deserves to be considered because it could shine some light on the investigation of life in the universe. In the book *Rare Earth*, Ward and Brownlee propose that intelligent life and animal life could be extremely rare in our galaxy and in the universe.[16]

Astrobiology is a "frontier" scientific-discipline with profound philosophical, social, and religious implications. For this reason, the Vatican Observatory has been involved in this field, organizing graduate schools and congresses on this important subject. As the Director of this institution and therefore member of the Pontifical Academy of Sciences, I proposed a Study Week meeting that gathered world scientists to undertake a collective discussion of Astrobiology, held under the auspices of the Pontifical Academy of Sciences in November of 2009. The Vatican authorities deemed the study on Astrobiology to be a perfectly sound and appropriate subject for study and discussion at the Pontifical Academy since it requires multidisciplinary collaboration. The following quote from Cardinal Giovanni Lajolo's address to the participants shows the appreciation of the Catholic Church for Astrobiology:

> It is a field which requires a range of all but the most profound of scientific knowledge, as well as highly refined research

16. See: Ward and Brownlee, *Rare Earth*.

techniques. Because it means often proceeding on the basis of scarce evidence and formulating hypotheses requiring strict verification, which in turn, can be diversely configured. It means resorting to results of research based on extreme aspects of possibility of life on Earth, and to study how to verify its presence on other planets or exoplanets. It means—at its limit—studying if and how one could verify the existence of extraterrestrial forms of intelligence and how to enter in contact with them. This is a task that demands scientific integrity . . . an intense and indispensable case of a vast multi-disciplinary research. In research . . . the scientist must also be allowed the possibility to walk paths which do not always lead to positive results . . .[17]

Exoplanets are beyond the reach of our space probes and the only option in the coming decades, if not centuries, is the remote study using spectroscopic analysis of the reflected or transmitted light by living organisms and observed through telescopes. The discovery of the first exoplanet orbiting the star 51 Pegasi, similar to our Sun, dates back to 1995.[18] Currently, the number of known planetary systems continues to grow. The exoplanets.org site is a database that contains useful information about exoplanets. At the time of writing, the number of confirmed planets is 3584. About two dozens are in a habitable zone of their parent star. A very important objective in the search for these exoplanets is to discover the existence of other Earth-like planets similar to our Earth, in mass and diameter, which orbit around its respective star that is similar to our Sun, within the habitable zone.[19]

The Kepler telescope has been specifically designed to be able to detect in a region of the Milky Way planets that are similar in size and density to our Earth and that are within the habitable zone. Using the Kepler telescope, astronomers have discovered the first Earth-like planet, Kepler 186f, in the habitable zone.[20] Recently astronomers announced that they had detected a planet orbiting Proxima Centauri in the habitable zone, the closest neighbor to our solar system.[21]

Considering the number of already-discovered exoplanets within our Milky Way Galaxy, it seems that the vast majority of stars in our galaxy

---

17. Impey et al., *Frontiers of Astrobiology*.

18. Mayor and Queloz, "A Jupiter-mass Companion," 355.

19. It's worth noting that there could exist habitable galactic zones, as in, the regions of the galaxy that host the formation of habitable worlds.

20. Bolmont et al., "Formation, Tidal Evolution, and Habitability," 3; NASA. Kepler. http://kepler.nasa.gov/news/nasake[plernews/index.cfm?FuseAction=ShowNews&NewsID=330.

21. Chang, "One Star Over."

are, at least potentially, able to have planets on which life may well have evolved. But we do not yet know if the phenomenon Earth-like planet is rare or common.

The characteristics of the universe must be such as to enable the emergence and evolution of the observer. This line of thinking is called the "weak anthropic principle." The "strong anthropic principle" states that the universe must have such properties that allow life to evolve from a certain point in its history.[22] Somehow the "strong anthropic principle" expresses a teleological view, a purpose, of the universe. The strong and weak formulations establish that any valid theory of the universe must be consistent with the existence of human beings or life in general.

Martin Rees puts on the spotlight our crucial position in the Universe: "The most crucial location in space and time (apart from the big bang itself) could be here and now."[23]

My first observation is that human thinking seems to oscillate between an Anthropocentric Principle (AP) and a Copernican Principle (CP). The AP places humanity in a privileged position, perhaps supporting anthropocentrism. The Ptolemaic system inherited by our religious forebears placed the Earth at the center. Might the AP return us to a human-oriented universe? Would the CP, which places the Sun at the center, relativize the human position in the universe? Does the CP de-center us on Earth? Does the physical relationship between Sun and Earth contribute to how we *Homo sapiens* should evaluate our relative worth to the cosmos?

A more recent version of Copernican de-centering makes our universe one of an infinite number of universes. This is what multiverse theory does to the geocentric mind. The Anthropic Principle might make our race of intelligent creatures on Earth feel special, whereas variants on the Copernican Principle, according to Richard Gott, say that our "location is not likely to be special."[24] One of the tasks of the astrotheologian, according to Ted Peters, is to draw out for religious sensibilities the role of geocentrism in human self-understanding.[25]

As Sara Seager, a world expert on exoplanets, has pointed out: "When and if we find that other Earths are common and see that some of them

---

22. Carr, "Cosmology and Religion," 148.

23. Rees, *OFH*, Chapter 1.

24. Gott, "Our Future in the Universe," 415. Perhaps we should clarify. The Anthropic Principle makes intelligent life special, whether on Earth or elsewhere. The Copernican Principle removes the special status of Earth in relation to other locations which may or may not bear intelligent life. The two principles do not sit on opposite ends of a single teeter totter.

25. Peters, "Astrotheology," *838–55*.

have signs of life, we will at last complete the Copernican revolution—a final conceptual move of the Earth, and humanity, away from the center of the Universe. This is the promise and the hope for exoplanets—the detection and characterization of habitable worlds."[26]

A "Copernican Revolution," in spiritual terms, means going out, leaving our own convictions, our well-known land and culture to change our mindset if necessary, not being auto-referential, as Pope Francis is encouraging the Catholic Church to be.

We can ask ourselves: What is the place of humankind in this huge Universe of billions of galaxies, each of them with billions of stars, and billions of worlds? This question opens our hearts and minds to another deeper question full of wonder as put by Psalm 8: "What are human beings that you are mindful of them, mortals that you care for them?" (Ps 8:4)

Life Is a Process and a Resilient Gift

The best scientific explanation that we have for the beginning of the Universe is the Big Bang model, which is confirmed by observations combined with speculations. Though there are many unknowns, our current understanding of physics allows us to reconstruct the history of the universe. We understand quite well the formation and evolution of galaxies, stars and planets. In this context the phenomenon of life is more remarkable than the other processes that give rise to planets, stars, galaxies, etc. The earliest hint of life on Earth is at 3.8 billion years before present. As far as we know it took about 9 billion years for the universe to produce life and about 3.8 billion years to produce modern humans who arose about forty thousand years ago.

As Paul Davies points out, "once life was initiated the universe would never be the same. Slowly but surely it has transformed Planet Earth. And by offering a route to consciousness, intelligence, and technology, it has the potential to change the universe."[27] Life is a process and although, in my opinion, it will be difficult to find life in actuality; we could find favorable conditions for the origin and development of life or we could find life in different stages of its evolution. Many scientists believe that life could arise wherever conditions permit. Therefore life would be in the natural order of things. Life would be written in the laws of nature. Thus we would be inhabitants of a biofrendly Universe. In this sense Angelo Secchi, in the

26. Seager, "Searches for Habitable Exoplanets," 231.
27. Davies, *The Fifth Miracle*, 13.

nineteenth century, believed that life filled the universe, and that life was associated to intelligence.

We also have learned that life is resilient. While we search for life out there, we need to understand life on Earth. Life has an extraordinary ability to adapt to the most extreme environments on Earth. Places once thought to be sterile, such as the Atacama Desert and the deep reaches of the Earth's crust, contain life. The range of adaptation of terrestrial extremophiles—organisms that thrive in physically or geochemically extreme conditions—implies that the traditional definition of the habitable zone within a planetary system is overly conservative.

As far as I know, the origin of life is still a mystery. Paul Davies in the Preface of his book *The Fifth Miracle*, explains the title in this way: "It derives from the biblical account in the book of Genesis, 'Let the land produce vegetation.' This is the first mention of life, and it seems to be the fifth miracle."[28] He also clearly states that he is not suggesting that the origin of life actually was a miracle.

I believe that our deepest experience is that life is given to us, it is a gift. Although I have said that life is resilient, it is also true that life is fragile and that we have to take care of it. In the Bible many times, life is referred directly to God as its origin and source, as the Creator, the Giver of life.

Two quotes at the beginning and end of the Bible illustrate this. In one of the accounts of Creation, the author of Genesis says: Then the LORD God formed man from the dust of the ground, and breathed into his nostrils the breath of life; and the man became a living being" (Gen 2:7). At the end of the book of Revelation, we read: "The Spirit and the bride say, 'Come.' And let everyone who hears say, 'Come.' And let everyone who is thirsty come. Let anyone who wishes take the water of life as a gift" (Rev 22:17).

Our Thinking Is Incomplete

Pope John Paul II reminded theologians that they have the duty to keep themselves regularly informed of scientific advances to take them into account in their studies or for introducing changes in their teaching if necessary.[29] Pope Francis, addressing the community of the Gregorian University, pointed out that "The theologian who is satisfied with his complete and conclusive thought is mediocre. The good theologian and philosopher has

---

28. Ibid., 24.

29. Pope John Paul II, Address to the Participants of the Plenary Session of the Pontifical Academy of Sciences.

an open, that is, an incomplete, thinking, always open to the mind of God and of the truth, always in development . . ."[30]

The challenges that science poses today should enter within the horizon of our philosophical and theological reflection. On one side, our philosophical and religious thinking cannot follow the latest scientific findings, or not yet verified hypotheses, on the other side we cannot remain attached or fixed to the "center." The dialogue with the modern world demands wide horizons, intellectual honesty, freedom of spirit, and, above all, not being afraid.

Today's astrotheologian ought not to begin with a dogmatic stance and then merely attempt to support it with whatever scientific evidence comes along. Rather, growing scientific knowledge fertilizes theological soil so that a new comprehension of God's creation can blossom. Isaac Newton's words illuminate what I mean: "I do not know what I may appear to the world, but to myself I seem to have been only like a boy playing on the sea-shore, and diverting myself in now and then finding a smoother pebble or a prettier shell than ordinary, whilst the great ocean of truth lay all undiscovered before me."[31]

Patience

The discovery of the existence of an ET civilization could happen tomorrow, or later in this twenty-first century, or perhaps never. Whatever might be the time period involved, the discovery of an ET civilization and then our potential "contact" with them will, inevitably, have a profound impact on our human philosophical, social, and religious comprehension of society and of the entire universe.

Giuseppe Tanzella-Nitti, astronomer and theologian, affirms: "The last word on the question of ET life does not come from theology, but from science. The existence of intelligent life on other planets beyond Earth is neither required nor excluded by any theological argument. For theology, as for all humanity, all we can do is wait, patiently."[32]

I would add: all we can do is wait patiently, looking forward to the surprises of God who "saw everything he had made, and indeed, it was very

---

30. Pope Francis, "Address to the Community of the Pontifical Gregorian University."

31. Brewster, *Memoirs of the Life, Writings, and Discoveries of Sir Isaac Newton*, vol. II. ch. 27.

32. Tanzella-Nitti, "Extraterrestre, Vita."

good" (Gen 1:31). God is not yet done making good things. The astrotheologian will be a patient theologian.

Questions to Theology

One task of Christian theology consists of a critical reflection on the contents of faith in order that the life of the believer be full of significance. A corollary task of astrotheology, then, consists of a critical reflection on the cosmos as interpreted by space science in order that the believer—including the believing scientist—may broadly and deeply assess the grand creation's significance. In this sense, theology is not limited to deductions from special revelation. The theologian, like the scientist, is dependent on empirical knowledge for broadening and deepening of his or her comprehension of reality. The astrotheologian right along with the astrobiologist relishes each new factual discovery, and each new factual discovery contributes to vision revision. This implies, among other things, that astrotheologians ought to speculate about some of the questions that arise with the potential discovery of ETI. The consideration of these questions could lead to a more comprehensive grasp of the contents of faith and their significance.

This entails, then, a readiness to revise in light of new discoveries. It could mean that some core Christian principles might require assessment. These are core principles that might be challenged by a potential discovery of ETI: uniqueness and centrality of humankind, uniqueness of the Incarnation, Christocentrism, original sin, and redemption. If we on Earth find we must share our universe with intelligent creatures living on other planets then, like trying on a new pair of shoes, the astrotheologian will have to see if these classic doctrines fit.

The Roman Catholic Church previously faced such challenges when Europeans encountered native residents living in the Americas. A potential discovery of ET life would pose a problem similar to the discovery of the Native Americans, about whom Pope Paul III (1534–1549) did not have difficulty in recognizing that they belonged to the descendants of Adam and Eve.[33]

Benedict XVI, addressing the participants in the Colloquium sponsored by the Vatican Observatory on the occasion of the International Year of Astronomy, engaged in some de-centering: "Modern cosmology has shown us that neither we, nor the earth we stand on, is the center of our universe, composed of billions of galaxies, each of them with myriads of stars and planets. Yet, as we seek to respond to the challenge of this Year—to

33. Ibid.

lift up our eyes to the heavens in order to rediscover our place in the universe—how can we not be caught up in the marvel expressed by the Psalmist so long ago?" (Ps 8:4–5).[34] The physical non-centrality of humanity can help us to be more humble and astonished that God takes care of us, small that we are. Physical de-centering translates into spiritual de-centering, and this is a logical consequence of the Copernican Principle.

Regarding the christological question—one incarnation on Earth or multiple, one for each planet bearing intelligent life?—Benedict XVI seems to side with a single incarnation. He stated, "Revelation tells us that, in the fullness of time, the Word through whom all things were made came to dwell among us. In Christ we acknowledge the true center of the universe and all history, and in him, the incarnate Logos, we see the fullest measure of our grandeur as human beings, endowed with reason and called to an eternal destiny."[35] Even the pope found he had to ask the question: one incarnation or many?

It is important to observe that the Magisterium of the Roman Catholic Church has no official teachings regarding ETI life. Because of my former position as director of the Vatican Observatory, media reporters easily found my phone number. I have been quoted many times related to this issue. In 2008 I gave an interview to *L'Osservatore Romano* where I basically responded to the journalist's questions in a very simple way that I will summarize here.

We do not know if life is a unique event within cosmic history or a universal and a quasi-inevitable phenomenon. I do not think that the existence of ETI would present a problem for the Catholic faith. Just as there is a multiplicity of creatures on the Earth, so there could be other beings, including intelligent ones, created by God. This is not in contrast with our faith, because we cannot set limits to the creative freedom of God. If we consider earthly creatures as brothers and sisters, as Saint Francis did, why should we not speak also of an ET family? They would still be part of God's good Creation.

I also proposed to borrow the Gospel parable of the lost sheep. The shepherd leaves the other ninety-nine sheep in the flock to go and look for the one who is lost. We can consider that in this universe there may be a hundred sheep, corresponding to diverse kinds of creatures. Human beings might be the lost sheep, the sinners who need the shepherd. God became human in Jesus to save us. Even if other ETI beings were to exist, they may not necessarily be in need of salvation. Perhaps they are not fallen. Perhaps

---

34. Pope Benedict XVI. Address of His Holiness Pope Benedict XVI.
35. Ibid.

ETI have remained in full friendship with their Creator. But if ETI are also sinners, then redemption would be possible for them. Our gracious God would make this happen. Jesus became God-in-the-flesh only once. The Incarnation is a unique event in the history of the whole universe, or even every other universe if the multiverse exists. As established by the Second Vatican Council: "By his Incarnation, He, the Son of God, in a certain way united himself with each man and woman."[36] We could extend this affirmation to our potential extraterrestrial family members. Whether one incarnation or many, somehow ETI would also have the possibility to enjoy God's mercy just as we have done.

Education and Public Outreach

The current search for ETI life poses questions that scientists could attempt to answer from a pure scientific and pragmatic point of view. Nevertheless an important investigation requires a cautious reflection from different perspectives and disciplines. A multidisciplinary approach is required to answer questions about the nature of life, intelligence, spirituality, as well as about values and scope of any civilization.

I believe that the researchers have an important social responsibility that consists of sharing their knowledge with their fellow citizens, especially contributing to the formation of the young generations. This interdisciplinary dialogue should not be limited only to experts, but should reach out to college and high-school students and the general public. School should be the place where the new generations can be trained in developing fine skills to participate in a multidisciplinary dialogue to discuss philosophical, social, and religious approaches to significant issues. College and high school faculty plays a key role in training new generations in multidisciplinary approach to discuss complex matters.

A multidisciplinary approach to the search of inhabited worlds could generate an entirely new and important field of human studies at the college and high school levels. These studies can help to motivate young people to explore more deeply what it means to be uniquely human and to develop a better understanding of who we are, from a truly "cosmic" and less narrow-minded perspective.

---

36. Pope Paul VI, *Gaudium et Spes*, 22.

Interreligious Dialogue

As we speculate about potential impact of confirming that we share our cosmos with other intelligent civilizations, we must ask about the future of religion on Earth.[37] Not only Christian theologians will be challenged, so also will the intellectual leaders of every great religious tradition.

When I was the Director of the Vatican Observatory, I was invited by the Iranian Embassy to the Holy See to visit Iran to meet my Muslim colleagues. As a result, I was personally responsible for organizing a workshop about the role of Astronomy in Christianity and Islam. I am, therefore, entirely conscious of both the challenges and the benefits of an interreligious dialogue. Based upon my direct experience, I would like to further explore the importance of this issue of the existence of other inhabited worlds in Christianity, Judaism, and Islam. I believe that is very important to seek collaboration on this unique subject of other Jewish and Muslim researchers, which I think to be crucial. The interreligious dialogue is urgently needed in our modern global civilization. In my view our survival as a species will depend on our ability to maintain a dialogue. For this reason education is extremely important.

Curiosity is a driving force to do science, to do research. Human beings are basically curious. We want to know how the universe works, the logic, the "logos" in the universe. This drive for knowledge has a basis in the nature of cosmos. Because there is rationality in the universe, we can do science. There is nothing better for religion than good science. People of faith have nothing to fear from science. None of us should be afraid of new results, new discoveries. Whatever the truth might be, we should be open to new results, once they are confirmed by the scientific community.[38]

With regard to the part played by theology in this context, Pope Francis's words can help to understand the role of religion today: "Religious classics can prove meaningful in every age; they have an enduring power to

---

37. For a more extensive treatment of the subject see Peters, *"Implications,"* 644–55; and Weintraub, *Religions and Extraterrestrial Life: How Will We Deal With It?*

38. On the positive attitude of the Catholic Church towards scientific progress, I find helpful this paragraph from Pope Francis' document *Evangelii Gaudium*, 243: "The Church has no wish to hold back the marvellous progress of science. On the contrary, she rejoices and even delights in acknowledging the enormous potential that God has given to the human mind. Whenever the sciences—rigorously focused on their specific field of inquiry—arrive at a conclusion which reason cannot refute, faith does not contradict it. Neither can believers claim that a scientific opinion which is attractive but not sufficiently verified has the same weight as a dogma of faith. At times some scientists have exceeded the limits of their scientific competence by making certain statements or claims. But here the problem is not with reason itself, but with the promotion of a particular ideology which blocks the path to authentic, serene and productive dialogue."

open new horizons . . . Is it reasonable and enlightened to dismiss certain writings simply because they arose in the context of religious belief?"[39]

Our Future

Martin Rees in his book *Our Final Hour* writes: "The wider cosmos has a potential future that could even be infinite. But will these vast expanses of time be filled with life, or as empty as the Earth's first sterile seas? The choice may depend on us, this century."[40]

Pope Francis, in his encyclical *Laudato Si'* on the care of our common home, affirms: "A fragile world, entrusted by God to human care, challenges us to devise intelligent ways of directing, developing and limiting our power."[41]

Until today interstellar travel has been proved to be quite difficult, it remains well beyond our technological capabilities, because of the technological requirements for engines, the enormous energy needed to accelerate spacecraft to speeds near the speed of light, and the difficulties of shielding the crew from radiation. We must be aware that Planet Earth is our only home in the solar neighborhood. Therefore we should take care of our common home.

Conclusion

To conclude I return to a very well known Gospel scene, the visit of the Magi. They were also part of the "People with Long Eyes" and, like Abraham, they left their land to pursue God's call through following the rising star. We can regard the Magi as wise and well-educated people. Today we could say that they were expert in many disciplines, they were part of the "Multidisciplinary People." Also they were intellectually honest, free to overcome cultural biases, and courageous enough to take the risk to leave their own certainties.

The roadmap to other Earths is a spiritual journey for all people of good-will not only for experts. Like the Magi, only if we have a free spirit, will we be overjoyed at seeing the star (Mt 2:10). "Count the stars, if you are able to count them," God told Abraham (Gen 15:5).

---

39. Pope Francis, *LS* n. 199.
40. Rees, *OFH* Chapter 1.
41. Pope Francis, *LS*, n. 78.

Bibliography

Adams, James Luther, and Jonathan Adams. *An Examined Faith: Social Context and Religious Commitment*. Minneapolis: Fortress, 1991.
Arber, Werner. "Preface." *Plenary Session on Evolving Concepts of Nature*. October 24–28, 2014. Vatican City: The Pontifical Academy of Sciences, 2014.
Aristotle, *Metaphysics*. Translated by W.D. Ross. http://classics.mit.edu/Aristotle/metaphysics.html.
Bolmont, Emeline, et al. "Formation, Tidal Evolution, and Habitability of the Kepler-186 System." *The Astrophysical Journal* 793 (2014) 3.
Brewster, David. *Memoirs of the Life, Writings, and Discoveries of Sir Isaac Newton*. Vol. 2. London: Thomas Constable, 1855.
Carr, Bernard. "Cosmology and Religion." In *OHRS*, 139–55.
Chang, Kenneth. "One Star Over, a Planet that Might be Another Earth." *The New York Times*, August 24, 2016. http://www.nytimes.com/2016/08/25/science/earth-planet-proxima-centauri.html?rref=collection/sectioncollection/science&action=click&contentCollection=science&region=rank&module=package&version=highlights&contentPlacement=1&pgtype=sectionfront&_r=1.
Chozick, Amy. "Hillary Clinton Gives U.F.O. Buffs Hope She Will Open the X-Files." *The New York Times*, May 10, 2016. http://www.nytimes.com/2016/05/11/us/politics/hillary-clinton-aliens.html?smprod=nytcore-iphone&smid=nytcore-iphone-share&_r=0.
Crowe, Michael J. *Extraterrestrial Life Debate from Antiquity to 1915: A Source Book*. Notre Dame, IN: University of Notre Dame Press, 2008.
Davies, Paul C. W. *The Fifth Miracle: The Search for the Origin and Meaning of Life*. New York: Simon & Schuster, 1999. Kindle Edition.
Dick, Steven J. *Life on Other Worlds: The 20th Century Extraterrestrial Life Debate*. Cambridge: Cambridge University Press, 1998.
Fantoli, Annibale. *Extraterrestri: Storia di un'idea dalla Grecia a oggi*. Rome: Carocci editore, 2008.
Frank, Adam, and Woodruff Sullivan. "Sustainability and the Astrobiological Perspective: Framing Human Futures in a Planetary Context." *Anthropocene* 5 (2014) 32–41.
Gott, J. Richard. "Our Future in the Universe." In *WU*, 400–424.
Impey, Chris, et al, eds. *Frontiers of Astrobiology*. Cambridge: Cambridge University Press, 2012.
NASA. Kepler. http://kepler.nasa.gov/news/nasake[plernews/index.cfm?fuseAction=ShowNews&NewsID=330.
Mayor, Michel and Didier Queloz. "A Jupiter-mass Companion to Solar-type Star." *Nature* 378: 6555 (1995) 355–59.
Peters, Ted. "Astrotheology." Chapter 72 in *The Routledge Companion to Modern Christian Thought*, edited by Chad Meister and James Beilby, 838–53. London: Routledge, 2013.
———. "The Implications of the Discovery of Extra-Terrestrial Life for Religion." *The Royal Society, Philosophical Transactions A* 369:1936 (February 13, 2011) 644–55. http://rsta.royalsocietypublishing.org/content/369/1936.toc.
Pope Benedict XVI. Address of His Holiness Benedict XVI to Participants in the Colloquium Sponsored by the Vatican Observatory on the Occasion of the

International Year of Astronomy. http://w2.vatican.va/content/benedict-xvi/en/speeches/2009/october/documents/hf_ben-xvi_spe_20091030_specola-vaticana.html.

Pope Francis. "Address to the Community of the Pontifical Gregorian University." April 10, 2014. http://w2.vatican.va/content/francesco/en/speeches/2014/april/documents/papa-francesco_20140410_universita-consortium-gregorianum.html.

———. *Evangelii Gaudium*. http://w2.vatican.va/content/francesco/en/apost_exhortations/documents/papa-francesco_esortazione-ap_20131124_evangelii-gaudium.html.

———. *Laudato Si'*. 2016. https://w2.vatican.va/content/dam/francesco/pdf/encyclicals/documents/papa-francesco_20150524_enciclica-laudato-si_en.pdf.

Pope John Paul II. Address to the Participants of the Plenary Session of the Pontifical Academy of Sciences. October 31, 1992. http://w2.vatican.va/content/john-paul-ii/it/speeches/1992/october/documents/hf_jp-ii_spe_19921031_accademia-scienze.html.

Radio Vaticana. http://it.radiovaticana.va/news/2014/05/12/papa_francesco_celebra_la_messa_a_casa_s_marta/1100388.

Rees, Martin. *Our Final Hour*. New York: Basic Books, 2003. https://itun.es/us/X9EXw.l.

Seager, Sara. "Searches for Habitable Exoplanets." In *Frontiers of Astrobiology*, edited by Christopher Impey et al, 231–49. Cambridge: Cambridge University Press, 2012.

Secchi, Angelo. *Le Soleil*. Paris: Gauthier-Villars, 1875.

———. *Le Stelle*. Milan: Dumolard, 1877.

Tanzella-Nitti, Giuseppe. "Extraterrestre, Vita." *Documentazione Interdisciplinare di Scienza e Fede*. http://www.disf.org/.

Ward, Peter D., and Donald Brownlee. *Rare Earth: Why Complex Life is Uncommon in the Universe*. New York: Copernicus, 2003.

Weintraub, David A. *Religions and Extraterrestrial Life: How Will We Deal With It?* Heidelberg: Springer, 2014.

Wikipedia. "Where No Man Has Gone Before." https://en.wikipedia.org/wiki/Where_no_man_has_gone_before.

Wikiquote. "Neil Armstrong." https://en.wikiquote.org/wiki/Neil_Armstrong.

———. "Vincent van Gogh." https://en.wikiquote.org/wiki/Vincent_van_Gogh.

4

Discovering ETI

What Are the Philosophical and Theological Implications?

ROBERT JOHN RUSSELL

There are infinite worlds both like and unlike this world of ours.

—EPICURUS[1]

Over the past half century, the interdisciplinary field of *Theology and Science* has undergone tremendous growth involving scholars from philosophy of science, philosophy of religion, the natural sciences, theology, ethics, history of science, and related fields.[2] Provocative and exciting conversations have been prompted by scientific topics such as quantum physics, relativity, physical cosmology, evolutionary biology, genetics, and neuroscience. Surprisingly underrepresented in this rapidly growing interaction, however, is

1. Epicurus, Letter to "Herodotus," 5.
2. For a scholarly introduction, see Barbour, *Religion in an Age of Science*; Murphy, *Theology in the Age of Scientific Reasoning*; Peacocke, *Theology for a Scientific Age*; Polkinghorne, *Faith of a Physicist*; Richardson and Wildman, *Religion and Science*. For a less technical introduction see Haught, *Science and Religion*; Peters, *Science and Theology*; Peters, "Science and Religion: An Overview," in *ER2*, 12:8180–92; Southgate et al., *God, Humanity and the Cosmos*. For a recent survey article with extensive references see my "Theology and Science: Current Issues and Future Directions" at www.ctns.org.

a focus on the philosophical and theological issues raised by the possibility of extraterrestrial intelligent life (ETI).[3] This is particularly curious since historians of science have shown that Christian theology contributed in significant ways to the assumption that ETI does in fact exist and that ETI belongs within the historical Christian concept of creation.[4] It is particularly timely, then, that these issues be adumbrated and addressed.

There are, in fact, a wealth of subtopics which could be discussed here as implications or questions raised by the discovery of ETI.[5] For the purposes of this short chapter, I will focus on six aspects of the scientific search for ETI and its philosophical and theological implications.[6] Let me preface my treatment with three preliminaries. First, I will speak from the context of Christian theology and welcome others to widen the circle of religious discussions of extraterrestrial life. Second, I will assume that for some forms of ETI the contact scenarios are reasonable. We need only address contact scenarios that are reasonable, not extraterrestrial life forms which remain unknown or with whom we may never communicate.[7] Third, my presentation of the philosophical and theological implications of the discovery of ETI employs the standard method for integrating science into theology, one which Ian Barbour calls a *theology of nature* and which I call *constructive theology in light of the natural sciences*. When the particular focus is on intelligent life in the universe, Ted Peters names it *astrotheology*.[8] I start with issues and implications which lie at the empirical end of the spectrum (even

---

3. See Peters, "Exo-Theology: Speculations on Extra-Terrestrial Life"; and "Extraterrestrial Intelligence," 779–80. For a recent survey of Roman Catholic views, see Vakoch, "Roman Catholic Views." See also McMullin, "Life and Intelligence Far from Earth"; Coyne, "The Evolution of Intelligent Life." For a recent survey of ethical implications see Randolph et al., "Reconsidering the Theological and Ethical Implications." Then see of course Ted Peters's essay in this publication.

4. Dick, *Plurality of Worlds*, esp. Ch. 2. "We cannot meaningfully speak of an absolute break between religion and the sciences after Copernicus and Descartes." Bolle, "Cosmology: An Overview," in *ER1*, 3&4:105–6.

5. In my opinion the discovery of ETI would not undermine the central claims of 'established religion', particularly Christianity, as many authors assume, although it would challenge a literal and inerrant reading of scripture in *any* religion. For Christians who take scripture as normative but not as literal and inerrant, there is plenty of room for a creative interaction between faith and science, including the discovery of ETI.

6. This is a revised and updated version published in Chela-Flores et al., *First Steps in the Origin of Life*.

7. See for example Dyson, *Disturbing the Universe*, ch. 19; Dick, "Extraterrestrials and Objective Knowledge," 47–48.

8. Peters, *GWF*, 701.

though we as yet have no direct evidence about them!) and move from there to implications which are more philosophical and theological.

1. Is Life in the Universe Rare or Abundant?

The relative abundance of life in the universe is one of the pivotal issues of our time, and a key question in the burgeoning new fields of exobiology and astrobiology. Projects such as the exploration of Mars now underway through NASA's Curiosity[9] and future plans for the exploration of Jupiter's moon Europa might well shed light on the question of pre-biotic and microbial life in our solar system. On the other end of the life-spectrum, projects which listen for signals using radiotelescopes might find evidence of advanced life in the nearby arms of our Milky Way galaxy. A notable example is the SETI Institute[10] in Mountain View, California. And since the announcement of the detection of the first exoplanet in 1996, 51 Pegasi b, by Michel Mayor and Didier Queloz, the search for extrasolar planets within a few thousand light-years from us has burgeoned.

Vatican Observatory researcher, the late William Stoeger, speaks for much of the scientific community. "Though we have yet to find an instance of extraterrestrial life, almost all experts agree that primitive life elsewhere in the universe is very, very likely, if not inevitable. They also consider conscious, even intelligent, life, though undoubtedly a much rarer occurrence, a definite possibility."[11] The cover of the May 3, 2013 edition of *Science* carried the striking title, "Exoplanets." The issue includes a sixteen page special section on the search for exoplanets and a four page report of the discovery of Kepler-62, a five-planet system with two small planets in the star's habitable zone. And the cover of the July, 2013 edition of *Scientific American* spotlights its key article, "The Dawn of Distant Skies," by Michael D. Lemonick. Here Lemonick begins as follows: "The galaxy is teeming with planets. Scientists are straining to peer into their atmospheres to seek signs of extraterrestrial life."

Still the question remains: Is life, and especially intelligent life, rare or abundant in our galaxy and our universe? Until the results of these projects are conclusive, we must be satisfied with theoretical attempts at estimating the relative abundance of life in the universe, and these attempts remain highly controversial. There are two sides to this controversy: those who

9. NASA. Curiosity Rover. http://www.nasa.gov/mission_pages/msl/index.html.
10. SETI, http://www.seti.org/.
11. Stoeger, SJ, review of *Vast Universe*, by Thomas F. O'Meara, in *Theology and Science*, 77.

affirm a rare or unique Earth versus those who are contact optimists. On the one hand, evolutionary biologist Stephen Jay Gould belongs in the rare Earth camp. He has stressed the unlikelihood of intelligent life elsewhere in the universe given the overwhelming role that chance has played in the evolution of life on earth.[12] And if life has evolved elsewhere, we should expect radical diversities in its morphologies, diversities reflecting the vastly differing evolutionary histories as shaped by strongly contingent events. Francisco Ayala has gone further and argued that life beyond the microbial stage is extremely unlikely even in our galaxy of several hundred billion stars, given the number of steps from simple living organisms to sentient creatures.[13] On the other hand, in the contact optimism camp we find Paul Davies[14] and Christian de Duve.[15] In a nutshell their argument is that since evolution is tightly governed by the laws of physics and biology, and since these laws led to the evolution of life on earth, they should lead to its evolution in abundance throughout the universe. Julian Chela-Flores has taken de Duve's point one step further, arguing that once life originates, the evolution of eukaryotes will lead to prokaryotes. From there the "convergence" of life towards intelligence is assured, although there may well be significant diversities in the morphologies which underlie intelligence.[16]

At this point, the empirical jury is out.

2. Will the Answer Influence Life's Value or Meaning?

Underlying the debate between the rare earthers and the contact optimists, however, is a more fundamental question: Is the meaning and value of life a function of its abundance in the universe? Or, is it essentially meaningful or meaningless *regardless* of what we find through the exploration of our solar system or through deep space via projects like SETI and the search for exoplanets? It is this question which has received considerable philosophical and theological reflection.[17]

12. Gould, *The Flamingo's Smile*.
13. For a delightful account in the author's own words, see Ayala, Closer to Truth.
14. Davies, "Teleology," 151–62.
15. de Duve, *Vital Dust*.
16. Chela-Flores, "The Phenomenon of the Eukaryotic Cell," 79–98.
17. Scarcity can indicate value: The discovery of a single palm tree in a vast desert does not mean that the tree is insignificant. Instead its scarcity makes it a tremendous discovery, for a hidden spring of life-giving water lies at its roots. I feel this way about Earth whether or not life is found elsewhere in the universe. Our planet is like the palm tree in what might in fact be a vast interstellar desert. If it takes a thousand million stars to produce the conditions for the possibility of a sea urchin, if it takes a billion years of

I will start with *arguments for meaninglessness*. Some scientists have suggested that biological life *per se* has little significance whether or not we are alone in the universe. They see life as essentially meaningless, a random product of physics and chemistry of no more significance than the wetness of water or the structure of Saturn's rings. Biological processes may take hold when really unusual conditions occur for the matter involved; but the universe, "at rock-bottom," is just endless mass-energy and curving spacetime.

Such cosmic nihilism is of course a *philosophical* interpretation of nature. It is not science *per se*, nor is it a position which can be proved by science. Rather, nihilism is a philosophical interpretation of nature widely propounded by such eminent scientists as Bertrand Russell[18] and Jacques Monod.[19] It is certainly the impression Steven Weinberg gave in his often-quoted conclusion to *The First Three Minutes*: "(H)uman life is ... just a more-or-less farcical outcome of a chain of accidents reaching back to the first three minutes ... The more the universe seems comprehensible, the more it also seems pointless."[20]

We will now turn to *arguments for meaningfulness*. Other scientists disagree with this nihilistic view, arguing instead for a philosophy in which the Big Bang followed by the evolution of life is, in some modest ways at least, a clue to a significant meaning of the universe. Paul Davies has depicted the evolution of life in terms of what he provocatively calls "teleology without teleology."[21] Here the emergence of what Davies calls the "order of complexity" is a genuine surprise, arising out of the "order of simplicity" described by the laws of physics. In a related way, William Stoeger has pointed to what he calls the "immanent directionality" of evolution. Stoeger's aim is to show that there is a directionality, perhaps even a teleology, immanent in nature that can be discovered through the natural sciences as they study the emergence of physical and biological structure, complexity, life, and mind.[22] More pointedly in his 1985 Gifford Lectures, *Infinite in All*

---

tinkering with genetic dice to produce a hummingbird, and if it takes a million years of scratching on bark and vocalizing intentions to produce a child who can reach out through human artifacts and chalkboard calculations and touch the edge of the visible universe, then life per se is surely "the pearl of great price" (Matt 13:45–46) and a clue to the theological meaning of the universe as a whole.

18. Russell, "A Free Man's Worship."
19. Monod, *Chance and Necessity*.
20. Weinberg, *The First Three Minutes*, 154.
21. Davies, "Is the Universe Absurd?" esp. 72–76; Davies, *Cosmic Blueprint*; Davies, "Teleology."
22. Stoeger, "The Immanent Directionality of the Evolutionary Process, and Its

Directions, renowned physicist Freeman Dyson explicitly rejects Weinberg's opinion, telling us instead he sees " . . . a universe of life surviving forever and making itself known to its neighbors across the unimaginable gulfs of space and time . . . Twentieth-century science provides a solid foundation for a philosophy of hope."[23] In sum, even without a philosophical teleology, many scientists turn their ears and listen to the faint voice of meaning rising up from what life is saying to us.

Who is right: those who argue against inherent meaning or those who argue for it? Again, the jury is out.

3. What Accounts for this Difference in Views?

Clearly the question of the relative abundance of life in the universe is an empirical one. The differences about the value and meaning of life based on the empirical facts, however, remain mainly philosophical and theological.

The philosophical difference—meaninglessness or meaningfulness—is not due merely to the presence or absence of data regarding alien life. It may stem in part from a deeper division between the way reductionist versus non-reductionist philosophies are used to interpret scientific discourse. If one assumes that the processes and properties characteristic of living organisms can be fully explained by physics and chemistry as Richard Dawkins does, there may be little, if any, basis for attributing meaning and value to life.[24] Non-reductionist arguments on the other hand, such as those deployed by Francisco Ayala,[25] Ernst Mayr,[26] and Charles Birch,[27] offer a basis within natural processes for attributing varying degrees of meaning and value to organisms with differing levels of complexity and organization.

These non-reductionist epistemologies, in turn, play a crucial theological role in a variety of views often referred to collectively as "theistic evolution."[28] Theistic evolutionists differ greatly from either creationists or intelligent design advocates. Theistic evolutionists begin with the sciences

---

Relationship to Theology," 163–90.

23. Dyson, *Disturbing the Universe*, 250; Dyson, *Infinite in All Directions*, 117–18.

24. Dawkins, *Blind Watchmaker*.

25. Ayala, "Introduction"; Ayala, "Reduction in Biology," 67–78; Ayala, "Can 'Progress' Be Defined?" 75–96.

26. Mayr, "How Biology Differs from the Physical Sciences," 67–78.

27. Birch, *A Purpose for Everything*; Birch, *Feelings*; Birch, "Neo-Darwinism," 225–48.

28. For an excellent introduction and overview see Peters and Hewlett, *Evolution from Creation to New Creation*.

of evolutionary and molecular biology; and they interpret the results using the theological categories of God's long term purposes and will for creation and for the human place within creation and redemption. This perspective of theistic evolutionists includes two central themes: *creatio ex nihilo* and *creatio continua*. 1) God as transcendent (meaning wholly other, not spatially distant) creates the universe out of nothing (*creatio ex nihilo*), holding it in existence at each moment and maintaining the regularities of its processes which we express scientifically as the laws of nature.[29] 2) God as immanent (meaning wholly present but not pantheistically identified with nature) creates the universe continuously in time (*creatio continua*), working "in, with, under, and through" the processes of nature,[30] as Arthur Peacocke and others phrase it. In short, God creates the complex diversity of living species by working in and through the natural processes whose very possibilities God created *ex nihilo*.

Scholars in theology and science have developed these themes in light of physical cosmology, quantum physics, chaos and complexity theory, evolutionary and molecular biology, anthropology, the neurosciences and cognitive sciences, etc. Arguably the most remarkable construction in the galaxy is the primate central nervous system: The number of connections between the neurons of the human brain is greater than the number of stars in the Milky Way. This staggering complexity makes possible the almost unimaginable feat of self-consciousness. We know ourself as a free, rational, and moral agent in the world. Thus on our planet, at least, we are privileged to discover a hint of what God's intentions might have been in creating a universe like ours, with its particular laws of physics. For when the evolutionary conditions are right as they have been on Earth, and as they may be elsewhere in our universe, God, the continuous, immanent, ongoing creator of all that is, working with and through nature, creates a species gifted with the "image of God" (the *imago Dei*) traditionally interpreted as including the capacities for reason, language, imagination, tool-making, social organization, and self-conscious moral choice, a species capable of entering into covenant with God and in turn with all of life.[31]

---

29. Arguments from "t=0" in Big Bang cosmology are often used to support this view, even though the theological claim is primarily about existence *per se* and not temporal origins. For recent references, see Russell, "Finite Creation Without a Beginning."

30. Note: I am assuming a *non-interventionist* view of God's immanent action here, one consistent with science. For details, see the CTNS/Vatican Observatory conference publications referenced in these Notes.

31. I return to the meaning of the *imago Dei* and recent arguments criticizing its traditional meanings in the companion chapter to this one, "Many Incarnations or One?"

Thus if it took the precise characteristics of this "fine-tuned" universe to allow for the possibility of the evolution of life, then even if life is scarce in the universe it is life as such that gives significance to our universe—and this obtains even if ours is only one of a countless series of universes, as some inflationary and quantum cosmologies depict. In short, I see life as the enfleshing of God's intentions through biological evolution which, in turn, is possible, in part, because of the physics God gives this universe. Together biology and physics are the ongoing expression of God's purposes in creating all that is. God thus offers to nature a special conscious experience of the God who acts immanently within nature as the transcendent ground of its being.

4. Will Intelligent Life Be Capable of Both Rationality and Moral Reasoning?

We turn here to the question of the biological origins of ethics.[32] If our human experience of moral capacity—like our capacity for rational thought and relationality—is a gift of God rooted in our biological nature and bequeathed us by God acting in and through evolution, does this suggest that wherever evolution results in creatures capable of rationality they would also be equipped by God with a capacity for relationship and moral reasoning?

Sociobiologists and, more recently, behavioral geneticists, have explored the biological basis of human social behavior in order to determine the relation between evolutionary and genetic constraints, on the one hand, and their cultural expression, on the other. Many scholars, such as E. O. Wilson, Richard Dawkins, and Michael Ruse, are unabashedly reductionistic. They interpret their scientific research in strictly deterministic and functionalist accounts of human behavior. Ruse, for example, has argued extensively that both the capacity and the content of human morality are entirely the products of evolution.[33] Ruse defines biological altruism as any cooperative behavior between organisms that increases evolutionary gain, while moral altruism refers to our considered choices to help others because it is seen as right to do so. For Ruse, moral altruism is a product of (non-moral) biological altruism with " . . . no objective foundation. It is just an illusion, fobbed off on us to promote 'altruism.'"[34]

---

32. This is *not* an attempt to derive ethics from biology. I wish to avoid the so-called 'naturalistic fallacy.'

33. See for example Ruse, *Sociobiology*; Ruse, *Darwinism Defended*; Ruse, *Taking Darwin Seriously*; Ruse, "Evolutionary Theory and Christian Ethics."

34. Ruse, "Evolutionary Ethics," 95, 100. As Rolston summarizes Ruse: "Ethics is

Geneticist Francisco Ayala takes a very different position. For nearly three decades he has argued against reductionism in biology,[35] disagreeing dramatically with Ruse over the evolutionary origins of human moral capacity. According to Ayala, evolution selected for intelligence in our ancestral hominid line, one of the many byproducts of intelligence is ethics. "Ethical behavior came about in evolution not because it is adaptive in itself, but as a necessary consequence of man's (sic) eminent intellectual abilities, which are an attribute directly promoted by natural selection."[36] Our ethical behavior, and the norms which we use to govern it, are not determined by our genes or our evolution. Instead ethics is open to cultural, philosophical, and religious sources. Similar arguments against reductionism have been developed by a wide range of scientists and philosophers, including Arthur Peacocke,[37] Ian Barbour,[38] Nancey Murphy,[39] and George Ellis.[40]

How does this bear on the question of ETI? Appealing to the principle of mediocrity, we are safe in predicting that the laws of physics and even the laws of biology would obtain anywhere and everywhere in the cosmos. "Whatever their (ETI) biologies, they would be based on the very same physics and chemistry as ours is, including the second law of thermodynamics," rightly observes Vatican Observatory astronomer Stoeger.[41] With this physical continuity in mind, I forecast that intellectual and even moral continuity would also obtain.

If what we have found out about life on Earth can be generalized, I expect that moral capacity will be present wherever life has evolved to the

---

not true, though it is functional. (But) ethics cannot be functional unless it is believed to be true in an objective sense, a false belief." Rolston, *Biology*, 8. Rolston is sharply critical of Ruse in *Genes, Genesis and God*.

35. See for example Ayala, "Introduction"; Ayala, "Reduction in Biology"; Ayala, "Can 'Progress' Be Defined?"

36. Ayala, "Ethical Behavior," 118.

37. Peacocke, "Reductionism"; Peacocke, *Creation and the World of Science*; Peacocke, *God and the New Biology*; Peacocke, *Theology for a Scientific Age*.

38. Barbour, *Issues in Science and Religion*; Barbour, "Ways of Relating Science and Theology," 21–48; Barbour, *Religion in an Age of Science*.

39. Murphy, "Theology in the Age of Scientific Reasoning"; Murphy, *Beyond Liberalism and Fundamentalism*; Murphy, "Supervenience and the Nonreducibility of Ethics to Biology"; Murphy, "Nonreductive Physicalism"; Murphy, "Supervenience and the Downward Efficacy of the Mental."

40. Ellis, *Before the Beginning*; Ellis, "The Theology of the Anthropic Principle," 367–406; Ellis, "Ordinary and Extraordinary Divine Action," 359–96; Murphy and Ellis, *On the Moral Nature of the Universe*; Ellis, "The Thinking Underlying the New 'Scientific' World-Views"; Ellis, "Reflections on Quantum Theory."

41. Stoeger, Book Review of Thomas F. O'Meara, *Vast Universe*, 83.

point of intelligence. By moral capacity, I mean that biological evolution will predispose extraterrestrial beings to moral discernment; but evolution will not determine the content or values invoked by extraterrestrial moral discernment. In sum, if we meet new friends living on other planets in the Milky Way, we can expect that they will have the capacity for moral reasoning. We will be able to talk about ethical matters with them.

5. What Are the Theological Implications for Christianity?

What sorts of responses might Christian theologians offer to the discovery of ETI with rational and moral capacities? Davies the physicist predicts it would "(shatter) completely the traditional perspective on God's relationship with man (sic)."[42] But theologian Peters finds "little or no credible evidence" for Davies's view.[43] Instead there is rich evidence in the history of Christian thought in support of a "plurality of worlds" with extraterrestrial life in the universe.[44] Contemporary theology, too, has been genuinely open to the possibility of rational and moral ETI. In support of this position Peters cites both Roman Catholic theologians and Protestant theologians, both historical and contemporary. In his research document, the *Peters ETI Religious Crisis Survey*, Peters reports that it is by and large the critics of religion who claim that the discovery of ETI would challenge religious faith, not those who espouse that faith! For more detail, see later chapters in this book.[45]

Personally I would find the discovery of ETI endowed with rational and moral capacities to be a wondrous exemplification of the intentions of God in creating a universe like ours, namely in order to achieve the evolution of creatures capable of bearing the *imago Dei* and capable of entering into genuine community and covenant with God through the divine

---

42. Davies, *God and the New Physics*, 71; Davies, "Transformations."

43. Peters, "Exo-Theology," 1. I agree with Peters (see my third footnote above). As Peters points out (p. 2; 7), it may have been the Protestant Fundamentalist reaction to UFOs in the 1970s that gave rise to the view of Christianity that Davies critiques. See also McMullin, "Life and Intelligence Far from Earth," 164-7.

44. As Peters shows, Greek atomists such as Democritus and Epicurus argued for many worlds, but Aristotle rejected their views. Thomas Aquinas sought to reconcile Aristotle with Christian tradition, defending a "one world" view for philosophical reasons. Others, such as John Buridan, Nicole Oresme, Albertus Magnus, the Jewish scholar Hasdai Crescas, and Nicholas of Cusa, favored the "many worlds" position. See Dick, *Plurality of Worlds*.

45. Peters with Froehlig, *The Peters ETI Religious Crisis Survey*.

initiative of grace. I am also persuaded by the philosophical arguments of non-reductionists that ethical choice and the contents of our moral codes remain at least partially a free variable. I would therefore expect that ETI will be morally accountable for its choices in some way even as humanity is. This leads directly to the challenge of moral failure, our final issue in this section of my chapter.

6. Will ETI Experience Moral Failure or Be Entirely Benign?

This question acknowledges a tragic dimension of reality at the heart of human existence here on Earth. Why do we act with a level of violence against our own kind and other species, with violence which far exceeds the needs of survival and far exceeds the level of violence of all other forms of life on Earth? Of course there are evolutionary precedents for such violence, particularly in mammals with whom we share a distant common ancestor. But the level of violence in humankind seems more like a qualitative rather than a mere quantitative difference in comparison with, say, tribes of warring chimpanzees. And when it comes to human culture, why do we lust after unlimited power and indulge in travesties such as racism, sexism, and speciesism? Put theologically, why do we sin? Why do we fail to love and serve God above all else and instead indulge ourselves in unbridled pride and inordinate sensuality?

One form of Christian response to this foundational question about the human condition has been to assert with Reinhold Niebuhr the paradox that sin is not an intrinsic part of human nature and yet it is an inevitable component of human behavior. Indeed, its only remedy is something that comes from beyond our human nature: the grace of God freely given us.[46] Declaring sin to be intrinsic would rob us of our individual and corporate responsibility on which our legal and political systems are built; failing to recognize its inevitability would lead to the false hope that we can free ourselves through one of countless self-help movements without depending radically on the grace of God. In sum, each of us inherits both the *imago Dei*, the image of God, and the inevitability of sin; and both seem to be unique to our species.[47]

---

46. Niebuhr, *The Nature and Destiny of Man: I*, VI–X.

47. It is interesting to compare the Augustinian view with contemporary cultural positions. For example, consider our two questions about sin: 1) is it in our nature? 2) is it avoidable by human effort alone? The Augustinian response is: 1) no; 2) no. A secular humanist would probably answer: 1) no; 2) yes. Interestingly, Carl Sagan seemed to

This traditional theological understanding conforms nicely to our contemporary scientific understanding of biological evolution, particularly with the philosophical theme of novelty within continuity. Thus, we humans as a species inherit diverse propensities from our pre-hominid past; yet, when crossing the evolutionary threshold to *Homo sapiens*, something strikingly new emerges. This newness is manifest both in the *imago Dei*—including our capacity for relationship, abstract thought, formal language, complex technologies, art, ethics, and science—and in the reality of human sin, including ruthless violence and our insatiable appetites for power and control. It is only through the grace of a loving God that our lives can be transformed into the fullness of what it truly means to be human. Conversely, the formation of authentic human personhood requires a lifetime of genuine wrestling with tough moral choices and the repentance of moral failures.

What then about ETI and the domain of moral failure: is it truly universal or is it limited to terrestrial history? I have suggested for scientific, philosophical, and theological reasons that the essential characteristics of human life are a genuine clue to the nature of life in the universe and not just an evolutionary fluke of the evolutionary processes on Earth. Siding with those who affirm meaning to life, I therefore expect that ETI will experience a kind of moral dilemma that in many ways resembles the moral quagmire of human experience, though obviously differing in its "moral morphology"—the personal and social form of ethics.

SETI's Jill Tarter has argued that extraterrestrial civilizations will be far older than ours, and to achieve such longevity, they will have had to overcome the temptation to warfare.[48] I tend to agree with her hopeful observation, but this optimism still leaves open the question of how such temptation is overcome? Just because an intelligent civilization has existed for a long time and overcome many obstacles, it does not follow that it will have overcome sin and its accompanying violence. Time alone does not heal, or save. Only God's grace can accomplish this.

---

take the position: 1) yes (the tripartite brain and its conflicts); 2) yes (through science).

48. Tarter claims that ET's longevity is inconsistent with them having "organized monotheistic religions" since they are the cause of warfare on earth. Clearly, indefinitely sustained warfare is inconsistent with cultural longevity, but her suggestion of a causal relation between monotheism and warfare would be hard to defend critically. Of course religion is often used as an excuse for violence by believers, but the real question is why humans are capable of such violence whether or not religion is the proximate cause. It also seems odd that she speculates favorably about ET having a "universal religion" with a "highly established code of ethics" centered on "the perpetuation of individuals" when these tenets are found in the monotheisms which she rejects. (Tarter, "Implications.")

My assumption is that ETI will experience an empowerment for their struggle to pursue the good, an empowerment by a source which transcends ETI's natural capacities. Put into theological language, I believe—perhaps I should say I predict!—that God will be present to the struggles of life everywhere, and that God's grace will redeem and sanctify every species in which reason and moral conscience are kindled.

This, in turn, leads to a seventh question for theology: will ET need redemption? I will defer the discussion of this question to my latter, companion chapter in this book.

Bibliography

Ayala, Francisco J. "Can 'Progress' Be Defined as a Biological Concept?" In *Evolutionary Progress*, edited by Matthew H. Nitecki, 75–96. Chicago: University of Chicago Press, 1988.

———. Closer to Truth. http://www.closertotruth.com/video-profile/Why-aren-t-Aliens-Already-Here-Francisco-J-Ayala-1-of-2-/188.

———. "Ethical Behavior as an Evolutionary Byproduct." In *Biology, Ethics and the Origins of Life*, edited by Holmes Rolston III, 113–36. Boston: Jones and Bartlett Publishers, 1995.

———. "Introduction." In *Studies in the Philosophy of Biology: Reduction and Related Problems*, edited by Francisco J. Ayala and Theodosius Dobzhansky. Berkeley, CA: University of California Press, 1974.

———. "Reduction in Biology: A Recent Challenge." In *Evolution at a Crossroads: The New Biology and the New Philosophy of Science*, edited by David J. Depew and Bruce H. Weber, 67–78. Cambridge: MIT Press, 1985.

Barbour, Ian G. "Five Models of God and Evolution." In *EMB*, 419–42.

———. *Issues in Science and Religion*. New York: Harper & Row, 1971.

———. *Religion in an Age of Science*. San Francisco: Harper & Row, 1990.

———. "Ways of Relating Science and Theology." In *PPT* 21–48.

Birch, Charles. *Feelings*. Sydney: University of New South Wales Press, 1995.

———. "Neo-Darwinism, Self-Organization, and Divine Action in Evolution." In *EMB* 225–248.

———. *A Purpose for Everything*. Mystic, CT: Twenty-Third Publications, 1990.

Bolle, Kees W. "Cosmology: An Overview." In *ER1* 3&4:100–107.

Chela-Flores, Julian. "The Phenomenon of the Eukaryotic Cell." In *EMB* 79–98.

Coyne, George V., SJ. "The Evolution of Intelligent Life on Earth and Possibly Elsewhere: Reflections from a Religious Tradition." In *MW*, 177–88.

Davies, Paul C. W. *The Cosmic Blueprint*. New York: Touchstone, 1989.

———. *God and the New Physics*. New York: Simon & Schuster, 1983.

———. "Is the Universe Absurd?" In *Science and Theology: The New Consonance*, edited by Ted Peters, 65–76. Boulder, CO: Westview, 1999.

———. "Teleology Without Teleology: Purpose through Emergent Complexity." In *EMB*, 151–62.

———. "Transformations in Spirituality and Religion." In *SETI*, 51.

Dawkins, Richard. *The Blind Watchmaker: Why the Evidence of Evolution Reveals a Universe without Design*. London: Penguin, 2006.
de Duve, Christian. *Vital Dust: Life as a Cosmic Imperative*. New York: Basic, 1995.
Dick, Steven J. "Extraterrestrials and Objective Knowledge." In *SETI*, 47–48.
———. *Plurality of Worlds*. Cambridge: Cambridge University Press, 1982.
Dyson, Freeman. *Disturbing the Universe*. New York: Harper & Row, 1979.
———. *Infinite in All Directions*. New York: Harper and Row, 1988.
Ellis, George F. *Before the Beginning: Cosmology Explained*. New York: Boyars/Bowerdean, 1993.
———. "Ordinary and Extraordinary Divine Action: The Nexus of Interaction." In *Chaos and Complexity: Scientific Perspectives on Divine Action*, edited by Robert John Russell et al. Scientific Perspectives on Divine Action Series, 359–96. Vatican City State and Berkeley, CA: Vatican Observatory / Center for Theology and the Natural Sciences, 1995.
———. "Reflections on Quantum Theory and the Macroscopic World." In *Quantum Mechanics: Scientific Perspectives on Divine Action*, edited by Robert John Russell et al. Vatican City State and Berkeley, CA: Vatican Observatory / Center for Theology and the Natural Sciences, 2001.
———. "The Theology of the Anthropic Principle." In *Quantum Cosmology and the Laws of Nature*, ed. by Robert John Russell et al., 367–406. Vatican City State and Berkeley, CA: Vatican Observatory / Center for Theology and the Natural Sciences, 1993.
———. "The Thinking Underlying the New 'Scientific' World-views." In *EMB* 251–280.
Epicurus. Letter to "Herodotus." *The Stoic and Epicurean Philosophers*, edited by Whitney J. Oates, 5. New York: Modern Library, 1940.
Chela-Flores, Julian, et al., eds. *First Steps in the Origin of Life in the Universe*. Proceedings of the Sixth Trieste Conference on Chemical Evolution. Dordrecht: Kluwer Academic, 2001.
Gould, Stephen Jay. *The Flamingo's Smile: Reflections in Natural History*. New York: Norton, 1985.
Haught, John F. *Science and Religion: From Conflict to Conversion*. New York: Paulist, 1995.
Mayr, Ernst. "How Biology Differs from the Physical Sciences." In *Evolution at a Crossroads: The New Biology and the New Philosophy of Science*, edited by David J. Depew and Bruce H. Weber, 67–78. Cambridge: MIT Press, 1985.
McMullin, Ernan. "Life and Intelligence Far from Earth: Formulating Theological Issues." In *MW*, 151–75.
Monod, Jacques. *Chance and Necessity*. Translated by Austryn Wainhouse. New York: Vintage, 1972.
Murphy, Nancey. *Beyond Liberalism and Fundamentalism*. Valley Forge, PA: Trinity International, 1996.
———. "Nonreductive Physicalism: Philosophical Issues." In *Whatever Happened to the Soul?* edited by Warren S. Brown et al. Minneapolis: Fortress, 1998.
———. "Supervenience and the Downward Efficacy of the Mental: A Nonreductive Physicalist Account of Human Action." In *Neuroscience and the Person: Scientific Perspectives on Divine Action*, edited by Robert John Russell et al. Vatican City

State and Berkeley, CA: Vatican Observatory / Center for Theology and the Natural Sciences, 1999.

———. "Supervenience and the Nonreducibility of Ethics to Biology." In *EMB*, 463–90.

———. *Theology in the Age of Scientific Reasoning*. Ithaca: Cornell University Press, 1990.

Murphy, Nancey, and George F. Ellis. *On the Moral Nature of the Universe: Theology, Cosmology, and Ethics*. Minneapolis: Fortress, 1996.

NASA. Curiosity Rover. http://www.nasa.gov/mission_pages/msl/index.html.

Niebuhr, Reinhold. *The Nature and Destiny of Man: I*. New York: Scribner, 1941. Reprint 1964.

Peacocke, Arthur. *Creation and the World of Science*. Oxford: Clarendon, 1979.

———. *God and the New Biology*. San Francisco: Harper & Row, 1986.

———. "Reductionism." *Zygon* 11.4 (December 1976) 307–34.

———. *Theology for a Scientific Age: Being and Becoming*. Minneapolis: Fortress, 1993.

Peters, Ted. "Exo-Theology: Speculations on Extra-Terrestrial Life." *CTNS Bulletin* 14.3 (Summer 1994) 1–9.

———."Extraterrestrial Intelligence." In *RPP*, 4:779–80.

———. *God—The World's Future*. 3rd ed. Minneapolis: Fortress, 2015.

———. "Science and Religion: An Overview." In *ER2* 12:8180–92.

Peters, Ted, ed. *Science and Theology: The New Consonance*. Boulder, CO: Westview, 1998.

Peters, Ted, and Julie Louise Froehlig. *The Peters ETI Religious Crisis Survey (2008)*. Counterbalance. http://www.counterbalance.org/etsurv/index-frame.html.

Peters, Ted and Martinez J. Hewlett, eds. *Evolution from Creation to New Creation: Conflict, Conversation and Convergence*. Nashville, TN: Abingdon, 2003.

Polkinghorne, John C. *The Faith of a Physicist: Reflections of a Bottom-up Thinker*. Minneapolis: Fortress, 1994.

Randolph, Richard O., et al. "Reconsidering the Theological and Ethical Implications of Extraterrestrial Life." *CTNS Bulletin* 17.3 (Summer 1997) 1–8.

Richardson, W. Mark, and Wesley J. Wildman, eds. *Religion and Science: History, Method, Dialogue*. New York: Routledge, 1996.

Rolston, Holmes, III. *Genes, Genesis and God*. Cambridge: Cambridge University Press, 1999.

Rolston, Holmes, III, ed. *Biology, Ethics and the Origins of Life*. Boston: Jones and Bartlett Publishers, 1995.

Ruse, Michael. *Darwinism Defended*. Reading, MA: Addison-Wesley, 1982.

———. "Evolutionary Ethics." In *Biology, Ethics and the Origins of Life*, edited by Holmes Rolston III, 89–112. Boston: Jones and Bartlett Publishers, 1995.

———. "Evolutionary Theory and Christian Ethics: Are They in Harmony?" *Zygon: Journal of Religion and Science* 29.1 (March 1994) 5–24.

———. *Sociobiology: Sense or Nonsense?* Dordrecht: Reidel, 1979.

———. *Taking Darwin Seriously*. Oxford: Blackwell, 1985.

Russell, Bertrand. "A Free Man's Worship." In *Mysticism and Logic*. London: Allen & Unwin, 1903.

Russell, Robert John. "Finite Creation Without a Beginning." In *Quantum Cosmology and the Laws of Nature: Scientific Perspectives on Divine Action*, edited by Robert

John Russell et al. Vatican City State and Berkeley, CA: Vatican Observatory Publications and Center for Theology and the Natural Sciences, 1993.

———. "Theology and Science: Current Issues and Future Directions." http://www.ctns.org/russell_article.html.

SETI. http://www.seti.org/.

Southgate, Christopher, et al., eds. *God, Humanity and the Cosmos: A Textbook in Science and Religion*. Harrisburg: Trinity Press International, 1999.

Stoeger, William R., SJ. Review of *Vast Universe*, by Thomas F. O'Meara. *Theology and Science* 11.1 (February 2013) 77–83.

———. "The Immanent Directionality of the Evolutionary Process, and Its Relationship to Theology." In *EMB* 163–90.

Tarter, Jill Cornell. "Implications of Contact with ETI Far Older than Humankind." In *SETI*, 45–46.

Vakoch, Douglas A. "Roman Catholic Views of Extraterrestrial Intelligence: Anticipating the Future by Examining the Past." In *SETI* 165–74.

Weinberg, Steven. *The First Three Minutes*. New York: Basic, 1977.

5

The Copernican Revolution That Never Was

Martinez Hewlett

> There are more things in heaven and earth, Horatio,
> than are dreamt of in your philosophy.
>
> —William Shakespeare, *Hamlet*, Act 1, scene 5

> Worlds on worlds are rolling ever
> From creation to decay,
> Like the bubbles on a river
> Sparkling, bursting, borne away.
>
> —Percy Bysshe Shelly

You step outside on a clear, cold, Taos night. You look up at a sky, filled with more stars than you can count. From this perspective, does it feel right to think that you are at the center of everything, of all that is?

In the middle of the second century of our era, Egyptian Claudius Ptolemy also looked up to the sky. He was impressed. He wrote a book, *Almagest*. In his own copy he penned a marginal note to himself. "I know that I am mortal by nature and ephemeral, but when I trace at my pleasure the windings to and fro of the heavenly bodies, I no longer touch Earth with my feet. I stand in the presence of Zeus himself and take my fill of ambrosia." Outer space had taken up residence in Ptolemy's soul.

Despite his desire to lift himself to the heavenly realms, Ptolemy's soul remained on Earth. Does this placement imply that Earth is located in the center of reality? It certainly looks that way. In the Western world, it was Ptolemy who eventually provided a kind of quantitative take on this seeming intuition, deriving a calculating tool that allowed observers of the night sky to make predictions about where the "movable stars" would be at a given time, or when the constellations formed by the "fixed stars" would appear on a certain date. Yes, it was also thought to be an accurate physical model of the cosmos. But at its heart, it was simply a tool ... a quantitative statement about observations.

Like all tools, the Ptolemaic model was subject to tweaking, to revisions, and, ultimately, to being discarded in favor of a better device. Philosopher of science Thomas Kuhn documents this series of events as the prime example of a revolution in science, as a shift in paradigms.[1] One day it was Ptolemy. The next day it was Copernicus. One day the Earth was in the center. The next day the sun was in the center. The Copernican Revolution was a scientific revolution about planetary revolutions.

Much changed. And yet, for all of the social upheaval that the Copernican heliocentric model precipitated, it did little to shift the conceptualization of how life somewhere other than on Earth might arise. Whether the Earth or the sun sits in the center of our solar system, human questions about other worlds and other sentient species in space continued to be asked in roughly the same form. But, to ask just how we might intelligibly imagine life appearing in other worlds like it did in our world was a task left to another revolutionary, Charles Darwin.

Although I am primarily an evolutionary biologist and secondarily a philosopher, in this chapter I will take up the first task of the astrotheologian as stated by Ted Peters in chapter 2: *Christian theologians along with intellectual leaders in each religious tradition need to reflect on the scope of creation and settle the pesky issue of geocentrism.* Our reflection here will be historical in character. We are drawing on the second source in the Wesleyan quadrilateral, tradition. The upshot of my argument here will be this: the tradition shows that both before and following Copernicus the question of geocentrism was raised, discussed, and debated. For the most part, the Copernican cosmology led to a rejection of geocentrism by both philosophers and theologians. Regarding views on our sharing this universe with intelligent neighbors on other planets, however, Copernicus's revolutionary

---

1. Kuhn, *Structure of Scientific Revolutions*.

cosmology changed relatively little. Why? To answer this we turn to the history itself.

The Ancients and Medievals and Off-Earth Life

I'm confident that every culture has had those who, while looking at the night sky, wondered if there was anyone else out there. Our records of these musings in the Western tradition go back at least to the ancient Greeks. There were two camps in this discussion. On the one hand, the atomists held that the universe was infinite and consisted of indivisible particles (atoms).[2] As such, other worlds like ours would be inevitable, since the entire cosmos consisted of identical materials. Democritus and his mentor, Leucippus, championed this cosmic pluralism, at least as a philosophical position.

But, Plato and Aristotle viewed things differently. Ultimately, the views of Plato and Aristotle would hold sway until the Copernican Revolution. Their position argued that other worlds were not possible. Aristotle makes the case in *De Caelo* (*On the Heavens*): any world other than ours must be made of the four elements, fire, air, earth, and water. However, each of these have, as their natural tendency or motion, the center of our world. Everything in reality is oriented around a center, a single center. Rocks, for example, fall down if left to their own motion. How could rocks exist anywhere else? Remember, this was a geocentric cosmos. Other worlds would have to be circling our own. Other worlds would be made of the same "stuff" as our own, an opinion held in common with Democritus. Still, no world with a center other than our own can exist.[3] Rocks fall down to Earth's center, not toward some other center. Thus, they denied the cosmic pluralism of the atomists. Centuries later, this became the opinion of early Christian philosophers and theologians, with one interesting exception.

---

2. Democritus's statement about the existence of other worlds is known through the work of Hippolytus of Rome (third century AD): "And he (Democritus) maintained worlds to be infinite, and varying in bulk; and that in some there is neither sun nor moon, while in others that they are larger than with us, and with others more numerous. And that intervals between worlds are unequal; and that in one quarter of space (worlds) are more numerous, and in another less so; and that some of them increase in bulk, but that others attain their full size, while others dwindle away and that in one quarter they are coming into existence, whilst in another they are failing; and that they are destroyed by clashing one with another. And that some worlds are destitute of animals and plants, and every species of moisture." The statement can be found in Robertson and Donaldson, *Ante-Nicene Fathers, Volume 5*, at Christian Classics Ethereal Library (http://www.ccel.org/ccel/schaff/anfo5.i.html).

3. Aristotle, *On the Heavens (De Caelo)*, 367.

St. Thomas Aquinas in the twelfth century inherited the Ptolemaic universe with the Earth as its center. Thomas was, in addition, an Aristotelian. This is not a trivial observation. Before the time of Thomas and his mentor, St. Albert the Great, Aristotle had been all but lost to the West. The acceptable philosopher for Christians had been Plato. The works of Aristotle were unavailable to a wide readership; but they had been preserved in two places: in the Greek world, in places such as Hagia Sofia; and in the world of Islam. At the request of Muslim intellectuals, the works of the Greek philosopher had been translated into Arabic. Eventually, great commentaries were written by Ibn Said (Avicenna) and Ibn Rushid (Averroes).[4] In the eleventh century, these works made their way back into the West in Latin, translated from the Greek and the Arabic.

For the more conservative churchmen, this new body of work bordered on heresy, differing in philosophical structure as it did from Plato. However, Thomas in Paris was attracted to the empiricism of the Aristotelian system. Make observations, and then use inductive reasoning to understand what was being observed. The Angelic Doctor, as Thomas was later to be known, is mistakenly thought of first as a theologian. This may result from the fact that most moderns encounter his great work, the *Summa Theologica*, at some point during their education. In fact, the greater body of his writings is philosophical in nature, many of them commentaries on a part of the Aristotelian corpus.

It is in this way that Thomas came to consider the issue of cosmic pluralism. He addressed the question of whether or not worlds other than ours exist. Thomas was not only an Aristotelian but also an ordained member of the Order of Preachers, a Dominican. Therefore, in classic Dominican fashion, he answered "yes and no."[5]

His answer was "no," if by other worlds and other beings we mean something exactly like ourselves. The argument for this answer uses exactly the same logic as did Aristotle. That is, any other world like ours would have to be made of the same four elements; and they would have to be oriented toward a center. Since all four of these have their natural location on the Earth with Earth's center, no other worlds like ours could possibly exist. Earth provides the single center for physical reality.

However, there is a fifth element, the quintessence or ether. This is what forms everything in the heavens above the sphere of the Earth, including the fixed stars and the movable "stars" such as Mars or Venus. Thomas

---

4. The loss of Aristotle's corpus to the Latin West, from the 6th to the 12th centuries, is discussed on the Stanford Encyclopedia of Philosophy web site (http://plato.stanford.edu/entries/aristotle/).

5. George, "Aquinas on Intelligent Extra-Terrestrial Life," 239–58.

argues that if beings exist on these quintessential bodies, they must also be made of quintessence. He says that this has some degree of logical support. After all, angels exist in these heavenly locales. So, after saying "no," Thomas considered what might be needed to grant a "yes." There just might be life in our sky.

That life in our sky would not be a duplicate of life on Earth; sky creatures would not have corruptible bodies like our inhabited by intellectual minds like ours. As we have just noted, the corruptible elements we are familiar with on Earth remain on Earth. This precludes the possibility of another earth-like planet with physical beings. So, the sky people Thomas speculates about would be similar to angels, adorned with quintessence.[6]

The Great Copernican Revolution . . . or Not

In our own era, some look at history through the lenses of what we might dub the *myth of the Copernican Revolution*. There is no doubt that a revolution regarding revolutions took place as a result of Copernicus's work. "The Copernican Revolution has expanded over a half-millennium, showing that our Sun is not the center of the universe, nor our galaxy, and the very presumption of a center is flawed," writes astrobiologist Chris Impey.[7] This is the revolution in scientific terms, physical terms. But, the so-called revolution for religion and culture has taken on mythical proportions.

According to this myth (more accurately: a legend), science plunged forward immediately and courageously while religion defensively resisted this progress in human knowledge. Allegedly, religious earthlings stood firm in their hubris: Earth is in the center! This hubris and planetary chauvinism over pride of place persisted even in the face of scientific evidence to the contrary. Geocentrism is a symptom; the disease is religious dogmatism. So goes the myth.

Here is the often repeated story in the words of Paul Davies. "In the sixteenth century, Europe was in thrall to the Roman Catholic Church, which clung to the ancient Ptolemaic view that the Earth lay at the center of the universe and—the natural corollary—that humankind represented the pinnacle of God's creation. By demoting our planet from the pivotal cosmic position, Copernicus initiated a trend—a principle of mediocrity—that continued for four centuries."[8] What has taken on a mythical role is the assumption that the church of Copernicus's era was stubbornly geocentric

6. See: George, "Aquinas on Intelligent Extra-Terrestrial Life," 239–58.
7. Impey, *Humble Before the Void*, 53.
8. Davies, *The Goldilocks Enigma*, 129.

and anthropocentric and unwilling to leave its dogmatic darkness for the light of scientific knowledge.

Philosopher Patricia Churchland's interpretation of history oozes with this myth. "Copernicus and Galileo threw that cosmology [Earth at the center] under the bus," she writes. "The very institution of the Christian Church was founded on the belief in the bodily resurrection of Jesus into an actual place—heaven. And this actual place is above the moon, and maybe even above the stars. If you believe something to be absolutely certain and foundational, it is profoundly shocking to find that your truth may be mushy, or worse, quite likely false . . . Some 400 years after Galileo's arrest, the Catholic Church conceded what every moderately well-educated person knew to be true: Earth is not the center of the universe."[9] If Churchland's version of history would be correct, then the scientific community along with "every moderately well-educated person" accepted heliocentrism about 1632, but the Roman Catholic Church conceded this only in 2032 (still a few years off). What Churchland does not include in her account is the founding of the Vatican Observatory in 1582 and the immediate employment of Galileo's telescope when it became available after 1610.

One of the items on our theological To-Do list in chapter 2, recall, is to criticize and correct mistaken images held within the scientific community regarding religious or theological matters. This myth represents a point where correction is called for. Let me explain why this myth of the Copernican Revolution is too simplistic an interpretation.

The heliocentrism of Copernicus generated much discussion during the era of Galileo, to be sure; but it took some time before it would become clear to scientists that heliocentrism would trump geocentrism. Why? For two reasons. First, the Copernican system did not seem coherent, at least at first. "Throughout the seventeenth century," writes Harvard astronomer and expert on Copernicus, Owen Gingerich, "the public acceptance of the Copernican cosmology was slow, far from unanimous, and based not on proofs but on the persuasion of what was increasingly seen as a coherent system."[10]

The second reason the Copernican Revolution proceeded so slowly was that the traditional Ptolemaic system continued to be functional. The geocentric system of Ptolemy works. That is, it's a good calculating tool. In fact, if you are on this planet and trying to carry out celestial navigation, geocentrism is essential. The only son in our family is a Naval aviator. He says that when he is flying out at sea at night and trying to find his way back

---

9. Churchland, *Touching a Nerve*, 14–17.
10. Gingerich, *God's Planet*, 46.

to the ship by dead reckoning, he'd better be Ptolemaic, not Copernican. Knowing that the Earth orbits around the Sun is of no help.

As an astronomical tool, however, the Ptolemaic model had picked up a few problems. The movable stars were the issue. Distant stars were thought to be fixed. Those stars that moved, were thought to be planets. Sometimes Mars could be observed to turn around and go in the opposite direction. This retrograde motion needed to be accounted for by epicycles. In philosophy of science the positing of epicycles is called an *ad hoc* hypothesis. One formulates an *ad hoc* hypothesis when you need to add something to make the observations fit your theory. As observational data accumulated these epicycles grew more and more complicated, and the calculating tool became more and more unwieldy. These are the elements that, as Kuhn argues, precipitate a scientific revolution and a paradigm shift. A better model of how nature works was needed. Hence, the Copernican Revolution.[11]

The Polish friar, Nicolaus Copernicus, was not an astronomer. He did not make observations. But he did know about this unnecessarily complicated star tool that was being used. It bothered his aesthetic sensibilities. Couldn't it be simplified? What if you put the Sun in the middle and had the planets and stars rotate around that body? Now the epicycles weren't so much of a problem. While it's true that this model had existed long before Copernicus, notably in both ancient Greece and not so ancient Islam, it is not at all clear that Friar Nicolaus knew this.[12] His great work, *De Revolutionibus Orbium Coelestium*, (*On the Revolution of the Heavenly Spheres*) includes the statement that the heliocentric model is "more pleasing to the mind" than the Ptolemaic model.[13] Today in science we might say that it is more "elegant."

Copernicus was already quite old when he undertook this work. Unfortunately, he died just days after its publication. Fortunately, he did not have to suffer from the social repercussions that his model would later have in the Christian West. The Ptolemaic calculating tool had become an agreed upon description of how the cosmos actually looked. In fact, it was even theologically relevant, having been buttressed by purportedly scriptural

---

11. Kuhn, *Structure of Scientific Revolutions*.

12. Aristarchus of Samos (310–230 BC) is thought to be the first to have proposed a heliocentric model in writing. Although the original of his thesis is not preserved, it is mentioned in some detail by Archimedes, in his short essay *The Sand Reckoner* (translated by Ilan Vardi, and found on the web at http://www.lix.polytechnique.fr/Labo/Ilan.Vardi/sand_reckoner.ps.). The Islamic astronomers of the Maragheh Observatory in Iran were critical of the Ptolemaic model, and considered other possibilities, including versions of heliocentrism. A good recent discussion of the Maragha Revolution, as it has come to be called, can be found in Saliba, *A History of Arabic Astronomy*.

13. Copernicus, *On the Revolution of the Heavenly Spheres*.

support. The Ptolemaic worldview had even influenced how one prayed. One prayed looking upward from down on Earth. Heaven was thought to be up, up where the stars could be seen. In short, Copernicus followed by Kepler and Galileo confronted Christian intellectuals and church authorities with a surprise: the world looks different from what you previously thought. From Mars or Venus, the Earth would look like it is up.

Yes, the change in scientific worldview created by Copernicus's heliocentrism led to changes in the religious worldview as well. But, just how revolutionary was this revolution? Radical? No, much less than radical.

A common mistake made by the myth of the Copernican Revolution tellers is to assume that the problem here was one of demoting ourselves from the center of creation, one of losing Earth-centrism. Astrophysicist Jeffrey Bennett provides an example of the standard mistake: "the central lesson of the Copernican revolution and nearly everything we have learned since has been that we are *not* central, after all . . . Our place in the universe is completely ordinary, which makes it reasonable to think our planet is quite ordinary as well."[14] This sounds right. It's common knowledge. However, we need to be cautious.

Such a contention oversimplifies what was at stake in Copernicus's era. Losing geocentrism did not worry us earthlings very much. To the contrary, one of the earliest objections to the Copernican model was that it appeared to elevate us.[15] Were we not we made of the corrupt four elements? Were not all of the other heavenly bodies made of the incorruptible fifth element? Were we now on a par with Mars and Venus? Were we of such lofty status that we could be moving in the same circles, so to speak, as these ethereal creations? The heavenly bodies had been thought to be ethereal, to be spiritually superior. The problem with the Copernican view was that it implied those of us on Earth would be elevated into equality with the heavenly realms. Such an elevation would violate the Christian presumption that human beings are strictly earthly.

This led initially to some religious rejections of Copernicanism. The Roman Catholic Church, already reeling from the Protestant Reformation, was in an understandably conservative mode of dogmatic retrenchment. Galileo Galilei was censured for teaching Copernican theory, even though his observations of the moons of Jupiter through his telescope provided some experimental support to the model. In truth, Galileo was punished more for his obstinacy about philosophy and theology than for his science.[16] What

14. Bennett, *Beyond UFOs*, 16.
15. Brooke, "Science and Religion," 1985–6.
16. Segre, "Light on the Galileo Case?" 484–504.

seems clear is that neither Christian theology nor Western civilization was clinging to geocentrism like Linus clings to his baby blanket.

Here is another part of the common mistake illustrated by the words of Bennett: "The Copernican revolution also opened the possibility of life among the stars."[17] As we have seen, the debate over the possibility of life among the stars began two millennia before Copernicus and had been a lively debate among Christian theologians through the middle ages. Certainly Copernicus offered a dramatic new model of the solar system for all of civilization to consider; but he did not precipitate a revolution regarding geocentrism or the question of extraterrestrial beings.

In sum, late medieval Christendom did not defend geocentrism against science because of an alleged pride of place accrued from living at the center of things. The picture painted by myth-makers of a recalcitrant Earth-centered church fending off attacks from intellectually honest scientists is a false picture. "The tale is a fabrication," says Dennis Danielson. "Reasonable people need not believe it."[18]

The Copernican Revolution Was Slow

There is no doubt that a revolution regarding planetary revolution took place, but its speed was slow like a calliope rather than speedy like the Indianapolis 500.

It was slow for both science and religion. The course of historical events has been complex and nuanced. One of the more interesting events of this period occurred in Rome. It was there, in the Campo di Fiore, that the Dominican friar Giordano Bruno, was burned at the stake. Modern interpreters of this tend to raise up Bruno as the first "martyr" for science. A statue of Bruno, the hood of his habit pulled low over his brow, now graces this piazza that for years was a platform for free speech. A closer examination of the case reveals that his execution was ordered more for his championing of a number of heretical positions on other doctrines, not just for his support of heliocentrism and cosmic pluralism.[19] The Vatican's objections were aimed at Bruno's theology, not his science.

Eventually, and in spite of theological objections, both the scientific community and the religious community began to accept the Copernican

17. Bennett, *Beyond UFOs*, 8.
18. Danielson, "Myth 6: That Copernicanism Demoted Humans," 58.
19. For two different views of Bruno's condemnation see the web site for the Galileo Project (http://galileo.rice.edu/chr/bruno.html) and the web version of the Catholic Encyclopedia (http://www.newadvent.org/cathen/03016a.htm).

cosmological model. This was not instantaneous. Some continued to argue that geocentrism should not be totally abandoned. Scientist Tycho Brahe, for instance, tried to merge the two models, with a system that had the planets rotating around the Sun, with the entire assemblage orbiting around stationary Earth. But, as observational evidence began to accumulate, and as the logic of the new system began to be appreciated, heliocentrism became the new paradigm.

This move by the scientists, especially in the face of continued pressure from the churches, did its part to precipitate the division between the two domains that would later characterize at least some of what came to be called the Enlightenment. The hegemony of the religious hierarchy over all things intellectual would be broken in the eighteenth century; and the secular academy would grow in importance. This is not to argue, however, that warfare broke out between science and faith. Most of the research scientists of the early modern era were theologically alert, even motivated to pursue natural knowledge to complement their theological knowledge. The skirmish between religious belief and scientific progress emerged later, during the final third of the nineteenth century, when Darwin's theory of evolution was subjected to public debate.

Off-Earth Life: Just More of God's Creation?

Coming out of the great advances of the Renaissance, with the new Copernican view of the cosmos, much speculation took place as to whether or not life existed on one of the other planets revolving around our Sun, or on planets yet to be discovered, orbiting other suns. Even René Descartes considered the possibility that people and animals might one day be seen on the moon, if telescopes became powerful enough to accomplish the task.[20] Of course, any form of off-Earth life would have had the same origins as that found on Earth; that is, the only model current at that time was special creation in the forms in which we see living things now, both here and theoretically on other astronomical bodies.

Among the most detailed Enlightenment explorations of this topic was the book by the French poet and philosophical commentator, Bernard le Bovier de Fontenelle, entitled *Entretiens sur la Pluralité des Mondes* (*Conversations on the Plurality of Worlds*), published in 1686.[21] The book, writ-

---

20. Descartes's comments regarding the possibility of animals or people on the moon can be found in correspondence and journal entries during his time in the Netherlands. This is mentioned in the biography by Rodis-Lewis, *Descartes*, 76–77.

21. The full text of *Conversations on the Plurality of Worlds* can be found at Google Books

ten in French rather than Latin, an unusual move for a work of that time, had, as its protagonist a noblewoman who is having a conversation with a philosopher concerning the Copernican model, Descartes, and the plurality of worlds.

The conversations take place over the course of six evenings. During this time, the philosopher and the marquise discuss heliocentrism as it relates to the earth spinning on its axis and the earth and other then known planets orbiting around the sun. They move on to speaking of the moon as a habitable orb, the nature of the inhabitants on the other planets, and even the idea of the fixed stars each being suns, with their own inhabited planetary systems.[22]

Throughout this discourse the influence of the Copernican model on astronomical thought is evident. However, the revolution is incomplete, in that there is no different model presented for the origin of living creatures, other than the one that had existed since antiquity: either the Greek idea of ideal forms that were eternal, or the Judeo-Christian tenet of Divine creation of kinds.

Fontenelle had his characters discuss the beings on the planets and their temperaments. Inhabitants of Mercury are "almost mad with vivacity," while of the inhabitants of Saturn, he says "If they are not very wise, answered I, they are at least, I suppose, very phlegmatic. Their features could not accommodate themselves to a smile; they require a day's consideration before they answer any question, and they would think Cato of Atica unmanly and frivolous."[23]

Interestingly, Fontenelle suffered no real criticism from theologians or the church. He was, it turns out, a devout Roman Catholic, but an adamant critic of superstitious practices within religions. He was a champion of science and the rule of logic, but not to the detriment of belief.[24]

While Fontenelle's work was seen by some as a popularization of science, it was taken seriously by members of the French Academy, to which he was admitted. In addition, the noted astronomer, Christiaan Huygens took up the challenge laid down by Fontenelle's *Conversations*. His book,

---

(http://books.google.com/books?id=VGoFAAAAQAAJ&pg=PA1&dq=Fontenelle +Bernard+inauthor:Fontenelle#PPP24,M2). An interesting, though somewhat dated discussion of Fontenelle's place within the French Academy during the Enlightenment can be found in Marsak, "Bernard de Fontenelle," 1–64.

22. Ibid.
23. Ibid.
24. Marsak, "Bernard de Fontenelle."

Cosmotheoros, was written as an explanation directed to his brother, Constantijn. It was finally published after both men had passed away.[25]

Huygens's approach had much in common with Fontenelle, but was much more scientifically based. In fact, Huygens addresses the inadequacies of the "French author's" attempt at the outset of his text.[26] He considers both the size and distance from the Sun of each planet, as well as the requirement for water so that life could exist in common with that found on the Earth. His discussion here has a lot of resonance with present day astrobiological thinking.

Huygens takes pains to deflect any theological criticisms of his work. He acknowledges that what we now call exoplanets are not mentioned in Scripture. He states "That it's evident God had no design to make a particular Enumeration in the Holy Scriptures, of all the Works of his Creation."[27] Further, he argues that such conjectures as life on other worlds can only add to our religious sensibilities: "And we shall worship and reverence that God the Maker of all these things; we shall admire and adore his Providence and wonderful Wisdom which is displayed and manifested all over the Universe."[28] Here again, the Copernican Revolution that so altered the course of the astronomical sciences, had no influence on the model of how life began or diversified.

Important for our work in this volume is William Derham who gave us the term, *astrotheology*, in his book of 1714, *Astro-Theology, or a Demonstration of the Being and Attributes of God from a Survey of the Heavens*. Derham divided history in to a series of paradigm shifts, or shifts in worldview of the type described by Kuhn. The Ptolemaic geocentric view of the cosmos was upset with the Copernican heliocentric view, Derham says. In addition, in Derham's own era, cosmology seemed to be moving beyond Copernicus to a third system. "The *New Systeme* supposeth there are many other Systemes of *Suns* and *Planets*, besides that in which we have our residence; namely, that every Fixt Star is a Sun, and encompassed with a Systeme of Planets." These planets "are *Worlds*, or places of *Habitation*."[29] Despite Derham's self-congratulation for dreaming up a new system of thinking about the cosmos, he in fact is continuing the pre-Christian and Christian traditions of imagining that we have neighbors in space. Further, he imagines that these space

---

25. Huygens, *Cosmotheoros*, 1698, in English translation online at http://www.staff.science.uu.nl/~gento113/huygens/huygens_ct_en.htm.
26. Ibid., Book I.
27. Ibid.
28. Ibid.
29. Derham, cited by Crowe, *ELD* 123-4.

neighbors are God's creatures like we are; and this supposed fact inspires us to appreciate the magnificence and grandeur of God's created world. We find no geocentrism or Earth-chauvinism here; but neither do we find a break with the traditional understanding of how life began or diversified.

Getting Ready for Darwin

This pre-Copernican and post-Copernican speculation regarding life off-Earth continued, of course, until the publication of Charles Darwin's great work in 1859, which we will consider repeatedly in subsequent chapters. Before that, however, there is one last drama in the pre-modern plurality of worlds debate that we must address. This is the Brewster/Whewell battle, played out in Britain in the decades just preceding *Origin of Species*.

William Whewell (pronounced Hew-ell) was one of the shining lights of Victorian England, probably more famous in his time than many of his contemporaries. An Anglican priest, he was also a Cambridge scientist, a philosopher of science, and a theologian. He was also a fairly liberal churchman. His adversary in this matter was David Brewster, a Scottish scientist. Although theologically trained in the Presbyterian tradition, Brewster did not become a preacher, but rather was adept in optics and thus became a member of the Royal Society.

Whewell was initially well disposed to the idea of extraterrestrial life. In fact, he had earlier given sermons extolling this idea as part of God's munificence. However, the publication of *Vestiges of the Natural History of Creation* by the Scottish journalist Robert Chambers[30] changed his position. The raw and unforgiving pre-Darwinian evolutionary view presented in *Vestiges* was antithetical to his theological stance. As a result, he published *Of the Plurality of Worlds*.[31] He argued that life exists nowhere in the universe other than on planet Earth. The heart of his objection was to the apparent randomness of Chamber's proposal, in opposition to his own beliefs in design and the law-like operation of nature. He did, however, raise

---

30. Chambers, *Vestiges*, available online at online.org.uk/content/frameset?itemID=A2&viewtype=text&pageseq=1. Chambers set out an organic explanation for the presence of life on earth, as well as for the occurrence of the all observed species, through an evolutionary process. It was in reaction to this model that Alfred, Lord Tennyson penned the line "nature red in tooth and claw" in his tribute poem, "In Memoriam A.H.H." The line later became associated with the Darwinian model, although it was written some fifteen years before the publication of *Origin of Species*.

31. Whewell, *Of the Plurality of Worlds*. Interestingly, Chambers's viewpoint in *Vestiges*, to which Whewell objected, was similar to that of modern theistic evolutionists. Chambers held that God created through the natural laws of the world.

scientific arguments. For instance, the observable planets are uninhabitable. At the time, no exoplanets were known. Finally, the history of life on Earth was only a small fraction of the geological age of the planet.

Brewster engaged the debate with his own book: *More Worlds Than One: The Creed of the Philosopher and the Hope of the Christian*.[32] His argument was based on a conviction in the unlimited creativity of God as well as the notion that humans could not be the highest expression of this creativity.

The resulting debate, carried out mainly in publications rather than in-person confrontation, was perhaps the last time one could argue about the creation of life and intelligence by direct divine intervention, so-called special creation, as the accepted scientific model. Within just a decade of this encounter, the publication of *Origin of Species* would shift the discussion to a methodological model where all life, even human, could be accounted for by the laws of nature.

Conclusion

Earth revolves around the sun. This knowledge was new with Copernicus. What was not new was the debate over other worlds somewhere in Earth's sky. This debate had gone on for two millennia before Copernicus, with arguments both for and against. After Copernicus, scientists and theologians continued to speculate about other worlds and about the possibility of creatures like us living on those other worlds.

This suggests two meanings for the one word, *geocentrism*. Physically, the pre-Copernicans were geocentrists. The post-Copernicans became heliocentrists. With regard to pride-of-place, however, not much changed with the advent of Copernicanism. Both before and following the Krakow cosmologist, many philosophers and theologians were straining their eyes through telescopes to see if their space neighbors were hanging out their laundry. Both before and after Copernicus, philosophers and theologians sacrificed geocentrism on behalf of the empyrean realm—the realm of quintessence or ether—in the highest heaven, which was home to God's creatures.

What we did not see prior to the middle of the nineteenth century, however, was an encounter with the question of life's genesis both on-Earth and off-Earth. What became new with Darwin has been the speculation that

---

32. Brewster, *More Worlds Than One*. The work begins with a theological justification, and then proceeds with a detailed scientific description of the then state of knowledge from astronomy. Brewster acknowledges that he is specifically challenging the work of Whewell.

a second genesis due to natural causes might take place on one or more of those worlds in the sky. The real revolution in Western thinking is not due to heliocentricism, but rather to the rise of reliance upon strictly natural explanations for natural phenomena. Generative and vibrant nature, whether here on Earth or on other planets revolving around other stars, only expands our vision of God's creation.

Bibliography

Archimedes. *The Sand Reckoner*. Translated by Ilan Vardi. http://www.lix.polytechnique. fr/Labo/Ilan.Vardi/sand_reckoner.ps.
Aristotle. *On the Heavens (De Caelo)*. Translated by J. L. Stock. Chicago: University of Chicago Press, 1952.
Bennett, Jeffrey. *Beyond UFOs: The Search for Extraterrestrial Life and Its Astonishing Implications for Our Future*. Princeton: Princeton University Press, 2008.
Brewster, David. *More Worlds than One: The Creed of the Philosopher and the Hope of the Christian*. Pdf version of the 1856 publication, http://ia700400.us.archive. org/19/items/moreworldsthanoobrewuoft/moreworldsthanoobrewuoft.pdf.
Brooke, John H. "Science and Religion: Lessons from History?" *Science* 282 (1998) 1985–6.
The Catholic Encyclopedia. http://www.newadvent.org/cathen/03016a.htm.
Chambers, Robert. *Vestiges of the Natural History of Creation*. http://online.org.uk/ content/frameset?itemID=A2&viewtype=text&pageseq=1.
Churchland, Patricia S. *Touching a Nerve: The Self as Brain*. New York: Norton, 2013.
Copernicus, Nicholas. *On The Revolution of the Heavenly Spheres*. Translated by Charles Glenn Wallis. Encyclopedia Britannica, 1939.
Crowe, Michael J. *Extraterrestrial Life Debate from Antiquity to 1915: A Source Book*. Notre Dame, IN: University of Notre Dame Press, 2008.
Danielson, Dennis R. "Myth 6: That Copernicanism Demoted Humans from the Center of the Cosmos." *Galileo Goes to Jail and Other Myths about Science and Religion*, edited by Ronald L. Numbers, 50–58. Cambridge: Harvard University Press, 2009.
Davies, Paul C.W. *The Goldilocks Enigma: Why Is the Universe Just Right for Life?* New York: Houghton Mifflin, 2006.
Democritus. In *Ante-Nicene Fathers*, edited by Alexander Robertson and James Donaldson. Vol. 5. Christian Classics Ethereal Library. http://www.ccel.org/ccel/ schaff/anfo5.i.html.
Fontenelle, Bernard le Bovier de. *Conversations on the Plurality of Worlds*. http://books. google.com/books?id=VGoFAAAAQAAJ&pg=PA1&dq=Fontenelle+Bernard+in author:Fontenelle#PPP24,M2.
The Galileo Project. http://galileo.rice.edu/chr/bruno.html.
George, Marie. "Aquinas on Intelligent Extra-Terrestrial Life." *The Thomist* 65 (2001) 239–58. https://www.unav.es/cryf/english/extraterrestriallife.html.
Gingerich, Owen. *God's Planet*. Cambridge: Harvard University Press, 2014.
Huygens, Christian. *Cosmotheoros*. 1698. In English translation. http://www.staff. science.uu.nl/~gento113/huygens/huygens_ct_en.htm.
Impey, Chris. *Humble Before the Void*. West Conshohocken, PA: Templeton, 2014.

Kuhn, Thomas. *The Structure of Scientific Revolutions*. Chicago: University of Chicago Press, 1962.
Marsak, Leonard. "Bernard de Fontenelle: The Idea of Science in the French Enlightenment." *Transactions of the American Philosophical Society* 49 (1959) 1–64.
Rodis-Lewis, Geneviève. *Descartes: His Life and Thought*. Ithaca, NY: Cornell University Press, 1998.
Saliba, George. *A History of Arabic Astronomy: Planetary Theories During the Golden Age of Islam*. New York: New York University Press, 1994.
Segre, Michael. "Light on the Galileo Case?" *Isis* 88 (1997) 484–504.
The Stanford Encyclopedia of Philosophy. http://plato.stanford.edu/entries/aristotle/.
Whewell, William. *Of the Plurality of Worlds*. Facsimile edition, edited by Michael Ruse. Chicago: University of Chicago Press, 2001.

PART 2

The Search for Extraterrestrial Life: Will We Meet Them?

6

Searches for ET Life in the Solar System

Exobiology, Astrobiology, and the Big Picture

MARGARET S. RACE

Isn't science supposed to be about logic and evidence? Sure, but when data are in short supply, scientists feel free to speculate.

—CHRIS IMPEY[1]

A book on Astrotheology is surely an invitation to think broadly about life and our place in the Universe—to speculate using Earth-based scholarly thinking and attempt to develop "a comprehensive and meaningful framework for understanding our human situation within an astonishingly immense cosmos."[2] As theologians consider the meaning and implications of a discovery of extraterrestrial (ET) life, they aren't the only ones who come face to face with complex questions and ideas. No matter what one's expertise or disciplinary training, we humans are all challenged by the continuous flow of scientific findings and technological advances that have arisen in this era of space exploration. And all of us are curious—even anxious—

1. Impey, *Living Cosmos*, 281.
2. Peters, "Astrotheology," 838.

to know what it would mean if and when life is discovered beyond Earth. Let this chapter be your invitation to explore the combined scientific and non-scientific implications of our search for and possible discovery of ET life—any kind, simple, complex or intelligent.

Consider what is likely to happen when headlines announce that ET life has been discovered. After reporters interview the scientists and engineers associated with the discovery, they'll probably have additional questions about its significance and meaning. Who better to turn to than familiar, trustworthy experts who can opine on the meaning and implications of the discovery from ethical, theological or philosophical perspectives? If you get such a call, it will be important to know the basics, if only to offset the frenzy of opinions and pronouncements likely to be communicated by the mass media and social outlets.

This chapter is intended to provide a brief introduction and overview about the extent of the known Universe, *where* we are searching for ET life, *how* we conduct the searches, what *evidence* we expect to find, how the evidence will be *interpreted*, and what *societal issues* may arise, upon discovery and perhaps afterwards. As one reads this book and ponders the astrotheological implications of finding intelligent cosmic cousins (or ETI), it is likewise important to be aware of the range of *other* issues linked to discovery—including details about simpler extraterrestrial life, which are probably not included in traditional theological or humanities discussions—at least not yet.

Astrobiological Perspectives about Life

For centuries, people have wondered if we are alone in the Universe. Until the invention of telescopes around the seventeenth century, questions about life beyond Earth were necessarily undertaken as *gedanken* or thought experiments based on whatever science, religious or cultural perspectives existed at the time. Likewise, before microscopes, it was inconceivable to ponder life smaller than visible. Today, humankind has the ability to search for life at all scales using an impressive combination of scientific approaches and technological capabilities. Our collective search efforts and research have yielded stunning advances in understanding all aspects of life—from pre-biotic chemistry to extremophiles living in wide-ranging environmental conditions, to the diversity of micro- and macro-scale life found on Earth over eons of time. When integrated with our likewise expanded knowledge of environments and processes in space, and the immensity of the Universe (or even *just* our Galaxy!), the general assumption among most scientists is that Earth life—in all its diversity—is unlikely to be the sole example of life in the Universe.

In its broadest sense, astrobiology is a relatively new, interdisciplinary field that brings innovative perspectives and methods to the way we address questions about life in the Cosmos. Astrobiologists, astronomers, and space explorers use scientific methods and new technologies to study the Universe, from its origin and diversity, to its current state and possible futures, across unimaginable expanses of time and space, drawing upon multiple disciplines and methods—all in attempts to learn whether we earthlings are alone or share the vastness of space with other living entities. While humankind has asked similar questions over the centuries, we are among the first generations with sufficient scientific and technical prowess to address the questions in experimentally reproducible and verifiable ways, both on Earth and beyond.

While the other chapters in this book center predominantly on comparative theological analyses and deliberations about life, it is useful to put the topic of extraterrestrial life into a broad scientific context from the start—particularly since not all discoveries of *other* life would mean the same things. Currently, our collective searches for ET life involve multiple disciplines and methods, diverse locations, and varied types of evidence and interpretations—scientifically, theologically, and otherwise. Because all legal, ethical, cultural, and even theological systems have developed based on life as we know it on Earth, any direct extrapolation from known life to other putative types of life should give us pause—particularly before pronouncing about its significance and meaning. Moreover, considerations about discovery go beyond our anthropocentric views of "humankind" and include questions about lesser relatives on Life's family tree as well—however er simple, different or unfamiliar they may be compared to us. Just because extraterrestrial life may be different than the intelligent variety, we should not presume that discovery would be devoid of societal meaning, concerns or questions.

Our Current Technological Reach in the Universe

Before describing the features of different searches and their presumed scientific interpretations, it is useful to get one's bearings—to map the space territory where we reside and explore. For readers unfamiliar with recent advances in the world of astronomy, it is good to start with updated information on the scale of our searches—and why scientists think that a discovery of extraterrestrial life is likely, and perhaps inevitable. This requires an understanding of Earth's place in both the Solar System and our galaxy, which are located in just a miniscule sliver of the known Universe.

Over the millennia, everything we know about life has developed from information drawn almost entirely from Earth, one of eight planets within our Solar System (only four of which are terrestrial or rocky in nature). During the past five decades of space exploration we've discovered many surprises. For example, there are more than 180 moons in our Solar System as well as literally billions of other orbiting bodies of various types and sizes (e.g., rocky asteroids, icy cometary bodies, meteors, dwarf planets, etc.)[3]

To appreciate the size of *just our* Solar System, it is useful to imagine taking a trip across it, just as the Voyager 1 spacecraft did when it launched from Earth in 1977. Even at rocket speed, it took Voyager until 2012 to reach the edge of the Solar System and enter interstellar space. Continuing at the same rate, it would take tens of thousands of years for Voyager to cover the distance to the nearest star (Alpha Centauri—the second star in our corner of the galactic neighborhood—is 4.3 light years away or 25 trillion miles away from Earth!).

Now consider that our sun is just one of billions of stars in our home galaxy, the Milky Way. It is within this galactic neighborhood of billions of distantly separated stars that all our current searches for ET life take place. It is estimated that even our most far reaching telescopic searches have spanned only a small portion of the area of the Milky Way.[4] Now consider that there are literally billions of galaxies that make up the universe, and imagine all the planets likely to be orbiting those billions of stars in each galaxy. Clearly, our searches for ET life and our technical capabilities are still quite limited. But not so limited that we can't think about the implications of discoveries of *other* life.

3. See NASA, Our Solar System, https://solarsystem.nasa.gov/planets/solar system.

4. If each star in the Milky Way were represented by a single page of a large atlas and you flipped through the pages at a rate of one per second (twenty-four hours/day, seven days/week), it would take ten thousand years just to count all the pages in Milky Way atlas. Put another way, if you were to travel *at the speed of light* from one side of the Milky Way to the other, it would take one hundred thousand years. (Ferris, "The Three Immensities," 140.)

Figure 6.1. Moon. (NASA photo)

How and Where We Search, and What Evidence We Collect

Just as our searches for extraterrestrial life are confined to a miniscule portion of nearby space, our interpretations about life will necessarily reflect our current and limited scientific knowledge about it. To be sure, some people espouse entirely science-supported explanations about life, while others rely exclusively on beliefs or faith-based pronouncements about the development and path of life and our world. But, as discussed by Arizona State University law professor Joel Garreau, "there is nothing unusual about human beings taking more than one path in their search for truth—science at the same time as religion, for example."[5] Nor is it unusual for people to interpret and make pronouncements about unknowns in the face of uncertainty. It has been done throughout history—and is likely to continue. So, when thinking about ET life, as well as the origin, complexity, diversity, and evolution of life and environments across time and cosmic scales, it is advisable to draw information from many perspectives.

If one examines how scientists study life on Earth, it obviously involves more than searching for one thing—like people or large organisms.

5. Garreau, "Environmentalism as Religion," 61–74.

Similarly, searching for life elsewhere requires analyzing diverse types of evidence—from living, dormant or fossilized organisms, to chemical precursors or building blocks of life, to biochemical markers, metabolic activity, environmental changes, "footprints," artifacts, signals, and more. Obviously, the types of inferences that can be made about life will depend upon what is discovered or detected, as well as where, when, and how. Understanding these distinctions is essential for meaningful discussions about the implications, significance, and meaning of finding ET life, wherever it is found, and regardless of one's starting perspective or expertise.

As described by SETI researcher Seth Shostak there are currently three major approaches used in searching for evidence of extraterrestrial life: 1) telescopic searches for signals or messages from extraterrestrial intelligence (SETI); 2) remote telescopic searches for extrasolar planets (or exoplanets) and associated spectroscopic characterization of atmospheric biomarkers; and 3) exobiological or astrobiological searches in the Solar System using a combination of space exploration and research on/from Earth. As briefly summarized below, the three search types are distinct in ways that can impact the interpretations about the meaning of any future discovery.[6]

1. Searches for ETIL

Searches for extraterrestrial intelligent life (SETI) have been conducted for decades by astronomers using a combination of telescopes and signal processing technology on Earth to seek detectable patterns or messages in the form of electromagnetic signals that may have been sent across interstellar distances within our galaxy, presumably by extraterrestrial civilizations and beings from afar. A verified signal detection would represent indirect contact with ETI and be interpreted as a technological manifestation of intelligence and consciousness, but in all likelihood, will not provide much information about the form or nature of the beings who sent the message, or whether they are similar to us biochemically or otherwise. It is also likely that these incoming signals may not translate readily into meaningful information. All we would know is that the "beings" exist in a particular location in galactic space, and are adept enough to have sent a receivable signal of some type (either deliberately or otherwise). Early ETI searches concentrated on detection of radio telescope signals, but new technologies and astronomical knowledge have resulted in expanded search methods, including searches

---

6. Shostak, "Current Approaches to Finding Life Beyond Earth," 9-22. For additional detailed information about the three search types, see: www.seti.org, http://planetquest.jpl.nasa.gov, and https://astrobiology.nasa.gov/.

for laser light pulses (optical SETI), detectable large-scale artifacts, or even waste heat.

The notion behind all searches for ETI is that *any* signal may be valuable as a sign of a possible intelligent civilization, regardless of whether there is any accompanying information on associated biochemistry or the type of life involved. Just as researchers can learn about past civilizations by translating or decrypting information from artifacts like the Rosetta Stone, Dead Sea Scrolls, or cave drawings, so too might linguists, anthropologists, and other social scientists review different information in their attempts to translate or interpret the message content. Because of the vast search distances involved, it will not be possible to know whether the technological civilization still exists—since the detected signals would have been received from sources tens, hundreds or thousands of light years away. Because the signal and evidence would be indirect in nature, there are no concerns about real time, direct interaction with ETI, and thus no chance of biological contamination or other immediate impacts or legal/regulatory concerns needing attention. Perhaps the biggest impact would be psychological—knowing that we are not alone, and wondering what that would mean.

2. Searches for Extrasolar and Habitable Planets

Searches are also underway to identify the existence of other planets and solar systems associated with stars other than our sun. Searches are accomplished using orbiting space telescopes (e.g., Hubble, Spitzer, Kepler telescopes, etc.) as well as ground telescopes to peer into just a small portion of nearby space in the Milky Way with its billions of stars. These exoplanet searches currently concentrate on areas tens to hundreds of light years distant and try to confirm the existence of other planets in assorted ways (e.g., by recording temporary decreases in the amount of light from a star when a candidate planet "transits" or passes in front of it.) Since 1995, when the first exoplanet was confirmed orbiting a sun-like star, the number has grown significantly. By the end of 2016, astronomers estimated over 3500 individual exoplanets in more than 2500 planetary systems, with nearly 600 multiple planetary systems confirmed. In addition, there are more than 4700 candidate planets awaiting verification. A recent statistical estimate suggests at least one planet around every star in the galaxy—which translates to around a trillion planets in our galaxy alone, many of them in Earth's size range.[7]

---

7. See: https://exoplanets.nasa.gov/exep/. Also https://exoplanets.nasa.gov/exep/indepth/ and http://exoplanetarchive.ipac.caltech.edu/.

Because the universe appears to be physically and chemically consistent throughout, scientists assume it may also be a biological universe, which presumably could be associated with indications of life elsewhere, such as atmospheric signatures of biological activity.[8] In addition to the ongoing telescopic hunts for planets beyond our Solar System, astronomers also use spectroscopic instruments to study temperatures and scan for atmospheric signatures of water vapor, oxygen, carbon dioxide, and other tell-tale metabolic indicators considered as reliable evidence of chemical processes associated with life as we know it (including possible microbial life). Of particular interest are those terrestrial planets located in the "habitable zones" of their stars—just the right size and distance from their suns to have environmental conditions suitable for liquid water, and possibly life, to persist. Ultimately, scientists hope to discover and characterize Earth-like planets around nearby stars, search for those with habitable conditions and eventually identify signatures of life indicative of possible inhabited places.

In a little over two decades of searching for extrasolar planets, the evidence for their existence is clear and incontrovertible—albeit, entirely indirect. Like the SETI searches, these discoveries are many light years away—meaning that the potential discoveries pose no apparent direct risks or impacts to Earth. Not surprisingly, there have been no concerns or controversies raised about searches for extrasolar planets or the meaning of their discovery. In fact, discoveries of new exoplanets are routinely announced by the astronomical and planetary science communities and space agencies. Discoveries are recorded and characterized as either confirmed or candidate planets—some of which are terrestrial and potentially habitable—but none of which can be inferred as inhabited until other methods and data can demonstrate that fact.

Exobiology, Astrobiology, and Searches for Extraterrestrial Microbial Life

In contrast to the aforementioned searches in distant galactic space, astrobiological and exobiological searches for evidence of ET life in the Solar System involve a greater variety of missions, methods, locations, technologies, scientific disciplines, types of evidence, and potential interactions with possible ET life and/or harmful contamination. The NASA Astrobiology roadmap, originally formulated in 1998, brought together assorted research and exploration activities to address three basic questions: How does life begin and evolve? Does life exist elsewhere in the universe? And what is the

8. Dick, *Twentieth Century Extraterrestrial Life Debate*.

SEARCHES FOR ET LIFE IN THE SOLAR SYSTEM 117

future of life, on Earth and beyond?[9] In many ways, the Roadmap provides the logical underpinnings that connects all three ET search types. However, if extraterrestrial life is discovered anywhere in the Solar System, it would be a real-time discovery, the only search type with the potential for direct contact, possible human interactions, and assorted risks to life as we know it—regardless whether the discovery occurs in space or back on Earth. In addition to the determining whether the "alien" is in the form of simple or complex life (as opposed to "intelligent"), astrobiologists may be able to study its biochemical nature as well, and determine whether and how it might be related to life as we know it—or if it represents a *second genesis* of sorts. Such speculations are based on considerable scientific research over the past several decades, especially in microbiology, genomics, cosmology, and astrobiology.

Exploration to study the Moon and other solar system bodies has involved over one thousand unmanned/robotic missions and ten manned Apollo lunar missions (Apollo 8 through 17)—comprising both one-way and round-trip missions beyond Earth orbit.[10] Together, these missions have collected both direct and indirect evidence of all types, returning data, planetary materials, and samples to Earth. NASA maintains archival laboratories of diverse extraterrestrial samples from many locations—including lunar materials (381.7 kilograms [842 lbs.] of soils, rocks, pebbles, and dusts) as well as diverse samples of meteorites, cometary and interstellar particles, cosmic dust grains, solar winds, and asteroidal particles and grains. Collected and returned materials are kept in appropriate containment facilities in order to maintain the scientific integrity and pristine nature of materials for further study. During the Apollo program, both samples and astronauts returning from the lunar surface were also placed in special quarantine facilities upon return to Earth, until extensive analyses determined there was no harmful backward contamination of concern.

While no definitive evidence of extraterrestrial life—simple or complex—has been discovered so far, humankind is already prepared to deal with its possible existence, at least in the Solar System. Under Article IX of the Outer Space Treaty,[11] planetary protection policies apply to both outbound and inbound spacecraft—with controls imposed on space missions depending on where they are targeted and what activities will be under-

9. *NASArm*.
10. Many other human missions have taken place, including those during the Mercury, Gemini, early Apollo, and Shuttle programs as well as expeditions to the International Space Station—but none these ever departed Earth orbit.
11. United Nations Treaty on Principles Governing the Activities of States in the Exploration and Use of Outer Space, Including the Moon and Other Celestial Bodies.

taken (e.g., flyby, orbiter, lander, subsurface drilling, etc). The Committee on Space Research (COSPAR) of the International Council for Science, oversees planetary protection, with the primary objective to develop, maintain, and promulgate planetary protection knowledge, policy, and plans to prevent the harmful effects of such contamination.[12] The strictest planetary protection controls apply to one-way and round-trip missions to bodies where conditions may be potentially supportive of life (e.g. Mars, with evidence of water ice and brines; Europa and Enceladus, both with salty oceans beneath their surfaces). Mission restrictions, spacecraft operating controls, and cleaning of various parts are required pre-launch to minimize *forward contamination* (transport of terrestrial contaminants and microbes on outbound spacecraft). For round trip missions, special controls and provisions are also required to prevent *backward contamination* (the return to Earth of uncontained planetary materials that could pose a threat to Earth and its biota).[13] Furthermore, missions returning to Earth from celestial bodies may also be subject to other laws and regulations imposed by launching and/or landing nations—such as controls on "importation of materials" or dealing with environmental, health and safety concerns.

In recent years, governmental launching agencies as well as commercial and private ventures worldwide have expressed increasing interest in expanding exploration and resource exploitation activities on the Moon and beyond. Accordingly, COSPAR and NASA are in the process of updating planetary protection requirements applicable to future human missions beyond Earth orbit.[14] Once again, there is considerable attention to ensuring that the missions will safeguard the human crew, the destination planetary surfaces (and possible ET life?), and Earth itself throughout all missions to nearby extraterrestrial locations—regardless whether ET life has been discovered yet or not.

What Happens When We Discover ET?

To be sure, the discovery of *any* type of life beyond Earth would be extraordinary—and have great significance. Those involved in astrobiology,

12. For more on COSPAR, see https://cosparhq.cnes.fr/about.

13. Over one hundred nations are signatories of the Outer Space Treaty, including all current space launching nations. Detailed information about international policies well as NASA Planetary Protection Requirements and controls for both robotic and human missions is available in the 'Documents' section of: https://planetaryprotection.nasa.gov/.

14. Race, "Planetary Protection Knowledge Gaps for Human Extraterrestrial Missions: Workshop Report."

astronomy and space missions have already considered various discovery scenarios and issues involving ET life of *any* type—simple and microbial included. Perhaps the most obvious questions associated with the discovery of ET life are *what does it mean? what should be done? by whom?* and *when?* Just as the methods and presumed life characteristics for the three search types are different, their discovery/response plans are as well.[15]

Radio-astronomers searching for ETI were the first to consider how to respond to a possible discovery. The SETI Principles, developed in 1989, articulate detailed action plans for use upon detection of a presumed signal.[16] The short term operational policy stipulates sharing and verifying the detected incoming signal, safeguarding the frequencies, notifying the international community, and announcing the discovery to the public. Moreover, the Principles indicate that no response should be sent until international consultations take place. No mention is made of ancillary questions such as: Who should respond on behalf of humankind? What if someone decides to send a message prior to international consultation? And what might the impacts of a response be upon future generations? Concerns have arisen within the astronomical and SETI communities about the prospects of premature active broadcasting that would announce our presence to ETI.[17] Some scientists, including physicist Stephen Hawking, have even suggested that messaging by rogue broadcasters is foolhardy and puts our entire planet at risk. In the meantime, proponents of active SETI— or METI (messaging)—continue to suggest that humankind initiate more deliberate messaging, even before detection of a signal (e.g., see *www.METI. org*). Recently METI.org announced plans to begin active broadcasts within the next several years. Currently, there are no guidelines or international policies to prevent or permit active SETI, and no deliberations have been undertaken except by a small elite group of scientists.

In contrast to searches for ETI, no searches for extrasolar planets or "discoveries" of potential habitable worlds have raised any concerns other than scientific and technical. Discoveries are routinely announced by space agencies and astronomers, and the focus is largely on cataloging and characterizing detected bodies. Even if a decision were made to send missions or signals to learn more about some far off terrestrial habitable planet, there would presumably be no indications of "contact" for extended periods of time—and little likelihood of determining anything about possible

15. Race, "Preparing for the Discovery of Extraterrestrial Life," 263–85.
16. International Academy of Astronautics. SETI Permanent Committee. "Declaration of Principles."
17. Grinspoon, "Who Speaks for Earth?"; also Vakoch, "In Defence of METI," doi:10.1038/nphys3897.

"inhabitants." No astronomical or other international body has raised any societal concerns about exoplanet searches, and none are anticipated.

While astrobiological searches for ET life in the Solar System involve the most active and diverse types of search scenarios, the only plans so far for a response to discovery are included in a Draft Protocol for handling returned Mars samples on Earth.[18] The Protocol emphasizes proper science testing and handling on Earth, including details on facilities, technological and administrative concerns; public communications; and emergency responses in the event of unplanned release from containment of possible ET life. The Protocol's plans mention that legal, ethical and social issues should also be considered as well—by including appropriate non-science expertise on an oversight committee(s) prior to any decision about release from containment. Subsequent review of Mars life detection efforts on Earth emphasized primarily biological contamination, planetary protection policy, and science capabilities predominantly.[19] Aside from discussions at these workshops, no comprehensive plans exist for making decisions about nearby ET life detection, either on Earth or on bodies in the Solar System. Presumably discussions would be undertaken by space agencies and research scientists, with likely international deliberations involving COSPAR.

Longer Term Questions about ET and Humankind

While theological analyses are clearly relevant when anticipating possible impacts on humankind associated with discovery of ETIL, it is equally important to contemplate issues linked with "lesser" types of life. The possible discovery of non-intelligent life may not lead to big philosophical questions, but it doesn't mean that the societal concerns or questions are unimportant. Because the remainder of this book focuses on topics related to ETI, it is appropriate to briefly highlight some of the societal questions that have already surfaced—particularly because many are outside traditional theological and religious considerations.

Some of the other issues raised about space exploration and activities would have far greater near-term impacts upon microbial or non-intelligent ET life—and perhaps even upon life on Earth. For example, provocative questions have been asked about the "rights" of indigenous ET life and what would constitute responsible human action regarding preservation, stewardship and environmental management of locations beyond Earth. Should widespread tourism, resource use, and commercial exploitation

18. Rummel, "A Draft Test Protocol."
19. Kminek, "Report of the Workshop," 1–5.

of other locations be allowed? Looking further ahead, who—beyond scientists, launching entities, and lawyers—should be involved in deliberations of whether to proceed with large-scale terraforming of planets, space settlements or colonization? Similarly, if human activities on other planets could result in ET microbes infecting or colonizing the human genome, who should decide what should be done—and to whom? Just as we consider possible impacts of emerging technologies, genomic interventions or climate change upon human futures, so too, it is advisable to examine the implications of our space activities from assorted perspectives, particularly if the activities might cause irreversible changes in evolutionary trajectories or the nature of life, on Earth or beyond.

The original NASA Astrobiology Roadmap and its subsequent revisions through 2008 included more than just science goals and objectives that focused on understanding the origin, diversity, and future of life. It also included a set of Principles as well—which, in many ways, anticipated the need for broad collaborative deliberations beyond just the science and technology communities. While the Roadmap aimed to provide a unifying sense of what it means scientifically to search for ET life, the basic Principles were considered fundamental to the overall Astrobiology program. Specifically, the Roadmap Principles indicated that Astrobiology

1. Is both multidisciplinary in content and interdiciplinary in implementation;

2. Encourages planetary stewardship through protection against forward and back contamination and recognition of ethical issues associated with exploration;

3. Recognizes a broad societal interest in its endeavors, especially in achieving a deeper understanding of life, searching for extraterrestrial biospheres, assessing the societal implications of discovering other examples of life, and envisioning the future of life on Earth and in space; and

4. Places a strong emphasis upon education and public outreach—and offers an opportunity to educate and inspire the next generation of scientists, technologists, and informed citizens.

Interestingly, the most recent revision of the Roadmap was published in a different format—an Astrobiology Strategy[20]—that emphasized science and exploration, but dropped the original basic operating Principles. Instead, there is a short appendix written by two authors (not a community

20. *NASAstr.*

deliberation) on how Astrobiology science may integrate with humanities and social sciences. It is a far cry from the earlier strong NASA statement about the importance of fundamental principles. Admittedly, it indicates that there are still many questions related to Astrobiology in fields like ethics, law, epistemology, history, social sciences, communications, and education. However, much detail is still lacking, even as individual authors begin to grapple with the issues.[21]

As we ponder the big science questions about life, it is appropriate to consider how the latest findings overlap with important questions in all fields from scientific and technological, to complementary societal, cultural, legal, theological, ethical, and philosophical specialties. As we revise our understanding about life, our planet, and our place in the Universe, it is appropriate likewise to ponder broad, new questions about meaning and how we relate to or interact with diverse types of organisms and environments, here and elsewhere.

Looking ahead, researchers in assorted non-science fields have an opportunity to take a more pro-active role in framing future collective questions and issues. At the very least, it would be appropriate to consider the meaning of astrobiology discoveries, plans, and activities upon *Homo sapiens* on Earth—and to extend non-science disciplinary thinking into our plans for the future. What *does* it mean to share the cosmos with space neighbors—whether intelligent, complex, or simple—near or very far? And who should be involved as we make decisions about our collective futures? Developing a clearer, updated "roadmap" of principles and non-science research issues would be a great start. It would be an important contribution to astrobiologists *and* all humankind as we endeavor to explore, extend, and understand our place in the universe.

Bibliography

Committee on Space Research (COSPAR). https://cosparhq.cnes.fr/about.
Des Marais, David J. NASA, *Astrobiology Roadmap* 2008. https://nai.nasa.gov/media/medialibrary/2013/09/AB_roadmap_2008.pdf.
Dick, Steven J. *The Twentieth Century Extraterrestrial Life Debate and the Limits of Science*. Cambridge: Cambridge University Press, 1999.
Ferris, Timothy. "The Three Immensities." *Forbes ASAP Magazine* (November 30, 1998) 140.
Garreau, Joel. "Environmentalism as Religion." *The New Atlantis* 28 (Summer 2010) 61–74. http://www.thenewatlantis.com/publications/environmentalism-as-religion.

21. Schwartz and Milligan, *ESE*.

Grinspoon, David. "Who Speaks for Earth?" *Seed*. http://seedmagazine.com/content/article/who_speaks_for_earth/.
Impey, Chris. *The Living Cosmos: Our Search for Life in the Universe*. New York: Random House, 2007.
International Academy of Astronautics. SETI Permanent Committee. "Declaration of Principles Concerning Activities Following the Detection of Extraterrestrial Intelligence." http://avsport.org/IAA/protdet.htm.
Kminek, Gerhard. "Report of the Workshop for Life Detection in Samples from Mars." *Life Sciences in Space Research* 2 (July 2014) 1–5.
NASA. Astrobiology at NASA. https://astrobiology.nasa.gov/.
———. *Astrobiology Roadmap*. https://nai.nasa.gov/media/medialibrary/2013/09/AB_roadmap_2008.pdf.
———. Astrobiology Strategy. https://nai.nasa.gov/roadmap/.
———. ExEP In Depth. https://exoplanets.nasa.gov/exep/inDepth/.
———. Exoplanet Exploration. http://planetquest.jpl.nasa.gov.
———. Exoplanet Exploration Program. https://exoplanets.nasa.gov/exep/.
———. Office of Planetary Protection. https://planetaryprotection.nasa.gov/.
———. Our Solar System. https://solarsystem.nasa.gov/planets/solarsystem.
NASA Expolanet Archive. http://exoplanetarchive.ipac.caltech.edu/.
Peters, Ted. "Astrotheology." In *The Routledge Companion to Modern Christian Thought*, edited by Chad Meister and James Beilby, 838–53. London: Routledge, 2013.
Race, Margaret S. "Planetary Protection Knowledge Gaps for Human Extraterrestrial Missions: Workshop Report." (2016) NASA DAA_TN36403. http://hdl.handle.net/2060/20160012793.
———. "Preparing for the Discovery of Extraterrestrial Life: Are we Ready? Considering Potential Risks, Impacts and Plans." In *LBE*, 263–85.
Rummel, John D. "A Draft Test Protocol for Detecting Possible Biohazards in Martian Samples Returned to Earth." (2002). NASA/CP-2002-211842. https://planetaryprotection.nasa.gov/summary/DraftTestProtocol.
Schwartz, James SJ and Tony Milligan, eds. *The Ethics of Space Exploration*. Heidelberg: Springer, 2016.
SETI Institute. http://www.seti.org.
Shostak, Seth. "Current Approaches to Finding Life Beyond Earth, and What Happens if We Do?" In *LBE*, 9–22.
United Nations Office for Outer Space Affairs. Treaty on Principles Governing the Activities of States in the Exploration and Use of Outer Space, including the Moon and Other Celestial Bodies. http://www.unoosa.org/oosa/en/ourwork/spacelaw/treaties/introouterspacetreaty.html.
United Nations Treaty on Principles Governing the Activities of States in the Exploration and Use of Outer Space, Including the Moon and Other Celestial Bodies. Article IX, U.N. Doc. A/RES/2222/(XXI) January 25, 1967; TIAS No. 6347, 1967.
Vakoch, Douglas A. "In Defence of METI." *Nature Physics* 12:890 (2016). doi:10.1038/nphys3897.

7

Exoplanets and the Search for Life Beyond Earth

JENNIFER WISEMAN

> The probability, therefore, is that each of these fixed stars is also a Sun, round which another system of worlds or planets, though too remote for us to discover, performs its revolutions, as our system of worlds does round our central Sun.
>
> —THOMAS PAINE[1]

Human beings have speculated on the nature of "other worlds" for as long as we could imagine the possibility. Sometime around 300 BC, Epicurus proclaimed "There are infinite worlds both like and unlike this world of ours ... We must believe that in all worlds there are living creatures and plants and other things we see in this world."[2] More than two millennia later, books and films likewise assume with creative fervor the existence of flourishing life on planets and moons in our solar system, and in star systems far beyond. Theologians and philosophers have likewise grappled for centuries with the implications of this imagined possibility of life beyond our own familiar world.

Yet for all science fiction and our limitless imaginations have to offer, they have always carried the caveat that no other inhabited, or even potentially habitable, worlds beyond our home planet have ever actually been

1. Paine, *The Age of Reason*.
2. Crowe, *ELD*, 4.

demonstrated, nor has there been any way to truly investigate and affirm this possibility scientifically. Until now.

We live in a most remarkable period of human existence. After thousands of years of recorded human history, human investigators have managed, within just the past few decades, to send robotic probes to other planets in our solar system, and to actually detect planets—thousands of them—orbiting stars outside of our solar system. The detection and study of these extra-solar planets, or *exoplanets*, at first seen as just a fringe activity of astronomers, is now one of the highest priorities of mainstream international space exploration.

How has this revolution come about? And is it leading to the detection of life beyond Earth? This chapter outlines the technical advances that have enabled this revolution in discovery and the resulting enrichment of contemplating the philosophical, and astrotheological, implications.

Exoplanets Everywhere!

In the early 1990s, the first confirmed detections of planets orbiting other stars were reported in scientific meetings and literature. The discoveries were not made by imaging these reported exoplanets. Rather, they were indirect detections, achieved through innovative technological advances and meticulous uses of telescopes and optical techniques.

Why not just point a telescope at a star and view any planets around it? That would be, in fact, an astronomer's dream, but in most cases it is very challenging and often impossible to image a planet around another star with current technology. A planet is very small and dim compared to its host stars, so much so that a Sun-like star would be a billion times brighter in visible starlight than the reflected light of an Earth-sized planet lost in its glare. While astronomers continue to work feverishly to develop techniques of blocking out such overwhelming starlight in hopes of imaging small planets nearby, other indirect techniques of finding exoplanets have proved to be very fruitful.

The vast majority of the first few hundred exoplanets discovered were found by detecting a periodic motion, or wobble, of the parent star. With the star and planets exerting a mutual gravitational pull on each other, the planets stay in orbit while the star wobbles to and fro. (In reality, the star and planets are all orbiting a point in the system known as a common center of mass, usually located within the volume of the star.) The stellar motion is usually measured by monitoring the back-and-forth shifting of the frequency, or color, of the light emitted from an element like hydrogen in the star's atmosphere. When starlight is spread out into its constituent colors with a spectrograph, similar to a prism, in a telescope, the pattern of emission from particular elements like hydrogen is very distinct, showing up only at certain colors along the rainbow. When an emitting object, in this case a star, is moving relative to the observer, the Doppler effect causes the light received to be shifted in frequency from that which was emitted. When the star moves away from the observer, the light received is reddened, or "red-shifted"; likewise approaching motion causes the received light to be "blue-shifted." Precise instruments in telescopes can measure this periodic shifting of color, allowing astronomers to deduce the amount of stellar motion along the line of sight and thus estimate the mass range of any planets orbiting that star that would be causing such a wobble with every orbit.

Measuring the wobbling motion of a star in the plane of the sky is also possible, with the careful technique of astrometry. But in more recent years, by far the most fruitful technique for finding and studying exoplanets is to look for transits. Transiting systems have one or more planets orbiting the parent star along our line of sight, so that periodically the planet(s) will orbit in front of the star, blocking out some of the starlight during this partial eclipse. Astronomers, once again, don't view the planet directly, but rather they monitor the total brightness of the star, which will appear to dim temporarily during the transit, and then return to full measured brightness. The amount of dimming tells the astronomer how large the transiting planet must be.

NASA's recent Kepler mission flourished with the transit technique of finding exoplanets, increasing the number of known exoplanets into the thousands over just a few years.[3] The Kepler space telescope stared toward the center of our Galaxy at a field of one hundred thousand stars, looking for periodic dimming caused by transiting planets in any of these star systems. While most stars in this crowded field are too far away for follow-up detailed study of their planets, the Kepler mission provided the first-ever statistics of how common exoplanets are. The harvest was surprisingly bountiful. The thousands of new planets identified, when extrapolated over the presumed random orientations of planetary systems relative to our sight-line, indicated that planets are not rare, they are abundant. Some stars have multiple planets, some have none, but according to the findings from Kepler, on average, every star has a planet. That means there are hundreds of billions of planets in our Milky Way galaxy alone![4]

Kepler data went further, allowing the sizes of these planets to be deduced. They span from smaller than Earth to larger than Jupiter, but the

3. NASA's Kepler and K2 Mission Pages: https://www.nasa.gov/mission_pages/kepler/main/index.html.

4. NASA Exoplanet Archive: http://exoplanetarchive.ipac.caltech.edu/.

majority seem to be slightly larger than Earth—so-called "super-Earths." Even though the Kepler telescope is no longer taking such a census of planets, the data already archived are so rich with information that scientists continue to make discoveries regularly from it.

While stellar wobbles and planet transits account for the majority of the thousands of exoplanet discoveries, there are still other ingenious techniques astronomers are using to find these other worlds. Gravitational microlensing, an effect of general relativity, is a rather exotic phenomenon of light from a distant star being detoured and magnified by space distortions caused by the gravity of a foreground star system the light is passing near on its way to our telescopes. Since the foreground and background stars generally are both in motion, they only align for a brief amount of time, resulting in a kind of flash of magnified light, from our perspective. If that foreground star has an orbiting planet, then there will be multiple components to that flash, corresponding to the microlensing of both the star and the planet. So by watching for such signals from magnified, lensed background stars, astronomers have identified still more planetary systems, and these need not be in an edge-on orientation for us to detect, like transiting systems.

Direct imaging of exoplanets is ultimately the technique of most interest, and though, as explained above, this is very difficult to do, astronomers have nevertheless been able to develop "light suppression" techniques, such as coronagraphs within telescopes, to block out bright starlight in a way that, for a few systems, has visually revealed planetary systems.

Figure 7.1. This artist's concept depicts one possible appearance of the planet Kepler-452b, the first near-Earth-size world to be found in the habitable zone of star that is similar to our sun. The habitable zone is a region around a star where temperatures are right for water—an essential ingredient for life as we know it—to pool on the surface. Scientists do not know if Kepler-452b can support life or not. (Image credit: NASA Ames/JPL-Caltech/T. Pyle.)

Are Exoplanets Habitable?

When a newly discovered exoplanet is reported, astronomers and the public all want to know if the new world could possibly support life. A basic measure of this possibility is the planet's position relative to the habitable zone around its parent star. The habitable zone or Hz is nominally the orbital region that is close enough to the star to allow the presence of liquid water on the planet (not just frozen ice), but not so close that the water would boil away. Since a connection to liquid water seems to be a basic requirement for all kinds of life on Earth, it is reasonable to expect, for a first exploration foray, that the most life-friendly exoplanets would be the ones with water.

Of course being in the habitable zone does not necessarily mean that the planet is inhabited with life, or that it even has liquid water. Furthermore, there are many other factors that would affect a planet's habitability, not just its distance from the star. Take Mars and Venus, for example: they are both in the nominal habitable zone around our Sun, but their atmospheric processes have long ago made life seemingly untenable, turning one into an oven-like greenhouse (Venus), and the other into a frozen dry desert (Mars). Atmospheric dynamics, orbit shapes, magnetic fields, bombardment from asteroids and debris, and proximity to stellar flares are among the many factors that influence the true habitability of a planet. In fact, the extreme harshness of the environment for many exoplanets has led some to conclude that the needed stability for life is unlikely, and therefore advanced life must be exceedingly rare.[5] Yet the abundance of exoplanets keeps the possibility alive, with others concluding that life must be plentiful.[6]

One unexpected twist in the approach astronomers are taking to finding habitable exoplanets has come with the realization that most stars in the galaxy are dwarf stars, redder and smaller than our Sun. So the hitherto default effort to seek out "Earth-like planets around Sun-like stars" is quietly being outpaced by an effort to find habitable planets orbiting red dwarf stars. The habitable zone around dwarf stars is much closer to the star, since the star is less bright. This means that any planets in that zone would have very fast orbits, circling their star in much less than an Earth year. Finding such planets should be easier in some ways, since those in an orbital plane along our line of sight would transit their stars frequently, and even direct imaging would be easier since the star itself is less bright. Yet even if these kinds of exoplanet systems are the most common, they may not be good candidates for habitable planets. It is known that red dwarf stars are

---

5. Ward and Brownlee, *Rare Earth*.
6. Darling, *Life Everywhere*.

very active, sending powerful flares regularly into their environments. With winds and charged particles ramming the planet regularly, a stable environment for life may not be possible to maintain.

While many such effects may make habitable zone planets around many types of stars uninhabitable, there are other bodies outside the habitable zone that nevertheless may be ideal homes for life. These are the *moons* of planets. We see this possibility even in our own solar system. Gas giant planets like Jupiter and Saturn harbor many moons with a dazzling breadth of character. While these planetary systems are nominally too cold for water and life, some moons, being tidally tugged and stretched by their interaction with their parent star, generate internal heat. Evidence is mounting that some of the ice-covered moons (e.g., Europa, Ganymede, Enceladus) have enough internal heat to keep water in liquid form under the surface ice. Could simple life exist in these ice-covered oceans? It's an intriguing possibility. The challenge is being able to sample such water to find out, but getting a probe to somehow drill through deep ice is not a simple concept. It's also not necessarily the only way to study the water of moons, but the point here is that even giant exoplanets in non-habitable zones of their stars may shepherd life-bearing moons.

Looking for Life in All the Right Places

We can send probes to sample dirt, atmospheres, and maybe even oceans in the bodies of our own solar system. But exoplanets are, for now, out of reach for humans to visit either in person or with probes. We are limited to "remote sensing" as a tool for exploring these alien worlds. So what kinds of evidence for habitability or life on an exoplanet could an astronomer perceive by using a telescope?

One way to know is by looking back at our own planet Earth from the platform of satellites and probes. From these, we can look at the imagery and the light spectrum from Earth's atmosphere and discern evidence of biological activity on our own planet. Astrobiologists are also building a wealth of information and predictions on the range of extreme conditions in which life could thrive, and the observable evidence of such biological activity.

From this we can predict that a life-bearing planet would likely show signs of biological activity, or "biomarkers" in the sense astronomers use the term, in the frequencies of light emitted or absorbed by their atmospheres. Telescopes with spectrographs can spread out the received light spectrum from an exoplanetary system, revealing the precise patterns of light frequencies that are emitted from, e.g., oxygen (which must be continually refreshed in atmospheres like Earth's by biological activity like photosynthesis in plants), methane (profusely emitted by Earth's livestock), and water vapor (evidence of water and habitability, though not necessarily of extant life).

Already, astronomers have detected water, sodium, and a variety of other features in the atmospheres of some exoplanets, by looking at transiting systems where light from the star passes through the limbs of the planet's atmosphere as it passes in front of the star. Elements and molecules in the planet's atmosphere absorb certain frequencies of the starlight passing through, making a discernable pattern in the telescope's spectrograph. From these preliminary studies on gas giant and super-Earth exoplanets, we already know that water vapor can be present. Astronomers are eager to apply these techniques to Earth-sized planets in the habitable zones of their stars, but need the next generation of more sensitive telescopes to be able to do this kind of analysis for most of these small bodies.

Astronomers will also need to realize that exoplanetary systems may be at a different stage of their evolution than Earth and Sun are at present. The snapshot analysis we may take of an exoplanet's atmosphere may reveal something similar to what was happening in Earth's atmosphere several billion years ago (or what our atmosphere will look like in the distant future). Three billion years ago, Earth's atmosphere was dominated by carbon dioxide and methane, with hardly any oxygen. Currently, thanks to the efficacies of oxygen-producing bacteria and plant photosynthesis, oxygen is much more abundant than methane and carbon dioxide in our own atmosphere. So there are many factors and possibilities that must be taken into account when astronomers begin to analyze light spectra and evidence for biosignatures from potentially habitable planets.

Advanced Life

This chapter has focused mainly on steps scientists are taking at the present to discern whether any of the plentiful exoplanets we are discovering are suitable for life or even harboring life, where that life would most likely be simple, such as monocellular life forms. The reason for this emphasis is that this is what most astronomers, astrobiologists, and planetary scientists around the world are focusing on, and the techniques of detection being developed will not necessarily differentiate between simple and advanced life biomarkers. Though the attraction of public discourse (and this book on astrotheology) to extraterrestrial life always focuses on the possibility of advanced life and civilizations throughout the cosmos, the current scientific efforts are focusing necessarily on the much more likely chance of encountering simple life forms. Even our own planet Earth harbored only simple life forms for the vast majority of its multi-billion year history of life. In fact, even some experts convinced of the possibility of advanced civilizations also predict that it is unlikely we will ever encounter them, given the vast separations in space and time between potentially communicating civilizations.[7]

This should not be discouraging for the search for extraterrestrial life. It should be emphasized that, to date, no life has been detected yet outside of Earth's environs, either in our own solar system or beyond. And yet the accelerating pace of related discoveries is making the detection of evidence for at least simple life beyond Earth quite possible in the coming years and decades, probably within the lifetime of some today. The detection of even simple life beyond Earth would be absolutely profound for humanity, indicating life has sprung up through "multiple Genesis events" throughout the universe.

That said, there is in fact a very concerted effort by some experts to look specifically for evidence of advanced, communicating civilizations in other star systems. The Search for Extraterrestrial Intelligence (SETI) program has involved many senior scientists, students, engineers, entrepreneurs, visionaries, and the greater public in efforts to detect electromagnetic signals that could only have been produced intentionally by intelligent life forms desiring to communicate. SETI has tuned their radio telescopes by making educated guesses as to what kinds of frequencies and patterns an alien intelligence might choose for communication. They also make special efforts to listen toward exoplanetary systems discovered by astronomers where habitable planets appear more likely.

No confirmed ETI signal has yet been detected. Yet the SETI teams remain undaunted and encouraged. They liken the search to an ocean of possibilities, where only a glass of water has yet been examined; they believe that patience and dedicated searching for many years is the necessary approach for such a bold endeavor.

Outlook

While encountering communication from an extraterrestrial advanced civilization is an unpredictable event, the complementary effort to find habitable, and perhaps life-harboring, worlds through astrobiology and exoplanetology is growing at an accelerating pace. NASA's Hubble Space Telescope is

7. Bennett, *Beyond UFOs*.

already analyzing the atmospheres of some super-Earth planets, looking for spectral signs of water. In the period when this book is published, NASA should be launching two space observatories with special capabilities for studying exoplanets: TESS (for studying exoplanets transiting nearby bright stars), and JWST (a large infrared observatory capable of studying the dusty planet-forming zones around stars). A still larger and more sensitive space telescope concept is currently being considered for launch in a decade or so that would allow direct imaging of Earth-sized exoplanets, blocking out the bright glare from the parent star and allowing detailed spectral analysis of the planets' atmospheres, uncovering many potential biomarkers. And very large telescopes on the ground will also make unique and complementary contributions to the characterization of exoplanets and their environments.

These very real scientific and technological leaps that are happening in our lifetime are setting the stage for reinvigorated conversations and contemplations of what it means for humanity to search for (and possibly find) life beyond Earth. Father José Funes, former Director of the Vatican Observatory with a chapter elsewhere in this volume, expressed this when asked why the Vatican hosted a recent scientific conference on astrobiology: "Why is the Vatican involved in astrobiology? Although astrobiology is an emerging field and still a developing subject, the questions of life's origins and of whether life exists elsewhere in the universe are very interesting and deserve serious consideration. These questions offer many philosophical and theological implications."[8]

Bibliography

Bennett, Jeffrey. *Beyond UFOs: The Search for Extraterrestrial Life and Its Astonishing Implications for Our Future.* Princeton: Princeton University Press, 2011.
Crowe, Michael J. *Extraterrestrial Life Debate from Antiquity to 1915: A Source Book.* Notre Dame, IN: University of Notre Dame Press, 2008.
Darling, David. *Life Everywhere: The Maverick Science of Astrobiology.* New York: Basic, 2002.
NASA. Exoplanet Archive. http://exoplanetarchive.ipac.caltech.edu/.
———. Kepler and K2. https://www.nasa.gov/mission_pages/kepler/Main/index.html.
Paine, Thomas. *The Age of Reason.* 1794.
"The Vatican Ponders Extraterrestrials." *The Guardian*, November 11, 2009. http://www.theguardian.com/world/2009/nov/11/vatican-extra-terrestrials-catholic.
Ward, Peter D., and Donald Brownlee. *Rare Earth: Why Complex Life is Uncommon in the Universe.* New York: Copernicus, 2004.

---

8. From "Vatican Ponders Extraterrestrials," *The Guardian*, November 11, 2009.

8

Yes, We Will Meet Them

The Drake Equation Tells Me So

HEIDI MANNING

Contact with extraterrestrial intelligence will have a profound effect on humanity. It will change forever the way we think about ourselves.

—JILL TARTER[1]

Are we alone? Is anyone else out there?
While staring into the dark sky and observing the myriad of tiny lights, humans have pondered these questions for centuries. Upon seeing so many twinkling stars and shining planets in the night sky, one wonders if there is anyone else out there. For decades now, we have been using our telescopes to listen and find out if anyone is out there trying to communicate with us, and occasionally, we have sent a message into space in case someone out there is listening and waiting to hear from us.
When we think of life in the universe, we often think in absolutes. We debate if it is either there or not. Based on the work of SETI Founder Frank Drake, I will show that extraterrestrial life is most likely out there, and

---

1. Tarter, "Contact: Who Will Speak for Earth?," 190.

we ought to change the debate to be on the implications of this life on our meaning and existence.

Throughout history, humans have considered the sky to be inhabited. Many civilizations view the sky as the home of the gods. It is the place we look up to when considering heaven. But beyond any deity, is there any intelligent being out there who might be like us, someone with whom we can communicate? A 1959 paper in *Nature* by Cocconi and Morrison[2] details the type of communication that could be anticipated and how we might search for such a signal. This paper concludes that we possess the means to detect the anticipated signal and the need to begin a search for communications from extra-terrestrial civilizations.

In 1961, at a meeting held at the National Radio Astronomy Observatory in Green Bank, West Virginia, Frank Drake proposed a mathematical equation that incorporated all the steps of stellar and biological evolution necessary to occur in order to have a civilization in our Milky Way galaxy capable of communicating with us. This equation is now known as the Drake Equation and was developed to provide a logical framework and justification that would guide the search for extra-terrestrial civilizations. By assigning values to the different variables in the equation, one can calculate the number of possible civilizations currently existing with whom we could communicate. The Drake Equation is $N = R^* f_p n_e f_l f_i f_c L$:[3]
Where,

N is the number of civilizations in the Milky Way Galaxy whose electromagnetic emissions are detectable,

R^* is the rate of formation of stars in our galaxy suitable for the development of intelligent life,

f_p is the fraction of those stars forming that have planetary systems,

n_e is the number of planets per solar system with an environment suitable for life,

f_l is the fraction of suitable planets on which life actually appears,

f_i is the fraction of inhabited planets on which intelligent life exists,

f_c is the fraction of intelligent civilizations that develop technology that releases detectable signs of their existence, and

L is the length of time such civilizations send detectable signs.

---

2. Cocconi and Morrison, "Searching," 844–46.
3. SETI Institute, www. seti.org or Drake xix.

Discussion of the Equation

The Drake equation estimates the number of communicating civilizations which could be present in our galaxy at this point in time, N. It incorporates all the conceivable steps necessary for such a civilization to exist assuming that the civilizations exist on a planet orbiting a star. The equation begins by determining the number of stars that form in our galaxy each year, R*. Multiplying that number by f_p determines the number of stars that form each year that are capable of having planets. To determine the number of habitable planets that form each year, the number of stars with planets orbiting them is multiplied by the average number of habitable planets that orbit a star, n_e. Now, because the conditions for life exist does not necessarily mean that life will arise there; so we need to determine the number of habitable planets that are actually inhabited by living organisms. To do this, the number of habitable planets is multiplied by the fraction of habitable planets on which life actually arises, f_l. Life exists in many forms from simple cell organisms to complex, intelligent beings. In order for us to communicate with a lifeform, it must be intelligent life. The number of planets with intelligent life is determined by multiplying the number of inhabited planets in the galaxy by the fraction of planets with intelligent life, f_i. There are many species of life on earth that might be considered intelligent (e.g., dolphins and chimpanzees). However, in order to communicate with us on Earth, the intelligent life form must have developed interstellar communication (e.g., radio signals being sent out). The number of intelligent communicating species in the galaxy is determined by multiplying the number of intelligent species by the fraction of those capable of communication, f_c. Finally, comes the most important variable, L, the length of time that an intelligence communicating will send out signals that we can receive and know of their existence. The question remains when an intelligent civilization is capable of interstellar communication, it is also able capable of destroying itself. How long can such a society exist?

Doing the Math

At the time that Drake developed this equation, only R* was known with any certainty, and it is estimated to be between five to twenty stars/year form in our galaxy.[4] In the recent decades, astronomers have made significant progress towards refining two more the numbers in the equation. Discoveries from exoplanet (planets around other stars) missions, particularly the

---

4. Plaxco, "Drake Equation Tutorial."

Kepler Mission, and its extended mission, K2, have yielded new information about how common it is for planets to form around other stars. Up until 1995, the only planets we knew in existence were the ones in orbit around our Sun. Scientific models predicted that planets should exist around other stars, but due to their small size and the vast distance to other stars, no planets outside of our solar system had ever been observed. In October 1995, 51 Pegasi b[5] was discovered, ushering in a whole new field of astronomical exploration that has helped to refine the numbers f_p and n_e in the Drake equation.

Because of the search techniques used, early searches for exoplanets were biased toward finding large planets orbiting very close to their host star. Most of the first exoplanets discovered used the Doppler Shift technique. This method indirectly detects the presence of a planet by observing the small shift or wobble of a star's orbit that results from a planet orbiting the star. When a planet orbits a star, the star is not stationary, but the two objects (the star and the planet) both orbit around a common center of mass. This results in the star wobbling slightly in its position, and this very slight movement of the star can be measured by observing the Doppler shift of light coming from the star that oscillates in a regular pattern. When the star wobbles towards us, the spectrum of light shifts slight towards the blue side of the spectrum. When the star wobbles away from us, the light shifts to the red side of the spectrum. If we observe the frequency of the starlight shifting back and forth from red to blue at a regular interval, we have indirect evidence of a planet orbiting that star. The time it takes the light to complete one oscillation indicates the distance the planet is orbiting from the host star, and the degree to which the spectrum is shifted is indicative of the mass of the planet. This technique has been very successful in discovering new planets around other stars and indicating how common it is for a star to have planets. However, it is not very efficient in finding planets that are potentially habitable, namely smaller worlds with solid surfaces located relatively far from their host star. Consequently, new techniques have been employed and new missions have been developed to search for habitable worlds. The Kepler Mission was specifically designed to search for smaller planets at a distance from their host star in the habitable zone.

The Habitable Zone

Habitability means that the necessary requirements for life are present. At the most basic level, life requires three things:

5. Mayor and Queloz, "A Jupiter-mass Companion," 355–59, doi:10.1038/378355a0.

1. The right chemicals (e.g., carbon, oxygen, hydrogen, and nitrogen)
2. A source of energy (e.g., solar energy, chemical energy from eating food, geothermal energy)
3. A liquid to transport energy into and out of the cells.

The necessary chemicals for life are ubiquitous in the galaxy. Since energy can come from a wide variety of sources, we assume that nearly every planet has an energy source that life could employ. Hence, the search for life is focused on finding a liquid for transporting energy in and out of cells. On Earth, all life uses water as the liquid for chemical transport. Water has many unique properties that make it an ideal liquid for life including the wide range of temperatures at which it is liquid, the unique property that its solid state (ice) is less dense than its liquid state (water), and that it is a polar molecule. Consequently, when we search for life, we generally search for places where water can exist in a liquid form. The region around a star in which liquid water can exist is called the habitable zone of that star. The quest for finding another intelligent, communicating lifeform starts with finding a planet in the habitable zone that is the right size.

The Right Size

When planets are forming from the nebular cloud of gas and dust, the very large planets with their huge mass tend to pull the gas toward it, creating a planet that does not have a solid surface but rather a giant ball of gas with a relatively small core. Without a solid surface, it is difficult to imagine a place on this planet where life could emerge and thrive. Therefore, our search for intelligent civilizations focuses on smaller planets similar in size to our Earth in the habitable zone of its host star.

The Kepler Mission has already discovered over three thousand planets[6] as it searches the hundred thousand stars that are our next-door neighbors in the Milky Way Galaxy using the Transit Detection Method. This method requires the telescope to continually stare at a star and look for slight changes in the intensity of the light coming from that star. If a planet is orbiting that star and if its orbit is directly aligned with our view of the star, the planet will pass directly between us and the host star. This is known as a transit. When a planet transits the host star, it momentarily blocks some of the light coming from the host star. Telescopes monitoring the intensity of the starlight will measure a slight decrease in the intensity. The magnitude of

6. NASA, Kepler, http://www.kepler.nasa.gov.

that decrease is indicative of the size of the planet and the distance at which it is orbiting its host star. The Kepler telescope has detected several stars with multiple planets orbiting them.[7] It has also discovered several planets in the habitable zone of their host star[8] and a few of these planets are about the right size to have a solid surface. This mission has made great strides in increasing our understanding of how likely planets are to form around stars (f_p) and the number of those planets that are habitable (n_e).

Another promising technique for indirectly detecting smaller exoplanets orbiting farther from their host star is microlensing.[9] Microlensing is a result of Einstein's Theory of General Relativity where light is bent when passing by an object with a large mass. The bending of light creates multiple images of a very distant star if a foreground star is located directly between the Earth and the distant star.[10] Changes in the intensity of light from the multiple images of the distant star can indicate the presence of a planet around the foreground star. Using this microlensing technique, astronomers now estimate the fraction of stars with planets to be nearly 100 percent.[11] Therefore, it is realistic to estimate the fraction of stars that have planets around them (f_p) in the Drake Equation to be nearly one. Because in science nothing is certain, 95 percent is a reasonable value to use. These discoveries have greatly increased our estimate of the number of planets in our galaxy.

The transit method (Kepler Mission) and the microlensing method have been successful in discovering planets located in the habitable zone of their host stars. The Kepler mission in particular is discovering earth-sized planets in the habitable zone around other stars. Suddenly the Earth does not seem so rare or unique. Using the discoveries made by these missions and estimating that there are two planets in our solar system within the habitable zone (Earth and Mars), a reasonable estimate for n_e is two.

The Other Variables

At this time, the remaining variables in the Drake Equation are known with far less certainty. Nevertheless, it is worth considering some reasonable estimates for the remaining four terms.

7. NASA Exoplanet Archive. http://exoplanetarchive.ipac.caltech.edu/.

8. NASA. Exoplanet and Candidate Statistics. http://exoplanetarchive.ipac.caltech.edu/docs/counts_detail.html.

9. Bennet and Hong, "Detecting Earth-Mass Planets," 660–64.

10. Einstein, "Lens-Like Action of a Star," 506–7.

11. Cassan et al. "One or More Bound Planets," 167–69, doi:10.1038/nature10684.

To estimate f_l, one must ponder how easily life can arise given the right conditions. How easily does chemistry turn into biology? The classic abiogenesis laboratory study was the Miller-Urey Experiment which created amino acids in an environment that replicated that of the early Earth.[12] Other abiogenesis studies have investigated the chemical mechanisms leading to an RNA world[13] from which life as we know it arose. For an estimate of a value for f_l, we must also consider the variety of environments in which we find life on Earth. These studies inform us of the wide and varied places in the galaxy where life might exist. The discovery of extremophiles (organisms living in extreme environments on Earth) has shown that living organisms thrive in many places we would consider very inhospitable. Life is everywhere on Earth. Living organisms have been discovered in the very cold and dark lakes under the ice sheets on Antarctica, in the sulfur-rich pools around the geysers in Yellowstone, near vents in the depths of the oceans, in highly radioactive water, and on rocks in the driest desert just to name a few. Wherever there is the slightest bit of water on Earth today, we find living organisms thriving and making use of a wide variety of energy sources. We also look back in the geological records and learn that life arose on Earth very soon after the Earth had formed and water was present. Evidence of photosynthetic bacteria has been found in the oldest rocks we have found on Earth.[14] Evidence of even earlier life has been discovered in 4.1 billion year old zircon.[15] This finding in particular has some interesting implications on the resiliency of life. Supported by the lunar samples brought back by the Apollo astronauts, we know that about 3.2 to 4.1 billion years ago the early Earth was pummeled by many comets and asteroids. This is a time known as the Late Heavy Bombardment.[16] During this period, these impacts delivered intense amounts of energy that would have melted at least part of the surface of the Earth.[17] Finding evidence of life on Earth 4.1 billion years ago, before the Late Heavy Bombardment occurred, indicates that life is very resilient and has survived the Hadean era on Earth or it indicates that life arose on Earth a second time. Either way, these discoveries signal

12. Miller and Urey, "Organic Compound Synthesis," 245–51, doi:10.1126/science.130.3370.245.

13. Copley, "The Origin of the RNA World," 430–43.

14. Summons et al., "2-Methylhopanoids as Biomarkers," 554; Des Marais, "When did Photosynthesis Emerge on Earth?" 1703–5, doi:10.1126/science.289.5485.1703.

15. Bell et al., "Potentially Biogenic Carbon," 14518–21, doi:10.1073/pnas.1517557112.

16. Tera et al., "Isotopic Evidence," 1–21.

17. Abramov and Mojzsis, "Microbial Habitability," 419–22, doi:10.1038/nature08015.

that life on Earth began rather easily and very early. Since we are finding life in so many different environments on Earth today and learning that it arose so soon after the conditions necessary for life existed on Earth, it is reasonable to conclude that the fraction of habitable planets on which life actually appears is quite high perhaps as high as 80 percent.

Earth is teeming with life. Much of it is simple, single-celled organisms, but many complex, multi-celled species have evolved too. However, even though these species are complex, very few would be considered intelligent species. The more intelligent species include whales, dolphins, chimpanzees, the great apes, and of course humans. But what is intelligence? Scientists have been debating and trying to quantify intelligence with some biological characteristic. One early measure Harry Jerison's studies developed is the encephalization quotient which correlates brain mass to body size.[18] Species with a larger-than-average-size brain relative to its body size are often found to be more intelligent.[19] More recently, Roth and Dicke found a better correlation between intelligence and the number of cortical neurons and conduction velocity.[20] Here on Earth, intelligence has evolved several times, and since intelligence has a distinct evolutionary advantage, it is reasonable that given enough time, intelligent species are likely to evolve on an inhabited planet. A reasonable estimate for the fraction of inhabited planets where intelligent life does eventually develop, f_i, is 75 percent.

No matter your definition of intelligence, not all intelligent species on Earth are capable of developing the technology that will allow them to communication with life on another planet. Humans are the only species that has developed this capability. Although species will communicate among themselves, interstellar communication is rare. While humans are now very intelligent and capable of interstellar communication, that has not always been the case. *Homo sapiens* have been on the Earth since roughly 200,000 BC–100,000 BC, and our direct ancestor *Homo habilis* evolved about 2.8 million years ago.[21] However, it is only in the past hundred or so years that we have been capable of sending and receiving radio signals; this is a small fraction of the time humans have existed. Consequently, a realistic estimate for f_c, might be relatively low. Ten percent would be reasonable based on what we know about the evolution of life on Earth.

18. Jerison, "Brain to Body Ratios," 447–9; Jerison, *Evolution of the Brain and Intelligence*.

19. Haug, "Brain Sizes," 126–42.

20. Roth and Dicke, "Evolution of the Brain and Intelligence," 250–57, http://dx.doi.org/10.1016/j.tics.2005.03.005.

21. Villmoare et al., "Early Homo at 2.8 Ma from Ledi-Geraru, Afar, Ethiopia," 1352–55, doi:10.1126/science.aaa1343.

Progress with Uncertainty

Recently acquired knowledge has logged the greatest progress regarding f_p, the fraction of those stars forming that have planetary systems. "In the past five years, the first reliable estimates of f_p have emerged," remarks Chris Impey at the University of Arizona. "About 10 percent of Sun-like stars have planets."[22] Kepler has already identified several hundred planets within a factor of two of the Earth in mass, about a dozen of which are located in the habitable zones of their parent stars. That is progress.

The most uncertain factor in the Drake Equation, however, is L, the length of time that an intelligent, civilization can exist once it has developed the capability of interstellar communication. "L is a measure of the civilizing or socialization process, and the variables that underlie it: biocultural coevolution and the interaction between the evolution of cognition and socialization," comments David Dunér at Lund University in Sweden.[23]

Can we project our experience on Earth spaceward? "Unfortunately," writes Garry Chick at Penn State, "for empirical information on how to estimate L, we have a sample of only one, ourselves."[24] So, let us remember that when humans began emitting radio signals that could be picked up by life elsewhere in the galaxy, we also developed the technology to destroy ourselves and the planet. This was the same era when nuclear and thermonuclear weapons became capable of destroying the entire Earth. Our use of fossil fuels is changing the climate of the planet, and if gone unchecked, it could sufficiently change the environment, making Earth no longer hospitable to humans. Between the threat of nuclear war and ecological disaster, Earth's civilization may not last the ten thousand years Frank Drake originally estimated.

The biggest question remains: how long will our interstellar communicating society exist? This is the most important factor in the whole equation. No matter the values chosen for all the other terms in the Drake Equation, a conservative estimate for L will give N<<1 and lead to the conclusion that we are alone. A very optimistic estimate for L will give N>>1

---

22. Impey, "Fraction of Stars," 81.

23. Dunér, "Length of Time," 242.

24. Chick, "Length of Time," 270. The length of time the window for communication is open is a most nuanced matter, as Douglas Vakoch suggests. "This assumption that any extraterrestrial intelligence that we detect would be much longer lived than our own civilization does not presuppose that there is any inherent tendency toward technology fostering survival. It may well be that civilizations attain the capacity for interstellar communication for only a brief period, say one hundred years, before they either annihilate themselves or lose interest in making contact with extraterrestrial civilizations." Vakoch, "Asymmetry in Active SETI," 477.

and draw the conclusion that the galaxy is teeming with life. Various people have provided rational reasoning for this value. Shermer has analyzed the lifetime of sixty civilizations that have existed throughout history and found an average lifetime of the civilization of 420 years. He also observed that more modern and technological civilizations have a shorter average lifetime of 305 years.[25] Even if one current civilization does fall, we have an amalgamation of knowledge across the world. Human knowledge has grown and is not a property of only one civilization or culture. Some analysts believe that a total nuclear war would not necessarily annihilate our entire species.[26] Unless the entire globe is destroyed, it is difficult to imagine that our ability to communicate with others off the Earth by sending and receiving radio signals would be lost forever. A much more optimistic value for L has been proposed by David Grinspoon.[27] He concludes that technology will help the civilization overcome all threats and the lifetime of the civilization will be infinite.

Lifetime is the primary factor in the equation and the one most uncertain. We do not know for how long we can survive. Will we destroy ourselves or will we figure out how to get past this technological adolescence? And if we get through this current stage, how much longer can we survive? Although intelligence is morally ambiguous, humans are capable of love and love trumps intelligence.[28] Love for one another will keep humans from self-destruction and our society will last for at least another hundred years. But nothing lasts forever. Every past successful society on Earth has eventually come to an end. Nature tells us that eventually, a post-human species will evolve. But in how long and will that lifeform also be capable of interstellar communication? There are always more questions yet to be explored. I believe that 500 years seems like a rational estimate for L.

Using these values estimated above (20, 0.95, 2, .8, .75, 0.1, 500) leads to a value of 1,100 communication-capable civilizations in our galaxy. Even with a much more conservative estimation of L of 200 years, the Drake equation still says 460 civilizations capable of communication should exist now.[29]

---

25. Shermer, "Why ET Hasn't Called," 21.
26. Martin, "Critique of Nuclear Extinction," 278–300.
27. Grinspoon, *Lonely Planets*.
28. Peters, "Outer Space and Cyber Space," 1–5, doi:10.1017/S1473550416000318.
29. Neil DeGrasse Tyson estimates one hundred current extraterrestrial communicating civilizations within the Milky Way by plugging similar numbers into the Drake Equation. Tyson, "Search for Life," 167.

One factor that the Drake equation does not account for is the possible galactic habitable zone.[30] All of the assumptions above assume that a planet orbiting a star at the proper distance would be habitable. However, it does not account for where the star is located in the galaxy. If a star is close to the center of the galaxy, it would be subject to intense radiation from the supermassive black hole[31] in the center of the Milky Way. Additionally, while the chemicals necessary for life might be spread throughout the galaxy, they would only be present around stars that are created from the ejecta of multiple supernovae. A very small, old star that formed early in the beginning of our galaxy would not have the heavier elements present necessary for life. So perhaps our estimate of two habitable planets per star is too high and 0.5 is a more reasonable value to account for the habitability within the galaxy. With this modification and assuming a lifetime of a civilization of 500 years, it still results in 290 civilizations existing in the galaxy now that are capable of communicating with us.

We are not alone. But with so many civilizations, why are we not communicating with at least one of them already? One consideration that must be accounted for is the vast dimension of the galaxy. The Milky Way galaxy is one hundred thousand light-years in diameter. If there are at least 290 civilizations out there, they may be spread throughout our immense galaxy. Since we have only been sending out radio signals for roughly one hundred years, only a civilization within fifty light-years of us would have had time to receive a signal from us and send a reply back. The time it takes for light to travel across the galaxy and the relatively brief time that humans have possessed the capability of interstellar communication greatly limits the region of the galaxy with whom we could communicate.

Our quest for understanding the origin of life on Earth and the search for more exoplanets in the habitable zone must continue so we know better where and how to look for any other possible civilization. We need to continue listening and looking for other life communicating with us. Additionally, we need to strive to preserve our society and our ability to communicate. Even though we have the ability to destroy ourselves and our environment, those with power need to live with restraint and in harmony. Only then will we survive, thrive, and continue the search for extraterrestrial intelligence, and only then can we be discovered by another civilization if there is indeed at least one other out there. As with anything, one can never be 100 percent certain of knowing the truth. That is where faith comes in. Like the faith one needs to believe in God, one needs faith to think that we are not alone. And

---

30. Gowanlock et al., "A Model of Habitability," 855–73, doi:10.1089/ast.2010.0555.
31. Schödel et al., "A Star in a 15.2-year Orbit,"694–96, doi:10.1038/nature01121.

why should we be? It seems rather self-focused to think that we must be the only beloved children of God. A human father can have many children and all are beloved. It is not unrealistic that an omnipotent, creating God could have many beloved children throughout the galaxy.

Yes, life is out there. Based on our current scientific knowledge, it is highly unlikely that we are alone. Just because we haven't communicated with anyone yet, does not change the end result that there is other life in the galaxy.

So onward goes the search.

Bibliography

Abramov, Oleg, and Stephen J. Mojzsis. "Microbial Habitability of the Hadean Earth During the Late Heavy Bombardment." *Nature* 459 (May 2009) 419–22, doi:10.1038/nature08015.

Bell, Elizabeth A., et al. "Potentially Biogenic Carbon Preserved in 4.1 Billion-year-old Zircon." *Proceedings of the National Academy of Sciences* 112:47 (November 24, 2015) 14518–21, doi:10.1073/pnas.1517557112.

Bennet, David P., and Sun Hong. "Detecting Earth-Mass Planets with Gravitational Microlensing." *The Astrophysical Journal* 472 (December 1, 1996) 660–4.

Cassan, Arnaud, et al. "One or More Bound Planets per Milky Way Star from Microlensing Observations." *Nature* 481:7380 (January 12, 2012) 167–69, doi:10.1038/nature10684.

Chick, Garry. "Length of Time Such Civilizations Release Detectable Signals into Space, L, 1961 to the Present." In *Drake* 270–97.

Cocconi, Giuseppe, and Phillip Morrison. "Searching for Interstellar Communication." *Nature* 184:4690 (September 19, 1959) 844–46.

Copley, Shelly D., et al. "The Origin of the RNA World: Co-evolution of Genes and Metabolism." *Bioorganic Chemistry* 35 (2007) 430–33.

Des Marais, David J. "When did Photosynthesis Emerge on Earth?" *Science* 289:5485 (2000) 1703–5, doi:10.1126/science.289.5485.1703.

Dunér, David. "Length of Time Such Civilizations Release Detectable Signals into Space, L, Pre-1961." In *Drake* 241–69.

Einstein, Albert. "Lens-Like Action of a Star by the Deviation of Light in the Gravitational Field." *Science* 84:2188 (1936) 506–7.

Gowanlock, Michael G., et al. "A Model of Habitability within the Milky Way Galaxy." *Astrobiology* 11:9 (2011) 855–73, doi10.1089/ast.2010.0555.

Grinspoon, David. *Lonely Planets: The Natural Philosophy of Alien Life*. New York: Ecco / HarperCollins, 2004.

Haug, Herbert. "Brain Sizes, Surfaces, and Neuronal Sizes of the Cortex Cerebri: A Stereological Investigation of Man and his Variability and a Comparison with some Mammals (Primates, Whales, Marsupials, Insectivores and one Elephant)." *American Journal of Anatomy* 180 (1987) 126–42.

Impey, Chris. "Fraction of Stars with Planetary Systems, fp, 1961 to the Present." In *Drake*, 71–89.

Jerison, Harry T. "Brain to Body Ratios and the Evolution of Intelligence." *Science* 121:3144 (1955) 447–9.
———. *Evolution of the Brain and Intelligence*. New York: Academic, 1973.
Martin, Brian. "Critique of Nuclear Extinction." *Journal of Peace Research* XIX:4 (1982) 278–300.
Mayor, Michel, and Didier Queloz. "A Jupiter-mass Companion to a Solar-type Star." *Nature* 378:6555 (1995) 355–9, doi:10.1038/378355a0.
Miller, Stanley L., and Harold C. Urey. "Organic Compound Synthesis on the Primitive Earth." *Science* 130:3370 (1959) 245–51, doi:10.1126/science.130.3370.245.
NASA. Exoplanet and Candidate Statistics. http://exoplanetarchive.ipac.caltech.edu/docs/counts_detail.html.
———. Exoplanet Archive. http://exoplanetarchive.ipac.caltech.edu/.
———. Kepler. http://www.kepler.nasa.gov.
Peters, Ted. "Outer Space and Cyber Space: Meeting ET in the Cloud." *International Journal of Astrobiology* (2016) 1–5, doi:10.1017/S1473550416000318.
Plaxco, Jim. "Drake Equation Tutorial." Astronomical Adventures. http://www.astrodigital.org/astronomy/drake_equation.html.
"Preface." In *Drake* xix.
Roth, Gerhard, and Ursula Dicke. "Evolution of the Brain and Intelligence." *Trends in Cognitive Science* 9 (May 2005) 250–7. http://dx.doi.org/10.1016/j.tics.2005.03.005.
Schödel, Rainer, et al. "A Star in a 15.2-year Orbit around the Supermassive Black Hole at the Centre of the Milky Way." *Nature* 419:6908 (2002) 694–6, doi:10.1038/nature01121.
SETI Institute. http://www.seti.org.
Shermer, Michael. "Why ET Hasn't Called." *Scientific American* (August 2002) 21.
Summons, Roger E., et al. "2-Methylhopanoids as Biomarkers for Cyanobacterial Oxygenic Photosynthesis." *Nature* 400 (1999) 554.
Tarter, Jill Cornell. "Contact: Who Will Speak for Earth and Should They?" In *ELU*, 178–99.
Tera, Fouad, et al. "Isotopic Evidence for a Terminal Lunar Cataclysm." *Earth and Planetary Science Letters* 22 (1974) 1–21.
Tyson, Neil deGrasse. "The Search for Life in the Galaxy." In *WU*, 146–69.
Vakoch, Douglas A. "Asymmetry in Active SETI: A Case for Transmissions from Earth." *Acta Astronautica* 68 (2011) 476–88.
Villmoare, Brian, et al. "Early Homo at 2.8 Ma from Ledi-Geraru, Afar, Ethiopia." *Science* 347:6228 (2015) 1352–55, doi:10.1126/science.aaa1343.

9

Yes, We'll Meet Them

A Scientific Argument for ETI

MARTINEZ HEWLETT

There is even increasing recognition of a new science, sometimes called exobiology—a curious development in view of the fact that this "science" has yet to demonstrate that its subject matter exists!

—GEORGE GAYLORD SIMPSON (1964)[1]

We have not observed life anywhere but on earth, but no natural fact is cosmologically more significant.

—THOMAS NAGEL[2]

In 1963, the U.S. National Aeronautic and Space Administration (NASA) was at the beginning of its great expansive phase. The Mercury flights had been successfully completed. The Gemini missions of two-man orbital launches were in process. In May of 1961, President Kennedy had fired the public imagination by proclaiming "I believe that this nation should commit

1. Simpson, "Nonprevalence of Humanoids," 769.
2. Nagel, *Mind and Cosmos*, 32.

itself to achieving the goal, before this decade is out, of landing a man on the Moon and returning him safely to Earth."[3] As a result, NASA began scaling up for the Apollo flights. The cost of this effort would exceed $20 billion.

It was against this backdrop that George Gaylord Simpson wrote and published the *Science* article from which the above quote is taken. His strong critique of the massive expenditure was based upon his view, as an evolutionary biologist, that the highly risky gamble of finding extraterrestrial life (the unspoken subtext of the missions) was not worth the outpouring of treasure that could be more beneficially used elsewhere. Writing almost three decades before the launch of the Hubble telescope, Simpson states that "there are no direct observational data whatever" to support the idea of planetary systems outside of our own.[4] He concludes, somewhat sarcastically, "I cannot share the euphoria current among so many, even among certain biologists (some of them now ex-biologists converted to exobiologists)" regarding the search for life beyond our own planet.[5]

In 2014, a half century after Simpson's debunking, NASA formulated its current astrobiology strategy which includes researching the transition from abiotic to biotic chemistry plus the search for life in off-Earth habitats.[6] Simpson represents the unique Earth or rare Earth position that scowls at such NASA efforts.

Over against Simpson I wish to argue: "Yes, we'll meet them." Over against Robert John Russell, who in another chapter of this book says the jury is out, I wish to argue the jury is close to a decision: "Yes, we'll meet them." I will offer a scientific argument that, in sum, looks like this: because our galaxy is now estimated to include between one hundred and four hundred billion exoplanets with at least seventeen billion of these Earth sized, some of those Earth-sized planets will play host to intelligent beings who may become our new neighbors in space. Church people can now enroll in the Bible-Welcomes-Aliens school of conjecture.

Curiously, it is skeptical Simpson's own body of work, being, as it was, a direct extension of the nineteenth-century cataclysm unleashed by Charles Darwin, that led to this spectacular outpouring of confidence that, if we simply look, we will find evidence of life elsewhere in the universe. What had happened between the debates of Whewell and Brewster concerning the extent of God's creative munificence and the publication of *Origin of*

3. President John F. Kennedy's address to a joint session of Congress, May 25, 1961. The section of his address related to space exploration can be found on the web at http://www.space.com/11772-president-kennedy-historic-speech-moon-space.html.
4. Simpson, "Nonprevalence of Humanoids," 770.
5. Ibid., 775.
6. *NASAstr*.

Species? After Darwin, science could now hold the conviction that, in effect, the Greek atomists were correct: the universe is the same throughout, in both substance and the law-like behavior of that substance.

The Revolution, as Seen from Down House in Kent

The journey northwest from the village of Down (now spelled Downe) in Kent to London in 1859, the year of the publication of *On the Origin of Species by Means of Natural Selection*, was somewhat longer than it is today. There was rail service to Down, but only horse and buggy out to the Darwin home. Nevertheless, Charles and his wife Emma had chosen this rural location because of its ease of access to the center of scholarship for his work.

This was a trip that Darwin made many times during his career. And yet, after his monumental voyage on HMS Beagle, it was in effect the longest journey he would ever take during the rest of his life. Despite this geographic limitation, the work that would emanate from the study in Down House would have repercussions in every corner of the globe.

As Darwin left on his earlier Beagle voyage, he was already coming to disagree with the currently accepted model for the origin and diversity of life: divine action. He had read and appreciated Rev. William Paley's *Natural Theology; or Evidence of the Existence and Attributes of the Deity*, which assumed that design in nature was the result of this action.[7] He had been exposed, however, to the work of Jean Baptiste Lamarck and his model for evolution of species.[8] He was also familiar with the writings of his own grandfather, Erasmus Darwin, whose 1796 volume *Zoonomia* included a view of deep time and evolutionary changes by adaptation.[9] During the Beagle voyage he was motivated by his reading of Charles Lyell's *Principles of Geology*, and the further expansion of the notions of geological age.[10] The final influential work for him was that of Thomas Malthus, from

---

7. Paley, *Natural Theology*; an online version of the 12th edition can be found at http://darwin-online.org.uk/content/frameset?itemID=A142&viewtype=text&pageseq=1.

8. Lamarck's works included *Recherches sur l'organisation des corps vivants*, *Philosophie Zoologique*, and the seven volume *Histoire naturelle des animaux sans vertèbres* (1815–1822). Online versions of these books are available at http://www.lamarck.cnrs.fr/?lang=en. One caveat: the works are posted in French, not English.

9. Darwin, *Zoonomia*, Section XXXIX. Darwin ultimately published a biography of his grandfather: Darwin, *Life of Erasmus Darwin*.

10. Lyell, *Principles of Geology*.

whom he got his idea for limited resources being a force on the adaptation of species.[11]

The five-year voyage ended and Darwin set up to work in London, near the center of intellectual activities. After his marriage to Emma Wedgewood, the couple sought a rural life, near enough to London, but removed from the city clutter. Thus, they found Down House and its surrounding acres. It was there that, over the course of some twenty years, his great theoretical achievement was completed.

On the Origin of Species by Means of Natural Selection was published on November 24, 1859. Almost immediately, repercussions were felt in both town and gown communities throughout Britain and beyond. Bypassing for now the discomfort and upheaval caused in certain religious circles, let's focus on the scientific reaction.

The intellectual *tour de force* that Darwin accomplished was evident. He effectively changed the entire discussion by giving the observed complexity of all living things a foundation in the law-like behavior of nature. And yet the reception was not instantly overwhelming. He lacked certain tools at his disposal, such as a coherent model for explaining inheritance. This was the focus of the critique by the engineer Fleeming Jenkin.[12] In Darwin's time, the accepted model for the transmission of characteristics from parents to offspring was blending inheritance. Jenkin argued that such blending would be certain to dilute out the influence of any beneficial traits, diminishing the power of natural selection to preserve these in subsequent generations. As a result, Darwin had no access to an interpretation of genetics that would support his model. His only recourse, in subsequent editions of *Origins*, was to fall back on Lamarckian arguments for characteristics acquired by large numbers of a population.

Within a decade of the appearance of *Origin of Species* Gregor Mendel would publish his paper on his breeding experiments with pea plants. His quantitative model for units of inheritance would not be recognized for another fifty years. But with the advent of the Mendelian genetic era at the beginning of the twentieth century, Darwin's model began to have some solid theoretical support. So it was that, by the middle of that century, we would read Julian Huxley's formulation of the "modern synthesis," uniting Darwinian evolution, Mendelian genetics, and an understanding of the molecular nature of genes and mutations.[13]

11. Malthus, *An Essay on the Principle of Population*, 17.

12. Jenkin, review of *The Origin of Species*, in *The North British Review*, 277–318. The text of the review can be found on the Web at http://www.victorianweb.org/science/science_texts/jenkins.html.

13. Huxley, *Evolution*.

The neo-Darwnian model became the paradigm for biological investigation that still remains firmly in place to the present. However, Darwin's explanatory theory deals with the origin of species, not life itself. In a letter to Joseph Hooker, Charles Darwin did speculate about what pre-cellular conditions might have been like:

> It is often said that all the conditions for the first production of a living organism are now present, which could ever have been present. But if (and oh! what a big if!) we could conceive in some warm little pond, with all sorts of ammonia and phosphoric salts, light, heat, electricity, &c., present, that a protein compound was chemically formed ready to undergo still more complex changes, at the present day such matter would be instantly devoured or absorbed, which would not have been the case before living creatures were formed.[14]

Darwin had formulated a theory that led to the logical conclusion that, if physical laws could be responsible for the complexity of life, surely these self-same laws could explain the origin of life itself.

Abiogenesis

Darwin's "warm little pond" imagery took on a theoretical dimension when Aleksander Oparin published his speculative considerations about the conditions necessary to spark the formation of organic molecules in the environment of the pre-biotic Earth.[15] His "primordial soup" suggested a possible avenue of experimental investigation.

It was Stanley Miller and Harold Urey who finally constructed the apparatus that would test this hypothesis.[16] Inside of a sealed glass vessel they placed water, methane, ammonia, and hydrogen to mimic the reducing atmosphere of the planet at that time. An electric spark was used to simulate a lightning discharge. After a time, they could identify the presence of many organic compounds associated with living systems. A recent re-creation

---

14. Charles Darwin, in a letter to Joseph Hooker, February 1, 1871. The letter is referenced in the Darwin Correspondence Project (http://www.darwinproject.ac.uk/darwinletters/calendar/entry-7471.html), but the complete text is not available electronically. The relevant passage can be found in the Internet Encyclopedia of Science, http://www.daviddarling.info/encyclopedia/D/DarwinC.html.

15. Oparin, *The Origin of Life*, originally published in 1924 in Russian. English translations appeared in 1938 and 1952.

16. Miller, "Production of Amino Acids," 528–29. Miller and Urey, "Organic Compound Synthesis on the Primitive Earth," 245–51.

and assessment of these experiments has been published,[17] confirming and expanding these results. Clearly, a primordial soup model can yield at least some of the building blocks of living systems. But how do we get living cells?

At least on our planet, the cell is the basic unit of structure for almost all living forms, exclusive of viruses. Cells first became evident to biologists with the advent of the microscope in the seventeenth century.[18] However, the question of the possible origin of living things from non-living matter (spontaneous generation) was not disproved until the work of Louis Pasteur in the nineteenth century.[19] *Omnia cellula e cellula* ("all living cells come from pre-existing living cells") became the maxim, popularized by Rudolf Virchow.[20]

The question of the origin of cellular life itself has been addressed more recently. For starters, Craig Venter and his colleagues have created a synthetic genome. The molecule, designed and assembled in the laboratory contained what the experimenters deemed would be the minimal necessary complement of genes to sustain a bacterial cell. The synthetic genome was transplanted into a recipient bacterial cell, from which the existing chromosome had been removed or inactivated. The transplanted genome was able to replicate and direct the metabolism of the now semi-artificial cell.[21]

Note that this is not the creation of a living cell from exclusively non-living starting material. You cannot take a collection of molecules, including a genome, surround it with a lipid membrane and get it to function as a living cell. All so-called "synthetic cells" such as the one described by the Venter group, begin with a cell that is already alive. Virchow remains unchallenged.

17. Johnson et al., "The Miller Volcanic Spark Discharge Experiment," 404.

18. The name "cell" was given by Robert Hooke, the inventor of the first microscope, in his 1665 publication. Upon viewing a section of tree bark (cork), he likened the small units that formed the structure to the monks' cells in a monastery. An e-book version of his work, *Micrographia*, can be found at Project Gutenberg (http://www.gutenberg.org/catalog/world/readfile?fk_files=3276692&pageno=1).

19. A concise discussion of this subject can be found at the National Health Museum web site, http://www.accessexcellence.org/RC/AB/BC/Spontaneous_Generation.php.

20. Virchow, *Die cellularpathologie in ihrer begründung auf physiologische und pathologische gewebelehre*, xv. A recent discussion of the contributions of the somewhat controversial Virchow can be found online in the journal *Lab Medicine:* Titford, "Rudolf Virchow," 311–12.

21. Gibson et al., "Creation of a Bacterial Cell Controlled by a Chemically Synthesized Genome," 52–56.

Life on Earth: Where Is it Found?

Perhaps a better question would be "where is it *not* found?" Most of us are certainly familiar with the wide variety of life in our local environment, wherever that might be. However, what is visible to us without instrumentation is only a very small sampling of the actual life forms on this planet. Microbial life constitutes the vast majority of cellular creatures to be found on Earth. We are learning more and more about the microbiome[22] of our planet through the Earth Microbiome Project,[23] as well as the microbiome of our own bodies through the Human Microbiome Project.[24] In fact, the number of microbial cells living on and in you is greater than your own cells by a factor of about one hundred.

Yet even this wide variety of forms exists in environments relatively close to the conditions we occupy. There are other places on the planet where life can be found, places where it is hard to reasonably expect it would be even close to compatible. Organisms that live in these environments are called "extremophiles." The range of conditions is mind-boggling, from high acidity to high alkalinity, from ultra low temperatures, to above that of boiling water. Life can exist in extremes of pressure and radiation, and in environments with essentially no water.

Lest we think that these extreme forms are all bacterial, consider the case of the water bear. This is an animal (phylum, *Tardigrada*), not too distant from the arthropods (mosquitoes and ticks). Water bears are small (from 0.1 to 1.5 mm, depending upon the species) and can be found in most environments on Earth. Very often, they are associated with moss or lichens, which serve as their food source.

Water bears are extremophiles in every sense of the term. When faced with limiting conditions, the animals eliminate more than 90 percent of their water and enter a dormant state. In this form they can withstand extreme lows (-200 °C) and highs (+150 °C) of temperature. They have been revived from dehydration after ten years, with some work suggesting viability after more than one hundred years. In the dormant state, water bears can withstand a thousand times higher doses of radiation than any other animals. Recently, water bears were launched into space and exposed to the extreme vacuum and radiation conditions found there. After ten days they

---

22. A microbiome is defined as all of the microbes in a particular environment, determined by measuring their genomes and their interactions.

23. Gilbert and Meyer, "Modeling the Earth Microbiome," 7.

24. Human Microbiome Project, at the National Institutes of Health web site, http://commonfund.nih.gov/hmp/.

were returned and tested for viability. Some even survived both the vacuum and ultraviolet radiation.[25]

All of these examples of life at the extremes play into the scenarios for what life might be like off-planet. I argue that we will meet ETI in large part because our universe is biophilic; it loves life.

Unique Earth versus Cosmic Biophilia

Not every one of Darwin's disciples would agree with me that the universe is biophilic, let alone designed in such a way that intelligent life is likely to evolve where conditions are fitting. Even Alfred Wallace, Darwin's contemporary, argued against the existence of ETI on behalf of the rare Earth or unique Earth position. Why? Because of contingency. The evolution of life from abiotic chemistry is itself a random chance event, not a principle built into natural processes. The subsequent evolution of species is similarly due to randomness, chance, and contingency. No laws of nature require a sequential march from the simple to the complex, from the stupid to the intelligent. What happened on Earth is not likely to be repeated on other worlds in space. So Wallace avers.

In the twenty-first century, Darwinian disciple Francisco J. Ayala follows Wallace in arguing for a rare Earth on the basis of contingency. Borrowing a phrase from Stephen Jay Gould, Ayala says that if we were to replay "life's tape" on another planet, we would not hear the same story that took place on Earth.[26] The course of Earth's evolution could not repeat itself elsewhere. Here on Earth the first two billion years saw only microbes on Earth. The eukaryotes were the first organisms whose cells have a nucleus containing DNA; and, he adds, there is nothing in the process that would make it likely that multicellular organisms would evolve like this elsewhere. Evolution could have stopped right there. No animals might have come into existence. Animals on our planet evolved only once. It was a chance occurrence. So, there is little likelihood that animals would evolve again if life's tape were replayed.[27]

But Wallace and Ayala are not the only musicians playing music for us to dance to. Biophilic melodies can be heard from other contemporary Darwinians. Cambridge biologist Simon Conway Morris, for example, whistles

---

25. The online magazine, *New Scientist*, has a report of this experiment, along with video material concerning water bears: Courtland, "'Water Bears' are First Animal to Survive Space Vacuum."

26. See Gould, *Wonderful Life*, 48.

27. See Ayala, "Evolution of Life," 57–77.

another tune: evolution would always follow a somewhat predictable track leading toward intelligent beings such as ourselves. "The emergence of human intelligence is a near-inevitability."[28]

The most biophilic melody can be heard in the music of Nobel Prize winning biologist Christian de Duve. "Life is the product of deterministic forces," he writes. "Life was bound to arise under the prevailing conditions, and it will arise similarly wherever and whenever the same conditions obtain. There is hardly any room for 'lucky accidents' in the gradual, multistep process whereby life originated. This conclusion is compellingly enforced when one considers the development of life as a chemical process."[29] When interpreting the Drake Equation, de Duve surmises that "the figure of about one million "habitable" planets per galaxy is considered not unreasonable. Even if this value were overestimated by several orders of magnitude, it would still add up to trillions of potential cradles for life. If my reading of the evidence is correct, this means that trillions of planets exist that have borne, bear, or will bear life. The universe is awash with life."[30]

I hum the contact optimism tune: the universe is awash in life. Yes, of course, nature must deal with randomness, chance, and contingency. Yet, when the conditions are right, then we can expect certain chemical processes to do their thing, so to speak. If an extraterrestrial planet orbits its respective sun in the habitable zone with liquid water and fitting physical conditions, then chemistry will become biochemistry and life will march forward looking for niches that will lead to intelligence. This understanding of evolutionary chemistry combined with the big numbers means this: yes, we will meet them.

The Search for Life Elsewhere in Our Solar Ghetto

Just as with those who came before us, the most obvious place for us to consider for the presence of off-planet life are our companions in this solar system. Speculation has focused most intently on the fourth planet, Mars. From astronomical observations dating back to the nineteenth century, evidence has pointed towards the presence of water at some time during that planet's history. For instance, the planet appears to have polar ice caps that grow and recede with the seasons. But these are conclusions reached from a distance. Obviously, the only way to know for sure is to go there.

28. Conway Morris, *Life's Solution*, xii.
29. De Duve, *Vital Dust*, xv.
30. Ibid., 121.

NASA is currently operating the Mars Exploration Program, with the express mission to look for the evidence of life, current or past, on our neighbor. The program assumes that water will be necessary, just as it is on Earth:

> To discover the possibilities for life on Mars—past, present or our own in the future—the Mars Program has developed an exploration strategy known as "Follow the Water." Following the water begins with an understanding of the current environment on Mars. We want to explore observed features like dry riverbeds, ice in the polar caps and rock types that only form when water is present. We want to look for hot springs, hydrothermal vents or subsurface water reserves. We want to understand if ancient Mars once held a vast ocean in the northern hemisphere as some scientists believe and how Mars may have transitioned from a more watery environment to the dry and dusty climate it has today.[31]

Mars exploration attempts have been ongoing since 1960, with the vast majority of the flights originating in the US. The former Soviet Union has had two successful missions in all of those years, while Japan and the European Space Agency each had one mission. Japan's was a failure, while the ESA's was a partial success.[32]

NASA's Mars exploration rovers Spirit and Opportunity (launched in 2003) were both geological missions, with the intent of seeking the presence of water in the mineral content of the Martian surface. Spirit is now inactive but, as of this writing, Opportunity remains functional and is still returning data.

The current excitement is over Curiosity, the Mars Science Laboratory rover that landed on the planet August 5, 2012. The experimental package aboard this vehicle is specifically aimed at searching for chemical evidence of life, either present or past, on the Red Planet. Instrumentation can even detect organic molecules in samples from the surface. Curiosity has the capability of drilling into rocks in search of molecules of interest. At present, two rocks have been examined in this way, although the complete data has not yet been made available. The first rock has yielded preliminary results,

---

31. Found on the NASA Mars Exploration Project web site: http://mars.jpl.nasa.gov/programmissions/overview/.

32. A table listing all missions can be found on the NASA website at http://mars.jpl.nasa.gov/programmissions/missions/log/.

suggesting that the Martian climate was at one time hospitable for at least microbial life.[33]

Other solar system planets are on the list to be explored. The Cassini probe, launched in 1997, has made fly-by of Jupiter and then flew on to Saturn, ultimately sending the Huygens probe to the surface of the largest Saturnian moon, Titan. Candidates for compatibility with life include Titan, and the Jovian moons Europa, Ganymede, and Calisto.[34]

The Search for Life on Extrasolar Planets

Until quite recently, the existence of planetary systems outside of our own was only a theoretical possibility. Then, in 1995, Michel Mayor and Didier Queloz at the University of Geneva identified the first exoplanet, 51 Pegasi b. A few months later Geoffrey Marcy and Paul Butler found two more exoplanets. The race to find planets was on.[35]

The launch of the Kepler Telescope accelerated the race. The Kepler Orbiting Observatory was launched on March 6, 2009, and became operational ("first light") on April 7 of that same year. The observatory is in solar orbit, trailing behind the Earth. The mission to locate extrasolar planets in a particular region of our own Milky Way galaxy began on May 12, 2009.

Kepler's instruments are designed to detect the motion of planets around stars, and to determine certain features of those planets. In particular, the mission looks for evidence of planets in what are called the habitable zones around stars, identifying so-called "Goldilocks planets." The detection is made by observing the effect of planetary orbits on the light emanating from a star. Changes in the intensity or wavelength of light indicates the presence of a planet making a transit across the face of the star as that planet moves in its orbit.

To date, Kepler has located 4500 candidate planets and 2300 confirmed planets.[36] Many of the candidates are Earth-sized. Many of these are located in the habitable zone of their star.[37]

---

33. Current news for Curiosity can be found on the NASA site at http://www.nasa.gov/mission_pages/msl/index.html.

34. Some of the data collected by Cassini can be seen at http://www.nasa.gov/mission_pages/cassini/main/index.html.

35. Lemonick, "Dawn of Distant Skies," 40–47.

36. The NASA Kepler information can be located at http://www.nasa.gov/mission_pages/kepler/main/index.html.

37. The Extrasolar Planets Encyclopedia is maintained online (http://exoplanet.eu) by Jean Schneider at the Paris Observatory. The current catalog can be viewed at http://exoplanet.eu/catalog/.

Astronomer Chris Impey offers his version of the big numbers that favor the existence of extraterrestrial life. "A conservative estimate might be a billion habitable spots—terrestrial planets in conventionally defined habitable zones, plus moons of giant planets harboring liquid water—in the Milky Way alone . . . That number must be multiplied by 10^{11} for the number of 'Petri dishes' in the observable cosmos."[38] Even though we have not yet met them, we will.

Conclusion: Yes, We'll Meet Them

Yes, we'll meet them. Based on the big numbers in these data sets, our galaxy is now estimated to have between one hundred and four hundred billion exoplanets, with at least seventeen billion of these Earth size.[39] Is it still reasonable to expect, given everything we have discussed in this chapter, that we are alone in the universe? No, it is not reasonable. Time to enroll in the Bible-Welcomes-Aliens school of conjecture.

Bibliography

Ayala, Francisco J. "The Evolution of Life on Earth and the Uniqueness of Humankind." In *Perch esiste qualcosa invece di nulla? (Why There Is Something Rather than Nothing?)*, edited by S. Moriggi and E. Sindoni, 57–77. Castel Bolognese, Italy: ITACAlibri, 2004.

Cassan, Arnaud, et al. "One or More Bound Planets per Milky Way Star from Microlensing Observations." *Nature* 481:7380 (January 12, 2012) 167–69.

Conway Morris, Simon. *Life's Solution: Inevitable Humans in a Lonely Universe.* Cambridge: Cambridge University Press, 2003.

Courtland, Rachel. "'Water Bears' are First Animal to Survive Space Vacuum." *New Scientist* September 8, 2008. http://www.newscientist.com/article/dn14690-water-bears-are-first-animal-to-survive-space-vacuum.html.

Darwin, Charles. Letter to Joseph Hooker, February 1, 1871. Darwin Correspondence Project. http://www.darwinproject.ac.uk/darwinletters/calendar/entry-7471.html.

———. Letter to Joseph Hooker, February 1, 1871. The Internet Encyclopedia of Science. http://www.daviddarling.info/encyclopedia/D/DarwinC.html.

———. *The Life of Erasmus Darwin.* London: Murray & Sons, 1879.

Darwin, Erasmus. *Zoonomia, or the Laws of Organic Life.* Vol 1. (1796). Section XXXIX. Project Gutenberg. http://www.gutenberg.org/files/15707/15707-h/15707-h.htm#sect_XXXIX.

Davies, Paul C. W. *The Eerie Silence: Renewing Our Search for Alien Intelligence.* Boston: Houghton Mifflin, 2010.

---

38. Impey, "First Thousand Exoplanets," 210–11.
39. Cassan et al., "One or More Bound Planets," 167–69.

de Duve, Christian. *Vital Dust: The Origin and Evolution of Life on Earth.* New York: Basic Books, 1995.
The Extrasolar Planets Encyclopedia. Exoplanet.eu. http://exoplanet.eu/catalog/.
Gibson, Daniel, et al. "Creation of a Bacterial Cell Controlled by a Chemically Synthesized Genome." *Science* 329 (2010) 52–56.
Gilbert, Jack, and Folker Meyer. "Modeling the Earth Microbiome." *Microbe Magazine* 7 (2012). http://www.microbemagazine.org/index.php?option=com_content&view=article&id=4399:modeling-the-earth-microbiome&catid=928&Itemid=1273.
Gould, Stephen Jay. *Wonderful Life: The Burgess Shale and the Nature of History.* New York: Norton, 1989.
Hooke, Robert. *Micrographia.* Project Gutenberg. http://www.gutenberg.org/catalog/world/readfile?fk_files=3276692&pageno=1.
Huxley, Julian. *Evolution: The Modern Synthesis.* London: Allen & Unwin, 1942.
Impey, Chris. "The First Thousand Exoplanets: Twenty Years of Excitement and Discovery." In *AHS* 201–12.
Jenkin, Fleeming. Review of *The Origin of Species*, by Charles Darwin. *The North British Review* 46 (June 1867) 277–318. http://www.victorianweb.org/science/science_texts/jenkins.html.
Johnson, Adam P., et al. "The Miller Volcanic Spark Discharge Experiment." *Science* 322 (2008) 404.
Kennedy, John F. Address to a joint session of Congress, May 25, 1961. http://www.space.com/11772-president-kennedy-historic-speech-moon-space.html.
Lamarck, Jean-Baptiste. *Recherches sur l'organisation des corps vivants* (1802). http://www.lamarck.cnrs.fr/?lang=en.
———. *Philosophie Zoologique.* 1809. http://www.lamarck.cnrs.fr/?lang=en.
———. *Histoire naturelle des animaux sans vertèbres.* 7 vols. 1815–1822. http://www.lamarck.cnrs.fr/?lang=en.
Lemonick, Michael D. "The Dawn of Distant Skies." *Scientific American* 309.1 (2013) 40–47.
Lyell, Charles. *Principles of Geology, Being an Attempt to Explain the Former Changes of the Earth's Surface, by Reference to Causes Now in Operation.* Vol. 1. London: John Murray, 1830.
Malthus, Thomas Robert. "An Essay on the Principle of Population: As It Affects the Future Improvement of Society, with Remarks on the Speculations of Mr. Godwin, M. Condorcet, and Other Writers." 1798. Library of Economics and Liberty. http://www.econlib.org/library/Malthus/malPop.html.
Miller, Stanley L. "Production of Amino Acids under Possible Primitive Earth Conditions." *Science* 117:3046 (May 1953) 528–29.
Miller, Stanley L. and Harold C. Urey, "Organic Compound Synthesis on the Primitive Earth." *Science* 130:3370 (July 1959) 245–51.
Nagel, Thomas. *Mind and Cosmos: Why the Materialist Neo-Darwinian Conception of Nature is Almost Certainly False.* Oxford and New York: Oxford University Press, 2012.
NASA. Astrobiology Strategy. https://nai.nasa.gov/roadmap/.
———. Cassini at Saturn. http://www.nasa.gov/mission_pages/cassini/main/index.html.
———. Curiosity Rover. http://www.nasa.gov/mission_pages/msl/index.html.
———. Kepler and K2. http://www.nasa.gov/mission_pages/kepler/main/index.html.

———. Mars Exploration. http://mars.jpl.nasa.gov/programmissions/overview/.
———. Mars Exploration. Historical Log. http://mars.jpl.nasa.gov/programmissions/missions/log/.
National Health Museum. http://www.accessexcellence.org/RC/AB/BC/Spontaneous_Generation.php.
National Institutes of Health. Human Microbiome Project. http://commonfund.nih.gov/hmp/.
Oparin, Aleksander. *The Origin of Life*. Russian orig., 1924. New York: Dover, 1938, 1952.
Paley, William. *Natural Theology, or Evidence of the Existence and Attributes of the Deity*. London: Faulder, 1802. http://darwin-online.org.uk/content/frameset?itemID=A142&viewtype=text&pageseq=1.
Simpson, George Gaylord. "The Nonprevalence of Humanoids." *Science* 143 (February 21, 1964) 769–75.
Titford, Michael. "Rudolf Virchow: Cellular Pathologist." *Lab Medicine* 41 (2010) 311–12.
Virchow, Rudolf. *Die cellularpathologie in ihrer begründung auf physiologische und pathologische gewebelehre*. Berlin: Hirschwald, 1858.

10

God's Self-Communication in a Cosmos Bound for Life

OLIVER PUTZ

The discovery of *any* sign that we are not alone in the universe could prove deeply problematic for the main organized religions, which were founded in the pre-scientific era and are based on a view of the cosmos that belongs to a bygone age.

—PAUL DAVIES[1]

It is an extraordinary thing that God would assume a material nature. Thus it is fitting that this should be a one-time event. And further it is appropriate that the nature that God assumed should be unique.

—MARIE GEORGE[2]

If physicist Paul Davies and philosopher Marie George are correct in their evaluations, the outlook for either Christianity or ETI is rather bleak. As far as Davies is concerned, contact would put the final nail in the religious coffin, bringing to completion the past five hundred years of secularization led by modern science. For George, the existence of ETI would not necessarily be incompatible with Roman Catholic magisterial teaching, but in light of biblical revelation and Thomistic theology highly improbable. Either way,

1. Davies, *Eerie Silence*, 188.
2. George, *Christianity and Extraterrestrials*, 90.

an undertaking such as *astrotheology* seems to be little more than an intellectual stillbirth: well intended in its attempt of reconciling theology and science, but ultimately blind to the obvious challenges that render the entire endeavor superfluous.

Nevertheless, I want to advance the admittedly outlandish thesis that there are legitimate theological reasons why we should *expect* sentient life to exist on other planets. The argument I will present swings on two theological hinges: First, it adopts a Scotistic doctrine of God which assumes that God creates to communicate Godself and that creation exists to receive this loving and unowed divine self-gift. Second, it espouses Karl Rahner's creative appropriation of Scotus in a theology of becoming, according to which God as transcendent cause moves creation toward the emergence of a range of self-conscious beings capable of experiencing the transcendent and, thus, of being in personal relationship with it. The universe is bound for life and once present, biological diversity will eventually lead to spiritual multiplicity.

Outlandish as my thesis may be, others in this volume have pointed out the historical fact that the idea of the plurality of worlds is not new to theological musing. In fact, as Arthur O. Lovejoy claims, the notion that life could exist elsewhere in the universe is based entirely on metaphysical and theological presuppositions.[3] So-called scientific arguments for ETI usually are arguments from analogy and, as such, neither scientific nor, for that matter, the most reliable research guide. Non-scientific proposals in Lovejoy's analysis commonly employ the *principle of plentitude*, that is, the premise that due to the nature of God, everything that can be will be, including ETI.[4] The history of the debate about extraterrestrial life abounds with examples of arguments in this vein—Giordano Bruno perhaps being the most famous one, Immanuel Kant a somewhat surprising other.[5] A more recent case in point is late Nobel laureate Christian de Duve's assertion that life is a cosmic imperative and, as such, "an obligatory manifestation of matter, written into

---

3. Lovejoy, *Great Chain of Being*, 99.

4. Ibid., 52.

5. Deeply influenced by neo-Platonism and Nicholas of Cusa, Bruno rejects the Aristotelian idea of a finite world and instead contends that the universe is boundless and filled with worlds like ours that are populated with people very much like us. Cf. Fantoli, *The Case of Galileo*, 49. For Kant see *Allgemeine Naturgeschichte und Theorie des Himmels*, 173. An early example for the use of the principle of plentitude is the *De rerum natura* of Roman Poet Lucretius, who adopts the atomism of Leucippus, Democritus, and Epicurus and claims that since space is infinite in every direction and matter abundant, other planets bearing life must exist. Cf. Lucretius, *On the Nature of Things*, 84–85. For a brief discussion, see Dick, *Biological Universe*, 12.

the fabric of the universe."[6] My own approach might be seen as a variation of this latter view, though I want to emphasize that it differs significantly from the principle's standard version. I do not think that God's omnipotence necessitates the creation of everything possible. Rather I want to argue that God's omnibenevolence and desire for self-communication in light of an evolutionary worldview suggests that creation moves toward multiple beings that can respond to the divine self-gift in an otherwise free process of becoming. In this sense, my proposal might be thought of as a weak version of the principle of plentitude.

The Self-Communicating God

As I see it, the underlying question of any astrotheological project today is what kind of God would desire to be in relationship with a plurality of beings? And what kind of God would create a cosmos that evolves toward that end?[7] If indeed life exists elsewhere, as astrotheologians surmise, God must have warranted it so or else not be the omniscient or omnipotent God of the Abrahamic traditions. But would not the idea of such a God who longs for a biological diversity reaching throughout the universe undermine central Christian faith commitments, foremost among them the doctrine of the incarnation? I believe that a Scotistic doctrine of God can help unravel this entanglement of theological ideas.

A God Who Loves

For medieval theologian and philosopher John Duns Scotus the question of ETI was hardly a concern, even though by the thirteenth century the possibility of life on other planets had been widely discussed.[8] Scotus was interested in the reasons for the incarnation, primarily in conversation with St. Anselm of Canterbury's satisfaction theory. Nevertheless, his doctrine of God offers an interesting perspective on how diversity might indeed be in the divine interest. God, as far as Scotus is concerned, creates out of love and for creation to exist in this love. Such a God might indeed warrant creatures

---

6. de Duve, "Life as a Cosmic Imperative?," 620. See also de Duve, *Vital Dust*.

7. Addressing this question in the context of ETI is merely one side of the issue. Elsewhere I have presented the same argument for nonhuman animals, some of which according to empirical studies are self-conscious and therefore, from the perspective of transcendental theology transcendentally conscious. See Putz, "Moral Apes," 613–24.

8. For historical overviews on the ETI debate see Dick, *Plurality of Worlds*.

capable of love to emerge on more than one average blue planet in a vast universe filled with potentially billions of habitable worlds.

Anselm saw the necessity for the incarnation in humanity's sinning against God by not surrendering to the divine will.[9] Such a transgression leaves a debt of honor, Anselm argues, that cannot be paid for by merely returning to the rightful path. Divine justice demands satisfaction, which requires amends to be made. But, considering the infinite magnitude of the offense, humans as finite beings are incapable of recompensing in an adequate fashion, which is why one who is both infinite and human must atone in humanity's stead.[10] Salvation can, thus, come only from and through a God-man possessing the divine ability to atone as well as the obligation *as* human being to do so. Hence, God incarnates in Jesus of Nazareth, who as the Christ saves through his death on the cross.[11] This, then, is the meaning of the incarnation: a sacrifice to God to atone for humanity's sins.

Like Anselm, Scotus accepts that humanity is saved in Christ, but he takes issue with reducing the incarnation to this objective. Instead, he insists on the absolute primacy of Christ and claims that God became human to reveal divine love. At the heart of Scotus's theology is the notion of predestination, which to him implies that God wills an intellectual creature to grace and glory.[12] This predestination has two aspects, namely God's intention from all eternity for a certain *telos* to be fulfilled, and the fulfillment of this intention in time. Since the eternal intention always precedes the temporal realization of the divine objective, Christ cannot be the response to a human need (i.e., redemption of sin).[13] For Scotus, this means that the incarnation would have taken place even if humanity had not sinned: "To think that God would have given up such a task had Adam not sinned would be quite unreasonable! I say, therefore, that the fall was not the cause of Christ's predestination and that if no one had fallen, neither the angel nor man in this hypothesis Christ would still have been predestined in the same way."[14] The primary meaning of the incarnation is not to redeem humanity, but to fulfill God's plan in which all of creation has its origin in the first place. The

---

9. Saint Anselm of Canterbury, *Cur Deus Homo*, Book 1, Chapter XI, 216.

10. Ibid., Book 1, Chapter XX, 240.

11. Ibid., Book I, Chapter XXV, 251.

12. Scotus, *Ordinatio*, I, d. 40, q. un, n. 4. This must not be confused with Augustine's theory of predestination or its later instantiation in for example Calvin's model. For these thinkers, what is divinely predestined is who is saved and who is not.

13. See Delio, "Revisiting the Franciscan Doctrine of Christ," 9.

14. Scotus, *Reportatio Parisiensis III*, d. 7, q. 4.

telos of creation is the communion of all persons with one another and with God.[15]

Herein, I believe, lies the key to a doctrine of God that opens possibilities for an argument for the necessary existence of ETI. A God who creates out of love and to communicate this love should find the communion of many more valuable than that of a few. To be sure, Scotus's God is immutable, untouched by the love of humanity, entirely self-sufficient while simultaneously completely selfless. God loves Godself and rejoices utterly in this love. And yet, it is because of this perfect love, which itself cannot but will perfection of love, that God creates the world. Scotus is convinced that God wants creatures to live this love and join in it with God. He writes: "First God wills good for himself as the end of all things; second, he wills that another be good for him. This is the moment of predestination."[16] Creation, then, is always already ordered toward creatures that can receive God's fullest love and glory and respond to it in terms. Since Scotus knows of only one being capable of not only receiving God's love, but also of responding to it consciously he concludes that God wants humanity as the "co-lover" with God. But what if there were more such beings?

God in Relationship

Anyone interested in adopting a Scotistic doctrine of God today would have to evaluate and possibly adjust it in light of contemporary natural science. For one, a modern Scotistic theology would have to embrace an evolutionary worldview, according to which sentient beings emerge out of the free process of cosmic evolution that is transcendentally conditioned by divine action.[17] A Scotistic theology would have to encompass a theology of becoming that takes scientific explanatory models like the Big Bang theory or the theory of evolution seriously. As I will try to show, Karl Rahner's idea of the self-transcendence of matter toward spirit offers a way for how we can envision this true becoming to unfold.

Second, as a consequence of its freedom, cosmic evolution cannot be seen as directed toward the human being as its pinnacle. To use the late

---

15. Ibid.

16. Scotus, *Reporationes Parisiensis III*, d. 32, q. 2.

17. It is important, here, to point out that by "evolving freely" I do not mean to insinuate "independently of God." On the contrary, as I will show, God as the transcendental cause of this evolutionary movement increases this freedom. Unlike in a zero-sum game, creaturely freedom, be it evolutionary or moral, gains with God's loving concurrence.

Stephen Jay Gould's infamous analogy: if we were to rewind the tape of evolution and play it all over again, no species looking exactly like us would emerge.[18] Humanity is no longer the center of the physical universe, and for theologians to insist that it remains so is a risky gamble. But if Scotus is right in his belief that God creates to self-communicate, and if his claim should hold even vis-à-vis scientific views on cosmic evolution, a being *very much like* the human being, capable of knowing (transcendental consciousness) and loving God, should emerge.[19]

Third, a God who wants to self-communicate, who predestines creation as recipients of this abundant love, and who by releasing creation into the free process of evolution (thus accepting the possibility of more than one co-lover to emerge) must want a multiplicity of spiritual beings. I believe that the mounting empirical evidence for self-consciousness in nonhuman animals on Earth supports this view. I shall return to this point momentarily, but let it suffice for now to say that from the perspective of transcendental theology possessing a self constitutes the necessary *and* sufficient precondition for being a recipient of God's self-communication. If humanity were but one of a multiplicity of transcendentally conscious beings, beings that are able to somehow know of the transcendent as their whence and wither, would it not enrich the unfolding of the divine self-communication? Would not more rather than fewer conscious species achieve the divine goal in an even fuller fashion than one lonely creature on a lonely planet amidst countless empty worlds? Personally, I think so. Hence, any modern scotistically informed doctrine of God should warrant the existence of other sentient beings—be they terrestrial animals or ETI.

Finally, another important adjustment a contemporary Scotistic doctrine of God would have to incorporate has to do with the notion of God's immutability. In the *Ordinatio*, Scotus presents his own version of the classic argument for the immutability of God that can be traced back to Augustine of Hippo and beyond, to the Greek philosophers. It goes like this: Since God is perfectly simple, that is, not composed of parts, which in turn would require a cause, God cannot be changed into any form received in Godself. As necessary being, God cannot be changed from being to non-being or vice versa. "Therefore God is said to be simply immutable in respect of any change, whether substantial or accidental."[20] The question that immediately

18. Gould, *Wonderful Life*, 48–50. Gould goes even further in his claim and argues that self-consciousness and intelligence would not evolve again. I share his view that *Homo sapiens* would not emerge twice, but—and that is my point here—I part with him on the matter of sentient intelligence.

19. This is what I mean by the *weak* principle of plentitude.

20. Scotus, *Ordinatio*, I, d. 8.2, n. 229.

presents itself to the modern mind is how such an immutable God could be a loving God.

I have argued elsewhere that a God who would not be affected by creation especially in the context of a loving relationship between self-conscious creatures and God seems hardly reconcilable with modern experience.[21] Instead, I adopt Abraham Heschel's notion of the "divine pathos," which "denotes, not an idea of goodness, but of a living care; not an immutable example, but an outgoing challenge, a dynamic relation between God and man."[22] English theologian John Taylor once remarked that to affirm that God is love implies that God is far more than self-sufficient.[23] In Taylor's opinion, God had to create the universe not because God needed anything beyond God, but because the nature of love demanded as much. By its nature, love tends to union, and without the other cannot be. Such a God in relationship is no longer Scotus's withdrawn, distant God of a pure agape, but one deeply moved by the life of every being. This God as fully transcendent ground of being creates the world out of love and in order to communicate Godself in and through that love, to which all of creation is invited. Creation, in turn, is predestined to move toward entering into love with God. Cosmic evolution moves toward a diversity of transcendentally conscious beings that can respond in terms to the divine invitation and by joining God in loving relationship affect God.

How are we to envision this process? And what does a being capable of entering into a loving relationship with God look like? For these questions, I turn to Karl Rahner.

God's Self-Communication in an Evolving World

In his theological project, Karl Rahner embraces Scotus's idea of a God who creates to communicate self in love.[24] But, as Harvey Egan observes, Rahner's is a *"creative* appropriation,"[25] and as such it offers some adjustments. For one, he makes the concept of the divine self-communication the center of his theology and interprets it in a transcendental analysis as the most intimate personal relationship between God and human being:

21. Putz, "Love Actually," 354–56.
22. Heschel, *The Prophets*, 289.
23. Taylor, *The Christlike God*, 195.
24. Rahner, "Die Christologie innerhalb einer evolutiven Weltanschauung," in *TI,ST* V:213, idem; 213; Rahner, "Intellektuelle Redlichkeit und christlicher Glaube," in *TI,ST* VII:69.
25. Egan, "Theology and Spirituality," 16.

What we call creation as the generation of non-divine being by God through an outward activity seems to be merely the establishment of the condition for the divine self-communication in gratuitous love, which *is* God himself. As the recipient's innermost constitutive element, God not only communicates created finite being distinct from Godself, but also Godself, thereby becoming giver and gift, and even the actual source of man's capacity to receive God as gift. Thus, the finite, by its very nature as finite, fulfills its being in God as the infinite mystery.[26]

God creates so that the communication of the divine self as love can unfold in those beings that God constitutes to receive the divine self-gift in each encounter of a concrete being, including the self.

Here a fundamental question arises that shapes Rahner's entire corpus. It takes its cue from Immanuel Kant's turn to the subject with its transcendental analysis of the necessary conditions for the possibility of human knowledge. Analogically, Rahner wants to know how the ontological structure of human cognition constitutes the precondition for any knowledge of God. Put differently, Rahner wants to know what about the human being enables it to receive the loving divine self-communication of which Scotus spoke and which Rahner reinterprets in light of modern thought, particularly of Martin Heidegger's philosophy of being.

A second, equally central question results from Rahner's acceptance of evolution as a given fact.[27] How, he asks, do we have to imagine such an evolutionary becoming that is dependent on God's activity to unfold? With that, Rahner obviously addresses an issue of paramount importance to all theology after Darwin and Lemaître, but more importantly to the question at hand, he also presents a theology of becoming that opens the door for the existence of ETI. Rahner is well aware of this and engages the issue repeatedly if only peripherally throughout his career. Taking his own proposal to heart, he concludes that while neither contemporary science nor theology can offer any data resolving the issue, the possible emergence of intelligent life on other planets cannot be dismissed as easily today as under a pre-scientific worldview.[28] And if they exist these spiritual extraterrestrials would have to be considered equally supernaturally oriented toward God as human beings. Rahner even goes so far as to say that "[g]iven the immutability of God in

---

26. Rahner, "Der dreifaltige Gott als Urgrund der Heilsgeschichte," 68. (This and all other translations from the German are mine.) For the self-communication of God as the central theme in Rahner's theology see Vorgrimler, *Karl Rahner. Gotteserfahrung in Leben und Denken*, 167.

27. Rahner, "Die Theologie der Erneurung des Diakonates," 183.

28. Rahner, "Naturwissenschaft und vernünftiger Glaube," in *TI,ST* XV:58.

himself and the sameness of the Logos one could hardly prove that multiple incarnations in different salvation histories are utterly unthinkable."[29] If indeed the evolution of self-conscious life as it played itself out in the case of humanity is an intrinsic property of the cosmos and thus warranted by God, the possibility that this has happened repeatedly throughout a vast universe is not at all unlikely.

Transcendental Consciousness: An Unreflected Awareness of the Divine

What for Karl Rahner characterizes the recipient of the divine self-communication is the capacity to experience God in virtually every instance of its conscious existence. In each act of knowing and willing, the recipient is brought before the underlying infinite mystery of its being in a fundamental revelatory event. This experience, however, does not yield a conceptual explanation but instead an unreflected, or as Rahner puts it, "anonymous knowledge" of the divine. Only in a subsequent consideration can the experience be put into concrete terms and, thus, become articulable.[30] By itself the revelatory encounter with God, which unfolds anonymously in the recipient's self-experience, affords a somewhat obscured awareness rather than an unambiguous knowledge of God. Given the primacy of the experience, one might think of it as a protoreligious state, where a subject is aware of the divine without necessarily being consciously aware of having the experience. Thus, the necessary condition for the possibility of any being—terrestrial or extraterrestrial—to be in loving, personal relationship with God is transcendental consciousness. Or, in Rahnerian speak: Any such being would be a transcendent being and, as such, a spirit in the world. Let me unpack this idea a bit further.

Underlying Rahner's phenomenological analysis of the experience of God is Heidegger's philosophy of being and its ontological distinction between beings (*Seiende*) as things that are and being (*Sein*) in which all things are grounded.[31] As far as Heidegger is concerned, it is being (*Sein*) that constitutes the concrete being (*Seiende*) and makes it accessible in the first place.[32] As *Dasein*, the human being asks necessarily about the mean-

29. Ibid., 59.

30. It is at this point that we can speak of a "theology" that informs religious practices, rituals, and beliefs.

31. Heidegger, "Grundprobleme der Phänomenologie," 22. For a discussion of the ontological difference see also Lotz, *Sein und Existenz*, 98.

32. In more recent readings, Heidegger is interpreted to understand being in a

ing of its being (*Sein*), which it encounters by merely existing.[33] Thus, the human being is as much the question of its being (*Seinsfrage*) as it is the one asking it. As Rahner points out, the natural sciences with their empirical anthropologies certainly offer important insights in an attempt to explain the human being, but since their inquiry is fundamentally divorced from the *Seinsfrage* per se, their answers fail to satisfy the *Dasein's* desire for self-understanding.[34] For in the end, as *Dasein* the subject does not seek to *explain itself ontically*, that is, in reductionistic, measurable terms, but *understand itself ontologically* as a way of being in the world and of relating to beings and to being. To this end, it will always transcend its self-experience and extend onto the infinite horizon of being-as-such. In every concrete, categorical experience, then, there is always also given a second, transcendental experience.[35]

Rahner calls this capacity to reach beyond the ordinary objects of one's sense experience onto the infinite horizon of being-as-such the *Vorgriff auf*

phenomenological-hermeneutical sense, where there is merely the current being of an infinite number of possible beings of an entity (Sheehan, "A Paradigm Shift in Heidegger Research," 190; von Sass, *Gott als Ereignis des Seins*, 299). Accordingly, the current being of an entity is what a *Dasein* takes this thing as. For example, a rock can mean to be a percussion tool when trying to crack a nut, a doorstopper when used to keep a door from shutting, or a paperweight when placed on top of a stack of papers. In this understanding, being is absent whenever there is no observing *Dasein*, so that being cannot truly constitute beings. This, of course, is close to Gadamer's distinction of *Welt* (world) and *Umwelt* (environment), with the former being only present to human beings (Gadamer, *Wahrheit und Methode*, 448.)

33. Heidegger, *Sein und Zeit*, 7. Heidegger defines *Dasein* here as that being (*Sei-ende*) which can ask about its being as such (*Sein*). At another place, Heidegger discusses the nature of "existence" and, thereby, defines once more what characterizes *Dasein* as the only concrete being that exists (as opposed to rocks or trees, which cannot pose the *Seinsfrage*): "What does 'existence' mean in *Time and Being*? It is the way of being of that concrete being which stands open for the openness of being in which it stands, by enduring it." (*"Was bedeutet 'Existenz' in* Sein und Zeit? *Das Wort nennt eine Weise des Seins, und zwar das Sein desjenigen Seienden, das offen steht für die Offenheit des Seins, in der es steht, indem es sie aussteht."* Heidegger, "Einleitung," 17, (my emphasis). I have translated *ausstehen* with *endure*, which, I believe comes closest to the German original. Another possibility would have been the use of the verb *to bear*. Either way, what for Heidegger is important is the fact that the *Dasein* cannot but be in the world and act in every moment of that existence. Human existence as the *Weise*, or way of being of the *Dasein* ends with death after which the *Dasein* has no longer to endure its existence and, thus, stands no longer open for the openness of being. Put differently, existence is the way of being toward death. For a helpful discussion of Heidegger's concept of being, see Lotz, *Sein und Existenz*, 97–119, as well as Lotz, *Martin Heidegger und Thomas von Aquin. Mensch-Zeit-Sein*, 30.

34. Rahner, *Grundkurs des Glaubens. Einführung in den Begriff des Christentums*, 38–39.

35. Ibid., 31–32.

esse, or pre-apprehension of being.³⁶ Since it is being-as-such (*Sein*) that renders concrete being (*Seiendes*) accessible, the *Vorgriff* has to be seen as the *conditio sine qua non* for the possibility of all sentient knowing and willing. Only against the *infinite* background of absolute being can the *finite* subject apprehend itself and any other concrete finite object. One could think of it analogically as how we can only recognize a clock as a distinct entity over and against the wall on which it hangs or a ship against the seemingly endless expanses of the sea.³⁷ Consequently, Rahner claims that by radically experiencing its own finitude, the sentient subject has already transcended this very finitude and by means of the *Vorgriff* reached upon the infinite horizon of absolute being.³⁸ At the same time, free choice becomes a possibility only insofar as the self experiences itself as a subject.³⁹ In its openness toward being, the sentient subject "is the transcendent being"⁴⁰ and "spirit in the world."⁴¹

But what is the whereunto of the *Vorgriff*? Or, asked more precisely, what is this being-as-such (*Sein*)? Heidegger's existential analysis, hinting at the concern of non-being that the human being experiences when faced with being in light of its own limited existence (being-toward-death), lets him conclude that being is nothingness.⁴² Whether this inevitably leads toward a negation of being, as Rahner interprets it, or to an approval of nothingness as being, where nothingness is not nothing, as Johannes Lotz reads Heidegger, is an open question.⁴³ Either way, Heidegger does not un-

---

36. Ibid., 44. Rahner uses the concept *Vorgriff* in reference to Heidegger, who introduces it in §32 of *Sein und Zeit* as part of the tripartite fore-structure of interpretation. Rahner of course redefined the concept in Thomistic terms, making it the necessary condition for the possibility of an experience of absolute being, which for Heidegger is impossible. The term has been translated into English in numerous ways, including *pre-apprehension, pre-grasp,* or *dynamic transcendence*, none of which captures its full meaning successfully. I, therefore, will use the German *Vorgriff* from hereon in.

37. Of course, the infinite horizon of being is beyond the this-worldly realm and as fully transcendent horizon must not be mistaken to be just another concrete being. Hence, the analogies given here are really just that, analogical.

38. Rahner, *Grundkurs*, 42.

39. Ibid., 49.

40. Ibid., 42.

41. Rahner, *Geist in Welt*, 146. *Geist*, or *spirit*, used as a descriptor of a subject that is infinitely open to being, Rahner develops out of Hegel's notion of spirit as the consciousness that experiences itself and other objects. Cf. Hegel, *Enzyklopädie der philosophischen Wissenschaften im Grundrisse*, particularly Dritter Teil, 274.

42. To be more exact, nothingness endures as being *"Aber dieses Nichts west als das Sein."* Heidegger, "Einleitung, 49. Nothingness shrouds being, yet does not oppose it or render it empty. One, therefore, might say that being is nothingness in the sense of not being concrete being, yet not as total absence of something—no-*thing*-ness.

43. Rahner, "Einführung in den Begriff der Existentialphilosophie bei Heidegger,"

derstand being as the subsisting transcendent. Being, he insists, is greater than all beings, including animals, works of art, angels, and even God.[44] As Lotz observes, here God is subordinated to being and thus never fully achieves the "is" of being.[45] For the Christian theologian Rahner, on the other hand, all being that has been, is, or ever will be is unified by and in the transcendent God, who according to Thomas Aquinas as absolute being is the subsisting being itself (*Deus est ipsum esse subsistens*).[46] Hence, he concludes that the *Vorgriff* goes unto God.[47] And as a result, the transcendental experience that is given in every categorical experience is ultimately a revelatory one.

The ramifications of Rahner's analysis are stunning indeed: Every conscious encounter with a concrete being in the world—be it one's eighteen-month-old son, Allegri's *Miserere*, or a cup of coffee—is potentially revelatory of God. Being free, a sentient subject like the human being can always ignore or flat-out reject the experience, but it cannot escape having it. As long as it experiences itself in the world vis-à-vis an other the subject will be brought before the infinite horizon of absolute being that is God. Self-experience and the experience of God are to Rahner corresponding dimensions that form a unity in difference, where the latter is the necessary condition for the possibility and moment of the former—and vice versa.[48]

It is important to remember here that this revelatory transcendental experience of which Rahner speaks does not provide us with a conceptual or thematic, but rather an anonymous knowledge of God. In the "original experience" (*Grunderfahrung*) of God, the divine is met as the infinite mystery, which communicates itself in silence and beyond which the *Vorgriff* cannot reach.[49] To use the analogy of the clock and the wall again, I do not notice the wall consciously when I look for the clock, even though I can only find the clock because it stands out from the background on which it hangs. Moving the unthematized background into the thematized foreground requires reflection on the underlying aspects of the experience and involves articulating the previously unspoken mystery using symbolic

---

340; Lotz, *Sein und Existenz*, 106.

44. Heidegger, *Über den Humanismus*, 19.

45. Lotz, *Sein und Existenz*, 109. But see also previous footnote for new readings of Heidegger.

46. Rahner, *Hörer des Wortes. Sämtliche Werke 4*, 94; Aquinas, *ST* P.I, q.4, a.2.

47. Ibid.

48. Rahner, "Selbsterfahrung und Gotteserfahrung," in *TI,ST* X:135–6. The term "corresponding dimensions" I borrow from Anton Losinger, who captures the relationship elegantly thus. See Losinger, *Der anthropologische Ansatz in der Theologie Karl Rahners*, 46.

49. Rahner, *Grundkurs*, 61.

representation and, ultimately, language. Of course, any attempt at capturing the absolute transcendent in categorical terms is bound to fail; for one, because our language is insufficient to articulate the ineffable, but secondly because any such categorical expression would be scrutinized immediately by the *Vorgriff*. Every reflection of the content of the transcendental experience renders God as a receding horizon that, while always present, remains fundamentally unattainable to the intellect. I may contemplate the ocean around my boat, but all I will find is that with each tentatively understood bit of water there is more endless blue I cannot comprehend. The transcendental experience places us, as Abraham Heschel puts it so hauntingly, in "unquenchable amazement" before "that which lies within our reach but beyond our grasp."[50]

Anonymous it may be, and yet this ubiquitous experience of the divine is to Rahner indicative of God's unsurpassable proximity in our lives and, thus, constitutes the very nature of the personal relationship between God and self-present subject. Rahner envisions this relationship in terms of God's self-communication, which to him is nothing else than the essence of supernatural grace as love.[51] It is the gratuitous unowed self-gift of God that instigates within the human being the desire for union with God as a *supernatural existential*[52] of human being. Self-communication, then, does not mean, as one might expect, that God discloses some finite truths about Godself, but rather that in an ontological act "God in his own reality makes himself the innermost constitutive element of the human being."[53] This existential manifests itself in the *Vorgriff*, which allows us to reach beyond and experience the transcendent mystery. In that sense, self-conscious beings are always already related most intimately to God and recognize God in an unreflected manner as their whence and wither. Moreover, they are as much recipients as they are event of the divine self-communication, since the latter unfolds only insofar as the subject is a conscious self in the world.

At this point, we can summarize what kind of beings should emerge from the cosmic evolutionary process, if indeed the goal of said process is the unowed, loving divine self-communication Scotus suggested: It would

50. Heschel, *Man Is not Alone*, 4–5.

51. Rahner, "Über das Verhältnis von Natur und Gnade," in *TI, TS*, I:334.

52. The term "existential" Rahner borrows from Heidegger, to whom existentials are the ontological, *a priori* structures of existence. In this sense, the question of the existentials of the *Dasein* is the transcendental analysis of *Dasein*. The orientation toward God is for Rahner a *supernatural* existential, because it has its origin in God acting gratuitously on the *natura pura* of the individual, who thereby is endowed with the *desiderium natural*, the desire for God in the beatific vision.

53. Rahner, *Grundkurs*, 122.

have to be a self-conscious subject, constituted as a spirit in the world to reach beyond categorical objects onto the underlying Holy Mystery of absolute being that is God. If ETI exists, it too would have to be an event of God's self-communication. But why expect such an event to occur more than once in more than one instantiation? And how does creation develop towards such a multiplicity of events of God's self-communication? This brings me to Rahner's theology of becoming.

Becoming through Active Self-Transcendence

Rahner engages the question of becoming in response to the potential challenges evolutionary biology poses to Christian theology.[54] Being convinced that theology must not ignore contemporary scientific explanations, he asks whether, assuming an evolutionary worldview is adequate, it can be incorporated into a Christian theology of hominization.[55] One of his major concerns is a theological response to the theory of evolution, which would not end up in the "moderate evolutionism" of *Humanis generis*, where the human body originates in pre-existent and living matter while the soul comes directly from God.[56] Rahner wants to avoid what he considers a dualist perspective in favor of what one might call a dual-aspect monist description of spirit and matter. Since for Christians all beings have their origin in God, all beings are united by similarities rather than divided by differences. The world as a whole has *one* origin, *one* self-realization, and *one* determination. Matter and spirit are not disparate entities, but two aspects of the same reality. What is evolving, then, is spirit-matter as a unity in difference.

In this metaphysical understanding, becoming is not merely the change of which the natural sciences speak when they describe newly emerging natural phenomena in a reductionistic bottom-up explanation that limits itself to an account of the reorganization of already existing parts. For Rahner, becoming marks a true *increase in being* (*Seinszuwachs*), where novel being possesses greater self-presence and is therefore more fully actualized. In this sense, becoming is the progressive actualization of self-consciousness. Underlying his proposal is the Aristotelian-Thomistic theory of potentiality and actuality, according to which a possibility of being (prime matter) is actualized through the informing action of a substantial form. Novel concrete being, thus, is a truly and essentially new substance endowed with a new informing principle. Overall, the evolution of spirit-matter moves toward

54. See also Putz, "Evolutionary Biology in a Catholic Framework," 316.
55. Rahner, "Naturwissenschaft und vernünftiger Glaube," in *TI,TS*, XV:41,45.
56. Rahner, "Die Hominisation als theologische Frage," 55.

greater self-consciousness, thus leading to a hierarchical organization of being.[57] Seen through this metaphysical lens, "evolution" is no longer understood in its biological sense as descent with modification leading to new species, nor the reorganization of elementary particles under influences of gravity and the second law of thermodynamics leading to the observable universe with its stars and planets. Rather, cosmic evolution now is a metaphysical process that is directed toward the emergence of self-consciousness and, thus, new beings that are in personal relationship with God.

Unlike biology, which cannot find any directionality in the evolutionary process, Rahner's metaphysical proposal assumes as much on the ground of the nature of being and the constant movement from the earliest form of spirit-matter to more complex life forms capable of conscious self-reflection. This movement, however, is not one toward human beings per se, but toward self-presence in material beings as events of God's self-communication. If we were to replay Gould's evolutionary tape, we would not again end up with *Homo sapiens*, but certainly with beings capable of receiving God's self-communication.[58] Evolutionary becoming for Rahner is the transition from one degree of actualized being to a greater one. This becoming is true self-transcendence, insofar as a being actively achieves its own fullness by surpassing itself. Matter transcends itself in order for spirit to be set free in the self-present being, which is why Rahner calls matter frozen spirit.[59] So far, so good, but does this active self-transcendence play out, and what is God's role in it?

Rahner's solution adopts and reinterprets the scholastic notion of *eductio e potentia materiae*, according to which the new form that is already present in the old being is drawn forth by finite causes, and combines it with the sixteenth-century concept of the *concursus divinus*, the divine concurrence, which views God as the transcendent constitutive that enables the finite being to actively assume fuller actualized being. Therefore, becoming in the world is always the bringing forth of a new substantial form through the actualization of that which is potentially already present in the subjacent by means of both creaturely and divine agency.

57. Scholasticism presented the *scala naturae* (great chain of being), which in a linear fashion leads to self-conscious corporeal beings (humans), on to the angels and finally God. As I will show, Rahner's proposal leads to a more organic, tree-like hierarchy. Moreover, the hierarchy is of a metaphysical, not biological nature!

58. As mentioned above, Gould argues that not only would evolution a second time around fail to produce *Homo sapiens*, but also self-conscious sentient intelligence in general. From Rahner's perspective, however, cosmic evolution unfolding in active self-transcendence is precisely moving toward the emergence of self-conscious sentient intelligence.

59. Rahner, "Die Hominisation als theologische Frage," 51.

To medieval thinkers, a new substantial form can arise under the efficient influence of external and internal agents acting on the old concrete being. What makes this idea problematic to the modern mind is for one that it has to invoke some type of vitalism to explain life and, secondly, that divine action, if at all needed, would be reduced to merely another efficient cause. That Rahner nevertheless embraces the *eductio* is due to the fact that it regards becoming as an increase in being while simultaneously implying that in becoming a finite being actively surpasses itself and assumes a new substantial reality. However, such an active self-transcendence cannot come about solely by the powers of the becoming finite being, or else the finite being would give more than it possesses, violating the principle of causality. To resolve the issue, Rahner casts the divine concurrence in a transcendental fashion, where the divine absolute being acts immanently in the finite being as the necessary condition of the possibility of its self-transcendence.

Rahner's solution is a synthesis of two diametrically opposed Catholic interpretations of efficacious grace that surfaced in the sixteenth century.[60] The problem is how a finite being can freely consent to divine grace so that said grace can bear its divinely intended effect. How can the freedom of the finite being be preserved when said freedom depends on divine grace? According to Dominican Domingo Bañez and his followers, God as ultimate cause moves every finite cause to its proper acts, which assures divine agency, but undermines personal freedom. In contrast, Jesuits Luis de Molina and Francisco Suárez argued that divine action cooperates with creaturely efficient causation to produce the overall outcome, thus guaranteeing individual freedom, but violating the principle of causality. By regarding divine action as the underlying necessary condition of active self-transcendence, Rahner opts for a both-and rather than an either-or. The notion of divine concurrence as necessary condition presumes that God as pure act pre-contains all reality and belongs to the constitution of the finite cause *as such*, without being its *intrinsic constituent*. God gives the UV radiation that brings about a gene mutation being, but God is not part of that radiation. Therefore it takes both, God as ultimate and UV radiation as efficient cause, to prompt a gene mutation that in turn actualizes the heretofore only potential new being of an organism.

60. The question of efficacious grace arose of course during the early days of the Reformation, when in particular Calvin suggested that human freedom was absolutely compromised in the Fall in which humanity lost both the image and likeness of God. Calvin famously adopted Augustine's idea of persevering grace, according to which those who were chosen by God were bestowed with grace they could not fall from. Calvin extended Augustine's views in his theory of divine election and predestination, where due to humanity's severe sinfulness and inability to choose what is right, God has to intervene with efficacious grace to assure the salvation of the chosen ones.

Rahner summarizes his philosophy of becoming in four ontological theses. First, becoming is always by its very constitution a self-exceeding and not replication. Second, this self-transcendence is made possible by the absolute being immanently at work within the becoming finite being as fundamental cause and ground of this becoming. Third, the essence of the self-transcending finite being does not limit what can become through this self-transcendence. It does, however, show that something more has to become, and, moreover, that the starting-point of this movement limits the possibility of what can become immediately of the self-transcending finite being. Fourth, these principles can be applied to the developing of matter towards spirit.

Tree of Being

As a theology of becoming, active self-transcendence, once one buys into it, may explain quite well the emergence of self-conscious terrestrial beings like us. The process has an intrinsic directionality toward self-presence as the necessary precondition of being an event of God's self-communication. So, if one adopts the Scotistic view that God creates to communicate God-self and relates it productively to the theory of evolution one might expect that self-consciousness should emerge from cosmic evolution sooner or later. But does this necessarily imply ETI?

One might want to argue with Marie George that on the grounds of a redundancy argument that one species alone would suffice to satisfy God's desire for self-communication. But in that case one would also have to explain why there is such a plentitude of other worlds that would consequently remain devoid of life. Why would God allow for an evolving cosmos to produce such a fantastic number of planets (between one hundred to four hundred billion in our galaxy alone)[61] and then not allow for life to emerge on them? As I pointed out above, that seems unlikely. Moreover, if one accepts active self-transcendence, as I want to, the movement toward self-present beings never ceases. It began in the moment of creation *ex nihilo* and has continued to this very day. But then, multiple events of God's self-communication should emerge—not only here on Earth, but all throughout the universe. Sentient life is indeed, as Christian de Duve suggests, inevitable. If indeed the evolution of self-conscious life as it played itself out in the case of humanity is an intrinsic property of the cosmos and thus warranted by God, the possibility that this has happened repeatedly throughout a vast universe is not at all unlikely. It should be expected.

61. Cassan et al., "One or More Bound Planets," 167–69.

Everywhere in the universe, the evolution of spirit-matter moves toward self-consciousness, yielding life from nonlife, spiritual multiplicity from biological diversity. Active self-transcendence leads toward a hierarchy of beings, with self-present beings in personal relationship with the divine. But this hierarchy is no longer the linear great chain of being of Scholasticism. It is an organic *great tree of being* with numerous branching points rather than individual rungs on a ladder.

Conclusion

In previous chapters, both Heidi Manning and Martinez Hewlett said, "Yes, we'll meet them." They provided scientific arguments to support this contention. They relied on the big numbers argument: given the vast number of exoplanets and galaxies, it would be unreasonable to think otherwise. Sooner or later we earthlings will meet them.

I agree with Manning and Hewlett that we should expect sentient life to exist on other planets, but I reach my conclusion through metaphysical and theological deduction rather than an empirical or scientific argument. My argument is based, on the one hand, on the Scotistic notion of a God who creates to self-communicate and who would warrant a multiplicity of recipients of this loving self-communication and, on the other hand, on the Rahnerian idea that cosmic evolution as a becoming of spirit-matter moves toward self-presence. Together, both premises imply that the emergence of sentient life throughout the universe is inevitable and warranted by God. Whether we will actually get to meet them will depend largely on either our or their abilities to develop spacefaring capabilities that allow us to travel the enormous distances between our respective worlds. But regardless of that, I think it safe to say that they are there. *Deus semper maior*.

Bibliography

Allgemeine Naturgeschichte und Theorie des Himmels. 1755. Reprint, Cologne: Könemann, 1995; 1st ed., .
Anselm of Canterbury. *Cur Deus Homo*. Translated by Sidney Norton Deane. Chicago: Open Court, 1962.
Aquinas, Thomas. *Summa Theologica* in Christian Classics Ethereal Library. http://www.ccel.org/ccel/aquinas/summa.toc.html.
Cassan, Arnaud, et al. "One or More Bound Planets per Milky Way Star from Microlensing Observations." *Nature* 481:7380 (January 12, 2012) 167–69.
Davies, Paul C. W. *The Eerie Silence: Renewing Our Search for Alien Intelligence*. Boston: Houghton Mifflin, 2010.

de Duve, Christian. "Life as a Cosmic Imperative?" *Philosophical Transactions of the Royal Society A* 369 (2011) 620.

———. *Vital Dust: Life as a Cosmic Imperative*. New York: Basic Books, 1995.

Delio, Ilia. "Revisiting the Franciscan Doctrine of Christ." *Theological Studies* 64 (2003) 3–23.

Dick, Steven J. *The Biological Universe: The Twentieth-Century Extraterrestrial Life Debate and the Limits of Science*. Cambridge: Cambridge University Press, 1996.

———. *Plurality of Worlds: The Extraterrestrial Life Debate from Democritus to Kant*. Cambridge: Cambridge University Press, 1984.

Egan, Harvey D. "Theology and Spirituality." In *The Cambridge Companion to Karl Rahner*, edited by Declan Marmion and Mary E. Hines, 13–28. Cambridge: Cambridge University Press, 2005.

Fantoli, Annibale. *The Case of Galileo: A Closed Question?* Notre Dame: University of Notre Dame Press, 2012.

Gadamer, Hans-Georg. *Wahrheit und Methode, Grundzüge einer philosophischen Hermeneutik*. 1960. Reprint, Tübingen: Mohr/Siebeck, 1990.

George, Marie. *Christianity and Extraterrestrials: A Catholic Perspective*. Lincoln: iUniverse, 2005.

Gould, Stephen Jay. *Wonderful Life: The Burgess Shale and the Nature of History*. New York: Norton, 1989.

Hegel, Georg Wilhelm Friedrich. *Enzyklopädie der philosophischen Wissenschaften im Grundrisse*. Berlin: Edition Holzinger, 2013; 1st ed., 1817.

Heidegger, Martin. "Einleitung." In *Was ist Metaphysik?* 1949. Reprint, Frankfurt: Klostermann, 2007.

———. "Grundprobleme der Phänomenologie." In *Gesamtausgabe*. Vol. 58. 2nd ed. Frankfurt: Klostermann, 2010.

———. *Sein und Zeit*. 18th ed. Tübingen: Niemeyer, 2001.

———. *Über den Humanismus*. 1951. Reprint, Frankfurt: Klostermann, 2010.

Heschel, Abraham J. *Man Is Not Alone*. 1951. Reprint, New York: Farrar, Straus & Giroux, 1976.

———. *The Prophets*. 1955. Reprint, New York: Harper Perennial Modern Classics, 2001.

Losinger, Anton. *Der anthropologische Ansatz in der Theologie Karl Rahners*. St. Ottilien: EOS, 1991.

Lotz, Johannes B. *Martin Heidegger und Thomas von Aquin: Mensch, Zeit, Sein*. Pfullingen: Neske, 1975.

———. *Sein und Existenz: Kritische Studien in systematischer Absicht*. Philosophie in Einzeldarstellungen. Ergänzungsband 2. Freiburg: Herder, 1965.

Lovejoy, Arthur O. *The Great Chain of Being*. 1936. Reprint, Cambridge: Harvard University Press, 1964.

Lucretius. *On the Nature of Things*. Translated by William Ellery Leonard. London: Dent, 1943.

Putz, Oliver. "Evolutionary Biology in a Catholic Framework." In *Teaching the Tradition: Catholic Themes in Academic Principles*, edited by John J. Piderit, SJ and Melanie M. Morey, 307–29. Oxford: Oxford University Press, 2012.

———. "Love Actually: A Theodicy Response to Suffering in Nature. In Dialogue with Francisco Ayala." *Theology and Science* 7.4 (2009) 345–61.

———. "Moral Apes, Human Uniqueness, and the Image of God." *Zygon* 44.3 (2009) 613–24.
Rahner, Karl. "Die Christologie innerhalb einer evolutiven Weltanschauung." In *TI,ST*.
———. "Der dreifaltige Gott als Urgrund der Heilsgeschichte." In *Mysterium Salutis* II, edited by J. Feiner and M. Lohrer. Einsiedeln: Benziger, 1967.
———. "Die Hominisation als theologische Frage." In *Das Problem der Hominisation: Über den biologischen Ursprung des Menschen*, by Paul Overhage and Karl Rahner, 13–42. Freiburg: Herder, 1961.
———. "Die Theologie der Erneurung des Diakonates." In *Diaconia in Christo: Über die Erneuerung des Diakonates*, 285–324. Freiburg: Herder, 1962.
———. "Einführung in den Begriff der Existentialphilosophie bei Heidegger." In *Sämtliche Werke*. Vol. 2. Freiburg: Herder, 1996.
———. *Geist in Welt. Sämtliche Werke 2*. 1939. Reprint, Freiburg: Herder, 1996.
———. *Grundkurs des Glauben. Einführung in den Begriff des Christentums*. Freiburg: Herder, 1976.
———. *Hörer des Wortes. Sämtliche Werke 4*. Freiburg: Herder, 1997.
———. "Intellektuelle Redlichkeit und christlicher Glaube." In *TI,ST*.
———. "Naturwissenschaft und vernünftiger Glaube." In *TI,ST*.
———. "Selbsterfahrung und Gotteserfahrung." In *TI,ST*.
———. "Über das Verhältnis von Natur und Gnade." In *TI,ST*.
Sass, Hartmut von. *Gott als Ereignis des Seins: Versuch einer hermeneutischen Ontotheologie*. Tübingen: Mohr/Siebeck, 2013.
Sheehan, Thomas. "A Paradigm Shift in Heidegger Research." *Continental Philosophy Review* 34 (2001) 190.
Taylor, John V. *The Christlike God*. 2nd ed. London: SCM, 2004.
Vorgrimler, Herbert. *Karl Rahner. Gotteserfahrung in Leben und Denken*. Darmstadt: Primus, 2004.

PART 3

What Will Happen When We Meet Them?

11

Extraterrestrial Life and Terrestrial Religion

A Crisis?

TED PETERS

Two things awe me most, the starry sky above me and the moral law within me.

—IMMANUEL KANT

My going to space strengthened my religion—to see everything, what we are part of, part of creation, to see whole nations and oceans and everything at a single glance, that just strengthened my belief.

—ASTRONAUT JOHN GLENN[1]

Frontier explorer Daniel Boone (1724–1820) is alleged to have said, "when I see the smoke of my neighbor's chimney, it's time to move on." Astrobiologists and SETI and METI researchers are just the opposite. "Let's strain our eyes to see how far we can see; and let's hope we can find a new neighbor in space!" Many among us are just itching to send a casserole to welcome extraterrestrials into our terrestrial neighborhood.

1. Levinson, "Interview with John Glenn," in Levinson and Waltemathe, *TFC*, 8.

Are all of us on Earth thirsty for confirmation that we share our universe with other beings, with extraterrestrial neighbors? Not according to some observers who fear that Earth's religious believers will suffer a crisis of faith. There are prophets of crisis among us who predict that confirmation of ETI will shatter traditional religious beliefs. Can this be true?

In this chapter we will examine the claims by some seers that terrestrial religions are vulnerable to a crisis because their alleged geocentrism (Earth-centeredness) and anthropocentrism (human-centeredness) are out of date, rendering them unable to adapt to a large universe shared with other intelligent races. I will subject this claim to analysis. In doing so, we will summarize some of the findings of *The Peters ETI Religious Crisis Survey*, which confirms that—though some religious believers tend toward geocentrism and anthropocentrism—a significant majority are enrolled in the *Bible-Welcomes-Aliens* school of conjecture.

The Prophets of Crisis

Why would we raise the question: would ETI provoke a religious crisis? Because there are social prophets among us who predict it. Some prophets-of-crisis tell us to worry about our religious traditions because of certain assumptions they make: they assume that religions born in a pre-modern age are geocentric and anthropocentric and vulnerable to disappointment at new scientific findings. Some of these crisis prophets rely upon survey data that seem to connect resistance to belief in ETI with conservative religious views. Let us look a bit more closely.

Physicist and astrobiologist Paul Davies is among those who tell us to worry about Earth's religions. Religions seem to be fragile, breakable. "The existence of extra-terrestrial intelligences would have a profound impact on religion, shattering completely the traditional perspective of God's special relationship with man."[2] Or, "undoubtedly the most immediate impact of an alien message would be to shake up the world's faiths. The discovery of *any* sign that we are not alone in the universe could prove deeply problematic for the main organized religions, which were founded in the pre-scientific era and are based on the view of the cosmos that belongs to a bygone age."[3] Religions are fragile and vulnerable to dissolution because their beliefs were formed in a pre-scientific age; and scientific knowledge regarding extraterrestrial intelligences could not fit into out-of-date religious worldviews. So the argument goes.

2. Davies, *God and the New Physics*, 71.
3. Davies, *Eerie Silence*, 188.

SETI's Jill Tarter supposes that the ETI who contact us will be more advanced than Earthlings. ETI will have either avoided religion altogether or have outgrown whatever organized religion they once had. This will precipitate a crisis on our home planet. "An information-rich message from these extraterrestrials will, over time, undermine our own world's religions."[4]

Some crisis prophets believe they can support this worry with data. Based upon a 2005 survey by the Center for Survey Research and Analysis, George Pettinico infers that "devout Christians in America are more likely than other Americans to hold very traditional views of humanity as the single culmination of God's creation and Earth as the one divinely chosen place for this culmination of creation to live and prosper. For many conservative Christians, the universe—despite its size and scope—exists for the benefit of humanity alone."[5] The survey asks, "*Do you believe that there is life on other planets in the universe besides Earth?*" Note how the survey question employs a word with a religious overload, *believe*, rather than a term one might expect here such as *think* or *surmise*. Among those who stated they do not "believe" in ETI, 45 percent attend religious services weekly; 57 percent monthly; and 70 percent rarely. This climb in percentage from apparently more devout to less devout seems to support Pettinico's conclusion that religious devotion is correlated with doubt about the existence of ETI. How does Pettinicio know that the explanation lies in conservative Christian theology that espouses geocentricity, with Earth existing for "the benefit of humanity alone"? Did he test for this belief? No, at least not according to the data he cited for his conclusion.

E. M. McAdamis takes a similar set of assumptions into his research regarding religious dispositions relevant to extraterrestrial life, although he uses the term "astrobiology" to refer to ET. He "contends that the more detached religious doctrines are from the centrality of the importance of humans in the universe, the more accepting they will be of astrobiological endeavors and evidence. Inversely stated, [his assumption] predicts that the more anthropocentric the religious doctrine, the more potential resistance to astrobiology is likely to result."[6] It may be difficult to find any religion which considers itself anthropocentric; most religions center their attention on God, gods, or supra-human realities. Might such an assumption skew the research?

McAdamis proceeds to examine nineteen terrestrial religious traditions; but his data neither confirms nor disconfirms decisively this particular assumption. Yet, he still offers a conjecture worth our taking note

4. Tarter, "SETI and the Religions of the Universe," 147.
5. Pettinico, "American Attitudes," 104.
6. McAdamis, "Astrosociology and the Capacity of Major World Religions," 339.

of. He concludes from his survey that terrestrial religion would survive the ETI storm. "Most of Christendom would be amenable to astrobiological evidence," and "the larger religious landscape of the world seems to be philosophically constituted to not merely survive astrobiological pursuits, but to be explicitly compatible with, or even validated by, evidence of the universe harboring life beyond Earth."[7] Despite his slightly misleading assumption regarding anthropocentrism, McAdamis's conclusion after pursuing research on nineteen religions including Christianity tends to refute the assumptions purveyed by Davies and others.

William Sims Bainbridge seems to operate with an assumption similar to that of McAdamis. He references a 1981 study of University of Washington students and concludes: "a major factor discouraging people from supporting attempts to communicate with extraterrestrial intelligence was their religion—in this case evangelical Protestants, exemplified by the Born Again movement."[8] He then analyzes in detail a 2001 survey conducted by the National Geographic Society along with the National Science Foundation. The survey contained many questions about UFOs and New Age beliefs. The survey did not zero in on religious devotion in relation to the existence of ETI. Bainbridge concludes, "although geocentrists were somewhat rare in the *Survey 2001* dataset, they are probably more common in the general population . . . In the history of western civilization, the geocentric viewpoint reflected the religious belief that human beings were central to God's plan for the universe, and this prejudice retarded the development of science."[9] In other words, we draw a conclusion about religion not based upon the data but rather based on what is not in the data. This is a curious method for pursuing science.

Bainbridge does note fittingly a reliable study performed by Douglas Vakoch and Yuh-Shiow Lee in 2000 which suggested that anthropocentric and religious individuals are less likely to affirm the existence of ETI.[10] In this case, the conclusion corresponds to the evidence and to the assumptions. In short, one must grant that some survey material does exist which suggests an inverse correlation between traditional religious devotion and affirmation of the existence of extraterrestrial life.

In contrast to the Vakoch and Lee survey, however, a survey conducted by Victoria Alexander suggests welcome acceptance on the part of religious

---

7. Ibid., 338.
8. Bainbridge, "Cultural Beliefs," 119.
9. Ibid., 137.
10. Vakoch and Lee, "Reactions to Receipt of a Message from Extraterrestrial Intelligence," 737–44.

believers to the prospect of someday meeting ETI.[11] So, it appears we have conflicting data. Matthew Shindell, who surveyed a large number of surveys "doubts that public reaction to the announcement of the discovery of extraterrestrial life can be predicted."[12] He adds, "I don't think these surveys necessarily ask the best questions."[13]

In sum, this survey of surveys is inconclusive. Prophets of religious crisis largely work without reliable data to support their assumptions and predictions. What all of this warrants, in my judgment, is further investigation to see whether prophecies of doom are realistic and just what might be transpiring in religious belief systems.

The Peters ETI Religious Crisis Survey

With the prophecies of crisis in mind, I constructed a survey instrument that zeroed in directly on the relationship between religious beliefs and the possibility of contact with ETI. Along with my Berkeley research assistant, Julie Louise Froehlig, I devised a survey: the *Peters ETI Religious Crisis Survey*.[14] The *Peters ETI Religious Crisis Survey* received more than thirteen hundred responses worldwide from individuals in multiple religious traditions. It became clear that the vast majority of religious believers, regardless of religion, see no threat to their personal beliefs caused by potential contact with intelligent neighbors on other planets. When we asked respondents to distinguish between their own personal beliefs and the beliefs of the religious tradition to which they adhere, anxiety rose just slightly; perhaps they fear their religious leaders might face a challenge. Still, religious adherents overwhelmingly registered confidence that neither they as individuals nor their religious tradition would suffer anything like a collapse.

We then asked respondents to forecast what would happen with religion in general, with religious traditions other than their own. What is startling, is that respondents who self-identify as non-religious are far more fearful (or gleeful?) of a religious crisis than are religious believers. In sum, those who self-identify as religious do not fear a challenge to their religious belief; but those who self-identify as non-religious prophesy that other people—those religious people!—will face a crisis in belief.

11. Alexander, "Extraterrestrial Life," 359–70.
12. Shindell, "The Public Response," 95.
13. Ibid., 100.
14. This data is drawn from Peters and Froehlig, *Peters ETI Religious Crisis Survey*. This chapter is a revision and update of Peters, "*Astrotheology and the ETI Myth*"; Peters, "*Would the Discovery of ETI Provoke a Religious Crisis?*" 341–55.

Let me share with you some of the data. In Figure 1 (Question 3), note the consistency of the dominance of the third bar, "disagree/strongly disagree." The short bars are "strongly agree/agree" and "neither agree nor disagree." This shows how Roman Catholics, evangelical Protestants, mainline Protestants, Orthodox Christians, Mormons, Jews, and Buddhists right along with the non-religious fear no threat to their personal beliefs.

The survey instrument permitted respondents to offer additional comments. Some reflected just what the crisis prophets would predict. One self-identified evangelical Protestant admits that confirmation of ETI would provoke a crisis but, curiously, not if ETI were a long distance away. "The actual interaction with advanced extraterrestrial life would create a crisis for my belief system. Finding some extraterrestrial life form in a far away planet would not." Another respondent, self-identified as a non-denominational Protestant, associates alien beings with Lucifer. "I believe that all extraterrestrial beings are fallen angels (demons, if you will). And whatever traits they have can be traced back to Lucifer." These comments came from the Agree / Strongly Agree pillars.

Figure 11.1: From the Peters ETI Religious Crisis Survey
(Ted Peters and Julie Froehlig)

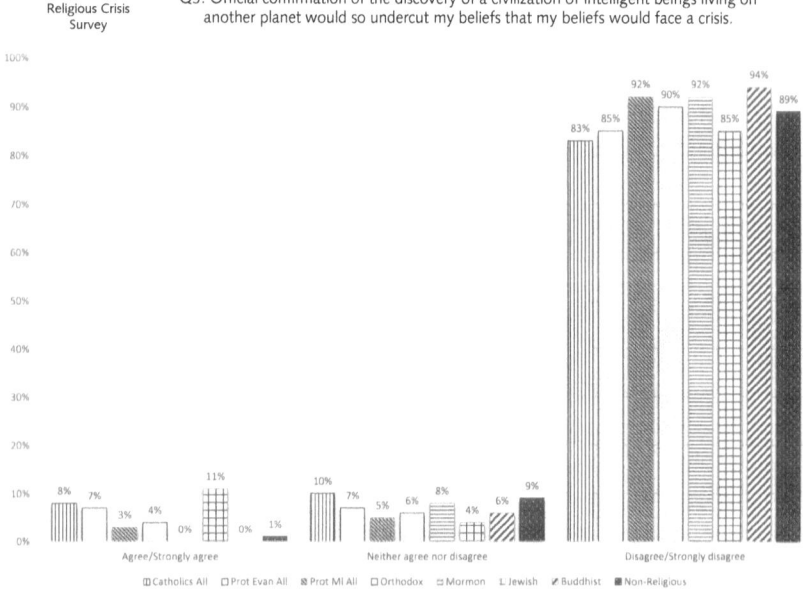

As already mentioned, the vast majority spoke positively about potential interactions with ETI. Among those who Disagree/Strongly disagree we hear a Muslim state, "Islamically we do believe that God created other planets similar to Earth." An evangelical Protestant sanguinely reports: "I can't see why the discovery of other life would affect our belief at all. God has made our world—and can make millions more, I suppose. Is Jesus the savior for all of them too, or did God do things very differently in those places? I'd be fascinated to find out, but not at all disturbed by it." A Roman Catholic foresees no crisis. "My religious viewpoints and practices would remain unchanged. The same God who made me is also capable of making extraterrestrials. His message of faith, hope, and love of neighbor goes beyond the borders of the known universe."

Some comments suggest more than mere grudging adaptation. They put out a welcome sign for space visitors. A mainline Protestant hopes that ETI would actually strengthen faith. "If life were discovered elsewhere in the universe, I think my faith in the absolutely mysterious and grace-giving God would actually be more confirmed than it is now. I would have to believe that God is involved not just on our planet, but in the universe in its entirety." So also does an evangelical Protestant: "Traditional Christian understanding teaches there are other intelligences in the universe who are more powerful than humans. Discovery that this teaching is confirmed strengthens Christianity, not weakens it." Another mainline Protestant wrote what has become my favorite comment: "I'd gladly share a pew with an alien."

Mormons already incorporate extraterrestrial entities in their theology; so it is not surprising to read one Mormon comment: "I believe that God, however he did it, created other worlds with other beings." Another Mormon foresees ETI as confirmation of beliefs already held. "First of all, my religion (LDS, Mormon) already believes in extra-terrestrials, and official doctrine and scripture even discusses names of extra-solar planets that are habitable for sentient life forms. If anything, an extra-terrestrial might even be looked at as confirmation of religious beliefs rather than something which would be though of as something to avoid."

Like the Mormons, Bahai followers already incorporate aliens into their cosmology. "The Bahá'í Writings contain many statements that implicitly and explicitly point to the existence of not only extraterrestrial life forms but to extraterrestrial intelligence as well."[15] It is interesting to note that the Bahai use the term, *exotheology,* rather than our term, *astrotheology.*

---

15. See the Bahai site: http://bahai-library.com/troxel_extraterrestrials_exo theology.

Eastern Orthodox Christians seem split. One comments, "I strongly disbelieve in the possibility of other intelligent life other than on earth. I think Christ came to release us from our sins on this planet and that is exclusive . . . But, if I were wrong and Christ can redeem other races, it would not change what I believe at all." In contrast, another Orthodox respondent comments, "I am constantly amazed at the ridiculous idea among some (a minority, I hope) in the scientific community that people of faith are ignorant and of low intelligence and therefore must have a 'God-of-the-Gaps' theology (i.e., God is used to explain all things which we cannot at this time explain scientifically) . . . I am a person of DEEP faith with a genius-level IQ, two doctorates, a hope that we will encounter intelligent life from outside of Earth, and a hope (but not necessarily an expectation) that any extraterrestrial intelligent life humanity encounters will be benevolent."

Numerous respondents mentioned they had read the works of British classicist and theologian, C.S. Lewis. One response is typical of a dozen or so: "There's an essay by C. S. Lewis (unfortunately published under various titles in various anthologies) that strongly argues against any particular significance for Christianity of any conceivable type of extraterrestrial. Whether or not one agrees, it is clear at any rate that extraterrestrials would not be disruptive to all religious belief."

Our survey did not test directly for geocentrism or anthropocentrism. Yet, these two items appear repeatedly among the voluntary comments offered by respondents. A Buddhist takes a stand against geocentrism: "I believe that anything is possible including life on other worlds. To think that in the infinity of the universe that we are the only intelligent life form in existence is ludicrous. I would only hope those beings would exhibit more wisdom than humans have in how they relate to their world and fellow beings." Another respondent self-identified as non-religious says almost the same thing. "I believe that we are not unique in the universe (it would be sheer hubris on our part, not that we are not a completely narcissistic species) but the universe is so large that contact among advanced civilizations is limited to neighboring planetary systems; and we may not have very advanced neighbors."

Opposition to geocentrism and anthropocentrism is common to respondents regardless of tradition. A Mormon exclaims: "Our universe is huge. So astonishingly huge that I find it absurd to think we are alone in this universe as a sentient life form." A Roman Catholic trumpets, "The world is too vast and wonderful and God's power is so limitless, that there must be more than little old us." One mainline Protestant explicitly rejects anthropocentrism. "God is God of all creation and all that is within it. The only way this should be a religious problem is if the true (though unstated) center of our worship is humankind." A Muslim similarly chastises anthropocentrism: "Only arrogance and pride would make one think that Allah made this vast universe only for us to observe."

This religious opposition to anthropocentrism is startling, when compared to the retrieval of anthropocentrism in today's scientific community. Of the many universes possible, says physicist Stephen Hawking, "only a very few would allow creatures like us to exist ... Although we are puny and insignificant on the scale of the cosmos, this makes us in a sense the lords of creation."[16] Why might a professional scientist defend anthropocentrism when an anonymous religious believer denies it? Perhaps some scientists are more likely to face a crisis of belief than the average religious believer.

Turning from the question of geocentrism and anthropocentrism, the matter of soteriology (work of Christ in salvation) came up in the survey. One evangelical Protestant thinks out loud, so to speak, about Christology and Soteriology. "From an evangelical Christian perspective, the Word of God was written for us on Earth to reveal the creator. We were created to bring glory to God. Why would we repudiate the idea that God may have created other civilizations to bring him glory in the same way? Christ as our Savior may be the method he chose to redeem us on Earth, but he could have used similar methods in other galaxies if he desired." A Roman Catholic follows the same thought experiment. "I believe that Christ became incarnate (human) in order to redeem humanity and atone for the original sin of Adam and Eve. Could there be a world of so called "extraterrestrials? Maybe. It doesn't change what Christ did." The general consensus among Christians of differing stripes is opposition to geocentrism or Earth chauvinism. Lutheran New Testament scholar and former Bishop of Stockholm, Krister Stendahl, attended a NASA symposium in 1972 and was asked his opinion on communicating with extraterrestrial intelligent life. "That's great," he said; "it seems always great to me, when God's world gets a little bigger and I get a somewhat more true view of my place and my smallness in that universe."[17]

In parallel fashion, a Buddhist commenting in the survey thinks out loud about the path to enlightenment. "As a Buddhist, it is clear that ALL

---

16. "... we can look forward to an age in which the understanding of life's mechanisms will be virtually total; at least, the principal systems will be understood molecule by molecule. From this total understanding will come—if we choose—total control ... It is not irresponsible—'sensationalist'—to suggest this. It is irresponsible to imply the opposite—that our power will always be too limited to worry about. We are entering the age of biological control, and we should gird our moral and political loins accordingly." Hawking and Mlodinow, *The Grand Design*, 9.

17. Stendahl, in *Life Beyond Earth and the Mind of Man*, 29.

sentient beings are subject to birth, old, and death and are, therefore, impermanent, subject to various forms of suffering and have no separate self. ET's would be, essentially, no different from other sentient beings, i.e., they would have Buddha Nature and would also be subject to karmic consequences of their actions. We might or might not be able to learn from them."[18]

Finally, from an evangelical Protestant we read: "I don't think they are out there. But if they are, that's cool."

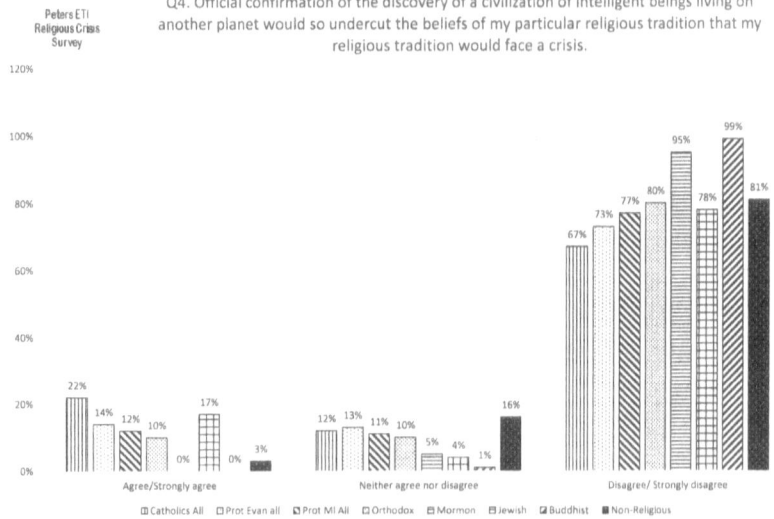

Figure 11.2: From the Peters ETI Religious Crisis Survey
(Ted Peters and Julie Froehlig, 2008)

When we turn away from one's own personal beliefs and ask about the beliefs of the respondent's religious tradition (Figure 2, Question 4), we notice a slight shift. The Disagree/Disagree strongly pillars still rise high, to be sure; yet not quite as high. Might this indicate that for some religious individuals who welcome confirmation of ETI a worry about their own religious tradition is evoked?

Some of the voluntary comments are illuminative. One Roman Catholic is not worried about his or her own faith; but the beliefs of the Catholic Church are in jeopardy because of its alleged anthropocentrism.

> The foundations of my religion (Catholic) and many others may be shaken by such a discovery because most human religions

---

18. One Buddhist writer who affirms the possibility of alien life also denies Christian anthropocentrism on Earth. "The belief that man was created specifically in the image of God has no significance in Buddhism." Kirthsinghe, "Buddhism, Biology, and Exobiology."

view human beings as special or privileged beings on the earth and in the cosmos. In Christian traditions humans seem to hold special favor with God. However, I see no reason God could not (or has not) create(d) other beings. It does not imply a lessening of God's love for us. In fact, come to think of it, it would be good for us to discover other beings—especially ones equally or more intelligent than ourselves—because it might knock down human arrogance towards other species right here on earth.

We find the same denominational anxiety in an evangelical Protestant. "I think the religious tradition with which I identify (Protestant Evangelical) is not prepared for the day we do make contact, but we need to start thinking this out and become prepared."

One theologian, Cynthia S. W. Crysdale, mentioned in an earlier chapter, similarly worries that Jews and Christians might not be prepared. "The Genesis accounts of creation presume that the Earth is central to God's purposes, most notably that of establishing a special relationship with mankind. The discovery of extraterrestrial life in any form would seem to contradict this core assumption of the Judeo-Christian tradition."[19] It would "seem" to conflict, she says. Although Crysdale herself would like to construct a theological vision that makes room for ETI, she is worried about the faith of the others who sit in her pew or who stand in her chancel.

A few Roman Catholic laypersons in the survey seem to be suspicious of their church's leadership, worrying that the hierarchy might find contact with ETI difficult to accept. This suspicion is curious, because the Holy See through the Jesuits sponsors the Vatican Observatory which, among other tasks, actively searches for extraterrestrial life.[20] A half century ago already, Francis J. Connell, C.S.S.R., dean of the School of Sacred Theology at the Catholic University of America during the 1940s and 1950s, spoke rather definitively: "it is good for Catholics to know that the principles of their faith are entirely compatible with the most startling possibilities concerning life on other planets."[21] In a homily on October 27, 2014, Pope Francis alluded to extraterrestrial life as having evolved in a manner consistent with God's universal plan of creation.[22] Of all institutions, the theological leadership of the Roman Catholic Church is the most likely to put out the welcome mat for alien visitors.

19. Crysdale, "God, Evolution, and Astrobiology," 220.
20. Ariel, "Vatican."
21. Connell, "Flying Saucers," 258.
22. Salla, "Pope Endorses Evolution of Alien Life & UFO Activity as Part of God's Plan," Exopolitics.org, October 28, 2014, https://exopolitics.org/pope-endorses-evolution-of-alien-life-ufo-activity-as-part-of-gods-plan/.

What might Muslims think? Even though our study provided a sample too small to be statistically significant, Islamic spokespersons have registered their judgments. "There are creatures not only in this world, but also in the space and the worlds beyond space. According to Islamic beliefs, there are more than eighteen thousand 'aliens'. There are genies, angels, and other creatures other than humans. UFOs [aliens] are in this category." These are the words of Mehmet Nun Yilmaz, head of the Department of Religious Affairs, Turkey.[23] This affirmation of other beings includes extraterrestrials, to be sure; but its reference to angels show that it entails much more. Important here is the observation that no evidence suggests a fragility to Islam if confirmation of ETI is established.

Now, we turn away from one's own personal beliefs and the beliefs of one's own tradition; we ask about forecasts for religions other than one's own. The *Peters ETI Religious Crisis Survey* asked respondents to forecast what will happen to the world's religions, those holding beliefs other than one's own. Here, something jolting is revealed. See Figure 3 (Question 5). Note how those self-identifying as non-religious are the ones who forecast a crisis in the world's religions. To say it in the first person: "my non-religious beliefs will not suffer a crisis, but other religious believers will have a problem."

Figure 11.3: From the Peters ETI Religious Crisis Survey
(Ted Peters and Julie Froehlig)

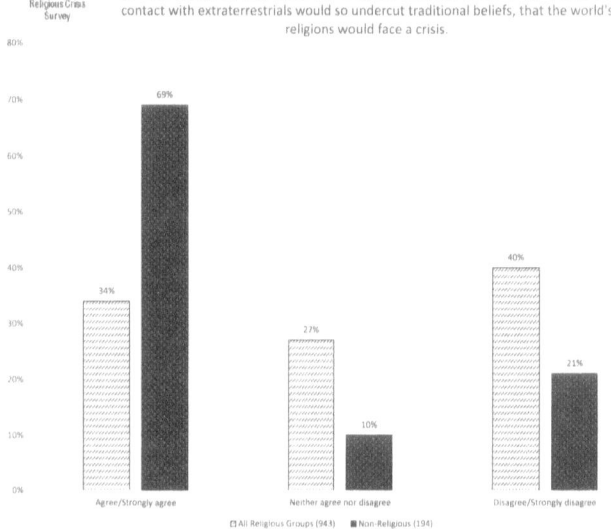

23. Yilmaz, cited in Sckerkarar, "History of UFO Sightings in Turkey," 9.

How should we interpret this graph in Figure 3? First, we can distinguish between one's own religious beliefs from the beliefs of others who differ. One Buddhist respondent expects easy acceptance of ETI by Asian religions but difficulty for the Abrahamic traditions. "Lumping together all the world's religions is a conceptual error. The religions of the book (the Abrahamic traditions) would have a very different set of reactions than the Asian traditions." Surprisingly, a Roman Catholic agrees. "I think Buddhism and Hinduism would be better equipped to face any encounter with other civilizations. The three monotheistic traditions, however, would enter into a serious crisis."

Second, and more dramatically, non-religious respondents are more likely to prophesy a crisis impending for those who are religious. The Agree/Strongly agree towers indicate this. According to the respondent comments, those who self-identify as non-religious or atheist tend to welcome confirmation of ETI on the grounds that they are self-described as open-minded. Presumably, religious people are not open-minded. "I hope we do find life beyond our planet. I have an open mind as to what shape it may take and what influence it will have." An atheist comments, "Discovery of ET would not affect my personal belief system because I am a stone atheist. I do search for ET and think we will come across some one day." Another hopes that ETI will mark a victory over religious anachronism. Contact with ETI is "bound to happen sooner or later, whether they're more advanced than us is uncertain, but it will certainly get rid of the notion that we're God's special little creatures made in his image . . . man I can't wait :)."

What might all this signify? It might signify that the prophecies of crisis are primarily the construction of non-religious persons who are making predictions about what will happen to religious persons. A decade prior to this survey Steven Dick made this observation: "In general, for Christians as well as for other religions, indigenous theologians see little problem, while those external to religion proclaim the fatal impact of extraterrestrials on Earth-bound theologies."[24] At the announcement of finding a Mars rock (ALH84001) with possible signs of life, Martin Marty wrote confidently, whether we find microbial life locally or intelligent life elsewhere, "theologians will adapt."[25] The *Peters ETI Religious Crisis Survey* tends to disconfirm the prophets of crisis and confirm the judgment of Dick and Marty.

After reviewing the data from the *Peters ETI Religious Crisis Survey* along with the earlier survey conducted by Victoria Alexander, Jeff Levin offers a surmise. "Perhaps the presumption of an inevitable religious crisis

24. Dick, *LOW*, 247.
25. Marty, "Life on Mars?," 23.

may reflect an irreligiousness or even hostility or condescension toward religion, on average, among physical scientists, engineers, military and intelligence officials, politicians, and federal bureaucrats, and influential media figures—in other words, the opinion leaders on this subject, by their own presumption . . . Prominent religious institutions would not collapse nor would major religious belief systems implode."[26] Perhaps condescension toward religious people leads to the negative prophecies.

If one's attitude toward religion is condescending, then perhaps the articulated beliefs of theologians become invisible. David Wilkinson, a hybrid physicist and theologian at Durham in the UK, frankly asks: "would the discovery of life elsewhere in the Universe so contradict the central beliefs of Christianity that it would bring it crashing back to the grave?" He answers: "for the vast majority of the Christian church, the existence of extraterrestrial intelligence is not a big deal."[27] Or, in the words of Notre Dame's Michael J. Crowe, "It is sometimes suggested that the discovery of extraterrestrial life would cause great consternation in religious denominations. The reality is that some denominations would view such a discovery not as a disruption of their beliefs, but rather as a confirmation."[28]

In summary, such survey evidence requires us to acknowledge that religious believers themselves do not fear that contact with ETI will undercut their beliefs or precipitate a religious crisis. It suggests that what passes for prophecies of crisis may be the product of what non-religious people say about religious people.

The *Peters ETI Religious Crisis Survey* deals with the general population. When we turn to our intellectual leaders—to our scientists and our theologians—the division remains constant. Constance Bertka has it right. "The contemporary dialogue amounts to scientists outside of the faith tradition predicting a famine for Christianity if extraterrestrial intelligent life is discovered, whereas theologians within the tradition are suggesting the possibility of a relative feast of ideas."[29]

Davies' Critique of the Peters ETI Religious Crisis Survey

Despite this evidence, Davies remains unpersuaded by my arguments. He finds it remarkable that so many respondents to the *Peters ETI Religious*

---

26. Levin, "Revisiting the Alexander UFO Religious Crisis Survey (AUFORCS)," 282.
27. Wilkinson, *Alone in the Universe?*, 116.
28. Crowe, *ELD*, 328–29.
29. Bertka, "Christianity's Response," 333.

Crisis Survey see no problem, no challenge to their respective religious beliefs. As we have noted, the survey results suggest that confirmation of extraterrestrial intelligence poses little or no challenge to previously developed religious understandings. Ancient worldviews will not be shattered if we discover that we share our universe with aliens. This appears remarkable to Davies because he is firmly committed to the belief that traditional religions will be undone by such knowledge. How can he handle evidence to the contrary?

Somehow, it seems, Davies feels he must explain away the evidence. How might he go about a refutation? How about an *ad hominem* description of the respondents? "Most of the comments had an air of sweeping the problem under the carpet. Very few of the Christian respondents tackled the theological minefield of the uniqueness of the incarnation and the species-specific nature of salvation. A handful did identify the conundrum, but no novel solutions were proffered."[30] The survey must be mistaken, says Davies, because the respondents failed to assess the severity of the theological problem they will face.

What concerns Davies theologically is the problem posed within Christian theology by the incarnation. Christians have long asserted that God became incarnate in Jesus of Nazareth, and this constitutes the Christ event in terrestrial history. The picture Davies draws of Christian theology is that Christian faith assumes a revelatory soteriology, that is, a species-specific revelation. And, he contends, this implies it would be necessary for Christians to affirm a new and unique incarnation on each planet for each species. Yet, a planet-hopping Christ appears absurd to Davies. What this may require us to ask is this: what do Christian theologians actually say about this matter? Does Davies draw an accurate picture? We will take this up in later chapters.

Geocentrism versus Anthropocentrism

As we can see from the *Peters ETI Religious Crisis Survey*, very few persons with religious commitment are likely to defend either geocentrism or anthropocentrism let alone both. Survey respondents tend to be impressed with the enormous scope of creation; and they find no grounds for promoting the status of humankind over against either God or ETI.

Our grip on what is at stake is strengthened when we remind ourselves of two related principles, the Cosmological Principle and the Copernican Principle. Most simply put, the *Cosmological Principle* is a scientific

---

30. Davies, *Eerie Silence*, 190.

principle which states that we can expect the laws of nature we discover here on Earth to apply equally everywhere in the cosmos. The *Copernican Principle* adds a cultural overlay, contending that neither planet Earth nor its human inhabitants should be privileged. Two Vatican Observatory astronomers tip toward the Copernican Principle to say: "there should be no location in space or time that is special or privileged in any way."[31] Because of the omnipresence of the laws of nature, Earth becomes de-centered, so it seems.

The manner in which Columbia University astrobiologist Caleb Scharf puts it deserves some attention. "The Copernican principle . . . states that the Sun, not the Earth, occupies the center of the heavens, and that a spinning Earth, as well as the other planets, circles around this fiery orb. It is a worldview that asserts that we are not the center of all existence; we are not special. In fact, we are as dull as they come."[32] What is at stake is more than merely the question of what is in the center. It is becoming "dull" that is bothersome. Note how Scharf glues together the physical features of our solar system with our belief system, as if the former determines the latter. Because Scharf worries that pre-Copernican religion experienced a big loss of meaning with the introduction of heliocentrism, he now tries to construct a new meaning to replace what was allegedly lost. What religion allegedly lost he will now replace with the help of science.

There is some historical evidence that the changing worldview in the sixteenth and seventeenth centuries partially disrupted traditional religious sensibilities. Poet John Donne (1572–1631), for example, reflects the disruption in his poem, "An Anatomy of the World." With the term "new philosophy" Donne refers to Copernican heliocentrism.

> And new philosophy calls all in doubt,
> The element of fire is quite put out,
> The sun is lost, and th'earth, and no man's wit
> Can well direct him where to look for it.
> And freely men confess that this world's spent,
> When in the planets and the firmament
> They seek so many new; they see that this

---

31. Consolmagno and Mueller, *Would You Baptize an Extraterrestrial?*, 275.

32. Scharf, *Copernicus Complex*, 8. Some, such as quantum physicist Ben Miller, try to retrieve the "special" status of Earth through their science. "Our universe is special because it's the one where the very fabric of existence is just right for life." Miller, *The Aliens are Coming*, 128. Rather than rely on geocentrism, Miller appeals to the Anthropic Principle to find his specialness. The problem of feeling "special" seems to be a problem for scientists, not for religious believers.

EXTRATERRESTRIAL LIFE AND TERRESTRIAL RELIGION 199

> Is crumbled out again to his atomies.
> 'Tis all in pieces, all coherence gone,
> All just supply, and all relation;[33]

Note how the "new philosophy" elicits anxiety because "all coherence gone." This anxiety is not due to a defense of geocentrism or anthropocentrism. Rather, it's due to the challenge to adapt to a new way of thinking after centuries of habit. Those who immediately incorporated the heliocentrism of Copernicus, such as the British Jesuits as early as 1577, saw as their first challenge one of locating heaven. No longer could we think of heaven as simply up in the heavens. This challenge to locate heaven did not in itself dissuade them from accepting heliocentrism, however. Recall that in an earlier chapter we demonstrated that the Copernican Principle is a myth, a false story told by astronomers to allege that medieval Christians were forced against their will by science to give up geocentrism. In sum, we must acknowledge historically that a cultural disruption was widely felt in Christendom for a period while Europe waited approximately a century for complete acceptance of the altered worldview, but no religious crisis of any scope occurred. Now, back to the treatment by Scharf.

Scharf assumes that the pre-Copernican worldview is totally lost and, along with it, the meaningfulness of life for religious people. Because of this alleged lacuna, he concludes that we need a new structure of meaning that undergirds human significance. Like a knight in shining armor, Scharf comes to the rescue! Science can provide what pre-Copernican religion lost, namely, human significance.

This time, according to Scharf, human significance will be based on science, not religion. How will he do this? Scharf begins by asking the existential question: why are we earthlings here and what is our significance? His only answer is a Promethean one: conquer space like Genghis Kahn conquered whatever real estate he could see. "Our reach deliberately extends far beyond the confines of this planet ... Our true cosmic importance may ultimately spring from this same surge for expansion, a vital signature that natural selection has written in our human genes. It is who we are. It is part of what makes us special. And it is how we can, if we choose, *make* ourselves significant."[34] In short, all by itself the universe has no meaning; and we on our dull planet cannot claim to be significant. So, if we want meaning we should make ourselves significant by expanding beyond our

---

33. Donne, "Anatomy of the World."
34. Scharf, *Copernicus Complex*, 231–32.

otherwise "dull" planet to fill the universe with our presence. By doing so, we become significant.

Can we take seriously what is being said here? No, for two reasons. First, as just mentioned, the Copernican Principle is, at least historically speaking, a myth that needs busting. It needs busting because relatively little anxiety over demoting Earth from its alleged central position arose, despite Donne's poem. Second, Scharf's attempt to establish a secular or humanized geocentrism to launch cosmic domination sounds imperialistic. Within an otherwise meaningless universe, Scharf's proposal is that we terrestrials create our own significance by expanding to what lies beyond Earth. Scharf asks us to surpass geocentrism by substituting geo-cosmic-domination. He asks us to embrace terrestrial manifest destiny—written by evolution into our genes—by manufacturing our own cosmic expansion.

This amounts to a glaring self-contradiction. Scharf reaffirms anthropocentrism by appealing to the establishment of our conquering human nature by our genetic destiny; and he reaffirms geocentrism by appealing to our planet Earth as the launching point for cosmic significance. The rest of the universe had better put up its guard: here come the earthlings!

This is a dangerous doctrine. Would any reasonable person see this as a step forward in the history of humility? Gone is the Copernican Principle, the principle of mediocrity, or any justification for human humility. In the name of science the Scharf doctrine proffers an ideology driven by self-divinization, manifest destiny, and domination. What we need here is a moral counter, an appeal to religious or moral reverence for humility to contain such a Promethean will to power.

The impact of outer space in the human soul should evoke a sense of humility. It is so frequently overlooked outside the theological community that Christians have always affirmed the immensity and transcendence of a loving God over against the humble status of God's beloved creation. Nothing that is a creature competes with the Creator for centrality. Christians along with many other traditions of faith are theocentric, not Earth-centric or human-centric. Historian of ideas Charles Taylor reminds us that "[T]he general understanding of the human predicament before modernity placed us in an order where we were not at the top. Higher beings, like Gods or spirits, or a higher kind of being, like the Ideas or the cosmopolis of Gods and humans, demanded and deserved our worship, reverence, devotion or love."[35]

Recall that the revolutionary work by Nicholas Copernicus, *De revolutionibus orbium coelestium* (On the Revolutions of the Celestial Spheres),

---

35. Taylor, *A Secular Age*, 18.

was published by the Lutherans in 1543. Did Lutheranism face a crisis over geocentrism or anthropocentrism? No, of course not. Robert Preus, a historian of Lutheran doctrine, makes this observation: "The notion of a Copernican revolution, according to which man somehow lost his importance because earth was no longer thought to be at the center of the universe . . . the Lutheran doctrine of God's immensity and transcendence over time and space would tend to make them quite unconcerned with this aspect of Copernicanism."[36]

In conclusion, we might say this: to attribute to pre-Copernican religious believers a naive geocentrism combined with anthropocentrism is not only historically misleading, it is unnecessarily patronizing. In addition, an attempt by post-Copernican scientists to construct human meaning by proposing expansion from Earth into space is more than simply self-contradictory; it proposes at the interplanetary level the same manifest destiny that has plagued geopolitics with international violence for millennia.

Neo-Geo-Centrism

It is reasonable to say that neither geocentrism nor anthropocentrism have held the status of cardinal doctrine in Christianity let alone in other religious traditions. Nor, when these two centrisms appear together, are they necessarily viewed as a single package. One can divide them. Ecotheologians divide them.

Many of today's *eco*-theologians are trying to affirm geocentrism over against anthropocentrism, presuming that cultural or religious anthropocentrism is responsible for environmental degradation and climate change. It should be easy to shift our allegiances from anthropocentrism to geocentrism, these theologians argue, because we have not traditionally affirmed anthropocentrism anyway. Because we do not have an investment in anthropocentrism, it should be easy to give it up on behalf of a holistic ethic that includes the entire biosphere on planet Earth. The health of our planet should provide our ethical goal, not the human exploitation or plundering of the planet. So the argument goes.

Eco-feminist Rosemary Radford Ruether provides an example of contemporary neo-geo-centrism. "God is not an old man outside the earth living in the sky, but rather a creative energy that is in and through the whole earth. This creative energy isn't a human being, male or female—rather it is within and underlying all beings (animals and plants), earth, air, and

---

36. Preus, *Theology of Post-Reformation Lutheranism*, 228.

water."[37] In developing her planetary naturalism here, Ruether is attempting to overcome anthropocentrism by placing the wellbeing of Planet Earth in its place. What is rendered invisible by Reuther, of course, is the dependence of Earth on its sun; and the sun's relationship, in turn, to the Milky Way and eventually all the stars.

Theologian and eco-ethicist John Cobb labels the neo-geo-centrism, *Earthism*. The advantage of Earthism is that it is inclusive; it values the health and wellbeing of the entire biosphere more than the so-called free market. Does this require that we choose between Planet Earth and the cosmic commons when considering the common good? On the one hand, Cobb says, yes. He conditionally supports *Earthism* as an inclusive spirituality to replace economic competition, and this excludes attention to off-Earth locations. "Even for Christians the whole of creation, meaning the universe with all its galaxies, is not a practical or fruitful object of devotion." Rather, for the foreseeable future our "penultimate devotion is the Earth, inclusive of all its inhabitants and especially the human ones. Earthism directs our energies most appropriately."[38] On the other hand, Cobb also says, no, we need not choose between Earth and space. Contact with extraterrestrial creatures might alter our neo-geo-centrism. "Should Earthlings ever encounter creatures from other parts of the universe, the idolatrous character of an unqualified Earthism would immediately become manifest."[39] In short, for practical reasons a qualified devotion to Earth rather than the market should guide the present generation, but we should prepare mentally for a future expansion of our concern to include outer space.

From the perspective of the astrotheologian, planet Earth is too small and too mediocre to place in the center of our worldview. Oh yes, it is important for the ecotheologian to promote neo-geo-centrism or Earthism for moral reasons. For the ecotheologian, geocentrism is of higher moral value than anthropocentrism. However, for the astrotheologian the cosmos is of still higher moral value than either anthropocentrism or geocentrism.

Apart from eco-ethicists, some neo-geo-centrists are actually trying to revive pre-Copernicanism. Robert Sungenis argues that Copernicus, Kepler, and Galileo were wrong and that the seventeenth-century Roman Catholic Church was right. The Earth does not move. Everything else moves around us. The Cosmological Principle is wrong; there is one privileged location in the cosmos, and that's planet Earth. Those who teach Big Bang cosmology

---

37. Ruether, "Empty Throne," 28.
38. Cobb, *Earthist Challenge*, 8–9.
39. Ibid., 179.

in the public school system belong to a giant conspiracy to suppress this truth.

Most scientists and theologians are tempted to simply turn a deaf ear to noises made by such atavists. Yet, skeptic Matthew Wiesner recommends, "Let's not ignore pseudoscience such as modern geocentrism; rather, let's discuss it and use it to teach and educate people about how to tell the difference between real science and junk science."[40]

My observation here is that the astrotheologian wants to transcend geocentrism and to affirm the entire cosmos as constituting the scope of divine creation. The atavistic neo-geo-centrist might object. The ecotheologian is not likely to object; but as of this date the cosmic dimension is simply not on the ecotheologian's radar screen.

Beyond the Crisis: The Astrotheologian and ETI

Georgetown University theologian John Haught maps three directions theologians might go when asked to speculate about the significance of extraterrestrial intelligent life. Firstly, if the theologian believes there is a *conflict* between science and religion, then we could expect the following: "the discovery of extraterrestrial intelligence (ETI) . . . would mark the end of the Abrahamic faith traditions. The narrowly provincial God of Judaism, Christianity, and Islam would seem so small as to be unworthy of worship. Secondly, if the theologian believes in a *contrast* between science and religion, then it "would not significantly alter our understanding of God's creativity, love, and providence if we find that ETI exists. To be frank, if SETI meets with success it will hardly make any difference to faith and theology that are already tuned in to the infinite creative extravagance of God." Thirdly, if the theologian is hoping for a *convergence* between science and religion, then "the discovery of an extraterrestrial world of living and intelligent beings elsewhere in our universe would . . . be a most interesting stimulus to theology."[41]

Haught himself follows the convergence path and, like others in the Bible-Welcomes-Aliens school of thought, he takes up the agenda which the ETI question stimulates. "The discovery of other intelligent worlds would be a powerful new incentive to radicalize monotheistic faith and confirm the fundamental unity of the cosmos . . . Theology's relevance to SETI lies most fundamentally in its conviction that all possible worlds have a common

40. Wiesner, "Modern Geocentrism," 53. See Sungenis, *Geocentrism 101*.
41. Haught, *Science and Faith*, 163–64.

origin in the one God. And by virtue of the omnipresence of this one God, we too would have an extended home in all possible worlds to which we might eventually travel."[42]

Despite the near unfathomable size of our cosmos, everything is connected. According to Big Bang theory, at the beginning all things were only one thing, a singularity. Only after the Big Bang banged did matter differentiate, the four forces become distinct, and the elements become established. Regardless of how distant an extraterrestrial civilization may be today, we share a common physical ancestry. We were birthed by the same stardust. Even though we cannot expect our space cousins to look exactly like us, we still belong to one family.

And we belong to one God. The one God of Israel worshipped by Jews, Christians, and Muslims, may seem like a little godlet when we read Old Testament stories. Still, the God of Israel was holy. That God was the creator of heaven and earth. Perhaps the size of heaven and earth in the minds of people living in the biblical world needs expansion in light of modern knowledge of the heavens and the many earths. But the concept of the divine creator remains consistent. Size does not matter. What else would we expect?

Recall one of our agenda items on the astrotheologian's To-Do list: *Christian theologians along with intellectual leaders in each religious tradition need to reflect on the scope of creation and settle the pesky issue of geocentrism.* It appears from the *Peters ETI Religious Crisis Survey* that pre-Copernican geocentrism is simply not a debilitating problem for religious people, regardless of their religious tradition. Those who accuse religious people of geocentrism tend to be the non-religious, not the religious. What we mean by the *pesky issue of geocentrism* might be the need to correct a misperception on the part of self-identified non-religious people.

The need to reflect on the scope of creation is a genuine need. Tacitly, our theologians have ignored outer space. Theologians have been geocentric, not because of Earth chauvinism but because geocentrism is a goal that takes us beyond parochialisms such as ethnocentrism, racism, sexism, and nationalism. Our religious leaders have prodded and goaded us to think beyond our ethnocentric or racial or national identities—beyond the we-versus-they tribal split—and to think in terms of a single planetary society. Theologians have striven to expand our consciousness to realize that the one God is the God of the entire planet and not merely of our own tribe or nation or religion. This expansion of scope has borne a sense of ethical urgency, because cultural myopia needs to be replaced with cultural pluralism

42. Ibid., 166.

and because ecological concerns require an international if not global mindset. Carrying out this task is intellectually exhausting for theologians.

Still, we ask more. The planet Earth could not be what it is without the sun's bathing it daily in energy and light. And the sun finds its place within the swirling flow of the Milky Way. And the Milky Way dances to the cosmic tunes of galaxies in motion. The scope of creation includes Earth, to be sure; but also much more. It is the task of the astrotheologian to expand the scope of our terrestrial thinking to include all that is terrestrial and, in a new way, to come to love the stars as God loves them.

Conclusion

Neil deGrasse Tyson presents the conventional challenge: "the discovery of extraterrestrial intelligence, if and when it happens, will impart a change in human self-perception that may be impossible to anticipate."[43] Impossible? Really?

We now know what people think will be their response to confirmation of the existence of ETI. If this future remains unpredictable, it would because those responding to the survey either told a lie or because they fall short in their self-understanding. It is possible that, despite what they have said, they will behave differently. However, to press such a scenario, one would have to demonstrate that people behave differently than what they predict of themselves. The burden of proof would be on the shoulders of the one who insists that we don't know when the evidence seems to suggest that we, in fact, do know.

Given what evidence exists, therefore, we can forecast a relatively favorable disposition on the part of Earth's religious believers toward aliens. As soon as confirmation of ETI is announced, we can forecast that church basements will be readied for a covered dish dinner to welcome aliens into our space neighborhood.

43. Tyson, *Space Chronicles,* 41. David A. Weintraub examines fifteen religious traditions and concludes that, with the exceptions of Orthodox Christianity and fundamentalism, "most religions appear theologically open to the possibility that intelligent extraterrestrial beings exist. Most religious belief systems are robust enough to accommodate the paradigm-busting news that the discovery of extraterrestrial life would represent." Weintraub, *Religions and Extraterrestrial Life,* 208.

Bibliography

Alexander, Victoria. "Extraterrestrial Life and Religion." In *UFO Religions*, edited by James R. Lewis, 359–70. Amherst NY: Prometheus, 2003.
Ariel, David. "Vatican: It's OK to Believe in Aliens." Associated Press, May 13, 2008. http://news.yahoo.com/s/ap/20080513/ap_on_re_eu/vatican_aliens.
Bainbridge, William Sims. "Cultural Beliefs about Extraterrestrials: A Questionnaire Study." In *Civilizations Beyond Earth: Extraterrestrial Life and Society*, edited by Douglas A. Vakoch and Albert A. Harrison. New York: Bergbahn, 2011.
Bertka, Constance. "Christianity's Response to the Discovery of Extraterrestrial Intelligent Life: Insights from Science and Religion and the Sociology of Religion." In *AHS*, 329–40.
Cobb, John. *The Earthist Challenge to Economism: A Theological Critique of the World Bank*. New York: St. Martin's, 1999.
Connell, Francis J. "Flying Saucers and Theology." In *The Truth about Flying Saucers*, edited by Aime Michel, 255–58. New York: Pyramid, 1967.
Consolmagno, Guy, SJ, and Paul Mueller, SJ. *Would You Baptize an Extraterrestrial?* New York: Image, 2014.
Crowe, Michael J. *Extraterrestrial Life Debate from Antiquity to 1915: A Source Book*. Notre Dame: University of Notre Dame Press, 2008.
Crysdale, Cynthia S. W. "God, Evolution, and Astrobiology." In *Exp*, 220–42.
Davies, Paul C. W. *The Eerie Silence: Renewing Our Search for Alien Intelligence*. Boston: Houghton Mifflin, 2010.
———. *God and the New Physics*. New York: Simon & Schuster, 1983.
Dick, Steven J. *Life on Other Worlds: The 20th Century Extraterrestrial Life Debate*. Cambridge: Cambridge University Press, 1998.
Donne, John. "An Anatomy of the World." *Poetry Foundation*. http://www.poetryfoundation.org/poem/173348.
Glenn, John. Interviewed by Paul Levinson. "Interview with John Glenn." In *TFC*, 8–11.
Haught, John F. *Science and Faith: A New Introduction*. New York: Paulist, 2012.
Hawking, Stephen, and Leonard Mlodinow. *The Grand Design*. New York: Bantam, 2010.
Kirthsinghe, Buddhadasa P. "Buddhism, Biology, and Exobiology." Chapter 10 in *Buddhism and Science*. http://books.google.com/books?id=fY9_X107du8C&pg=PA72&dq=Buddhism+and+exobiology&hl=en&sa=X&ei=IL4GUYnlIOi8igKp54DwBw&ved=0CCoQ6AEwAA#v=onepage&q=Buddhism%20and%20exobiology&f=false.
Levin, Jeff. "Revisiting the Alexander UFO Religious Crisis Survey (AUFORCS): Is There Really a Crisis?" *Journal of Scientific Exploration* 26.2 (2012) 273–84.
Marty, Martin E. "Life on Mars?" *The Lutheran* 9.10 (October 1996) 23.
McAdamis, E.M. "Astrosociology and the Capacity of Major World Religions to Contextualize the Possibility of Life Beyond Earth." *Physics Procedia* 20 (2011) 338–52. http://www.sciencedirect.com/science/article/pii/S1875389211006006.
Miller, Ben. *The Aliens are Coming*. New York: The Experiment, 2016.
Peters, Ted. "Astrotheology and the ETI Myth." *Theology and Science* 7.1 (February 2009) 3–30.
———. "Would the Discovery of ETI Provoke a Religious Crisis?" In *AHS*, 341–55.

Peters, Ted, and Julie Louise Froehlig. *Peters ETI Religious Crisis Survey*. Counterbalance. http://www.counterbalance.org/etsurv/index-frame.html/.
Pettinico, George. "American Attitudes about Life Beyond Earth: Beliefs, Concerns, and the Role of Education and Religion in Shaping Public Perceptions." In *Civilizations beyond Earth: Extraterrestrial Life and Society*, edited by Douglas A. Vakoch and Albert A. Harrison, 102–17. New York: Berghbahn, 2011.
Preus, Robert D. *The Theology of Post-Reformation Lutheranism*. 2 vols. St. Louis: Concordia Publishing, 1970–1972.
Ruether, Rosemary Radford. "The Empty Throne: Reimagining God as Creative Energy." *Tikkun* 29.3 (2014) 28, 65.
Salla, Michael. "Pope Endorses Evolution of Alien Life & UFO Activity as Part of God's Plan." Exopolitics.org. October 28, 2014. https://exopolitics.org/pope-endorses-evolution-of-alien-life-ufo-activity-as-part-of-gods-plan/.
Scharf, Caleb. *The Copernicus Complex: Our Cosmic Significance in a Universe of Planets and Probabilities*. New York: Scientific American / Farrar, Straus & Giroux, 2014.
Sckerkarar, Esen. "History of UFO Sightings in Turkey." *MUFON UFO Journal* 443 (March 2005) 9.
Shindell, Matthew. "The Public Response: Reviewing Public Policy Survey Data on Science, Religion, and Astrobiology." In *Workshop Report: Philosophical, Ethical, and Theological Implications of Astrobiology*, edited by Connie Bertka et al. Washington, DC: AAAS, 2007.
Stendahl, Krister. In *Life Beyond Earth and the Mind of Man*, edited by Richard Berendzen. Washington, DC: NASA Scientific and Technical Information Office, 1973.
Sungenis, Robert. *Geocentrism 101*. Catholic Apologetics International, 2014.
Tarter, Jill Cornell. "SETI and the Religions of the Universe." In *MW*, 43–50.
Taylor, Charles. *A Secular Age*. Cambridge: Harvard University Press, 2007.
Troxel, Duane. "Intelligent Life in the Universe and Exotheology in Christianity and Bahai Writings." Bahai Library Online. http://bahai-library.com/troxel_extraterrestrials_exotheology.
Tyson, Neil deGrasse. *Space Chronicles: Facing the Ultimate Frontier*. New York: Norton, 2012.
Vakoch, Douglas A., and Yuh-Shiow Lee. "Reactions to Receipt of a Message from Extraterrestrial Intelligence: A Cross-Cultural Empirical Study." *Acta Astronautica* 46 (2000) 737–44.
Weintraub, David A. *Religions and Extraterrestrial Life: How Will We Deal With It?* Heidelberg: Springer, 2014.
Wiesner, Matthew P. "Modern Geocentrism: A Case Study of Pseudoscience in Astronomy." *Skeptical Inquirer* 39.1 (2015) 50–53.
Wilkinson, David. *Alone in the Universe?* Crowborough, UK: Monarch, 1997.

12

Jewish Theology Meets the Alien

Norbert M. Samuelson

> Hear, O Israel: The Lord our God, the Lord is one.
> —Deuteronomy 6:4

When Judith Resnik was named as an American astronaut, some Israeli rabbis asked what she would do about Shabbat in space. She herself, did not care. The question was purely academic. My understanding, which may not be correct, is that she was a secular Jew who was completely non-observant.

The answer that I recall from the time in the Philadelphia Jewish press was that of the then chief rabbi of the Israeli army, Shlomo Goren. He said that in space the time of the Sabbath is measured from Jerusalem, so that when it is Shabbat in Jerusalem it is Shabbat in space, and when it is not Shabbat in Jerusalem it is not Shabbat in space. I don't know if his decision either then or now was considered binding on halachic Jews (i.e., Jews who to some extent, accept the authority of judgments by traditional rabbis, both in the past and in the present), but the question is important.

When Judith Resnik went to space, she left familiar Earth for an alien environment. That word, *alien,* garners multiple meanings. In what follows, this term will be applied to space, to an extraterrestrial entity, to a Jew, and even to God. As God positively creates everything, everything in the cosmos negatively defines God. The creature negatively defines the creator. This means that the God of Israel, who is the God of the cosmos, is the most alien, i.e., the paradigm alien of all aliens.

In Heaven as on Earth?

As there are some differences in Jewish law in and outside of Jerusalem, it is reasonable to assume that Jewish law outside of the Earth's sphere need not coincide with Jewish law on Earth. As, for example, traditionalist Jewish communities add an extra day to the festivals outside of Jerusalem but they do not do so within the walled city of Jerusalem, there may be parallel questions about outer space. An obvious question is, how long is a day on another planet? For example, is a day on Jupiter the same as a day on Earth, more or less than twenty-four hours, depending on the time of year, or is it slightly more than ten hours given that planet's average periodic rotation around the sun? Most significantly for liturgical reasons, when is it Shabbat on Jupiter and how long does it last? (The problem is the same for Rosh HaShana, Yom Kippur, and the three "pilgrim festivals" of Pesach, Shavuot [the feast of weeks], and Sukkot [the feast of booths], and there is the same problem with every object in space that rotates around a sun. By extension, just as it is not clear what a day is in space, it similarly is not self-evident what a person is in space. For example, need a "person" to be a person have a carbon based body, or could he-she-it have a body composed primarily from some other element, say silicon? Certainly looking like a *Homo sapien* need not be by anyone's standards sufficient or even necessary to be one. (The *Zohar*, for example, states that the entire cosmos has the shape of a human being, and gives this claim as an interpretation of what Genesis means when God says "Let us [God and whatever] make humankind [a human, *adam*] in our image [*b'tsalmo*], according to our likeness [*kidmuteynu*]" [Gen 1:26]).

Scientists generally ignore the distinction between persons and non-persons as too "fuzzy" or "soft" a way to distinguish people from non-people in biology. But for Jewish law, let alone for Jewish theology, the difference is critical. A "person" is a "living" being (*chayah*), and at least in classical (i.e., medieval) Jewish theology a person is an entity that has a "soul" (*nefesh*). Now a "person" was assumed to be responsible for her moral judgments presumably because he-she had a soul, i.e., a non-physical agent who determines morally-evaluated actions (so-called "acts of will.")

Whatever a soul is, it is what (e.g., breath) differentiates a living being from a non-living being, for everything alive (vegetable, animal, and human) is said to have a "soul." Note "everything" means everything and not just a human being (*Homo sapiens*), and everything that has a soul is in principle subject to moral judgment as appropriate to its-his-her species. So, for example, vegetables are given a limited range of divine commandments (e.g., "be fruitful and multiply!") and they can be morally judged within the domain of those commandments, but commandments that lie beyond their domain have no applicability (e.g., do trigonometry!). So you can judge a pumpkin to be a good or bad pumpkin, but you cannot judge a pumpkin to be a good or bad mathematician, or logician, or even to be morally responsible. Moral responsibility minimally assumes choice, choice assumes volition, and there is no reason to assume that vegetables have volition.

In short, whether or not creatures are subject to moral judgment depends on whether or not their physical behavior is voluntary. Otherwise, be they (in rabbinic language) merely bodies (e.g., rocks) or merely souls (e.g., angels) they are not subject to ethical evaluation. Furthermore, the range of moral responsibility is limited to creatures who are embodied souls (or, to say the same thing, insouled bodies). In other words Jewish law and Jewish ethics extends to but only to all embodied souls. Minimally this category includes all animals and human beings. Can it also include creatures who are neither human nor animals, i.e., to creatures not native to the planet Earth?

At the most general level, Jewish law classifies human beings into two sets of categories: (1) males and females; and (2) Jews and non-Jews. These categories are mathematical in the sense that the lines between them are not rigidly drawn, or at least not necessarily so. However, while Jewish law does not recognize a hard-and-fast line in nature between males and females (for it does recognize that some people lack clear sexual differentiation, e.g., people who are androgynous or hermaphrodite), and the same could be said for the Jew/Gentile bifurcation. Even should a born Jew convert to a different religion, he or she continues in Jewish law to be considered as a Jew. In current Jewish law, gender is a rigid classification marker.

How might this classification tradition within Judaism apply to the relationship of earthlings to extraterrestrial intelligence? We can ask: is the distinction between an earthling and an extraterrestrial entity a mathematical distinction, in the way I am using the term "mathematical"? That is, is the human versus non-human a binary, rigid classification? Would we say that, according to Jewish law, a human is defined by two markers: (1) a carbon-based entity (2) tied by genetic history to the planet Earth? If so, does an extraterrestrial being (i.e., a being with consciousness and volition that has only one but not both of these markers) become, by definition, an alien?

The Jew as an Alien

There is a long folk history of Jews being thought of as a distinct, non-human not-quite-natural species. For the most part the tradition is distinctly

European and Christian. For example, some medieval Europeans thought that Jews were androgynous, more precisely that Jewish males have both a vagina and a penis. Furthermore there is also a literary tradition that claims that Eastern European folklore about the vampire is a projection of fears that the Jewish people are vampires. Jews like vampires are sophisticated, worldly foreigners of no specific or local ethnic identity, who speak the native language like foreigners from some mysterious land, who possess extra-human, magical powers, and who drink blood (perhaps a faulty generalization from misapprehending Jewish circumcision ceremonies).

Consider the evil scientist who advises the ruler in Fritz Lang's *Metropolis*. Lang's city is divided into two levels—a bottom level of slave laborers who make possible the idyllic life of the ruling class in the upper stories of the metropolis. Eventually there is a lower class rebellion by the physical workers against their upper class ethereal managers that threatens the very survival of the polity. Salvation and happiness come only at the end when the workers and masters unite through (Christian?) love and execute the scientist whose technological skill enabled the ruler to realize his plans.

It is said that Adolf Hitler, to Fritz Lang's great surprise, loved the movie, and it is not hard to imagine why. The scientist, like the Jewish people, was a foreigner. So he lives outside the city (i.e., in a Jewish "ghetto") under his own law and customs. In fact the general ideas for governing the city are not those of the "magician," i.e., of the scientist whose science is "black magic" (*kabbalah*?) provides the how for the rulers why. (In other words, the Jew was "only following orders.") Note also that the five pointed star of alchemy on the wizard's door reflects the six pointed star on the scientist's door, the *"magen david"* (shield of David), which universally became an identity sign for the Jew and which here, on analogy with the story of the Exodus from Egypt, identifies the city's evil engineer as a Jew.

For most of the Christian European Middle Ages Jews were viewed like Lang's alchemist. They were strangers (aliens) in a strange land (an alien world). They had their own distinctive language, customs, and mores, which caused them to be judged in virtue of their religious lifestyle as alien by those who lived next door to them. Furthermore, the Jews saw themselves as their neighbors saw them, viz., as foreigners, as strangers from a strange land.

As Jews were aliens in a land entirely different than their own, so at least the theologians recognized that their deity, the God of Israel, was even more foreign. In fact, as the sole creator of the universe, God was viewed as absolutely foreign. No one expressed this view more sharply than did Judaism's most influential theologian, Moses Maimonides (1135–1204 CE).

God as The Alien

Moses Maimonides begins his major code of Jewish law with an enumeration of what was generally assumed, but rarely spelled out, the 613 commandments of the first five books of the Torah, which according to rabbinic belief were dictated "mouth to mouth" by God at Mount Sinai. As listed by Maimonides, the first law is the first verse of the so-called Ten Commandments, "I am the Lord your God" (Ex 20:2), which Maimonides interprets to be the obligation to affirm and believe that God exists. The second commandment is the so-called *"Shm'a,"* Deut 6:4, which says "the LORD our God, the LORD is one."[1] It is taken to assert that God is one, whatever that means, and its assertion is universally considered by all Jews to be the most fundamental affirmation of Jewish faith. It is the first commandment that Maimonides discusses in his collection of "the roots (*'iqqarim*, i.e., foundational principles or dogmas) of the religion of our master Moses, peace be unto him"[2] that a "man" (*adam*, a human being, and not just a Jew) that comes "before anything else," i.e., this commandment is addressed by God at least to all human beings and not just to the Jewish people.

Maimonides calls this book "The Book of knowledge (*mada*)," whose subject matter includes (the affirmation of) the oneness (*yichud*) of the name (i.e., of God) and the prohibition against the worship of the stars (*'avodat kokhavim*, i.e., idolatry). Note that the term *"mada"* is closer in meaning to the term "science" than to the general term for "knowledge"; only medieval science is very different than modern "science." As Maimonides and his contemporaries used the term, it expresses some substantive claims about nature such as those found in Aristotle's *Physics* and *Metaphysics*, whose professed truth was taken more or less for granted by medieval philosophers. Like modern "science" the claims of medieval natural science were taken to be true by a consensus of almost all educated people, and were judged to be self-evident to almost all rational minds. But unlike modern science the beliefs called "scientific" by most medieval philosophers were not empirical. The beliefs said to be scientific were all considered to be necessarily true even though most of them were empirical claims, which, though empirical, purported to command consensus by all so-called "rational people," i.e., all animals said to think logically, i.e., human beings.

---

1. Here the NIV is quoted. It is interesting to note that this verse is the biblical proof text cited by the authors of the Nicene Creed for what arguably is the strangest affirmation (at least from the perspective of logic) of the Christian Trinity. As it functions in this most Christian of Christian creeds, what the *Shm'a* says is, "Here Israel (1) the Lord (2) our God (3) is one."

2. Maimonides, *Guide of the Perplexed*, 94–95.

Note that, on Maimonides's terms, while belief in God's unity is a first principle of Judaism, it does not merely define Jewish belief. Rather it defines what any rational human being would believe. As such the belief in God's oneness is not considered by either Maimonides or any Christian or Muslim philosophers to be uniquely Jewish. However, Maimonides explanation of what constitutes this Jewish belief is so radically original that it constitutes a distinctive Jewish (at least Maimonidean) belief. It is so radical for his time (as it is for our time as well) that I can claim that the creator of the world is with respect to the world absolutely alien.

Maimonides presents this root belief (i.e., this dogma) in two very different formats. First, it is the subject of part I, chapters 50 through 53 of the work Maimonides addressed to all professing Jewish philosophers, *The Guide of the Perplexed*; second, it is the subject matter of chapter 1, *halacha* 7 of the *Mishneh Torah* addressed to all believing Jews, be they trained philosophers or not.

Maimonides presents the following summary of what he says about God's oneness in *The Guide* in the final paragraph of chapter 52.

> He, may He be exalted, is one in all respects; no multiplicity should be posited of Him; there is no notion that is superadded to His essence; the numerous attributes possessing diverse notions that figure in the Scriptures and that are indicative of Him, may He be exalted, are mentioned in reference to the multiplicity of His actions and not because of a multiplicity subsisting in His essence, and some of them, as we have made clear, also with a view to indicating His perfection according to what we consider perfection.[3]

God is one in every sense of the term "one." He is innumerably one in the sense that there exists one but only one deity in the universe, but he is also one in the sense of qualitative as well a quantitative oneness. As God is one in the sense that there is one and only one deity, so in absolute terms the God is unlike any one or thing else that can be called a god. Two reasons are given here in this selection. One has to do with grammar and the other has to do with ethics. First, with respect to grammar, if a predication can apply validly to two different subjects, then there must be one respect in which the two are alike (which permits the application of a single term to describe both of them), and there must be another respect in which they are not alike, which means they would be complex and not absolutely simple, i.e., not one. Second, with respect to ethics, God is the paradigm to which all moral valuations are based. Only God is absolutely good, and as such God

3. Ibid., 119.

functions as the moral ideal towards which all things strive. God's oneness as such expresses a single ideal in comparison to which everything is judged to be good or bad. ("Good" means comparatively closer to be divine; "bad" means comparatively more remote.)

Now as such there is no quality of God that is comparable to any quality of a creature. Furthermore, since nothing is like God there is nothing with reference to which comparisons can be made. Hence, comparisons between God and anything are comparison between what the creator and his creatures do, i.e., their actions, and not between what they are. Hence, properly speaking, you can say that God creates, reveals, and saves, but not that he is the creator, the revealer, and the savior. The former said of these terms express actions only and not qualities.

What Maimonides says in the *Mishneh Torah* makes explicit what the *Guide* has explained.

> This God is one. He is not two or more. But one, unified in a manner that surpasses any unity that is found in the world, i.e., He is not one in the manner of a general category which includes many individual entities, nor one in the way that the body is divided into different portions and dimensions. Rather, He is unified (*yichud*), and there exists no unity similar to His in this world.[4]

God is not only one numerically; he is one absolutely in every sense of the term, i.e., he is absolutely simple, and as such he is utterly unique, i.e., he is absolutely strange. The reason is the dogma that he and he alone created the universe.

For Maimonides the created world came about through eternal forms becoming materialized (or, what is to say the same thing) spatial-temporal matter becoming informed or intelligible. It exists because it is situated in space and time, but it is also intelligible because it has a form. Hence everything created is complex, i.e., not simple, no matter how simple it is, simply because it is real (i.e., spatially and temporally located), and it is intelligible (i.e., because it has form). The one exception to this judgment is God, who as the sole creator of everything, has no form or matter. For our purposes the critical inference is that he must not be informed. This radical judgment becomes known as Maimonides's "Negative theology."

---

4. Maimonides, *Mishna Torah*, 144.

Conclusion

The method of negative theology implies that anything and everything in the cosmos—earthlings and extraterrestrials included—are not God. Perhaps the first principle of a Jewish astrotheology would be this: we earthlings and our extraterrestrial neighbors belong together, because it is God who is the real alien.

As Franz Rosenzweig says in Part III of his *Star of Redemption*, think of something. What you are thinking about is not God.[5] Furthermore, that necessarily it is not God defines whom God is by not being Him. As God positively creates everything, everything in turn negatively defines God. In other words, the God of Israel who is the God of the entire cosmos is the most alien, i.e., the paradigm alien being, of all aliens.

Bibliography

Maimonides, Moses. *Guide of the Perplexed*. Translated by Shlomo Pines. Chicago: University of Chicago Press, 1963.

———. *Mishna Torah*. Translated by Eliyahu Touger. New York and Jerusalem: Moznaim, 1989.

Rosenzweig, Franz. *The Star of Redemption*. Translated by William W. Hallo. 1971. Reprint, Notre Dame: University of Notre Dame Press, 1985.

---

5. Rosenzweig, *Star of Redemption*.

13

Islamic Theology Meets ETI

MUZAFFAR IQBAL

The religious view of the order of nature must be reasserted on the metaphysical, philosophical, cosmological, and scientific levels as a legitimate knowledge without necessarily denying modern scientific knowledge.

—SEYYED HOSSEIN NASR[1]

One of the tasks assigned to the astrotheologian is to assess the scope of God's creation in light of a possible future encounter with intelligent creatures elsewhere in space. This chapter explores questions arising from a potential encounter with extra-terrestrial intelligence (ETI) from an Islamic perspective by situating such an encounter within the broader cosmological scheme presented by the Qurʾān. It explores the following questions: Is there room for the existence of ETI within the creation theme of the Qurʾān? If yes, what are the specific modalities of their existence? What challenges would such an encounter bring to the Muslim conception of the Divine? What would it mean for Muslim understandings of the End of Time, Resurrection, and Judgment?

---

1. Nasr, *Religion and the Order of Nature*, 6.

Praise Be to God, the Lord of the Worlds

Practicing Muslims proclaim *Praise be to God, the Lord of the Worlds* at least seventeen times a day, because this first verse of the first sura of the Qurʾān, *Sūrat al-Fatiḥā*,[2] is recited as part of every cycle of the five obligatory daily prayers. The verse is also commonly uttered by Muslims in the course of social interactions, whether as a response to everyday inquiries ("how are you?") or to express gratitude to God (the word *ḥamd*, "praise," also encompasses "thanks" *[shukr]*). Given that the clock timings of the daily prayers shift across the seasons, and given that Muslims live across all regions of the earth, one can reasonably assume that someone on earth is praising God at all the moments of day and night, offering thanks not only for restful sleep and nourishing sustenance but also for his being *Lord of the Worlds*—a phrase that is understood to include all realms of the unfathomable and wondrous cosmic span, and which might include yet-unencountered intelligent beings living somewhere other than the planet Earth.

Semantically, this understanding is anchored in the polysemous Arabic term *ʿālamīn*, which denotes, among other meanings, "intelligent creatures." As the plural of the plural noun *ʿālam*, it encompasses all creatures, whether humans know them or not.[3] It should also be noted that the Qurʾān addresses itself to two sentient beings, the humans and the *jinn*, both of whom are held accountable for their beliefs and actions.[4] Thus, in some ways, the basic questions asked by astrotheology already have an immediate referent in the form of the Jinn, who are invisible beings created from fire (Q 55:15). The Jinn have free will and live on earth in a world parallel to that

---

2. Often inadequately and erroneously translated as "chapter," *sūra* (pl. *suwar*) is a distinct unit which identifies a complete section of the Qurʾān. There are 114 *suwar* in the Qurʾān, ranging from three (Q 108) to 286 verses (Q 2).

3. *Al-ʿālamīn* appears 73 times in the Qurʾān. For a comprehensive range of meanings, see Lane, *Arabic-English Lexicon*, sub ʿ-l-m.

4. The Arabic word *jinn*, from the root *j-n-n*, means to hide or conceal. Their existence is known through the Qurʾān and authentic Prophetic sayings: *Indeed We created man from dried sounding clay and We created the Jinn before that from the smokeless flame of fire* (Q 15:26–27). The Prophet said, "The Angels were created from light and the Jinn from smokeless fire" (*Ṣaḥīḥ Muslim*). The Jinn were created for the same purpose as human beings, for God says: *I did not create the Jinn and mankind except to worship Me* (Q 51:56). Jinns, like humans, can believe or disbelieve. One sura of the Qurʾān is called al-Jinn and it begins: *Say [O Muhammad]: It has been revealed to me that a group of Jinn listened and said, 'Indeed we have heard a marvelous recitation. It guides unto righteousness so we have believed in it, and we will never make partners with our Lord'* (Q 72:1–2). In many aspects of their world, the Jinn are similar to human beings. They eat and drink, marry, have children, and die. They will be present with humanity on the Day of Judgment and like humanity will be judged and committed to either Paradise or Hellfire.

of humankind. But since this chapter deals with specific questions related to intelligent beings inhabiting spaces other than the planet Earth (a planet which we have become accustomed to assuming is the sole locus of intelligent species), our discussion is limited to the following questions: What would happen if we were to actually encounter beings endowed with intelligence who live on other planets? What would such an encounter do to our fundamental beliefs regarding creation and resurrection? How would it change our understanding of the Divine? How would it alter our understanding of our own place in the cosmos? Would it yield a "crisis of faith"?

To be sure, these are now real questions, as opposed to the speculations of previous centuries when planetary explorations in search of extraterrestrial intelligence (ETI) were only the stuff of science fiction. Now we have an actual spacecraft made by humans entering the orbit of Jupiter, the largest planet in the solar system, and beaming images back to earth.[5] It is also important to recognize that while the questions arising from a possible—even anticipated—encounter with ETI may be common to all faith traditions, the answers may not be similar. The methodology of exploring these questions is certainly not the same even across the three Abrahamic faiths, for the architectonics of knowledge are different in these three religious traditions.

In the case of Islam, the primary questions of ETI may be considered through the vast interpretive literature on the Qurʾān, which Muslims believe to be the speech of God (*kalām Allāh*), verbally revealed to the unlettered Prophet over a period of twenty-three years (609–632 CE) through the medium of the archangel Jibrīl (Gabriel), and handed down to successive generations in its original form through an uninterrupted human chain. A related source is the sayings and actions of the Prophet, which have been codified in the Hadith literature. Besides these two core sources and the massive commentary traditions to which they have given rise, discourse on ETI questions may also be anchored in the disciplines of *kalām* (often rendered as "scholastic theology") and of *falsafa* (philosophy).

The Creator and the Created in the Qurʾān

The Qurʾān directs every inquiry to God, envisaged to be independent of history, and inasmuch as He is as He has described Himself. As in the other Abrahamic faiths, God is the supreme Creator, *Lord of the heavens and the*

---

5. NASA's spacecraft Juno arrived at Jupiter on July 4, 2016. Juno is programmed to travel about 5 million miles (8 million kilometers) away from the solar system's largest planet; for details, see http://www.nasa.gov/mission_pages/juno/main/index.html.

earth and all that is between them; to Him alone is worship due (Q 1:2; 5:28; 7:54; 13:16; 19:65; 37:4–5; 38:65–66). His creation encompasses the entire cosmic order—from the particles of sand in the desert to distant galaxies. The Qurʾān does not detail its cosmogonic scheme, but declares the creation of this hierarchical and layered cosmos a *fait accompli*. It asks rhetorically: *Who created seven heavens one upon another?* (Q 67:3) and *Have you not regarded how God created seven heavens one upon another?* (Q 71:15). The pivotal word in the latter verse, *ṭibāqa*, is explained by the Andalusian exegete Abū ʾAbd Allāh Muḥammad b. Aḥmad al-Qurṭubī (d. 671/1273) as "one above the other, each heaven encompassing (*muṭabbaqa*) the other, like domes (*ka-l-qibāb*)."[6] It is a cardinal article of Muslim creed that all existent beings came into existence through the command of God, Who, when He desires to create something, simply says, *"Be; and it is"* (Q 2:117; 3:47, 59; 6:73; 16:40; 19:35; 36:82; 40:68).

Allah—The Supreme Name of the Creator

The word "Allah," "the most definite of all definites" (*aʾraf al-maʾārif*), as the master-grammarian Abū Bishr ʾAmr b. ʾUthmān b. Qanbar Sībawayh (d. 180/796) explains, is exclusively reserved for the exalted Originator (*al-Bārī taʾālā*).[7] The Qurʾān pointedly asks: *Do you know any who could be His namesake (lahu samiyyan)?*[8] He is the Creator of all that exists; no one other than Him is a Creator in any real sense. *Indeed, I am Allah—there is no divinity save Me* (Q 20:14); *No sight can perceive Him while He encompasses all sights; He is Subtle, All-Aware* (Q 6:103). He is *the Lord of east and west; there is no divinity but Him* (Q 73:9); He is the Real (*al-Ḥaqq*) (Q 20:114); *the Eternally Self-Sufficient*. Allah is Absolutely Unique, One, and incomparable (Q 2:163; 4:171; 5:73; 6:19; 16:22; 18:110; 21:107; 41:6; 42:11; 112:4). He has no partner, no helper (Q 2:22; 2:165; 14:30; 34:33; 39:8; 41:9) and like Him there is nothing (Q 42:11). To Him belong the most Beautiful Names (Q 59:24). He has no opponent or rival (Q 6:19; 15:96; 17:22, 39, 42; 21:22; 23:91, 117; 51:51). He possesses all the attributes of perfection (Q 59:23; 62:1). He begot neither a son (Q 2:116; 4:171; 6:100–101; 9:30–31; 10:68; 17:111; 18:4; 21:26; 25:1) nor a daughter (Q 6:100; 16:57; 37:149; 43:16; 52:39); He has no mate (Q 6:101; 72:3); He is beyond duality (Q 16:51) or trinity (Q 4:171; 5:73). He has always existed and He will be when there is nothing else—*He is the First and the Last, the Outwardly Manifest and the*

---

6. *Tafsīr, sub* Q 71:15.

7. Māwardī, *Nukat, sub* Q 1:1; Rāghib, *Mufradāt, sub* ʾ-*l-h*; Rāzī, *Lawāmiʿ*, 79–81.

8. Q 19:65.

Inwardly Hidden, and He has knowledge of all things (Q 57:3). He is changeless (Q 2:255; 3:2; 20:111; 112:2).

Proofs of His Existence through ETI?

Would an encounter with ETI provide further proofs for the existence of God? After all, Fakhr al-Dīn al-Rāzī (543–606/1148–1209), the philosophically inclined influential commentator from Rayy (in present-day Iran), comments that the creation of humans is among the most convincing and pervasive Qurʾānic arguments for the existence of the Maker (*al-Ṣāniʾ*) (referring primarily to Q 2:21, 258; 20:50; 26:78; 96:1–2).[9] Likewise, the Khurasānī sufi commentator Zayn al-Islām Abū-l-Qāsim ʾAbd al-Karīm b. Hawzān al-Qushayrī (376–465/ca.986–ca.1073) comments that Q 56:58–59 (*Have you ever considered that [fluid] which you emit? Do you create it, or are We its Creator?*) comprises a fundamental Qurʾānic argument for the existence of the Maker, because human creation—precipitated by two drops gathered together in the womb, where they undergo various stages of division and regeneration and unite in a specific form—occurs neither through the conscious work of male and female parents (who lack the requisite knowledge and capability), nor spontaneously through the sperm and ovum on their own (these being mere liquids without knowledge or power). This, al-Qushayrī contends, cannot but establish the existence of the Pre-Eternal Maker (*al-Ṣāniʾ al-Qadīm*), the Omniscient King, Who is the Creator.[10]

If we were to extend these and other Qurʾānic arguments for the existence of a Creator to new realms populated by ETI, would this yet-unencountered form of life increase our faith in God or lead to a crisis of faith? We may infer an answer from what al-Qurṭubī notes in his exegesis of Q 10:101, that the Prophet was instructed by God to tell disbelievers: *"Ponder on whatever there is in the heavens and on earth!"* The critical word here is *whatever*. It allows us to extend reflection to ETI, because the insistent Qurʾānic exhortation to reflect on the creation of God is not earth-bound or even earth-centered. It draws on a rich lexicon that employs several words to refer to the Divine act of creation, strictly speaking to bring out existents from non-existence. The most frequently used word is *khalq*, which is the volitional act by which God bestows existence upon particular objects. The uniformity of the act of creation treats all created beings as containers or bodies (*hayākil*), created along definite patterns, subjected to accidents, and consisting of a reality that perishes. It is possible to conceive of ETI

---

9. *Tafsīr*, Muqaddima, faṣl 3, al-kalām fī majmūʿ tafsīr hadhih al-sūra.
10. *Tafsīr*, sub Q 56:58–59.

following the same general pattern, whereby the mere existence of ETI will not produce a crisis of faith. But what of the possibility of decentering our beloved terrestrial realm, and our perceived superior position therein?

The Terrestrial Realm

The earth, mentioned 461 times in the Qurʾān, has an important place in the Qurʾānic creational hierarchy, but there is no verse that suggests (let alone requires) a geocentric model. Premodern Islamic scientific tradition mostly followed the Aristotelian geocentric cosmology, because for the scientists and philosophers of that era, Aristotle had proved "with impeccable philosophical rigor, not only that the whole universe was spherical, but also that the Earth was at its center. And if one did not have an Earth there, one had to assume an Earth as the fixed point of any moving sphere, besides being the ultimate point of heaviness of the universe."[11] The resulting cosmological models, especially in the tradition of Ptolemaic astronomy, were, however, subjected to detailed scrutiny by Muslim scientists, leading to the emergence of new genres of writings—called *shukūk* (doubts)—addressing specifically the totality of those Ptolemaic problems which appeared dubious. In time, "every serious astronomer felt that he had to take part in this enterprise" of systematic and detailed scrutiny. Specific objections to Aristotle's model were raised as early as the eleventh century by al-Bīrūnī (d. 1048) and several others in the next three centuries—eventually leading to an important reform in previous cosmological models proposed. It was Naṣīr al-Dīn al-Ṭūsī (d. 1274) who finally developed the famous Ṭūsī Couple, a theorem he first proposed in his *al-Tadhkira fī-l-Hayʾa* (*Memoir on Astronomy*) (*ca.*1260). The success of this theorem was such that it was used by "every serious astronomer that followed Ṭūsī, including the Renaissance astronomers such as Copernicus and his contemporaries."[12] Other significant contributors to the new cosmological models included by the astronomer Ibn al-Shāṭir (d. 1375), the timekeeper at the Umayyad Mosque in Damascus.

What is significant for our purpose is the fact that for Muslim astronomers and theologians, these new models were devoid of any theological concerns: whether or not the earth was the center of the cosmos made no difference to their faith. Thus, when heliocentrism prevailed following the Copernican revolution, it was a non-event for Muslim theologians. The heliocentric model did not produce a crisis of faith in Islam and it did not spur

11. Saliba, *Islamic Science and the Making of the European Renaissance*, 120.
12. Ibid., 158.

debates of the kind which allegedly emerged in Europe in the sixteenth and the seventeenth century.[13]

ETI with a Superior Soul and Body?

But what if we encountered ETI with a soul and a body superior to the human constitution? In Islamic philosophical and scientific traditions, existents (*al-mawjūdāt*) are conceived as things constituted of matter (*al-mādda*) and thus anything which has a body (*al-jism*), whether residing on earth or elsewhere, can be uniformly studied insofar as it undergoes change (*al-taghayyur*). Muslim philosophers such as Ibn Sīnā (d. 1037) and the already quoted exegete and theologian Fakhr al-Dīn al-Rāzī held that the fundamental principles governing natural sciences are derived not from these sciences, but from meta-science. This means that while the cosmogonies of existence and fine details of the working of physical systems are studied through natural sciences, the very existence of things cannot be studied on the basis of their own existence. In other words, if ETI with souls and bodies exist, they would have their own constitutive elements specific to them, but they could still be studied under the same metaphysical principles which are used for the earthly natures (*al-ṭabī ʾa al-ariyya*)—which al-Rāzī describes as:

> requiring settledness (*al-istiqrār*) but on the condition that this [earthly] body (*al-jism*) is found existing in its natural place (*makānihi al- ṭabiʿī*) which is the earth (*al-arḍ*), while the motion [of this body] toward it [the earth] is on the condition that this body is outside its [natural] place. The existing faculty (*al-quwwah al-mawjūdah*) for this effect (*al-athar*) [whether of settledness or motion] is a faculty having neither consciousness nor apprehension (*idrāk*) at all of the effect, and furthermore this effect is a single effect (*athar waḥīd*) occurring in a single manner (*wāqiʿan ʿalā ṭarīqatin wāḥidatin*).[14]

If this definition is applied to ETI, one could conceive of their alternate locus of existence and replace "earthly" with an adjective better pertaining to

---

13. This is not say that there was a total absence of such debates, for there were some such in the Muslim world in the nineteenth century; but, as Ahmad Dallal has noted, "they seem to have been mechanical transpositions of earlier Western controversies brought to the Muslim world in missionary schools." (Dallal, *Islam, Science, and the Challenge of History*, 163).

14. Quoted in Adi Setia, "Fakhr al-Dīn al-Rāzī on Physics and the Nature of the Physical World."

their abode. But would it be possible to frame their intelligence and their relationships with each other and with God in a similar fashion? Would it be possible to think of ETI as having a soul (*nafs*), which, in humans is a motive principle (*mabdaʾ al-ḥaraka*)? If yes, could their souls be divided into classes or a hierarchical structure (as with the human, vegetative, celestial, and animal souls)? These are important questions for Muslim thinkers, but they cannot be fully answered without an actual encounter with ETI, which will determine the applicability of the fundamental principles of this theoretical framework.

Status of Humans after the ETI Encounter

Would the status of humanity suffer if we were to encounter superior extraterrestrial intelligent beings? Clearly, it all depends on what is meant by "superior." As an adjective, "superior" denotes something higher and better, whether in rank, constitution, or relative to a broader cosmological scheme, whether in intellect or other manifest characteristic or faculty such as physical strength. None of these would be of much consequence to the Divine honoring of humankind (an honoring that is marked through revelation and responsibility). However, if we were to encounter intelligent beings that are superior to human beings in their relationship with the Creator, then humankind would be demoted in rank—because so far as Islam is concerned, the honoring of humanity is based on the capacity to recognize the Creator. While all existent beings are cognizant of their Creator and extol his glory,[15] human beings have a special position in the hierarchy of creation: *We have honored* (*karramnā*) *the Children of Adam and carried them on land and sea, and provided them with pure things, and preferred them greatly over many of those We created* (Q 17:70). The great encyclopedic exegete Abū Jaʾfar b. Jarīr al-Ṭabarī (224–310/839–ca.922) explains the "honoring" in this verse as being mastery over other creatures, who have been made subservient to humankind.[16] Other exegetes explain this honouring to be due to knowledge granted to humans. For instance, Abū al-Layth al-Samarqandī (d. 373/983) quotes several scholars in his explanation:

---

15. Q 17:44: *The seven heavens and the earth, and whosoever in them is, extol Him; nothing is, that does not proclaim His praise, but you do not understand their extolling. Surely He is All-clement, All-forgiving.* Also Q 24:41: *Hast thou not seen how whatsoever is in the heavens and in the earth extols God, and the birds spreading their wings? Each knows its prayer and its extolling; and God knows the things they do.* Likewise, Q 62:1 and Q 64:1.

16. al-Turkī et al., eds. *Jāmiʿ al-bayān ʿan taʾwīl āy al-Qurʾān*, sub Q 17:70.

Al-Ḍaḥḥāk [(d. ca.102/720)] said, [this "honouring" is] "through intellect and discrimination (tamyīz)." It is said that Allah Most High created the earth's plants and trees, and placed therein a spirit; for as long as its spirit dwells within it, it grows and increases by itself; when it withers, the spirit leaves it, and its growth and increase is cut off. He created animals, and placed an augmentation of spirit within them, by means of which they seek out their own provision, and are able to hear sounds; and He created the sons of Ādam, and placed within them [another] augmentation of spirit, by means of which they apply their intellects, distinguish [between things], and know. [Moreover], He created the Prophets, and placed within them [a yet greater] augmentation of the spirit, by means of which they hear the angels, and by means of this receive revelation, and have knowledge of the Hereafter. That is, [man has been preferred] to the jinn, devils and animals. It is related of Ibn ʾAbbās that he said "they have been preferred over all of creation, other than a group amongst the angels, namely Jibrīl, Mīkāʾīl, Isrāfīl, and the like of them"; and it is related of Abū Hurayra that he said "in the sight of Allah, the believer is more distinguished than the angels that are in His Presence."[17]

For the mystics, this honouring is understood to be through the granting of spiritual insight. Al-Sulamī (325–412/936–1031), for instance, quotes his great Sufi predecessor al-Junayd al-Baghdādī (d. ca.296/908) that this verse means, "We honoured the children of Ādam with understanding [of spiritual realities] bestowed by Allah."[18] Fakhr al-Dīn al-Rāzī writes at length:

> The human soul is uniquely singled out because of intellective ability or power of intellection (al-quwwa al-ʾāqila), which perceives the essence of things as they are in themselves (ḥaqāʾiq al-ashyāʾ kamā hiya). It is within [the capacity of] this intellection that the light of the knowledge of Allah Most High becomes manifest, and the illumination of His Grandeur shines forth, and it is this that enables [human beings] to become acquainted with the secrets of the two worlds of Creation and of the Command (ʾālamay al-khalq wa al-amr), and to encompass the different divisions of the creatures of Allah, the spirits and bodies, as they are in-themselves. This [intellective] power is the result of the fecundation of the holy substances, and the incorporeal spirits of the Divine (al-jawāhir al-qudsiyya wa al-arwāḥ al-mujarrada al-ilāhiyya). This power is incomparable with respect to the

---

17. al-Samarqandī, *Tafsīr al-Samarqandī al-musammā Baḥr al-ʿulūm*, sub Q 17:70.
18. al-Salmī, *Ḥaqāʾiq al-Tafsīr*, sub Q 17:70.

vegetative and animal powers, both in nobility and distinction; and if this is the case, it is clear that the human soul is the most noble of the souls existing in this world.[19]

Since the exegetical tradition already employs ontological taxonomies, given its familiarity with the Qur'ānic notion of superiority (*tafḍīl*)—the classical division among "degrees of existents" (*marātib al-mawjūdāt*), "degrees of being" (*marātib al-wujūd*), and "degrees of creatures" (*marātib al-makhlūqāt*)—one can speculate that an encounter with ETI would immediately be placed in this matrix and the resultant categorization would be based on the same criteria as used for other created beings.

In short, if we were to encounter ETI of superior physical abilities, who are not endowed with higher spiritual qualities, that would not demote humans from their creational rank, but an encounter with ETI superior in intelligence and spiritual receptivity indeed would. The latter scenario would mean that God, who is *the Exalter of Ranks* (*darājāt*) (Q 40:15), created a cosmos in which there exists a creation superior to humankind. It should be noted that Q 17:70 also states that humans are *preferred greatly over many of those We created*, but it does not grant an absolute or exclusive superiority to humankind. The honouring, such as it is, is due to a creation's ability to recognize the Creator, while decrease in such honour is due to an existent's lack of receptivity: *Surely the worst of living creatures in Allah's sight are those that are deaf and dumb and do not understand* (Q 8:22). Najm al-Dīn al-Kubrā (540–618/1146–1217) explains that they are those who "do not listen to the words of the Real, with acceptance and [a submitting] heart, and they are dumb with regard to the words of the Real, and to speaking with the Real."[20]

End of Time, Resurrection, and Judgment

Islamic eschatology presents a grand scenario of a sudden and decisive moment when all existents will reach their end—an absolute annihilation, uniformly encompassing the tiniest atom and the vast galaxies, leaving nothing but *the face of thy Lord*.[21] If there are beings on planets other than Earth, they would be included in this universal return to the Creator. This will be followed by Resurrection and final judgment. Would ETI be resurrected, and if so what would be their final fate? Is there paradise and hell for ETI?

19. *Tafsīr, sub* Q 17:70.
20. al-Kubrā, *al-Taʾwīlāt al-najmiyya fī-l-tafsīr al-ishārī al-ṣūfī, sub* Q 8:22.
21. Q 28:88.

It is not beneficial to speculate, but the basic Qurʾānic framework of accountability (*taklīf*) may suggest a fruitful way to frame this question. God imposes obligations on his creatures which subject them to his command; their final fate depends on how well they discharged these obligations. Thus, if we encounter ETI who are granted the resources and freedom to be governed by his Law of obligation and accountability (i.e., they are *mukallaf*), then, strictly speaking, there would be resurrection and judgment for them, just as for humans. This would in turn raise the same philosophical questions which exist regarding humankind: (i) What proof do we have that God has imposed any obligation on us? (ii) Why does God impose obligations on us? (iii) Are God's obligations possible for us to fulfill, and, if not, can God oblige someone to do the impossible?

The answers provided by the three schools of Islamic theology may also apply to ETI. In turn, (i) Muʾtazilīs and Māturīdīs agree that we know our obligations through the exercise of reason (*bi-l-ʾaql*) and through revelation (*bi-l-sharʾ*), while Ashʾarīs hold that only the latter source is effective; (ii) Ashʾarīs and Māturīdīs do not admit knowledge of the divine motives (*ʾilal*), while Muʾtazilīs believe either that God does whatever is best (*aṣlaḥ*) for humans, or that (for those following Abū Hāshim al-Jubbāʾī (d. 321/933)) God imposed his obligations to facilitate human fulfillment of the general duties to which He subjected them, and thus to remove every excuse for their not doing so.[22]

Conclusion

Some Christians interpret their scripture in such a way that the *Bible-Welcomes-Aliens* to Earth. For the Muslim, the Qurʾān does not exactly welcome aliens to our planet, although the Qurʾān certainly does not put up any "no trespassing" signs. Theologically, what is important is that extraterrestrial intelligent creatures are, like us, creatures of God.

To be sure, the speculations of this chapter are based on an analogy with our human situation. We ask: what would be the true nature of the Divine obligations imposed on ETI? In response, one can quote the Islamic adage: God knows best.

---

22. For a basic outline of Islamic theology, see http://www.muslimphilosophy.com/ip/rep/H009.htm.

Bibliography

al-Kubrā, Aḥmad b. ʾUmar Najm al-Dīn. *al-Taʾwīlāt al-najmiyya fī-l-tafsīr al-ishārī al-ṣūfī*, edited by Aḥmad Farīd al-Mazīdī. Beirut: Dār al-Kutub al-ʾIlmiyya, 1429/2009.
al-Salmī, Abū ʾAbd al-Raḥmān. *Ḥaqā ʾiq al-Tafsīr*, edited by Sayyid ʾImrān. 2 vols. Beirut: Dār al-Kutub al-ʾIlmiyya, 1421/2001.
al-Samarqandī, Abū al-Layth Naṣr b. Muḥammad b. Ibrāhīm. *Tafsīr al-Samarqandī al-musammā Baḥr al-ʾulūm*, edited by ʾAlī Muḥammad Muʾawwaḍ, et al. 3 vols. Beirut: Dār al-Kutub al-ʾIlmiyya, 1993.
al-Turkī, ʾAbd Allāh b. ʾAbd al-Muḥsin, et al., eds. *Jāmiʾ al-bayān ʾan taʾwīl āy al-Qurʾān*, 26 vols. Cairo: Dār Hajar, 1422/2001.
Dallal, Ahmad. *Islam, Science, and the Challenge of History*. New Haven: Yale University Press, 2010.
Islamic Theology. http://www.muslimphilosophy.com/ip/rep/H009.htm.
Lane, Edward William. *Arabic-English Lexicon*. 1863. Reprinted, Cambridge: The Islamic Texts Society, 2003.
Māwardī. *Nukat*.
NASA. Juno. http://www.nasa.gov/mission_pages/juno/main/index.html.
Nasr, Seyyed Hossein. *Religion and the Order of Nature*. Oxford: Oxford University Press, 1996.
Rāghib. *Mufradāt, sub ʾ-l-h*.
Rāzī. *Lawāmiʾ*.
Saliba, George. *Islamic Science and the Making of the European Renaissance*. Cambridge, MA: MIT Press, 2004.
Setia, Adi. "Fakhr al-Dīn al-Rāzī on Physics and the Nature of the Physical World: A Preliminary Survey." *Islam and Science* 2.2 (2004) 161–80.

14

Toward a Constructive Naturalistic Cosmotheology

STEVEN J. DICK

> Religious naturalism . . . describes a person's interpretive, spiritual, and moral responses to our understandings of nature.[1]

Science, particularly in the form of astronomy and cosmology, continues to reveal more and more about our place in the universe. For almost a century since Edwin Hubble discovered extragalactic space and hinted at the expanding universe, we have known that life on Earth is part of the vast unfolding of cosmic time and space, over 13.8 billion years according to the latest observations of spacecraft such as Hubble, COBE, WMAP, and Planck. And the search for life *beyond* Earth, once the stuff of science fiction, is now a robust research program with a well-defined Roadmap.[2] The science of astrobiology—and there is no longer any doubt it is a science, simplistic slogans about "a science without a subject" notwithstanding—is funded by NASA and other institutions to the tune of tens of millions of dollars of ground-based research, not to mention the hundreds of millions spent on space-related missions. Biogeochemists study extremophile life on Earth, biologists study the origins of life, a bevy of spacecraft have orbited or

---

1. Goodenough and Deacon, "Sacred Emergence of Nature," 864.
2. Des Marais et al., "The NASA Astrobiology Roadmap," 715–30. For an entrée to the large literature of astrobiology see Sullivan and Baross, eds. *Planets and Life*. On the history of astrobiology see Dick and Strick, *The Living Universe*. And on the critical issues in the field see Dick, "Critical Issues."

landed on Mars, others have found oceans on Jovian and Saturnian moons as well as organic molecules on Titan, and the Kepler spacecraft has discovered thousands of planets beyond the solar system—all just a prelude to future studies. Recent Congressional hearings on biosignatures and complex life beyond Earth indicate astrobiology is a hot topic in the policy arena.[3] And international interest is also strong, particularly within the European Space Agency.[4] Although no life has yet been found beyond Earth, the search for such life as part of the natural unfolding of cosmic evolution shows no signs of abating.

Given that these scientific results bear so heavily on our place in the universe, it is important to examine their societal implications. And in no area of human endeavor are these results likely to have broader impact than in theology, even if the impact is not immediate. It has been fifteen years since I first elaborated principles of what I called "cosmotheology," as part of a Templeton Foundation meeting on the theological implications of the new universe, and ten years since I revisited the subject for a German audience—evidence that interest in the subject is international, if not yet global.[5] In the intervening decade—precisely because of the new results in science—the problem of adapting theologies to current knowledge has only grown more urgent. Books, symposia, and discussions on the subject now appear with increasing frequency, most recently Thomas F. O'Meara's *Vast Universe: Extraterrestrial Life and Christian Revelation*, David Wilkinson's *Science, Religion, and the Search for Extraterrestrial Intelligence*, Guy Consolmagno's *Would You Baptize an Extraterrestrial?* and David Weintraub's *Religions and Extraterrestrial Life: How Will We Deal with It?* All of this activity—engaged in by scientists and theologians alike (Thomas O'Meara is a Dominican and Guy Consolmagno a Jesuit)—is only the latest manifestation of a controversy that has been building over the last five hundred years, since the heliocentric theory of Copernicus made the Earth a planet and the stars potential Earths. The history of the implications of cosmic evolution for theology, particularly its extraterrestrial life aspect, has been written elsewhere in considerable detail.[6] Here I intend to take a closer look at the

---

3. U. S. Congress, House Committee on Science, Space, and Technology, Hearings on "Astrobiology: Search for Biosignatures," and Hearings on "Astrobiology and the Search for Life in the Universe."

4. Fridlund and Lammer, "The Astrobiology Habitability Primer," 1–4. The entire issue is devoted to European Space Agency work in this area.

5. Dick, "Cosmotheology"; Dick, "Kosmotheologie – neu betrachtet," 156–72, online in English at http://bdigital.ufp.pt/bitstream/10284/778/2/287-301Cons-Ciencias%2002-5.pdf.

6. Dick, *LOW*; Crowe, *ELD*; Crowe, "History of the Extraterrestrial Life Debate,"

foundations, principles, and necessity for cosmotheology as part of broader efforts in what has variously been called exotheology, astrotheology, and astroethics.[7] In doing so I hope to clarify my own position by comparison with others, and to indicate what the future may hold.

Foundations and Principles of Cosmotheology

As Einstein began with the assumption that the speed of light is a fundamental constant independent of the motion of the light source, with all sorts of seemingly strange consequences such as time dilation and shrinking objects in his resulting special theory of relativity,[8] I begin with the assumption that the supernatural does not exist—with all sorts of results that many in established religions will consider strange, but I consider both enlightening and liberating. So entrenched has the idea of the supernatural become in Western civilization that many will ask how anyone can defend the radical assumption of its absence. But it is actually the supernaturalists who need to defend their position, since the existence of a reality beyond the natural world, while a legitimate question, is the extraordinary claim—one that supernaturalists have been defending for the last few thousand years. The idea of the supernatural arose early in the history of civilization, and undoubtedly existed even before the rise of civilization as a response to forces humans did not understand.[9] We now understand most of those forces, to such an extent that many theologians have given up on "god of the gaps" arguments, precisely because science has filled the gaps at an increasing pace. The idea of the supernatural is still with us only because the Abrahamic religions have adopted it as dogma over the course of thousands of years. To those outside established supernatural religions, the existence of a realm beyond the natural world would seem to be no more necessary than Aristotle's terrestrial-celestial dichotomy, an idea that survived intact only through the Middle Ages. I realize this is no small claim given the importance of the supernatural in terrestrial history. But many historical ideas are contingent and have had to be discarded; the absence of a supernatural realm inside or

---

147–62; and Dick, "Cosmic Evolution."

7. For example, Lamm, "The Religious Implications of Extraterrestrial Life," 107; Peters, "Exo-Theology: Speculations." 1–9, expanded in Lewis, *The Gods Have Landed*; Peters, "Astrotheology: A Constructive Proposal," 443–57, where he lays out elements of an "astroethics" program.

8. Isaacson, *Einstein*, 118–22.

9. Armstrong, *History of God*.

outside the universe is the core of my argument and the foundation for any naturalistic cosmotheology.

What then, is cosmotheology? *Cosmotheology* is simply a theology that takes into account what we know about the universe based on science. It is therefore a naturalistic cosmotheology, but it is not coextensive with scientism because it does not imply that science is the only way to understand the world.[10] Its first principle is that humanity is in no way physically central in the universe. This has been proven beyond the shadow of a doubt in a continuous series of "de-centerings," beginning with the Copernican removal of the Earth from the center of the solar system in the sixteenth century, followed by Harlow Shapley's proof around 1920 that the solar system is on the periphery of the Milky Way Galaxy, and Edwin Hubble's proof in the late 1920s that our Galaxy is only one of many in an almost infinite space. Some have argued that the Copernican de-centering was actually beneficial in terms of human dignity, since the Earth, even though central in the Aristotelian and medieval cosmos, was considered the dregs of creation (hell was, after all, down below somewhere), until Copernicus placed our planet in the realm of the heavens.[11] This may be true, but surely it is beyond doubt that the subsequent physical de-centerings hardly elevated humanity's conception of its physical place in the universe.

One may well argue that physical de-centerings do not matter that much; it is biological, cognitive, and moral status that is at stake in assessing our real place in nature. This brings us to the second principle of cosmotheology, that (if astrobiological endeavors are successful in finding intelligent life) any such theology must take into account the probability that humanity is not central biologically, mentally or morally in the universe. The word "probability" is crucial here, since we have not yet found a single instance of life beyond Earth, much less intelligence. That is what the science of astrobiology is exploring, but all indications are that life will eventually be found, in abundance for microbes, and perhaps more rarely for intelligence. But in such a vast universe "rare" is a relative term. It is certainly true that having found exoplanets by the thousands, some of them *Earth-sized* and in the habitable zones of their parent stars, does not necessarily mean these are *Earth-like* planets. That will need to be determined by examining biosignatures in their atmospheres and other means. But it has been interesting to watch the skeptics over the last twenty years first deny that there are any other planets beyond the solar system,

10. For a spirited, robust, but light-hearted defense of scientism see the book by Duke University philosopher Alex Rosenberg, *The Atheist's Guide to Reality: Enjoying Life without Illusions*.

11. Danielson, "The Great Copernican Cliché," 1029–35.

then emphasize that they are only uninhabitable gas giants, then that they are only Earth-sized and not Earth-like. There is a trend here; surely it is only a matter of time before Earth-like planets are found, and (it seems to me) only a matter of time before life is found. In short, thus far the general principle of the uniformity of nature's laws has held with respect to the existence of exoplanets, and the expectation is confirmed that what is true of Earth is true of other places in the universe. Although this has still to be proven in the case of life, the trend is clear, despite the great diversity of planetary systems found. No one would expect a solar system exactly like ours; as with life, so with planets: diversity is the coin of the realm.

If this universal production of life is true, the third principle of cosmotheology holds that we must take into account the probability that humanity is near the bottom in the great chain of beings in the universe. This follows from 13.8 billion years of cosmic evolution, and the fact that planets and life could have formed billions of years before our Earth originated 4.5 billion years ago. The extreme youth of our species, which has only in the last few thousand years emerged to the point that it can contemplate its place in the universe, is surely a sobering fact when placed in the context of cosmic evolution. Again, it is true (as Ted Peters and others have pointed out) that "progress" in social development, and even continuous linear complexity in biology, are not assured in a universal context. But certainly it is undeniable (without invoking any goal-oriented or teleological principles) that life on Earth is more complex after 3.8 billion years of evolution, and that culture is more highly developed than it was when modern *Homo sapiens* originated some two hundred thousand years ago. What millions or billions of years of biological and cultural evolution have produced in the wider universe is certainly open to debate, but to claim that most intelligent civilizations would not be more advanced than we are seems to border on nihilism.

Fourth, cosmotheology must be open to radically new conceptions of God, not necessarily the God of the ancient near East, nor the God of the human imagination, but a God grounded in cosmic evolution. It is entirely possible that beings have evolved in the natural course of the universe with many of the traits we attribute to God, including omnipotence, omniscience, and so on. It is even possible such beings have meddled in human affairs, though I hasten to add there is no evidence of this, and certainly no evidence that a figure such as Jesus Christ was the son of God or divine in any way, uplifting as he may have been to his numerous followers. Whether one wishes to call such a superior being "God" is also open to discussion, but an expansive theology might do so.

Fifth, cosmotheology must have a moral dimension, extended to embrace all species in the universe—a reverence and respect for life in any form. This principle—a challenge even on Earth—gives rise not only to astrotheology but also to the related field of astroethics.[12] It is often stated that morality stems from theology and the existence of God, with the unfortunate implication that any non-believer has no basis for morality. By contrast astroethical principles stem from this reverence for life in all its manifestations, the product of the creativity of cosmic evolution, whether terrestrial or extraterrestrial. Cosmic evolution can also serve as a framework for human moral orientation in other ways.[13]

Sixth, although human destiny has often been couched in divine terms, as in Reinhold Niebuhr's *The Nature and Destiny of Man* (1941) on Pierre Lecomte du Nouy's best-selling *Human Destiny* (1947), or, indeed as in the entire Christian theology, it need not be linked to the supernatural. Rather, it can be linked to the process and endpoint of cosmic evolution. If cosmic evolution ends with humans and we are alone in the universe, our destiny involves stewardship of our pale blue dot and perhaps spreading, nurturing, or even creating, life in the universe—all pathways filled with ethical considerations. If cosmic evolution results in a biological universe— one in which life and intelligence is common, our destiny is to interact with this life in all its myriad possibilities, invoking a quite different set of ethical considerations raised in the fifth principle.[14]

Cosmotheology and Religious Naturalism

By now it should be clear that cosmotheology fits comfortably within the tradition known as religious naturalism.[15] In the words of its premier historian, Jerome Stone, religious naturalism "affirms a set of beliefs and attitudes that there are religious aspects of this world which can be appreciated within a naturalistic framework. There are some events or processes in our experience that elicit responses that can appropriately be called religious. These

---

12. For astroethics, see: *ELU*; Schwartz and Milligan, *ESE*; Gilmore, "Space Exploration"; and Peters, "Ten Ethical Issues in Exploring Our Solar Ghetto."

13. Kaufman, "The Epic of Evolution," 175–88. This issue of *Zygon* is dedicated to cosmic evolution in a religious context. See also Peackocke, Lupisella and others cited below.

14. Dick, "Cosmic Evolution," 45–49; Vidal, *The Beginning and the End*; McKay, "Does Mars Have Rights?" 184–97. Randolph and McKay, "Protecting and Expanding," 28–34.

15. "Religious naturalism . . . is not so much a religion as it is a philosophical proposal for the recreation and redefinition of what counts as religion in light of the ascendency and productivity of science." Braxton, "Modern Cosmology and Religious Naturalism," 124.

experiences and responses are similar enough to those nurtured by the paradigm cases of religion that they may be called religious without stretching the term beyond recognition."[16] In short, religious naturalism denies that an ontologically distinct and superior realm including God, the soul, and heaven is required to give meaning to the world. Nor does it identify with a pantheism that identifies God with Nature, or even a Spinozan version of pantheism (sometimes called panentheism), in which the universe is a subset of God. Rather, meaning is derived from our knowledge of the natural world, from the creativity and beauty of nature. This natural God is compatible with the concept of Einstein, for whom God "does not play dice" nor concern himself with the fate and actions of men. But Einstein's God "appears as the physical world itself, with its infinitely marvelous structure operating at atomic level with the beauty of a craftsman's wristwatch, and at stellar level with the majesty of a massive cyclotron."[17]

Stone distinguishes three types of religious naturalists: those who conceive of God as the creative process within the universe; those who conceive God as the totality of the universe; and those who do not speak of God and yet can still be called religious due to the feelings of reverence and awe the universe inspires. Even those who fall in the first two types and use the term "God" do so in a rigorously naturalistic way, not invoking a supernatural realm. It seems to me there is no need to use the loaded term "God" for a naturally creative process or for the universe itself. Stone's third type of religious naturalism is the more common usage, and naturalistic cosmotheology falls comfortably within that category.[18] This view resonates with the principles found in biologist Ursula Goodenough's *The Sacred Depths of Nature*, but draws its inspiration and principles from astronomy rather than biology. It also resonates with complexity theorist Stuart Kauffman's radical views in *Reinventing the Sacred*, where he proposes a natural divinity that draws its sacred quality from the creativity of the universe itself—and which in his view can still be called God.[19] An astronomically inspired cosmotheology, a biologically inspired view of the sacredness of nature, Kauffman's insights into the complexity of nature, and even the broader environmental movement, all end up at the same place: with a reverence for the creativity of the natural universe and an evolving understanding of our place in it. Nor is it a universe necessarily reducible to

---

16. Stone, *Religious Naturalism Today*, 1.

17. Clark, *Einstein*, 38; Einstein, "Religion and Science," 36–40.

18. Stone, *Religious Naturalism Today*, 6. The quote is from Murray, "Religious Humanism Yesterday, Today and Tomorrow," 84.

19. Goodenough, *Sacred Depths of Nature*; Kauffman, *Reinventing the Sacred*.

physics; Kauffman, for example, argues vigorously for a scientific worldview that embraces the reality of *emergence* for life, meaning and value, a natural process but not predictable because not subject to natural law.

These principles in turn resonate with many aspects of the new humanism, with its openness to "wonder and mystery and transcendence in a naturalistic framework." They emerge in part from what Stone identifies as the two major roots of religious naturalism in America, the Columbia school centered on Columbia University, and the Chicago school, where the Meadville Lombard Theological School plays a major role.[20] Such principles are already being incorporated into some religions. Although humanism transcends the boundaries of any religious denomination, the Meadville Lombard Theological School, for example, trains Unitarian-Universalist ministers, many of whom espouse some form of religious naturalism. Although small, the Unitarian tradition dates back centuries, and in a broader sense naturalism not only has deep roots, but also has played a major role in history. Indeed Matthew Stewart has recently argued convincingly that many of those who played a major role in the founding of the American republic—including Thomas Jefferson, Benjamin Franklin, Thomas Paine, and Ethan Allen—believed only in "nature's God," rendering ironic, or at least problematic, the vociferous claim that the United States was founded as a Christian nation.[21]

Religious naturalism is, of course, a controversial position in more ways than one. The renowned historian of science John Greene, for example, cautioned in no uncertain terms that naturalism is just another worldview, no more privileged than supernaturalism or anything else to serve as the source of value and meaning in human life. He pointed out that scientists such as Ernst Haeckel, Julian Huxley, Ralph Burhoe, and E. O. Wilson "all typify the scientist-ideologue bent on saving society by promulgating a new ethics and a new religion claiming the sanction of science." The pattern is "depressingly familiar," in Greene's view: "When will scientists and others learn that naturalism is a philosophical point of view with no more claim to the status of science than any other philosophical viewpoint, whether Marxian, Freudian, Russellian, Whiteheadian, or whatever. Scientists have as good a right to expound their philosophical, ethical, and religious views as anyone else, but they have no right to palm these off as the findings of science."[22] In my view this attitude is both instructive and dead wrong. Instructive in the sense that a science-oriented theology is far from the sole

---

20. Stone, *Religious Naturalism Today*, 5.
21. Stewart, *Nature's God*.
22. Moore, *History, Humanity and Evolution*, 404.

source of meaning and value in life. And dead wrong in the sense that science has *at least as much* claim to infuse theological thought as any other world view, and perhaps more, since being grounded in the natural world gives it a pillar of support that supernatural religions do not have, numerous attempts at natural theology notwithstanding. Exactly how to draw meaning from the findings of science is open to interpretation, and not just by scientists. Thinkers such as Stuart Kauffman have no problem finding meaning and value in the creativity of the universe. Philosopher-scientist Mark Lupisella has described a long-term worldview that "can be characterized as a morally creative cultural cosmos—a post-intelligent, post-technological universe that enters the realm of conscious evolution driven largely by moral and creative pursuits,"—in other words a worldview in which meaning and value may be bootstrapped from the universe. Many others are espousing similar views at an increasing pace.[23]

Naturalistic cosmotheology may be seen as one strand in a galaxy of astrotheologies proposed over the last few centuries. How does it differ from the others? The idea of an "astrotheology" dates back at least to 1715, when the English clergyman and natural philosopher William Derham penned a book by that name whose subtitle indicated its purpose: *Astro-theology: Or a Demonstration of the Being and Attributes of God from a Survey of the Heavens.*[24] In other words, this was an exercise in natural theology, an attempt to prove the existence of God from his created work. The term "cosmotheology" originates with the philosopher Immanuel Kant, who referred to it in 1781 in his *Critique of Pure Reason* (without endorsing it) as a "transcendental theology" method of "inferring the existence of a Supreme Being from general experience," rather than a natural theology method of inferring the nature of a Supreme Being from the particulars of nature.[25] Both of these views, of course, are far from our concept of a naturalistic cosmotheology, and in fact they are far even from the modern view of most non-naturalistic astrotheologies.

How, then, does naturalistic cosmotheology compare to more modern versions of astrotheology? As early as 1994 the theologian Ted Peters took up the challenge of what he then called "exo-theology," arguing that the discovery of extraterrestrial intelligence would present no significant challenge to theology. Even though that is seriously questioned by some theologians such as former Vatican Observatory director George Coyne,[26]

23. Lupisella, "Cosmocultural Evolution," 322.
24. Derham, *Astrotheology*.
25. Kant, *Critique of Pure Reason*, sec. VII, "Critique of All Theology based upon Speculative Principles of Reason," 364–65.
26. Coyne, "The Evolution of Intelligent Life," 187–88.

I tend to agree with Peters that after a period of perhaps wrenching change (depending on the nature of contact), theologies would expand to include the new view, and that this could in fact be an enriching experience for theology. Peters, in fact, strongly argues that the latest science, including evolution, should be incorporated into theology, totally in agreement with cosmotheology as laid out here. On the other hand, when Peters concludes in an essay two decades later that "contact with extra-terrestrial intelligence will expand the existing religious vision that all of creation—including the 13.7 billion year history of the universe replete with all of God's creatures—is the gift of a loving and gracious God," he is speaking within a supernatural tradition that makes little sense to those outside of it—unless the concept of God is expanded to mean the natural universe itself. But is admittedly hard to see the universe as "loving and gracious."[27]

Similarly, when the Anglican priest and biochemist Sir Arthur Peacocke argues in a thoughtful, farsighted, and beautifully written article that any theology "will be moribund and doomed if it does not incorporate the perspective [of the epic of cosmic evolution] into its very bloodstream," even including the significance of Jesus Christ, I can fully agree. Yet when he concludes that "humanity is incomplete, unfinished, falling short of that instantiation of the ultimate values of truth, beauty, and goodness that God, their ultimate source, must be seeking to achieve to bring them into harmonious relation to Godself," he is speaking a language that is foreign to a religious naturalist.

Again, in his forward-looking book *Thank God for Evolution* Michael Dowd embraces the evolutionary epic, but only within the Christian tradition that, according to him, gives it meaning. And the Methodist theologian and physicist David Wilkinson, in arguing that the discovery of extraterrestrials would not diminish us in the eyes of God, bases his position on biblical revelation, citing "divinely initiated redemption, an action of a gracious God on behalf of a fallen cosmos."[28] This classical theism too is foreign to a religious naturalistic cosmotheology, which finds no need to embrace the concept of a fallen cosmos.

All of these efforts are, like those of Thomas Aquinas in another turbulent era, attempts to reconcile new scientific knowledge within the Christian tradition, rather than attempts to ask the more basic question of whether the Christian tradition still makes sense in the modern world. Whether they are steps along the way to religious naturalism is a dubious hope; for one thing,

---

27. Peters, "Exo-Theology," 1–9, expanded in Lewis, *The Gods Have Landed*.
28. Peacocke, "The Challenge and Stimulus of the Epic of Evolution to Theology," 92; Dowd, *Thank God for Evolution*; Wilkinson, *SRSEI*.

religious naturalism does not offer salvation from a transcendent God, and never will, even in the form of advanced extraterrestrials. As Carl Sagan said in his book *Pale Blue Dot* referring to the Earth as seen from the Voyager 1 spacecraft, "our posturings, our imagined self-importance, the delusion that we have some privileged position in the Universe, are challenged by this point of pale light. Our planet is a lonely speck in the great enveloping cosmic dark. In our obscurity, in all this vastness, there is no hint that help will come from elsewhere to save us from ourselves."[29] A religious naturalist finds the equivalent of salvation in other ways, through social justice, good works, and making this world a better place to live—all in a cosmic context that defines our place in the universe.

A Difference in Worldview

One way of understanding the enduring differences in religious naturalism and religious supernaturalism is in terms of worldviews. The construction of worldviews and their influence on our thinking are deep philosophical problems.[30] What you believe is interesting, but why you believe it is even more so. If we could understand why people believe what they do, we might begin to tackle some of the world's problems, and not just in the religious domain. In this respect, science and theology are two worldviews, with different epistemologies, different sources of knowledge, and different aims except in the broadest sense of human attempts to understand our place in the universe. The main epistemology of science is empiricism—which admits of theory and observation, with all their complexities and problems, and yet which taken together have revealed so much about the universe around us. The main epistemology of theology, on the other hand, is revelation, faith, and a heroic attempt at empiricism via what has been known over the last four centuries as natural theology. But as we have emphasized, natural theology is no longer considered a good theological argument; even most theologians admit the complex universe no longer requires a God to explain its inner workings, intelligent design biologist Michael Behe's idea of irreducible complexity in "Darwin's black box," and atheist philosopher Thomas Nagel's mysterious teleological principles as applied to mind and consciousness, notwithstanding.[31] This leaves revelation, which also reduces to faith—faith that somehow God revealed his or her word to the authors of the Bible and that we should be following its dictums thousands of years lat-

29. Sagan, *Pale Blue Dot*, xv–xvi.
30. Vidal, *Beginning and the End*, 3–57.
31. Behe, *Darwin's Black Box*; Nagel, *Mind and Cosmos*.

er. Faith, then, is the difference between science and theology. Some people have it, some people don't. If your personal epistemology does not include faith, the practice of religion and the weaving of supernatural theologies will seem strange activities indeed. But if your worldview includes faith as a source of knowledge, it makes perfect sense, though it must be pointed out that with all the religious diversity found on Earth today, one person's faith is another person's heresy.

It is axiomatic that problems arise when one worldview interacts with another, in this case the religious and the scientific. Science and religion can indeed seek common ground—and I applaud efforts to engage a dialogue between science and religion. But when religious beliefs result in attempts to stifle the teaching of science in the form of evolution or anything else, it is dangerous indeed. Intelligent Design might well be taught in religion class, but not in a science class. Similarly, while it is easy to denigrate religion and theology, it remains true that religion has been the source of much good in the world, even if that must be weighed against the numerous religious wars and enormous personal anguish it has engendered, and that continues today. Religion means a great deal in the lives of many people, and freedom of religion is a foundational concept no less than academic freedom in science. A little humility is in order both for religious and scientific fanatics, each of whom should realize that their respective worldview does not explain everything of value in the universe.

Some may well object that the religious naturalist worldview really qualifies as cosmophilosophy rather than cosmotheology. But if theology is defined broadly as that which gives meaning and value to life in a cosmic context, then even a naturalistic cosmotheology is indeed a theology, albeit one without God, at least the standard God. In its emphasis on evolutionary becoming, cosmotheology resonates with Alfred North Whitehead's process theology, without seeing God as the beginning and source of all possibilities. It also resonates with the Jesuit Teilhard de Chardin's evolutionary cosmology, but without the teleological end, the "Noosphere" he identified with Jesus Christ. And the sixth principle of cosmotheology makes clear that human destiny may be couched in terms of natural cosmic evolution rather than supernatural divinity.

Following the Consequences: Cosmotheology and the Speed of Light

I began this essay with the core principle that the supernatural does not exist, promising to follow that premise wherever it led, in the same way

Einstein assumed the speed of light was constant, despite the fact that it led to radical and non-intuitive concepts such as time dilation and shrinking objects as part of special relativity. Where does the absence of the supernatural leave us? For starters it leaves us in a much better position with respect to a explaining a variety of problems that have generated elaborate arguments among theologians and bewilderment among the populace.[32] There is no need to explain "why bad things happen to good people." They happen not because of a vengeful or inattentive God, but because that's the way the universe is—hardly loving and gracious when it comes to the forces of nature. The universe has the potential for great beauty and good, and yet the cosmos out of which we originated is a harsh place that may inflict pain and suffering on humans, who, as a product of cosmic evolution, have themselves generated evil throughout history. There is no need for a Creator God, even though we cannot yet fully explain the origin of the universe or the origin of life. And there is certainly no need for a judgmental God in a world that overflows with judgment grounded in different worldviews large and small. There *is* a need for a loving and compassionate God, in the best tradition of all theologies. But this can be expressed naturally in the way we treat our fellow humans each and every day. Indeed, Karen Armstrong has made the point that, stripped of the supernaturalism and other accoutrements, compassion is at the core of all religions, even if the ideal is not always met.[33]

A naturalistic cosmotheology also leaves us with a number of questions. From an historical, psychological, and sociological point of view: how did the idea of the supernatural originate in human societies, by what mechanisms did it evolve to its present unassailable position in the human mind via the majority of major religions, and how does it maintain its hold? From an astrobiological point of view: what are the chances extraterrestrial intelligence would have evolved similar points of view? Put another way, is supernaturalism a universal or contingent feature of high-order consciousness and intelligence? From the point of view of astroethics: aside from a general reverence for all forms of life, including compassion, what moral principles can be drawn from a universe that, while it may seem harsh and indifferent, indisputably has ingrained in its core, laws of nature that are biofriendly? And last but not least, from a religious point of view: can such a naturalistic worldview satisfy human needs as much as supernaturalistic worldviews? If so, by what mechanisms? If not, why not?

---

32. Kushner, *When Bad Things Happen to Good People*; Arnold, *Why Do Bad Things Happen*, etc.

33. Armstrong, *Twelve Steps*, 6.

Surely an extraterrestrial anthropologist or theologian visiting Earth and studying the beliefs of its inhabitants would find curious the religious ideas many Earthlings take for granted, passed down through generations and often accepted without thinking too much about them. Specific beliefs associated with religion have clearly grown out of history: in the Christian tradition the doctrine of the Trinity of God dating to the fourth century; a baby filled with "original sin" that needs to be washed away, dating substantially to Augustine of Hippo in fifth-century Africa; a supernatural God who incarnates a son; the Virgin birth of this son—all these ideas and more have accreted over the centuries and given rise to innumerable religious wars over the most minute details, even as they have arguably formed the basis for Western civilization. Similarly with Islam, whose Sunni and Shia sects have waged wars large and small based on their version of the true successor to Muhammed in the seventh century, from which their respective religious principles flow. Whether in these Abrahamic religions or some other tradition, these are also principles that give many people meaning and solace to endure the difficulties of life. And they are one source of a morality and worldview that arguably and in important ways have held together the social fabric for millennia, even as its darker centrifugal forces sometimes tear that fabric apart.

By contrast, an extraterrestrial might well recognize moral principles grounded in the shared worldview of cosmic evolution. But can a cosmotheology, grounded in the natural rather than the supernatural, contribute to uniting the social fabric without the deleterious consequences of supernatural religions? As many astrotheologians and scientists now accept, the epic of evolution is a shared worldview that has brought humans where we are today, a concatenation of elements spewed from a supernova, congealed through eons of evolution, now contemplating the universe and our place in it, while seeking meaning and value even as that evolution continues. In the modern view of cosmic evolution—Genesis for the third millennium—the concatenation of elements may have produced innumerable other intelligences throughout the universe. They likely also seek meaning and value in the universe, and while we will undoubtedly find new extraterrestrial modes of seeking such meaning and value, we all share the same universe.

"Dust thou art, and into dust thou shalt return," according to one popular source of terrestrial theological wisdom. "Stardust thou art, and into stardust thou shalt return" is the inspiring vision offered by cosmic evolution. It is a unifying vision good enough for increasing numbers of people as we enter the third millennium. Whether it is good enough for the majority of Earthlings and Extraterrestrials, over the long or short term, remains to be seen.

Bibliography

Armstrong, Karen. *A History of God: The 4,000 Year Quest of Judaism, Christianity and Islam*. New York: Ballantine, 1993.
———. *Twelve Steps to a Compassionate Life*. New York: Knopf, 2011.
Arnold, David. *Why Do Bad Things Happen to Good People?: Answers to One of Life's Greatest Moral Questions*. New York: Creation House, 2009.
Behe, Michael. *Darwin's Black Box: The Biochemical Challenge to Evolution*. New York: Free Press, 1996.
Braxton, Donald M. "Modern Cosmology and Religious Naturalism." In *RCRS* (2012) 124–34.
Clark, Ronald W. *Einstein: The Life and Times*. New York: Avon, 1972.
Coyne, George. "The Evolution of Intelligent Life on Earth and Possibly Elsewhere: Reflections from a Religious Tradition." In *MW* (2002) 177–88.
Crowe, Michael J. *Extraterrestrial Life Debate from Antiquity to 1915: A Source Book*. Notre Dame: University of Notre Dame Press, 2008.
———. "A History of the Extraterrestrial Life Debate." *Zygon* 32 (1997) 147–62.
Danielson, Dennis. "The Great Copernican Cliché." *American Journal of Physics* 69 (2001) 1029–35.
Derham, William. *Astrotheology: Or a Demonstration of the Being and Attributes of God from a Survey of the Heavens*. London: Innys, 1715.
Des Marais, David J., et al. "The NASA Astrobiology Roadmap." *Astrobiology* 8 (2008) 715–30.
Dick, Steven J. "Cosmic Evolution: History, Culture and Human Destiny." In *Cosmos and Culture: Cultural Evolution in a Cosmic Context*, edited by Steven J. Dick and Mark Lupisella, 25–59. Washington, DC: NASA, 2009.
———. "Cosmotheology: Theological Implications of the New Universe." In *MW*, 191–210. http://www.metanexus.net/essay/many-worlds-cosmotheology.
———. "Critical Issues in the History, Philosophy, and Sociology of Astrobiology." *Astrobiology* 12 (2012) 906–27.
———. "Kosmotheologie—neu betrachtet." In *Leben im All: Positionen aus Naturwissenschft, Philosophie und Theologie*, edited by Tobias Daniel Wabbel, 156–72. Dusseldorf: Patmos Verlag, 2005. In English: http://bdigital.ufp.pt/bitstream/10284/778/2/287-301Cons-Ciencias%2002-5.pdf.
———. *Life on Other Worlds: The 20th Century Extraterrestrial Life Debate*. Cambridge: Cambridge University Press, 1998.
Dick, Steven J., and James E. Strick. *The Living Universe: NASA and the Development of Astrobiology*. New Brunswick, NJ: Rutgers University Press, 2004.
Dowd, Michael. *Thank God for Evolution: How the Marriage of Science and Religion Will Transform Your Life and Our World*. New York: Viking, 2007.
Einstein, Albert. "Religion and Science." In *Ideas and Opinions*, 36–40. New York: Bonanza, 1954.
Fridlund, Malcolm, and Helmut Lammer. "The Astrobiology Habitability Primer." *Astrobiology* 10 (2010) 1–4.

Gilmore, Michael. "Space Exploration." In *ESTE* (2005) 4:1831-35.
Goodenough, Ursula. *The Sacred Depths of Nature*. Oxford: Oxford University Press, 1998.
Goodenough, Ursula, and Terrence W. Deacon. "The Sacred Emergence of Nature." In *OHRS* (2006) 853-71.
Impey, Chris, et al., eds. *Encountering Life in the Universe*. Tucson: University of Arizona Press, 2013.
Isaacson, Walter. *Einstein: His Life and Universe*. New York: Simon & Schuster, 2007.
Kant, Immanuel. "Critique of all Theology Based upon Speculative Principles of Reason." *Critique of Pure Reason*. http://www2.hn.psu.edu/faculty/jmanis/kant/critique-pure-reason6x9.pdf.
Kauffman, Stuart A. *Reinventing the Sacred: A New View of Science, Religion and Reason*. New York: Basic, 2008.
Kaufman, Gordon D. "The Epic of Evolution as a Framework for Human Orientation in Life." *Zygon* 32 (1997) 175-88.
Kushner, Harold S. *When Bad Things Happen to Good People*. New York: Schocken, 1981.
Lamm, Norman. "The Religious Implications of Extraterrestrial Life: A Jewish Exotheology." In *Faith and Doubt: Studies in Traditional Jewish Thought*, 105-59. Jersey City: Ktav, 1971.
Lewis, James R., ed. *The Gods Have Landed: New Religions from Other Worlds*. Albany: SUNY Press, 1995.
Løgstrup, Knud Ejler. *The Ethical Demand*. Notre Dame: University of Notre Dame Press, 1997.
Lupisella, Mark. "Cosmocultural Evolution: The Coevolution of Culture and Cosmos and the Creation of Cosmic Value." In *Cosmos and Culture: Cultural Evolution in a Cosmic Context*, edited by Steven J. Dick and Mark Lupisella, 321-59. Washington, DC: NASA, 2009.
McKay, Christopher P. "Does Mars Have Rights? An Approach to the Environmental Ethics of Planetary Engineering." In *Moral Expertise*, edited by Don MacNiven, 184-97. New York: Routledge, 1990.
Moore, James R., ed. *History, Humanity and Evolution: Essays for John C. Greene*. Cambridge: Cambridge University Press, 1989.
Murray, William. "Religious Humanism Yesterday, Today and Tomorrow." *Religious Humanism* 35 (2000) 55-90.
Nagel, Thomas. *Mind and Cosmos: Why the Materialist Neo-Darwinian Conception of Nature is Almost Certainly False*. Oxford: Oxford University Press, 2012.
Peacocke, Arthur. "The Challenge and Stimulus of the Epic of Evolution to Theology." In *MW*, 89-118.
Peters, Ted. "Astrotheology: A Constructive Proposal." *Zygon* 49 (2014) 443-57.
———. "Exo-Theology: Speculations on Extra-Terrestrial Life." *CTNS Bulletin* 14.3 (1994) 1-9.
———. "Exo-Theology: Speculations on Extra-Terrestrial Life." In *The Gods Have Landed: New Religions from Other Worlds*, edited by James R. Lewis, 187-206. Albany: SUNY Press, 1995.
———. "Ten Ethical Issues in Exploring Our Solar Ghetto." *Journal of Astrobiology and Outreach* 25.3 (2016). doi:10.4172/2332-2519.1000149. http://www.escience

central.org/journals/ten-ethical-issues-in-exploring-our-solar-ghetto-2332-2519-1000149.php?aid=70370.

Randolph, Richard O., and Christopher P. McKay. "Protecting and Expanding the Richness and Diversity of Life, an Ethic for Astrobiology Research and Space Exploration." *International Journal of Astrobiology* 13 (2014) 28–34.

Rosenberg, Alex. *The Atheist's Guide to Reality: Enjoying Life without Illusions.* New York: Norton, 2012.

Sagan, Carl. *Pale Blue Dot: A Vision of the Human Future in Space.* 1994. Reprint, New York: Random House, 1994.

Schwartz, James SJ, and Tony Milligan, eds. *The Ethics of Space Exploration.* Heidelberg: Springer, 2016.

Stewart, Matthew. *Nature's God: The Heretical Origins of the American Republic.* New York: Norton, 2014.

Stone, Jerome A. *Religious Naturalism Today: The Rebirth of a Forgotten Alternative.* Albany: SUNY, 2010.

Sullivan, Woodruff T., III, and John A. Baross, eds. *Planets and Life: The Emerging Science of Astrobiology.* Cambridge: Cambridge University Press, 2007.

United States of America Congress, House Committee on Science, Space, and Technology, Hearings. "Astrobiology: Search for Biosignatures in our Solar System and Beyond." December 4, 2013. http://science.house.gov/hearing/full-committee-hearing-astrobiology-search-biosignatures-our-solar-system-and-beyond.

———. "Astrobiology and the Search for Life in the Universe," May 21, 2014. http://science.house.gov/hearing/full-committee-hearing-astrobiology-search-biosignatures-our-solar-system-and-beyond.

Vidal, Clement. *The Beginning and the End: The Meaning of Life in a Cosmological Perspective.* Heidelberg and New York: Springer, 2014.

Wilkinson, David. *Science, Religion, and the Search for Extraterrestrial Intelligence.* Oxford: Oxford University Press, 2013.

15

"ET, Call Church!"

Astrosemiotics and Shared Spirituality

MARK GRAVES

Regardless of how varied the communication between persons may be, it always involves the risk of one person daring to lay him or herself open to the other in the hope of a response. This is the essence of communication and it is the fundamental phenomenon of the ethical life.[1]

—KNUD EJLER LØGSTRUP

The distances between stars let alone galaxies far exceed distances originally conceived as astronomical. In the event that we identify a "nearby" inhabited exoplanet one hundred thousand light-years distant, will we be able to communicate with whomever lives there? Even if we would send messages at the speed of light, it would take one hundred thousand years for the exocivilization to answer the phone after we have dialed the right number. That planet's inhabitants might have gone extinct in the meantime. If interactive communication is to take place, timing will be important.

1. Løgstrup, *The Ethical Demand*, 17.

Frank Drake's Greenbank equation incorporates the right timing. One of the components of Drake's equation is the fraction of civilizations that develop a technology to release detectable signs of their existence into space. Implied is the assumption humans can detect that signal as communication. If extraterrestrial intelligence (ETI) communication is indistinguishable from background natural phenomena, humans could not detect the signal. Conversely, without certain cognitive and technical capacities, ETI is unlikely to receive and understand human communication.

We have seen in previous chapters how both scientists and theologians are concerned about the rational capacity of aliens living on other planets. Some theologians identify the *imago Dei* with the human capacity for reason; and Enlightenment philosophers in the Kantian tradition make reason a prerequisite for dignity. In this chapter we will take up the matter of interplanetary communication. In taking up communication, we must imagine with greater specificity just which rational or cognitive capacities might be requisite.[2]

Three rational or cognitive capacities which appear necessary for interplanetary communication will be explored in this chapter.[3] First, abstract thinking combined with symbolic language. The ability to model nature adequately enough to create technology for interplanetary signaling likely requires an ability for abstract thought and the ability to communicate those abstractions to others in what for humans consists of symbolic language. The field of *semiotics* examines communication in its most general form, and the tools of semiotics can categorize the signals that humans (and perhaps ETI) might understand regardless of the source.[4]

In semiotics, symbols refer to some object through socially constructed convention. Conventions are arbitrary: the English first-person plural *we* has no physical connection or similarity to the French affirmative declaration *oui*. Conventions can also refer to a complex shared interpretation of numerous people spread over millennia, long separated from their original

---

2. My employment of semiotics here is partially dependent on cognitive science. "Cognitive science is the science that seeks to relate the psychological functions of information processing (in thought, emotion, intention, volition, valuation, agency) to the physically measurable signals from the human body (measures from the brain, heart, skin, eyes, breath, bodily posture, bodily movements)." Spezio, "The Cognitive Sciences," 285.

3. Additional ETI capacities may be required, including having cognition at all, but for extraterrestrial *intelligence* I bracket those implied cognitive or intellectual capacities as well as any evolutionary or embodied assumptions to focus on the cognitive capacities specifically required for communication.

4. Cf, Coe et al. who consider evolutionary theory to analyze interstellar communication. Coe et al., "ET Phone Darwin," 214.

physical actions (e.g., religious doctrine, stock price, or republic). One can imagine intelligent communication without symbolic language (e.g., among chimpanzees or dolphins or possibly through music), and ETI language likely differs from human language, but the development of advanced technology appears to require the ability to manipulate abstractions of nature and to communicate those abstractions to others in a complex way.

Second, in addition to the cognitive and linguistic skills necessary to communicate with symbols, interplanetary communication likely requires the capacity to imagine an absent and unknown communication partner. Our newly discovered extraterrestrial partners would have to have developed a theory of mind. *Theory of mind*, according to Frans de Waal, "refers to the ability to know what others know, that is, to attribute mental states such as intentions, goals, and knowledge to others."[5]

We earthlings enjoy theory of mind: we can imagine ETI might exist and also how they might think about communicating with us.[6] Without the ability to conceive of other minds, ETI would have no reason to intentionally attempt interplanetary communication. Terrestrial humans might discover ETI who do not have a theory of mind, but such ETI could not *communicate* with us.

A third capacity is the ability to create shared interpretations against which one can relate one's own interpretation. Communication has the Latin root *commūn(is)*, which means to make common—the same root as communion. In communicating with another person, one attempts to create a shared interpretation of some experience or abstraction. Astronomers know about billions of stellar objects and events, even though no individual astronomer knows everything; and an astronomer can compare their interpretation of an observation against the collective understanding (through literature and database searches and peer review). Science depends upon these shared interpretations of the natural world. So do most human social endeavors, including traditional institutional religion and a person's spiritual experience of something *more*. Communing with the other—or, for Christians, with Christ—requires some type of shared communication, even if nonverbal.

---

5. de Waal, "Apes Know What Others Believe," 39. Astroethicist Kelly Smith would argue that without a certain level of socially functioning intelligence we could not value let alone communicate with alien entities. Smith, "The Trouble with Intrinsic Value," 261–80.

6. For an early discussion of interstellar communication, Cameron, *Interstellar Communication*. For current investigations, see Vakoch, *Communication with Extraterrestrial Intelligence*.

In this chapter, I tackle two of the four tasks for astrotheologians listed in the second chapter. I expose and examine extra-scientific assumptions (the Third Task) and prepare the public for the eventuality of extraterrestrial contact (Fourth Task). Most of the extra-scientific assumptions I highlight are how humans interpret, understand, and communicate about the world in which we live. Although not overtly religious, those assumptions are influenced by long-standing views of the person intertwined with religious values about the human person. Because theology attends to the interpretation of human existence, I can use theology to highlight nuances of how humans and ETI might interpret each other, which my scientific tools would have missed. In exposing assumptions and clarifying the way humans and/or ETI might interpret each other, I identify human cognitive structures that ETI may or may not have. The search for ETI (SETI) generally assumes ETI have those cognitive structures even though evolution on Earth suggests that those abilities are relatively unlikely to evolve. Rather than assume that the intelligence of ETI depends upon human-like cognitive and linguistic abilities, examining less anthropocentric models of intelligence may open up new avenues for the search. Finally, if we do discover ETI capable of interplanetary communication (or they discover us), their capacity to develop technology suggests aliens might have a cognitive capacity to interpret nature that we would consider intelligent, symbolic, and imaginative.

Semiotics also makes explicit the process of interpretation that the American philosopher of religion Josiah Royce used to develop his understanding of spirituality; and I use Royce's philosophy of community to describe how humans and ETI might have shared spirituality. Shared spirituality depends upon sharing interpretations, such as human interpretations of humans and of ETI, ETI interpretation of ETI and of humans, and each interpretation of the shared experience of discovering another exoplanetary civilization. The three cognitive capacities appear necessary for interplanetary communication and also appear foundational for humans to develop religion. Although ETI might have interplanetary technology without having religion, they could not have such technology and also communicate with humans using symbolic language without the capacity for collective symbolic interpretation that humans would understand as having a spiritual dimension.

In that vein, I suggest one way to interpret and value ETI that might lead to a harmonious shared spirituality (and identify why it might not). In Royce's philosophy, stable communities—which he describes in a positive way as true communities—depend upon a principle of loyalty. This loyalty or commitment by the members is aimed not only at the community-specific

"ET, CALL CHURCH!" 249

vision, beliefs, etc., but also to a universal principle of loyalty. For harmony among these "true" communities (grounded for him in the early Christian church), each community must commit to the right of other communities to commit (or remain loyal) to their vision, belief, etc., even if they conflict with one's own. The only requirement is that they also remain loyal to the principle of loyalty (which prevents false or oppressive communities). By affirming the right of ETI to hold their beliefs yet also insisting they affirm the right of others to independent and possibly conflicting beliefs, humans take an ethical stance leading toward harmonious shared spirituality.

The human ethical stance may prove essential to human-ETI relations. Although ETI may require cognitive capacities comparable to humans in order to communicate with humans through interplanetary technology, aliens may only have a capacity for spirituality without having developed ethical and theological frameworks adaptable to unknown others. (Consider European medieval responses to Islam or the evangelization and colonization of indigenous populations in the Southern hemisphere.[7]) ETI capacity for technology development may easily exceed that of humans, but somewhat scarily, what if their ethical framework is less inclusive than humans? What if their ethics never adapted to pluralism and they consider humans as savages requiring colonizing? In that case, an appropriate ethical stance and readiness to commit to harmonious shared spirituality may prove to be in everyone's best interest.

For communication between humans and ETI, I consider three cases: (1) ETI receiving human signals, (2) ETI transmitting signals humans can receive, and (3) shared ETI-human communication in some form. Then, I examine possibilities for a shared spirituality and reflect upon that theologically.

First Case: Human Signals and Meaning

Imagine extraterrestrial beings on a planet close enough to eventually receive signals sent from Earth during the twentieth century. At first, they might pick up modulations in radio waves that appear to vary over time in ways unlikely to naturally occur. Signals from earth would significantly differ from noise or simple regularities in a way we would describe as having *information*. Information in this technical sense (of Shannon information) refers to the capacity to transmit different alternatives, which we measure

7. For a historical and linguistic examination in a SETI context of early colonial activities, see Kuznicki, "Inscrutable Names of God," 203.

mathematically using "bits" (now a familiar term to measure computer memory or the capacity of a computer network). One could measure early radio waves similarly, and the amount of information in a transmission would appear significantly greater than naturally occurring stellar objects. That information exists even if no ETI were to receive it, but if they were to *receive* information from Earth that reception would be information in a different sense of the word—the signal would *mean* something in addition to what the content of the signal refers. ETI might interpret the signal as meaning, "we are not alone," "yet another alien civilization is messing up our subspace communication," or "hope the planet Earth can save our dying civilization."

As ETI eventually receives signals from Earth's transmissions of the early twentieth century, they would likely have several challenges to learn about humans. In picking up early radio transmissions, they might isolate signal from noise, but how would they interpret that signal? Even if they had the same five senses as humans, which sense should they use? Humans primarily depend upon vision to navigate our world, but many animals use smell, so those could be plausible first attempts. An exceptionally perceptive ETI project manager in their space program might realize that a likely first use of Earth's new technology might be for entertainment, and thus smell would be the most likely sense to communicate over radio waves (as everyone on the ETI project and all of Earth's beagles would readily affirm). How would an ETI engineer attempt to decode smell signals that the animals from Earth might transmit? After attempting various decodings without success, a few fringe thinkers might consider sound, but the rational and practical engineers would probably consider that intrinsically noisy medium of limited usefulness and thus unlikely. After attempting to decode smell, sight, and taste and also attempting a complex mathematical project to decompose the signal's degrees of freedom across various abstractions in case Earth creatures might have a different sense, some hard working engineer late after the second sun sets on the horizon might try sound. After the ETI sound engineer decodes the signal using the results of prior failed attempts to narrow the range of parameters, he/she/it might eventually "hear" a human speak.

What is the human saying? How long would it take ETI to realize Earth has multiple languages? What are the humans talking *about*? Further analysis might identify words and phonemes. A particularly long transmission might give an ETI scientist enough regularity to calculate word frequency. (If ETI only had one language, would they even have the concept of "language" much less the discipline of linguistics?) Eventually, they may be able to group the words into languages. But with just a bunch of words, how could they ever discover what the words mean?

When television images begin to arrive on an exoplanet, ETI has some new information. After the process of decoding Earth television signals, ETI can begin associating images with sound. A human speaking would show what we look like. A concert would show how we make music. An analysis of *The Three Stooges* might at first confuse them greatly.

Eventually correlations between Earth images and the words in a transmission may give ETI some connections between them. Stories about forests, national parks, and logging might have pictures of trees. Stories about tropical storms, surfing, and ships might have images of the ocean. When American education television programs of 1960s and 1970s begin to reach extraterrestrial receivers, then aliens may learn direct co-occurrence relationships between words and images; but without any instruction, learning what the words really mean would remain elusive. Imagine someone unfamiliar with baseball attempting to understand "ground rule double," "double play," and "infield fly rule" without instruction. To understand any game such as baseball, one needs to learn the objectives, roles, objects, and rules, and sometimes the rules depend upon a history not apparent in any particular game.

The Austrian philosopher Ludwig Wittgenstein describes language in terms of a game. For language to have meaning beyond the mathematical information carried by its signal, one must interpret the language in a particular context—a particular group who understand not only the language rules of grammar and word meaning but also how the language is used. Even in a small group of friends, a family, or a research field, aspects of language have complex and particular meanings that differ from those outside that group.

Semiotics generalizes human language to any medium or sense modality and examines the process of meaning formation through an organism's apprehension of the world through signs. The American philosopher Charles S. Peirce developed semiotics with several organizations of signs. The simplest organization of Peirce's semiotics consists of three kinds of relations to objects in signs: icon, index, and symbol. An *icon* signifies by resembling its object like a painting or a map. It possesses a quality that resembles or duplicates those of the object. An *index* represents its object through an existential connection between itself and the object. For example, a fingerprint not only resembles the ridges of a fingertip, it signifies the existence of a particular finger. An index may also signify by a causal relationship, such as a thermometer or weather vane. A *symbol* represents its object through a convention that governs how the symbol will be used: A symbol refers to an object by social convention without direct similarity (as in an icon) or existential or causal connection (as in an index). For example, the word *cat* is connected to an animal through conventional English usage, and thus one can categorize human language as symbolic language.[8]

On the plaques attached to the spacecraft Pioneer 10 and Pioneer 11 launched in the 1970s, some thought went into assuring the visual signs were iconic and indexical (though I found no evidence that Carl Sagan or others involved explicitly drew upon semiotic theory). A line drawing of the Pioneer spacecraft iconically represents spacecraft and is used to communicate shape and size of human figures. An icon of the hyperfine transition of hydrogen also creates an indexical relationship between its wavelength and the units of measure used elsewhere on the plaque. The relative position of the Sun to the center of the Milky Way galaxy and fourteen pulsars uses indexical signs (like a fingerprint) to communicate Earth's location iconically. The iconic map of the Solar System has an indexical line of the Pioneer's planned trajectory through the Solar System with an arrowhead symbolically indicating direction (and inadvertently referring by human social convention to the cultural activity of using a bow and arrow).[9]

Symbols require interpretation. The social group that defines a language also defines possible ways the words and more complex abstractions such as metaphor might be interpreted. The cognitive linguist George Lakoff pushes past the obvious metaphors in language and claims that most (if not all) language is metaphoric to some degree unless describing physical reality. These metaphors influence how we interpret meaning in symbolic language. For Lakoff and his philosopher coauthor Mark Johnson, the essence of metaphor is to understand and experience one kind of thing in terms of another. One understands Argument Is War in terms of attack, defense, counterattack, victory, defeat, strategy, and the like, and an intense verbal battle can even trigger emotional responses appropriate for physical danger. Some metaphors directly connect to one's body—an argument that "makes one's blood boil" does relate to actual body temperature, blood pressure, and heart rate.[10]

In addition to interpreting the metaphor *Argument Is War*, one could also use the metaphor *Argument Is Game-Playing*, and lay one's cards on the table. More complex symbol interpretation may function similarly to the

---

8. Note, Paul Tillich and Carl Jung define symbol and sign differently from Peirce: for them, a symbol partakes of the reality to which it points and a sign simply points to it.

9. Gombrich, "Visual Image," 255–58; and Gombrich, *Image and the Eye*, 150–51. For further analysis of possible signs, see Vakoch, "Signs of Life Beyond Earth," 313.

10. Lakoff and Johnson, *Metaphors We Live By*, 4–5, 15.

basic linguistic interpretations. What about interpreting the symbol love? One could use Love Is War: "She fled from his advances" or "He won her hand in marriage." One could use Love Is a Physical Force: "I could feel the electricity" or "They lost their momentum." Another metaphor is Love Is a Journey: "We hit a dead end and may have to go our separate ways" or "Look how far we have come." Forget about ETI communication for a second, imagine the issues that might arise between two humans when one of them interprets love using the metaphor Love Is A Journey and the other uses Love Is War.

Might our alien listeners be literalists? Consider ETI who has the ability to manipulate symbols, but lacks the ability to understand metaphor (e.g., similar to a computer). ETI might identify regularities in human communication but would not have the capacity to learn what they *mean*. A super-literal interpretation of symbolic language is challenging to imagine, but ETI might have some limited capacity for basic metaphoric reasoning without developing the many layers of modern language (e.g., as early linguistic humans may have gradually developed metaphoric reasoning and communicated that through culture).

Even strictly literal communication depends upon implicit knowledge. How much analysis must ETI do to understand a story about water pollution caused by paper production, if they do not drink water from rivers and live on a planet without trees? Like a game, language has interconnected rules, various objectives, and reference to objects and events that may have occurred in the distant past. How much historical and cultural knowledge is needed to understand the sentence, "In the face of plummeting stock prices, the president's speech to investors hit a home run"? One must understand the obvious metaphoric connection to baseball, the no-longer-novel metaphor that stock prices have a height and can rise and fall, and the implication that some unmentioned event triggered the stock price fall. But if one were to consider "president" in its original meaning of one who presides over a meeting, one would misunderstand the meaning of the sentence. The president presides over something unmentioned in which the investors invest. Furthermore, in the context of an example sentence in an academic text, does the sentence refer to the president of the United States after many stock prices plummeted or to the president of a corporation after its stock price fell? In an American business context where the CEO, Chairman of the Board, or CFO (Chief Financial Officer) of a corporation might have plausibly addressed its investors, what does it mean that the president of the company spoke instead?[11]

---

11. One may also interpret religious symbols metaphorically. The contemporary

Given the complexity in understanding language, it is amazing that humans ever manage. But humans appear to learn new meanings in a novel context fairly easily and can use simple meanings to create complex abstractions almost automatically. Other animals find human language impossible even though the intelligence of some animals is otherwise comparable to humans. Humans have a few cognitive capacities that make language use relatively simple, and for animals (or ETI) without those capacities, human-like language is essentially impossible.

Human beings have the ability to abstract and interpret symbols and to manipulate them using language while keeping in mind how others interpret them. For ETI to interpret the signals from Earth, they likely require a similar ability to abstract the regularities of nature and consider differing interpretations. For ETI to receive and understand the meaning of human signals from Earth, they likely must be capable of symbolic communication with humans (even if also capable of other means of communication). ETI may use that capacity to interpret regularities of nature that led humans to develop science, mathematics, and technology capable of interplanetary communication, in which case, they may wish to signal us.

Second Case: ETI Communications Coming Our Way

The 2016 movie, *Arrival,* provides an imaginary test case in the semiotics of terrestrial-extraterrestrial communication. Before interstellar space ships arrive on our planet, however, more than likely we will receive a radio signal.

At that point we will need to distinguish between accidental and intentional communication. Although possible ETI accidentally signaled Earth without considering who might be listening (as humans did through radio and television broadcasts in the early twentieth century), I consider the case where ETI emits a signal with the intention of interplanetary communication and listens for interplanetary signals. Otherwise, if they only accidentally signaled Earth, we could only learn they exist, but we could not communicate with them.

What cognitive capacities are necessary for ETI to transmit signals to humans? Extraterrestrial signals are challenging to imagine and characterize, but communication requires a receiver. Given the somewhat unlikely, but very important assumption that ETI can communicate with humans,

---

theologian Sallie McFague makes explicit the role of metaphor in attempts to reason about God. McFague develops a metaphorical theology that uses a variety of metaphors to try to understand God's characteristics. McFague, *Metaphorical Theology.*

one can characterize signals of ETI by what humans can receive. I argue that releasing detectable signs of existence into space would very likely also require the capacity to communicate in a way characterized by semiotics as symbolic.

Consider if ETI (a) were sufficiently intelligent to quickly grasp general patterns in the world and (b) could communicate about the world to others, but (c) lacks the ability to abstract those general patterns into symbols. One can imagine very intelligent chimpanzees or dolphins who can quickly solve complex concrete problems, but cannot manipulate abstractions such as the number 999, a social contract among equals, or the gravitational constant. The capacity to abstract the world into natural numbers (not to mention zero, infinity, algebra, or calculus) requires an ability to abstract generalities in nature to stand-alone abstractions that one can manipulate. Similarly, the ability to conceptualize social relationships into political or civic abstractions requires a specific capacity beyond the ability shared with other animals to participate in power hierarchies. Humans in groups not only choose a leader, they develop a variety of political theories. Without that ability, ETI would likely lack the capacity to socially organize the study of nature sufficient to manipulate it as technology. (In many ways, an ETI without the ability to create social abstractions could not form a "civilization" as Drake envisioned it.)

Iconic and indexical communication is probably insufficient for developing interplanetary technology. Given the prevalence of iconic and indexical communication among species on earth and that out of several million years of intelligent animals, only humans have symbolic language, the likelihood is relatively low that intelligent life without symbolic language will develop technology capable of releasing detectable signals. Without symbolic language, ETI could neither understand human language nor develop technology to transmit or receive interplanetary signals.[12]

---

12. There is a small gap in my argument here. ETI requires the capacity to generalize its world into abstractions that semiotics would identify as symbols and communicate them. But human language also has grammars that allow for a potentially infinite number of sentences. I write this sentence for an astrotheology book chapter while sitting in a coffee shop in San Francisco on June 6, 2013. That prior sentence has probably never been previously uttered, and one easily generate trillions of similar sentences by changing dates, locations, places, activities, etc. ETI requires the ability to create general abstractions and communicate them, but they might have languages with simpler grammars. (E.g,, Noam Chomsky identifies four levels of formal grammars.) In that case, ETI might be able to develop interplanetary technology with something similar to symbols but without having human-like symbolic language. E.g., in Peirce's philosophy he describes the "generals" of nature, which are more dependent upon reality than symbols are, and ETI might be able to communicate with these generals in some way to develop a type of science. However, ETI would still need some way to

Conversely, given the short period of time between when the evolution of symbolic language in humans began and the current stage of technology, the likelihood of creatures having a capacity for symbolic language and *not* developing advanced technology appears small. The time between culturally identifying the abstraction of numbers and civil structures and then developing modern science was relatively short. Once ETI has the capacity beyond iconic and indexical communication to manipulate symbols, they may quickly use those symbols to create technology (if they have some need such as evolutionary pressure on their survival or flourishing).

What capacity is necessary for ETI to listen for interplanetary communications? For ETI to intentionally transmit and listen to interplanetary signals, a likely assumption is that they have an ability to imagine us. Not humans in particular, of course, but an ability to imagine that something "out there" exists. For them to listen suggests they can conceive of an unknown other. We humans consider the unknown other in a variety of possible ways including desire for connection, fear of out group members, hope of relational fulfillment, etc. Without speculating on the particular forms of ETI mentality, it appears likely ETI who can intentionally send signals can conceive of another who might be affected by their actions and change beliefs to know they exist.

Because we enjoy theory of mind, we humans are good at imagining other's beliefs about something even if those beliefs differ from one's own. As mentioned above, social cognitive scientists call this ability *theory of mind*. Theory of mind refers to the ability of humans to form distinct ideas, or "theories," about the minds of others. Although somewhat controversial, the evidence appears to suggest that only humans older than a few years of age (four to seven years) have theory of mind. Other animals and the youngest humans have various precursors, such as the ability to distinguish between accidental and intentional acts or the ability to share attention or follow eye gaze, but cannot reason about the distinct beliefs of others.

Experiments on theory of mind often involve a false-belief task where a subject watches an experimenter hide an object, such as a ball, under one of two boxes as a second person observes. The second person leaves the room, and the experimenter switches the location of the ball. Where does the subject believe the second person will look when that person returns? Under the second box where the subject knows the ball rests, or under the first box where the subject knows the second person saw it placed? Adult humans expect the second person will look under the first box, where that person believes the ball was placed; but human infants, very young children,

---

create new abstractions to imagine and develop advanced technology.

and most (if not all) nonhuman primates indicate (through a variety of experimentally verified nonverbal cues) that they believe the second person will look where the ball now actually rests, not where the second person thinks the ball rests. Those that lack a theory of mind cannot separate their own beliefs about the world from the different beliefs another person might hold.

Both symbolic language and something like theory of mind are necessary to listen for interplanetary symbols. Consider three alternative scenarios. For the first scenario, consider ETI who has (a) symbolic language, (b) the ability to imagine an unknown other, and (c) a society with the organizational capacity adequate for developing a technology for interplanetary signaling. Their civilization would have developed to the point the ETI would develop shared understandings of many particular symbols. They would have needed to imagine the technology before developing it—probably a long series of technological advancements.

Second, consider ETI with the cognitive capacity to create abstractions but that (a) never developed oral language that humans acquired with a descended larynx and conscious control over breathing (for controlled vocalization) nor (b) developed the ability to combine symbols in a complex grammar. Imagine a hominid such as Neanderthal who have large brains like *Homo sapiens* but must gesture or grunt to communicate. They could create sophisticated tools but would likely lack the ability to develop advanced mathematics, theoretical science, or linguistics. Such an ETI might develop long-distance ways of communicating indexical relationships (such as a signal fire to communicate that an enemy is approaching) but would not listen for symbols defined through another group's convention.[13]

Third, consider ETI with extensive capacity to communicate using symbols but lacks the ability to model the beliefs of others. Although difficult to imagine how such a creature might have evolved, one can approximate such creatures using artificial intelligence as it currently (or soon might) exists. Such an ETI would lack the ability to imagine whether another might exist and would not listen for communications from unknown others.

All three forms of ETI have symbols. The first can share those symbols with others and cognitively model different ways in which the other could interpret the symbol. The second is limited in the ability to share those symbols, and the third lacks the ability to understand how they might be

---

13. In this example, I assume Neanderthals had the ability to create symbols, but not fully developed vocalization or grammatical abstraction. Some recent evidence suggests Neanderthals may have some of those abilities. In that case, earlier hominins would be a better example. Barney et al., "Articulatory Capacity of Neanderthals," 88–102.

interpreted. The second and third both lack the ability to develop shared interpretations.

ETI capable of developing interplanetary communication likely requires symbols and the ability to conceive of unknown others. ETI must be capable of creating new symbols that refer to human meanings and then acquire what those symbols mean. One could use independently discovered properties of the universe as a source of objects about which humans and ETI could develop symbols with a shared interpretation, but both groups must have the capacity to learn the distinct interpretations of the other in order to begin to reconcile those interpretations.

Third Case: Humans and ETI Sharing Interpretations

Consider the case where aliens and humans can directly communicate with each other through symbols. As a simpler alternative to directly visiting the other planet, perhaps ETI broadcasts a complex signal that when received and decoded by Earth scientists contains sufficient instructions to build a device capable of instantaneously transmitting simple signals across stellar distances. (Maybe they suggest a creative use of a consistent, distant source of entangled particles or vibrations across folds in space-time.) What do we signal to them? How do we construct a shared language? The instructions may give us some information about ETI. Diagrams would suggest they have vision. Measurements may suggest something about their size and dexterity. Challenges in interpreting their instructions may suggest cognitive differences. How do we expect they would interpret the first signal from us? Do we signal "hello," a scientific question, a picture, our location, or something else?[14]

What can we imagine they might know? If aliens have developed technology capable of transmitting and receiving radio signals, they almost certainly would have noticed general patterns that human scientists have identified as frequency, noise, energy, etc. Their science might differ from that of humans, but they would most likely have noticed the same regularities and relationships in nature we have discerned.

Nature has regularities—Atoms clump together in certain ways; Energy occurs in specific forms. But the human senses limit our perception of those regularities. Birds and bumblebees perceive the regularity of polarization in sunlight that humans cannot naturally perceive. Human cognitive structures limit the ways we can conceive them. Bats have the ability to

14. See, for example, Vakoch, "Earth Speaks."

echolocate using bounced sound waves and may conceive of the world and their place in it differently than humans.[15] If an intelligent species similar to Earth's shark were to develop science, they would not need Maxwell's equations to model the relationships between electric and magnetic fields, because sharks' ability to sense electromagnetic fields directly would preclude them from originally conceiving electric and magnetic fields as separate.[16] Even if aliens were to have similar biology to Earth's, they would likely have very different ways of conceptualizing science.[17]

One might argue that mathematics captures nature's regularities without biological or cultural dependence. But mathematics is still a language, and philosophers of mathematics characterize the variety (and often infinite variety) of ways mathematics could develop.[18] In addition, many physical problems and most biological systems defy analytic solution (e.g., because of chaotic processes highly dependent on initial conditions). In the human brain, various regions of the parietal lobe appear essential for our ability to do mathematics, and ETI may have a similar ability to create abstractions about regularities and manipulate them in complex grammatical ways. The ability to develop these types of languages may be a precursor to developing technology capable of signal transmission or reception across stellar distances.

As an aside, one may wish to clarify Drake's fraction f_i of planets that develop intelligent life. I suggest we distinguish between extraterrestrial civilizations with bare intelligence which depends upon indexical representation from an extraterrestrial civilization which depends upon symbolic abstractions for civil roles and structures. On Earth, chimpanzees and dolphins are intelligent; but they could not develop civilizations. The fraction f_c of civilizations that develop a technology that releases a detectable signal into space may also need clarification about what civilization requires. One needs to be specific about how the four components of intelligence, symbolic language, civilization, and advanced technology map to two fractions

---

15. Nagel, "What Is It Like to Be a Bat?" 435–50.

16. In addition to vision, sharks also sense electromagnetic fields through the *ampullae of lorenzini*. Thanks to Oliver Putz for suggesting this example.

17. Even among humans, cultural differences impact the development of science. Douglas Vakoch identifies the "Incommensurability Problem" as a cognitive mismatch between humans and ETI and suggests that differences in human and ETI science might best be overcome by developing iconic signs for extraterrestrial messages that refer to phenomena of nature. Vakoch, "Constructing Messages to Extraterrestrials," 697–704.

18. For a discussion of alternatives in the SETI context, see ibid.

f_i and f_c instead of assuming that intelligence and civilization are synonymous.[19] For example:

- The fraction of planets with life that develop intelligent species may be moderately high (it just may take several hundred million years).
- The fraction of intelligent species that develop symbolic language is very low (on Earth).
- The fraction of species with symbolic language that develop civilizations is likely high. (It happened in less than one hundred thousand years with humans and perhaps much less.)
- The fraction of species with symbolic language and civilizations that develop advanced technology may also be very high (depending upon motivational factors).

Imagine ETI with an ability (a) to create, manipulate, and communicate symbols and (b) to conceive of unknown others. They might create technology to communicate and listen for others. What would motivate them to do so? What longings would they have? Is there a capacity necessary for wanting more out of one's life? These aliens would have shared interpretations among themselves. Would they play with those interpretations in poetry or push them in art? Would they attempt to improve the accuracy of their interpretations of the world through science? How would they deal with conflicts between different beliefs or desires?

Another cognitive capacity that appears necessary to understand human beings is our ability to conceive of oneself as other. How would another conceive of me? Cognitive scientists call this capacity "mind reading." It has nothing to do with clairvoyance, but with the ability to read what another person might think about a third person's thoughts (including oneself).[20] What does Mary think of what John thinks about Sue? What do I think you the reader think about ETI thinking about us? One might get lost in a sentence with too many redirections, but the film genre of romantic comedy plays with slight distinctions between what two people think about what another thinks of them. Television serial dramas (such as soap operas) can build complex stories about various different beliefs that others know about each other and how those distinct beliefs change. Would ETI without a capacity similar to mind reading understand those human broadcasts?

---

19. *Drake* xix; or Drake, SETI Institute, "Drake Equation."

20. More precisely, the ability to read what another person might think about a third person's thoughts may be called second-order mind reading.

Human emotional development depends upon the ability to separate one's beliefs and desires from others, especially in the presence of strong emotion or attachment; and for most people, compassion is a spiritual aspiration rather than a reflexive act. ETI might have greater capacity to think about multiple distinct beliefs and desires and how they might relate to each (as a particularly adept politician might) but would still miss something of human experience without the capacity to conceive of oneself as other. Would an alien understand racism without the capacity to understand how he or she is perceived differently from another person by a third? Could ETI develop the virtue of courage without understanding how another might conceive of their actions as cowardly or foolhardy? Would ETI develop theology without the ability to wonder what an unknown deity thinks of one's actions and beliefs?[21]

I contend that ETI with symbolic language and the ability to conceive of an unknown other has the technological capacity for interplanetary communication, but without the ability to wonder about humans wondering about them, they might not build it. Even among humans, those who do not imagine ETI wondering about us do not participate in SETI. Perhaps Drake's equation needs a fractional component to capture not only the civilizations capable of advanced technology but also those who hope to find exoplanetary others enough to build such technology. What if the galaxy were populated with numerous civilizations, and we were the only one who cared? To communicate with extraterrestrial beings, earthlings must find not only a technologically capable civilization but also one that wishes to commune with us—a civilization desiring to find exoplanetary others to learn how they look through our eyes. Such a civilization may not have developed institutional religion, but they would have the capacity to understand human longing for something *more*.

The possibility of conceiving of as-yet-unknown others is what enables us humans to imagine deities and other unknown spiritual forces. The ability to create shared interpretations enables humans to wonder how those religious *Others* might interpret the world. Although ETI would not necessarily have considered unknown others in our religious terms, I argue they must have the capacity to learn this about us if communicative sharing is to take place. My argument is that the linguistic capacity required to communicate with humans and the cognitive capacity required to *develop* interplanetary

---

21. ETI might have another motivation such as an immediate and pressing need to find a habitable planet, but then they are searching for a physical place or perhaps a "biological" environment, not searching for an extra-planetary intelligence with which to communicate.

technology is sufficient to *understand* human spirituality. This is step one toward developing a shared human-ETI spirituality.

Sharing Spirituality

Spirituality orients one toward something more than one's immediate context. Spirituality directs one beyond one's individual self and concerns. In this metaphor of *Spirituality Is Direction*, does one follow a particular path or wander around lost? How does one discern one's path? Does one have to begin following the path to really see it? How does one choose the path? Does the path choose the individual? Where does one find guidance and direction?

Religion defines an orientation through its beliefs, rituals, practices, and social structures. For Christianity, the Wesleyan quadrilateral (described in chapter 2) provides four sources of direction: Scripture, tradition, reason, and experience. Each of those four sources guides the Christian in many ways. One thing the four have in common is they help the individual interpret the world in a particular way. As one reads and interprets Scripture, one learns concepts, symbols, patterns of moral behavior, and stories of significant figures who model (and fail to model) Christian faith. The tradition (or history) of one's religious denomination includes doctrines, creeds, significant human exemplars of faith, and ways of interpreting Scripture and what happens in the world. Interpretation is an aspect of reason, and one can learn ways to reason that are more inclusive, nuanced, and rational using values from the other sources of Scripture, tradition, and experience. Experience requires both a bare encounter with one's world and an interpretation within the cognitive structures defined by one's culture, religious tradition, and language.

Outside of religion, many tight-knit social groups may have their shared ways of interpreting reality and, therefore, a type of emergent spirituality. Families, not-for-profit corporations, sporting clubs, businesses, universities, towns, etc., each have a particular way of interpreting the world and the activities of its members. This shared way of interpreting is a beginning of spirituality. Within highly cohesive groups, the shared interpretation among members influences individual behaviors and interpretations, i.e., the shared interpretation can cause something to happen: one acts in a way different than one would without the shared interpretations.

The timing for when one stands up or sits down in a church worship service depends upon a shared interpretation in a tradition that one may not be able to fully explain. One may independently look up at the stars, but identifying planets and constellations depends upon shared (scientific and historical) interpretations. When one looks at Mars, how one interprets that experience depends upon a long history of others looking and interpreting that sight, which one might subsequently share with another. Does one identify the sight with a god of a religious tradition (e.g., the Roman god of war)? Does one recall recently viewed images from the Mars rover? Does one marvel at the distance light traverses from the Sun to reflect off Mars and travel to Earth? Or, as in the case of many ancients, does one believe light is tied to vision and thus one is *seen* by Mars?

When a group of individuals with high social cohesion share a common interpretation, that group forms a community. The group will likely share many beliefs and goals and disagree on other ones, but as I use the term *community*, it has some common vision, value, or concern. I suggest that one shared value for any ETI with which we find ourselves communicating is the value of interplanetary communication. Humans and ETI likely share the vision of some unknown other existing out there. By committing to a shared cause or vision, the members of a community create the unity that holds the community intact. The individuals work and communicate to enact, articulate, and interpret the shared vision rather than simply to maintain that social structure. M. Scott Peck defines community in terms of deep respect and true listening for the needs of others in community.[22] In classical cognitive terms, a community shares a heart, mind, and will that results in social cohesion rather than self-interest.

Drawing upon the early twentieth-century, pragmatist philosopher of religion Josiah Royce, I suggest that visions that include a commitment to the right of others to commit to alternative visions support a harmonious spirituality. The inclusion of alternative vision supports both the diversity of interpretations within a community and the harmonious interaction with other communities that have possibly conflicting views. To form a shared spirituality with ETI, humans must value the difference that ETI might bring to our interpretations of others and our world.[23]

In community, one commits to a cause that acknowledges the right of others to commit to their cause, with the one restriction that the other cause also include the same principle of commitment to commitment, or what Royce calls "Loyalty to Loyalty." In addition to the particularities of one's cause, one also commits to the principle of commitment. Humans might commit to the cause of promoting human flourishing, and ETI might commit to the cause of promoting their own flourishing. The groups might

22. Peck, *Different Drum*.
23. Royce, *Philosophy of Loyalty*.

independently form cohesive social groups but would not form Roycean spiritual communities without committing to the right of the other group to have their own cause. One's dedication and commitment to the principle of commitment demonstrates support and obligation to the loyalty of those in other communities even when those causes differ.

Royce argues convincingly that Loyalty to Loyalty provides harmony and suffices to distinguish ethical or "true" causes to which one might aspire. Royce's approach divides possible visions or causes into two camps: those that support Loyalty to Loyalty and those that do not. Two groups can directly oppose each other's vision as long as they acknowledge the right of the other to hold and work toward the opposing perspective. By affirming the right of others to become loyal to other causes that include Loyalty to Loyalty, one indirectly affirms the right of others to live freely and pursue their own vision, and excludes causes that might harm, enslave, or disempower others for the simple reason that those injuries would also affect their right to become loyal to their cause. Harmful causes lack Loyalty to Loyalty. One's Loyalty to Loyalty supports the loyalty of others even if one fails to directly support their cause. However, one cannot support another's predatory cause because one has also committed to the principle of Loyalty to Loyalty and the predatory cause would undermine that commitment. Thus, in Loyalty to Loyalty, one respects another's loyalty, avoids unnecessary conflict in the interest of harmony, and resists the other's cause to the extent it undermines Loyalty to Loyalty. Such behavior defines a range of ethical causes and increases harmony across diverse social structures regardless of the particular cause, which one might not fully know.

Royce explores the relationship between individuals and community and distinguishes among various types of communities. Of particular relevance is the significance of sharing diverse interpretations. Diversity is essential to making a group a community. The individual cannot melt or merge completely with the community but must retain distinction for the community to exist as relationships between the distinct individuals. Otherwise the community would degenerate into a group with a single perception of the world—incapable of the differences needed to form diverse interpretations. When each person attempts to interpret each other's interpretations in the context of the shared vision, Royce calls that ideal a community of interpretation.[24]

---

24. For Royce, communities also include shared lives and atoning love where the community is willing to forgive any repentant member and has the creativity to posit an atoning act that makes the world better than if the member's betrayal had never occurred. Royce, *Problem of Christianity*.

In the shared interpretive process, the community gains a collective interpretation of events that no single individual may hold. The community may reach a general consensus even though every individual lacks some detail or perspective. As a whole, the community can interpret events in a way that does not reduce to the interpretations of the individuals. That interpretation will continue to develop as additional members join and leave the community, and the prior communal interpretants will influence future interpretations. The community's interpretive process has causal power and shifts how individuals interpret events. Because the community's diverse perspectives and historical memory enrich the interpretive process beyond the capacity of any individual, the individual may adapt to the community's interpretation to better live out the individual's commitment to the shared vision.

Because the interpretive process may outlive individual members in a long-standing community, and influences the actions and tendencies of others, one can fruitfully consider the shared "Interpreter of a community" as having agency. The interpreter of a community distributes its actions over the members of a community and orients itself toward the shared vision. It helps maintain social cohesion and responds when members stray from their commitment. In living out the shared vision, the community orients itself toward discerning the decisions of the community's Interpreter.

The dynamic Interpreter has the role that Josiah Royce characterized as Spirit. The interpreter of a community emerges from the mental processes of individuals, and the emergence of the interpreter occurs in Royce's community of interpretation when each person interprets the vision in the context of the others' minds in the community. One does not just interpret the vision as an individual; one interprets the vision in the context of other minds in the community. What does SETI mean to those who fear the unknown? How do those who are economically disadvantaged feel about spending resources on SETI? What does astrotheology mean to those who put their hope for extraterrestrial salvation rather than human religious wisdom? Those diverse interpretations influence the spirituality of the project. How do others see us? What would ETI think of our search? One continues to commit to one's cause, such as SETI—within the constraints of Loyalty to Loyalty. Different communities have the right to differing causes. Harmony emerges from the full interpretation of what a cause means.

Community has a spirit that emerges from the interactions among its members as they interpret one to another. The interpreter spirit is more than the sum of the individual interpreters. This use of spirit resonates with an awareness of the "spirit" of a family, city, nation, church, organization, or the like. For Christian theology, communities of interest include the family, local congregation, all Christians, and all humans. For astrotheology, one can consider a community forming from humans and ETI sharing a vision of finding exo-civilizations and interpreting eachother's right to thrive through commitment to the principle of Loyalty to Loyalty.

As mentioned in the introduction to this chapter, ETI may not be as inclusive as the human capacity for inclusivity. Or, they might find offensive the extent to which humans fail to fulfill our principle of Loyalty to Loyalty. However, if aliens share the capacity to enter into self-reflective dialogue about how we view each other, we would have the capacity to begin forming a shared spiritual community across interplanetary distances. Together, we could begin the exploration for additional exocivilizations toward the end of creating a diverse community of civilizations that could give true meaning to a *universal* shared spirituality.

Conclusion

In summary, for interplanetary communication to take place, ETI requires the ability to communicate using symbols and to conceive of an unknown other. In addition, ETI requires the ability to reflect upon oneself as another's other (which is foundational for human spirituality) in order to develop the technology for interplanetary communication. ETI may develop such technology without developing organized religion, but if they develop such technology and can communicate with humans, aliens would necessarily have the capacity to understand human spirituality.

Although at first one might be surprised by the similar cognitive abilities required for SETI (in both directions) and religion, in hindsight, there are several commonalities. Both depend upon a type of communication with something not of Earth and a hope that initiates and sustains the communication attempts even when no immediate response is forthcoming. The various religious beliefs and rituals around UFOs and ETI echo the plurality of historical religions and suggest deep human longings apparently not fulfilled for some by traditional religions. It actually speaks well of existing religions that more people do not seek salvific meaning in symbols of outer space, alien life, ETI, or galactic civilizations. Human longing for home may give a heaven-like connotation to the first discovered habitable planet. Discoveries of alien life may stir passions for relationship. ETI in exocivilizations would enable humanity to give itself new meaning and may even redefine what it means to be human. Theology can help us interpret those new discoveries and distinguish our human longings from the immediate novelty that touches them.

Discovering exocivilizations will add new meaning to humanity, as does every major discovery, and will affect Christian and other religions' understanding of Creation. The significant possibility that many aliens may not be capable of interplanetary communication with humans may disappoint some; but this possibility also increases our responsibility as stewards of a much larger garden. If we discover intelligent alien life, what does our capacity for ethical reasoning demand as our moral response? Any decision affects what it means to be human. If we also meet exoplanetary others capable of a shared spirituality, who we are as humans will affect how we are met. Discovering new ways to relate to others and ourselves will not eliminate our relationship with a Creator God or Savior Christ; and contact may very well help us better understand those religious relationships. Discovering we are not alone may teach us that, on a deeper level, we share a universal relationship not limited by human conceptions and with which we have the privilege to commit.

Bibliography

Barney, Anna, et al. "Articulatory Capacity of Neanderthals, a Very Recent and Humanlike Fossil Hominin." *Philosophical Transactions of the Royal Society B: Biological Sciences* 367:1585 (2012) 88–102.

Cameron, Alastair Graham Walter. *Interstellar Communication: A Collection of Reprints and Original Contributions, Physical Investigations of the Universe.* New York: Benjamin, 1963.

Coe, Kathryn, et al. "ET Phone Darwin." *Civilizations Beyond Earth: Extraterrestrial Life and Society*, 216–26. New York: Bergbahn, 2011.

de Waal, Frans B. M. "Apes Know What Others Believe." *Science* 354:6308 (October 7, 2016) 39–40.

Drake, Frank. SETI Institute. "The Drake Equation." http://www.seti.org/drakeequation.

Gombrich, Ernst Hans. *The Image and the Eye: Further Studies in the Psychology of Pictorial Representation.* Oxford: Phaidon, 1982.

———. "The Visual Image." In *Media and Symbols: The Forms of Expression, Communication, and Education*, edited by David R Olson, 255–58. Chicago: National Society for the Study of Education, 1974.

Kuznicki, Jason T. "The Inscrutable Names of God: The Jesuit Missions of New France as a Model for SETI-Related Spiritual Questions." Chapter 12 in *Civilizations Beyond Earth: Extraterrestrial Life and Society*, edited by Douglas A. Vakoch and Albert A. Harrison. New York and Oxford: Bergbahn, 2011.

Lakoff, George, and Mark Johnson. *Metaphors We Live By.* Chicago: University of Chicago Press, 1980.

Løgstrup, Knud Ejler. *The Ethical Demand.* Notre Dame: University of Notre Dame Press, 1997.

McFague, Sallie. *Metaphorical Theology: Models of God in Religious Language.* Philadelphia: Fortress, 1982.

Nagel, Thomas. "What Is It Like to Be a Bat?" *Philosophical Review* 83.4 (1974) 435–50.

Peck, M. Scott. *The Different Drum: Community-making and Peace.* New York: Simon & Schuster, 1987.

"Preface." In *Drake* xix.

Smith, Kelly C. "The Trouble with Intrinsic Value: An Ethical Primer for Astrobiology." In *Exp* 261–80.

Spezio, Michael L. "The Cognitive Sciences." In *RCRS* (2012) 285–95.

Royce, Josiah. *The Philosophy of Loyalty.* New York: Macmillan, 1908.

———. *The Problem of Christianity. Lectures Delivered at the Lowell Institute in Boston, and at Manchester College, Oxford.* New York: Macmillan, 1913.

Vakoch, Douglas A. *Communication with Extraterrestrial Intelligence.* Albany: SUNY Press, 2011.

———. "Constructing Messages to Extraterrestrials: An Exosemiotic Perspective." *Acta Astronautica* 42.10 (1998) 697–704.

———. "Earth Speaks." SETI Institute. http://earthspeaks.seti.org/.

———. "Signs of Life Beyond Earth: A Semiotic Analysis of Interstellar Messages." *Leonardo* 31.4 (1998) 313.

PART 4

Jesus Christ on Earth and Elsewhere

16

One Incarnation or Many?

TED PETERS

What is not assumed is not healed, but what is united to God is saved.
—GREGORY OF NAZIANZUS, *TO CLEDONIUS AGAINST APOLLINARIS*, EPISTLE 101.7

Incarnation does not sanctify history, but qualitatively transforms it.
—MICHAEL VON BRÜCK[1]

Please consider this proposition: *If multiple societies of extraterrestrial intelligent beings on exoplanets exist, we can predict that God will or already has provided a species-specific incarnation for each planet parallel to God's incarnation in Jesus Christ on Earth.* The position I take is the negative. I disagree with this proposition.

Is the Christian Ship Likely to Sink?

Is the Christian ship likely to sink? Do we have a hole in the bottom of the boat, a hole bored by the concept of incarnation? If we find ourselves sharing our cosmos with extraterrestrial neighbors, will the flood waters rush in and sink the Christian religion? Paul Davies certainly thinks so. Because Davies

1. Von Brück, *Unity of Reality*, 240.

so forcefully poses the challenge, let us cite him once again: "The existence of extra-terrestrial intelligences would have a profound impact on religion, shattering completely the traditional perspective on God's relationship with man. The difficulties are particularly acute for Christianity, which postulates that Jesus Christ was God incarnate whose mission was to provide salvation for man on Earth. The prospect of a host of 'alien Christs' systematically visiting every inhabited planet in the physical form of the local creatures has a rather absurd aspect."[2] The idea of multiple incarnations seems "absurd" to Davies.

Even some theologians reflect a sense of absurdity, but for a slightly different reason. Process theologian Norman Pittenger challenged the cosmic significance of an earthly Christ: "How can the Christian gospel, concerned with the salvation of men in this world, have any universal significance when we know that there may well be intelligent life on other planets."[3] Oxford hybrid biologist and theologian, Arthur Peacocke, feared that we are flirting with nonsense. "What can the cosmic significance possibly be of the localized, terrestrial event of the existence of the historical Jesus? Does not the mere possibility of extraterrestrial life render nonsensical all the superlative claims made by the Christian church about his significance?"[4] Because a planet-hopping Christ seems absurd to Davies and a cosmically significant Christ seems nonsensical to Pittenger and Peacocke, it appears that traditional belief in incarnation makes the Christian faith vulnerable to criticism if not even falsification. It appears theologians have three options in the face of contact with ETI: (1) explore the cosmic significance of salvation won on Earth by the historical Jesus Christ; (2) explore a planet-hopping Christ which becomes incarnate for each extraterrestrial race of intelligent beings; or (3) throw in the towel and quit pursuing absurd or nonsensical theology. In this chapter, I will elect the first option: exploring the cosmic significance of salvation for the cosmos won on Earth by the historical Jesus Christ.

The term *planet-hopping Christ* is obviously pejorative, unnecessarily pejorative. Theologians who defend the idea of multiple incarnations—including Robert John Russell—do so with intellectual integrity. Our intention

---

2. Davies, *God and the New Physics*, 71. One Roman Catholic theologian seems to illustrate the problem Davies raises. On the basis of the Chalcedonian formula in which the incarnate Jesus Christ is truly divine and truly human, there can be only one true incarnation and not multiple. And, because "human" refers to earthlings, there can be no non-human subjects of divine redemption on other planets. "I conclude that either a part of the traditional Christian doctrine or belief in the pervasiveness of extraterrestrial intelligence (ETI) has to be abandoned by rational theists." Weidemann, "Did Jesus Die for Klingons Too?" 130.

3. Pittenger, *Word Incarnate*, 248.

4. Peacocke, "Challenge and Stimulus of the Epic of Evolution to Theology," 103.

here is to clarify respectfully the first two answers to the one question: would the existence of extraterrestrial intelligent beings require multiple incarnations of the divine Logos or would the incarnation of God in Jesus Christ on Earth suffice for the cosmos? We're not throwing in any towels.

Curiously, this is not a new question. It arose within Christian theology early in the Copernican-Galilean era. Theologians speculated about other worlds among the stars, asking about the implications of alien life. Thommaso Campanella (1568–1634), when defending Galileo in *Apologia pro Galileo*, speculated: "If the inhabitants which may be in other stars are men, they did not originate from Adam and are not infected by his sin. Nor do these inhabitants need redemption, unless they have committed some other sin."[5]

In our own post-Copernican era, this question places us in a dilemma. Here is the question: if intelligent beings live on extraterrestrial planets, would the God of Jesus Christ need to provide a separate incarnation for each of those species? If a theologian answers *yes*, then critics would dub the idea of a planet-hopping Christ absurd. If a theologian answers *no* and affirms that the incarnation of Christ on Earth is efficacious for the entire cosmos, then it appears that a pre-Copernican Earth chauvinism is at work. Ouch.

What we see entangled here is a combination of multiple speculations. As we proceed, we will need to sort out just which anthropology fits which christology? If aliens are rational and conscious as Martinez Hewlett suggests in an earlier chapter, then might they require their own divine incarnation as a revelatory event? If aliens are sinful, might they require their own atonement? If aliens are not sinful, might there be some other reason for a species-specific incarnation? Might we be mistaken, in the first place, for transferring our human traits to aliens, who by definition are not human? We are working here with the equivalent of intellectual spaghetti. Now, we will try to separate out each individual noodle.

Are Aliens Persons?

If, by definition, extraterrestrial aliens are not human, then what traits might they hold in common with us? How can we get a hold of the equivalent of theological anthropology—understanding the human condition that corresponds to the person and work of Christ? The question of alien personhood pops up at this point.

---

5. Campanella cited in Crowe, *ELD*, 12.

Earthlings in general, and Christian earthlings in particular, are frequently criticized for planetary chauvinism, dubbed by many as *geocentrism* or *anthropocentrism*. Allegedly, we think we belong at the center of activity, or the top of the ladder of importance. We belong to the in-group, and our very word *alien* suggests that extraterrestrials belong to the out-group. The moral problem with chauvinism is that it denies personhood to outsiders. The pride associated with every chauvinism blinds us to the moral status of the outsider. How might this affect a christological assessment?

Theologian Andrew Burgess poses the challenge. He describes Earth chauvinism in terms of a "cosmic hubris." Terrestrial believers assume that the Creator of the universe was incarnate on our own bit of interstellar debris and maintains a special relationship to the human species."[6] Burgess joins others in identifying this chauvinism with an outdated pre-Copernican cosmology. He writes: "As long as someone is thinking in terms of a geocentric universe and an earth-deity, the story has a certain plausibility. . . . As soon as astronomy changes theories, however, the whole Christian story loses the only setting within which it would make sense. With the solar system no longer the center of anything, imagining that what happens here forms the center of a universal drama becomes simply silly."[7] Whether such hubris and chauvinism derive from a pre-Copernican cosmology or not—Martinez Hewlett in another chapter demonstrated that this tale of de-centering is a myth to be discredited—we need to press the anthropological question: would we consider aliens as persons or not?

Singapore philosopher Roland Puccetti asks this question: would ETI be considered as persons? Yes, he answers. According to Puccetti, "despite secondary biological differences they [extra-terrestrial intelligent beings] would certainly qualify for person-status, since they would be both capable of assimilating a conceptual scheme and the sort of entity to which one can quite reasonably ascribe feelings. Since these are . . . the essential requirements for constituting a moral agent, to say that extraterrestrials are persons is the same as to say they are moral agents."[8] An alien would have a physical body and a mental mind, making the alien a moral agent. Therefore, says Puccetti, the alien must be considered a person.

Having established the personhood of ETI, Puccetti then engages in a push-me pull-you number on Christian theology. On the one hand, he argues that multiple incarnations ought to be expected. On the other hand, he tries to show that the concept of person in incarnational theology lacks

6. Burgess, "Earth Chauvinism," 1098.
7. Ibid.
8. Puccetti, *Persons*, 106.

coherence. "Except on the extreme kenotic view—that between His Incarnation and Ascension the Son of God more or less lost His divine attributes and was temporarily nothing *but* man—Christianity has no conclusive theological reasons for opposing multiple incarnations in which non-human rational corporeal beings would have their natures hypostatically united to that of the same Son of God."[9] Putting himself into the place of a Christian theologian, which he is not, Pucetti argues that if no change in the Godhead is involved in Christ's earthly incarnation, and if what Christ does is take human nature up into the divine life, then it would follow that this could be done multiple times, each with a different life form. If two natures, one finite and the other infinite, are compatible in one person, then why not several times on different planets?

Then, like a military surge, Puccetti attacks by firing big numbers. He elects to use 10^{18} for the number of likely life bearing planets in the universe. The incarnate Son of God would be a corporeal person, to be sure; and this means he could be only one place at one time. Yet, the number of places to visit would be overwhelming. "There would be on the order of 680,000,000 to 3,400,000,000 incarnations occurring simultaneously from now to the extinction of life on all such stars. Allowing for earlier incarnations reduces this figure a bit, but not enough to make any real difference."[10] The impression the reader gets is that this would be absurd. The numbers are too big. The concept of an incarnate person seems to collapse under its own incoherence. To believe in a planet-hopping Christ would be embarrassing to a theologian.

This appears to mark a victory in Puccetti's attack on religion. "What I have been trying to do . . . is to show how a correct analysis of the person-concept combined with the not unreasonable belief in extraterrestrial natural persons actually undermines belief in God."[11] In sum, the very existence of persons on other planets undermines an Earthling's belief in God.

Former Notre Dame philosopher Ernan McMullin is not persuaded by the Puccetti argument. He finds it ironic that Puccetti demands that corporeality be included in his concept of person. The Latin term, *persona*, from which we derive "person" in its modern usage, originally applied to the three persons of the divine Trinity. The concept did not originate in reference to corporeal beings.

McMullin goes on to identify some shaky presuppositions in the Puccetti argument, one of which is a warning about relying on big numbers.

9. Ibid., 136.
10. Ibid., 139.
11. Ibid., 143.

The Drake equation is strictly speculative, recall, not empirical. "One has to be wary here of a fallacy induced by the contemplation of large numbers . . . It is one thing to discover one or a small number of ETI sites based on the interpretation of incoming radiation. It is another thing entirely to establish, on the basis of a theoretical analysis of the multiplicity of processes involved in the appearance and survival of intelligent life, that the number of centers of such life in the universe is of a certain order or even that it is, in very general terms, extremely large."[12] McMullin does not believe this is a slam dunk refutation of Puccetti, even if he has reduced the force of Puccetti's argument.

Just how should we think of personhood? The first thing to note is that personhood is a moral category. We connect personhood with dignity. This means each person is a moral end. Immanuel Kant spoke for the entire Enlightenment when he declared that a person should always be treated as an end, and never as merely a means. If such a thing as Earth hubris characterizes us, then we need to become self-aware: we dare not treat an alien person as a means instead of an end. This is the moral dimension of personhood.

There is also a rational component to personhood, at least in ancient theology and modern philosophy. For ancient Boethius a person is an individual substance with a rational nature.[13] Modern Kant said that a person becomes a moral end if he or she is rational. Rationality defines a human person as a person. In saying so, the eighteenth-century Kant is continuing a tradition that derives from a long held belief within Christian theology; and this theological appreciation of human rationality derives largely from Plato in ancient Athens. Rational capacity is decisive for twentieth-century Catholic Karl Rahner, for example. He holds that beings living among the stars who are intelligent and free are "not distinguished in an important way by where they are located in the cosmos . . . [but rather by] their intellectual subjectivity determining the reality of space and time."[14] Rationality is what places a creature in special relationship to its creator.

Not all contemporary Christians are sanguine about this identification of rationality with dignity let alone personhood, however. The reason is that many human persons are unable to pass a rationality test. Very young children or seniors with brain dementia are either pre-rational or post-rational.

12. McMullin, "Life and Intelligence Far From Earth," 167.

13. Boethius, *Liber de persona et duabus naturis*, III. "Following Boethius," writes Oliver Crisp, "I have reasoned that a person is a fundamental substance of a rational nature . . . the natures of Christ are to be understood as fundamentally concrete substance-like things, not as fundamentally rich properties, though natures may include properties or predicates." Crisp, "Desiderata," 40.

14. Rahner, "Sternenbewohner: Theologisch," 1061–62.

Yet, they are still persons; and they are still to be treated as a moral end. The absence of the rational capacity does not justify putting them to death, or treating them in any way with diminished dignity.

The Vatican—especially popes John Paul II and Benedict XVI—devoted considerable passion to defending human dignity, especially the dignity of pre-rational and post-rational persons. They rest their case for human dignity on the presence in each person of the divine image, the *imago Dei*. Each human person deserves to be treated with dignity due to the presence of God's image, not due to the presence of rationality.

As we speculate about alien life, we cannot help but ask whether aliens are rational. Certainly, the assumptions at work in the ETI myth presuppose that rational aliens have advanced in science and technology and even in morality. Still, theologians might ask: will aliens exhibit the image of God? How will the criterion of rationality aid us in discerning this?

One Incarnation for Each Race of Little Green Men?

Let us work with the assumption that the extraterrestrials in question are conscious and rational. Even so, the possibility is open that our scientists will discover microbial life within our solar system. Such microbial life would not likely be conscious or rational. Microbial life would belong to God's creation, to be sure; but for the sake of argument let us concentrate on rational beings who are capable of understanding and thinking. Extrasolar planets may house such rational creatures who, like *Homo sapiens* on Earth, share our world with other living creatures of varying degrees of thinking capacity. Our interest for the time being will be on extraterrestrial rational beings capable of understanding and communicating. With this assumption in mind, we can see how the question of a planet-hopping Christ gets raised and why critics take the opportunity to ridicule Christian belief in revelation and redemption.

Davies takes this opportunity. He contends that Christian theologians are hopelessly trapped in the absurd commitment to positing a planet-hopping Christ, that is, a God who becomes incarnate repeatedly to accommodate various species. Davies lays down the gauntlet with an allusion to little green men. "Theologians and ministers of religion take a relaxed view of the possibility of extraterrestrials. They do not regard the prospect of contact as threatening to their belief systems. However, they are being dishonest. All the major world religions are strongly geocentric, indeed homocentric. Christianity is particularly vulnerable because of the unique position of Jesus Christ as God incarnate. Christians believe that Christ died specifically to save humankind. He did not die to save little green men."[15] Even though theologians say that they will welcome extraterrestrial neighbors, they are "dishonest," says Davies. Theologians are dishonest because they fail to admit to the vulnerability of their Christology.

John Polkinghorne does not feel vulnerable. He picks up the gauntlet without any sign of theological distress. "There must surely be many sites in the universe suitable for the development of some form of life . . . Theology does not altogether know what to think about extraterrestrial possibilities. God's creative purposes may well include 'little green men' as well as humans, and if they need redemption we may well think that the Word would take little green flesh just as we believe the Word took our flesh."[16] Perhaps Dr. Polkinghorne should be given a lie detector test to see whether he is dishonest.

Let's look at bit more closely at what is going on here. The cipher "little green men" comes from UFO reports in the early 1950s. It is not a scientific term. Still, it points to our concern for alien intelligence. This makes it a reasonable question for us to ask: might divine incarnation need to be species-specific? Might God find a way to communicate the divine presence to every domicile of rational creatures? Notre Dame's Thomas O'Meara is blunt: extraterrestrial "incarnations would correspond to the forms of intelligent creatures with their own religious quests."[17] Neither Polkinghorne nor O'Meara appear to feel vulnerable; nor do they appear to be dishonest. They quite unabashedly argue that the idea of multiple incarnations is reasonable.

Also in the camp of many incarnations we find philosopher of religion John Hick. "God would become incarnate more than once—and indeed, in principle, an indefinite number of times—for the sake of separate groups of people . . . on other planets of other stars."[18] Each planet with intelligent creatures would host a visit from the second person of the divine Trinity. Despite Davies' charge that extraterrestrial Christology would be "absurd" and that theologians who engage in extraterrestrial speculation are "dishonest," such speculation within the religious community is moving ahead with no signs of vulnerability.

15. Davies, "Transformations," 51.
16. Polkinghorne, *Science and the Trinity*, 177.
17. O'Meara, *VU*, 48.
18. Hick, "Response to Hebblethwaite," 192.

Are There Churches on Exoplanets?

The question is more important than the answer. The question is this: if extraterrestrial intelligent beings exist, would we expect to see multiple incarnations of the second person of the Trinity or would the saving work of Jesus Christ on Earth suffice for the entire cosmos? Theologians line up on two sides: some say yes, God would find a way to engage rational creatures everywhere in the universe. Others say no, God's redemptive work in the history of Earth suffices for the entire scope of God's creation. Either one of these two answers would seem to deny the tenability of the Davies position.

Paul Tillich belongs in the multiple-incarnation camp. In his influential *Systematic Theology*, Tillich addresses directly the question of the efficacy of redemption on earth for life on other planets. How should we

> understand the meaning of the symbol 'Christ' in the light of the immensity of the universe, the heliocentric system of planets, the infinitely small part of the universe which man and his history constitute, and the possibility of other 'worlds' in which divine self-manifestations may appear and be received ... The function of the bearer of the New Being is not only to save individuals and to transform man's historical existence but to renew the universe ... The basic answer to these questions is given in the concept of essential man appearing in a personal life under the conditions of existential estrangement. This restricts the expectation of the Christ to historical mankind.[19]

Having affirmed here the efficacy of redemptive action by God in Jesus Christ, Tillich speculates about additional possibilities. "At the same time, our basic answer leaves the universe open for possible divine manifestations in other areas or periods of being. Such possibilities cannot be denied ... Incarnation is unique for the special group in which it happens, but not unique in the sense that other singular incarnations for other unique worlds are excluded."[20] Multiple incarnations would be reasonable to Tillich, even though to date we have no proof that such a thing has happened. Tillich follower Durwood Foster similarly contends that "the love of God manifest in Jesus Christ has surely not remained unknown wherever there is spiritual receptivity."[21] What seems to be operative here is the assumption that the function of incarnation is primarily revelatory, an event in which God's

---

19. Tillich, *ST*, 2:95.
20. Ibid., 96.
21. Foster, *God Who Loves*, 125.

grace and love become known to creaturely consciousness. What about sin or estrangement?

Tillich goes on to speculate that ETI might be in a situation similar to us here on Earth. They might be fallen, what Tillich calls "estranged" from God. This is where the Christian vocabulary regarding sin kicks in with the words, *sin* and *evil*. "Sin is the estrangement between God and humans instigated by human defiance or abnegation. Evil is the disorder within humans individually and among them collectively," writes Mark Heim.[22] Might alien civilization exist within a condition similar to ours: fallen, estranged, subject to sin? Does the extraterrestrial situation need redemption? Redemption is more than mere revelation. Redemption constitutes the *work* of incarnation, the *work* of reconciliation, not merely the revelatory presence of God.

It appears to me that a theologian with a revelational or exemplarist emphasis in Christology is more likely to sympathize with the idea of multiple incarnations than is a theologian with a high Christology, with the idea that Jesus Christ performed the decisive work of atonement. The Bible is filled with accounts of divine revelations to humans: the appearance of God in the burning bush to Moses; call visions to prophets such as Jeremiah and Isaiah; communication through dreams; along with "signs and wonders." Jesus provides one more in a long list of revelation moments. Even though Jesus was revelatory of God, to be sure; God is not without a means of self-revelation in countless ways, with or without Jesus. It would be easy to extrapolate this to off-Earth worlds, it would seem to me; so that God's revelatory work could occur virtually anywhere to virtually any individual or civilization. If by *incarnation* a theologian means that in Jesus we find a revelation of a truth about God or an example as to how God's creatures should live, then such a Christology would be amenable to multiple incarnations of the one and same second person of the Trinity. The Logos could become a Christ many times.

When he takes up the issue of many worlds, Tillich seems to rely upon this revelational or exemplarist type Christology, according to which the incarnation makes present the eternal creative and redemptive power of God. Even if aliens live in an estranged situation, God as the ground of their being would be at work with reconciling love, according to Tillich. "If there are non-human 'worlds' in which existential estrangement is not only real—as it is in the whole universe—but in which there is also a type of awareness of this estrangement, such worlds cannot be without the operation of saving power within them . . . The expectation of the Messiah as the bearer of the

---

22. Heim, *Depth of the Riches*, 60.

ONE INCARNATION OR MANY? 281

New Being presupposes that 'God loves the universe', even though in the appearance of the Christ he actualizes this love for historical man alone."[23] God acts one way in the historical world of human beings and in a parallel way for the extraterrestrial worlds of non-humans. Incarnations of the one Christ could be multiple.

Lewis Ford would modify this. Ford is a card carrying process theologian of the Whiteheadian school. "Salvation is not just limited to men but applies to all intelligent beings wherever they may dwell," he writes.[24] Intelligent non-human creatures count, according to Ford. Embracing the concept of evolution and applying it to other worlds yet to be discovered, Ford asserts that God is always and everywhere drawing the evolutionary process toward greater complexity and higher value. "We may define God as that dynamic source of values which lures the evolutionary process to an ever-richer complexity productive of increasing freedom and intensity of experience. As such, God is necessarily operative in the development of every life and in every culture, whether terrestrial or extraterrestrial."[25] The Ford position may not require multiple incarnations of the second person of the Trinity, because the lure of God toward more intense experience is built into God's action throughout creation. What Ford shares in common with Tillich and Foster is the receptivity of knowledge of God on the part of intelligent creatures whether terrestrial or extraterrestrial.

Even though it does not matter to Ford whether ETI are persons, it is not clear here whether Tillich or Foster would consider ETI as persons in the sense that Puccetti does. Tillich uses the term "non-human." Yet, Tillich considers the possibility of estrangement and awareness of estrangement, which would put them into the category of moral agents. So, perhaps with Tillich, ETI would have to be non-human persons. And, because Jesus the Christ appeared for historical human persons, an additional incarnation for these non-human persons would be in order.

Yet, Tillich does not need to follow the trail to this destination, in my judgment. Such a species-specific incarnation would not be necessary, given Tillich's assumption that God's creative presence and grace is available throughout the creation. All Tillich really needs to affirm is that God's saving work would be present, regardless of whether in incarnate or some other form.

David Wilkinson is cautious about following the logic of the multiple incarnations school of thought. The logic of the multiple incarnation

23. Tillich, *ST*, 2:96.
24. Ford, "Theological Reflections on Extra-Terrestrial Life," 2.
25. Ford, *Lure of God*, 63.

position has implications, and four of these implications are worrisome. First, "to drive a wedge between the cosmic Christ and the human Jesus does begin to open the door to the view that Jesus was just a good man used by God." It undercuts the creation-wide implications of the eternal logos taking up residence in the temporal Jesus. Jesus isn't just a nice guy God used for his generation alone.

The second worrisome implication is this: "if god's nature is to reach out in love in embodied form, why should there not have been multiple incarnations in different cultures on the Earth?" God is certainly manifest in multiple cultures, but Christian theologians have restricted the idea of incarnation to one and only one instance, Jesus in Israel. "Jesus is still held to be supreme," says Wilkinson.

Thirdly, God is not limited to revelation through incarnation alone. "The Bible is full of other images of God communicating, including through visions, through awe at the natural world, through angelic visitations, through burning bushes, through dreams, through the written word, through prayer, and through prophets." Should God wish to reveal the divine self to races of extraterrestrial beings, methods other than incarnation are available.

Fourth and most importantly, the multiple incarnation position would make sense only if the function of incarnation is revelation. If incarnation also entails salvation, then more is at stake. "The incarnation is about both revelation and salvation," Wilkinson makes clear.[26] If the soteriological work of Jesus Christ in terrestrial history is efficacious cosmically, then this must include all creatures within God's creation, terrestrial and extraterrestrial alike.

Wilkinson's cautions indirectly support the opposite position: the sufficiency of the soteriological work of the terrestrial Christ for the whole of reality, ETI included.

Does One Incarnation Suffice for the Entire Cosmos?

Let us turn to the other camp, where spokespersons assert that the single event of redemption in Jesus Christ on Earth is cosmically valid. As a preface, we note that a major theme in Christian theology is to affirm that the atoning work of Jesus Christ is cosmic in scope. All of creation finds its redemption in this historical event within creation's history. The gospel "has to do with the cosmic redefinition of reality," writes Lutheran Peter Berger. "The Christian Gospel is about a tectonic shift in the structure of the

26. Wilkinson, *SRSEI*, 158–59.

ONE INCARNATION OR MANY? 283

universe, focused on the events around the life of Jesus."²⁷ What might this imply for the ETI question?

In *Nostra Aetate*, the Second Vatican Council enunciated a similar position, namely, that God's universal grace is ontologically tied to God's historical action on planet Earth in the event of Jesus Christ. This warrants preaching, the sharing of news about this event. "Christ in His boundless love freely underwent His passion and death because of the sins of all men, so that all might attain salvation. It is, therefore, the duty of the church's preaching to proclaim the cross of Christ as the sign of God's all-embracing love and as the fountain from which every grace flows."²⁸ Vatican II did not address directly the question of extraterrestrial life, yet we could surmise that God's grace would be effective on other planets even though this grace expresses the love of the very same God revealed in Jesus Christ on Earth.

Australian Lutheran theologian Mark Worthing draws out the implications of the cosmic Christ by advocating the single-incarnation view. "If there is other intelligent life in the universe then God relates to it through Christ—the same Christ through whom God reconciles us to Godself. I do not believe Christian theology can posit a multiplicity of Christs and remain Christian theology."²⁹ Members of the single incarnation camp rely on an assumption voiced succinctly by Peter W. Marty, "We ought to think of the work of Jesus Christ as cosmic in scope."³⁰

During an interview, Jesuit L.C. McHugh was once asked: what would be the relation of intelligent beings inhabiting a far corner of the cosmos to Jesus Christ? McHugh responded saying, such creatures "would fall under the universal dominion of Christ the King, just as we and even the angels do."³¹ McHugh even employs the term *exotheology*, which corresponds roughly to our term here, *astrotheology*.

J. Edgar Bruns, former president of Notre Dame Seminary in New Orleans and a New Testament scholar, wrote that "the significance of Jesus Christ extends beyond our global limits. He is the foundation stone and

---

27. Berger cited in Thuswaldner, "Conversation with Peter L. Berger," 19. The cosmic Christ belongs in the camp with other higher Christologies. "Since no one else will ever be God incarnate, he is logically indispensible; and since no one else will ever die for the sins of the world, he is materially decisive. For high Christologies, Jesus is an exclusively unique person who accomplished an exclusively unique work." Hunsinger, "*Salvator Mundi*," 58.

28. "Declaration on the Relationship of the Church to the Non-Christian Religions," 667.

29. Worthing, "Possibility of Extraterrestrial Intelligence," 83.

30. Marty, "Who Gets Saved?" 3.

31. McHugh, "Life in Outer Space?" 29.

apex of the universe and not merely the Savior of Adam's progeny."[32] The position of these two Roman Catholics, McHugh and Bruns, implies that, should alien beings be discovered, missionaries would be called for much as they were when Europe discovered the Western hemisphere.

Among those holding to a single incarnation we find Munich Lutheran theologian, Wolfhart Pannenberg. Pannenberg, like Tillich, would agree that God's redemptive as well as creative work would be present in other worlds. Still, Pannenberg comes closer to saying that the saving work of Jesus Christ on Earth is efficacious for the entire universe.

> It is hard to see . . . why the discovery of nonterrestrial intelligent beings should be shattering to Christian teaching. If there were such discoveries, they would, of course, pose the task of defining theologically the relation of such beings to the Logos incarnate in Jesus of Nazareth, and therefore to us. But the as yet problematic and vague possibility of their existence in no way affects the credibility of the Christian teaching that in Jesus of Nazareth the Logos who works throughout the universe became a man and thus gave to humanity and its history a key function in giving to all creation its unity and destiny.[33]

Pannenberg can express confidence in the universal efficacy of Jesus Christ because his incarnation is the incarnation of the universal Logos, the principle by which all of creation is generated and held together. Like thoughts originating in our mind, the Logos originated in God. And as our thoughts come to expression in speech, when God speaks the Son becomes differentiated from the Father, and the world with all of its particularity comes into existence. This is the divine nature of the Son, the universal ground of all finite reality.

We are stepping back from redemption to the doctrine of creation here. Pannenberg associates creation with the Trinity, not with the Father alone. It is through the Son as the Logos that the existence of the entire finite creation is wrought. "If the Logos is the generative principle of all the finite reality that involves the difference of one thing from another—a principle grounded in the self-distinction of the eternal Son from the Father—then with the advent of ever new forms differing from what has gone before there comes a system of relations between finite phenomena and also between these phenomena and their origin in the infinity of God. As the productive principle of diversity the Logos is the origin of each individual creature in

32. Bruns, "Cosmolatry," 286.
33. Pannenberg, *ST*, 2:76.

its distinctiveness and of the order of relations between the creatures."[34] Note that the Logos establishes both individuality and the relations between individuals. Perhaps this is the condition that makes it possible for the Logos to become a single individual, Jesus of Nazareth, while still expressing universal finite reality.

Because each and every personal life on another planet would have been created through the same divine Logos, one need not assume a lack of connection between what happens on Earth and what happens in even the furthest reach of outer space. The temporal and spatial finitude of Jesus of Nazareth distinguishes him as an individual when set next to all other sentient beings either on Earth or elsewhere. One need not postulate some special metaphysical track that each species of intelligent beings would follow, warranting their own incarnation. In Tillich's words, the incarnation in Jesus is the "concrete universal."[35]

As we untangle the strands of theological spaghetti, we should separate out the question of rational capacity from the question of estrangement or need of redemption. If rational creatures living on another planet are curious and capable of understanding, then one could easily posit that God would communicate the divine reality simply to share communion with creatures. Both Tillich and Pannenberg would admit that the divine logos or divine reason maintains the same structure everywhere in the cosmos, so rational creatures would be by nature attuned to the presence of God, whether incarnate in flesh or merely apprehensible through mind. If we add to the rational capacity the condition of estrangement—fall into sin—then the situation becomes more complex. As human reason on Earth becomes distorted by sin, so also we could expect that extraterrestrial reasoning might be similarly distorted. The work of incarnation would be more than merely revelation. It might include atonement and reconciliation, the gracious overcoming of estrangement.

On this point, McMullin unpacks with a bit more detail what was compact in Pannenberg.

> If the incarnation and the death of Christ be seen as the response to human sin, and if that sin itself be seen as a contingency that

34. Ibid., 62.
35. Brother Guy Consolmagno would concur. "Even though the life of Jesus occurred at a specific space-time point, on a particular world line (to put it in general relativity terms), it also was an event that John's Gospel describes as occurring in the beginning—the one point that is simultaneous in all world lines, and so present in all time and in all space. Thus, there can only be one Incarnation—though various ET civilizations may or may not have experienced that Incarnation in the same way that Earth did." Consolmagno, "Would You Baptize an Extraterrestrial?"

might well not have occurred, then the question about an ETI race might be, Did its first progenitors fall when challenged? If they did, then the Son of God might become incarnate in their nature and die to expiate their sins. If they did not . . . then an Incarnation there would not be called for. If it be supposed in addition, however, that Christ's death on Earth was a unique event, sufficient of itself to restore the balance disturbed even at the cosmic level by the sin of Adam, then a further Incarnation on other inhabited planets would be unnecessary.[36]

Note the conditional argument: if the extraterrestrial condition is characterized by fallenness, then it follows . . .

This points to an already existing issue within Christian thinking: is the reason for the divine incarnation in Christ a response to human estrangement? Or, would God have become present in Christ regardless of the situation of sin? Might God engage in self-communication simply out of love for the world, perhaps as a continuing expression of God's world-perfecting work? With this as an existing terrestrial question, we can see how it might have implications for speculation about extraterrestrial creatures. Let us turn to the question of creaturely estrangement; and then we will turn to divine motives for incarnation.

Why Incarnation? The Fix-a-Broken-Creation Model

We can fantasize about ETI. We could imagine that they are better than we are. In C.S. Lewis's novels, *Out of the Silent Planet,* and *Perelandra,* he could imagine a race of people who had not fallen into sin. Their planet remains as ours was in the Garden of Eden—that is, intelligent creatures do not live in a state of estrangement from their creator. Many of today's scientists and UFO true believers fantasize that ETI are more virtuous than we are, because they are more highly evolved. Until we meet them, of course, we simply will not know for certain.

Robert John Russell, in contrast, predicts that alien beings will be found to exist in a non-Edenic situation. Most likely, ETI will find themselves in the same situation of ambiguity that we find ourselves. "I predict that when we finally make contact with life in the universe . . . it will be a lot like us: seeking the good, beset by failures, and open to the grace of forgiveness and new life that God offers all God's creatures, here or way out there."[37] The physical features of reality are constant regardless of where one lives in this

36. McMullin, "Life and Intelligence Far from Earth," 172–73.
37. Russell, "What Are Extraterrestrials Really Like?" 66.

universe; so it is reasonable that extraterrestrial beings experience the same struggles with the same failures and achievements we have come to know.

Now, with this in mind, let us address the question already internal to theology: what warrants incarnation? Had Adam and Eve not fallen, would the Son of God still appear in incarnate form? Is it the world's fallen state that creates the need for redemption? Or, might the incarnation be due strictly to God's self-communication in creation? Tillich seems to assume that our situation of estrangement calls out for an incarnate visitor from the ground of being. Fallenness calls out for redemption. Pannenberg, in contrast, makes the incarnation independent of the fall. "The incarnation cannot be an external appendix to creation nor a mere reaction of the Creator to Adam's sin."[38] God's presence in Jesus Christ adds the grace of redemption upon the grace of creation. The latter is a completion of the former. Tillich and Pannenberg represent two models of terrestrial incarnation, the fix-a-broken-creation model and the divine-self-bestowal or incarnation-anyway model.[39]

In the minds of most Christian theologians incarnation and redemption seem to belong together in the fix-a-broken-creation model. The church long remembers Athanasius (296–373) saying of Jesus Christ, "He was made man that we might be made God."[40] Eastern Christians both ancient and contemporary hold that in the incarnation Jesus Christ recapitulated all that is human, healed it, and set us on a course toward deification, *theosis*. Such deification had been God's original plan in creation, to be sure; but because of human sinfulness God found it necessary to take redemptive action. "The Fall demands a change, not in God's goal, but in His means," writes Russian Orthodox theologian, Vladimir Lossky; "For the atonement made necessary by our sins is not an end but a means, the means to the only real goal: deification."[41] According to this line of thinking, the Christ event fixes a broken creation; it lifts up what has fallen.

Why Incarnation? The Divine-Self-Bestowal or Incarnation-Anyway Model

In contrast, some Western theologians have argued on behalf of the divine-self-bestowal model because the incarnation was destined due to the internal

38. Pannenberg, *ST*, 2:64.
39. The *incarnation anyway* designation is employed by Adams, *Christ and Horrors*, 174; van Driel, *Incarnation Anyway*, 4–5; and Crisp, *Revisioning Christology*, 23.
40. Athanasius, *Incarnation of the Word* §54.
41. Lossky, *Orthodox Theology*, 110–11.

dynamics of trinitarian life. Even without sin, Earth would have played host to the incarnation anyway. Accordingly, "the Christ-event is not thought of as an addition to creation," writes Denis Edwards. "It is not primarily a corrective for a creation that went wrong. It does not come about simply as a remedy for sin, although in the light of sin, it is certainly a radical act of forgiveness, healing, and liberation . . . God's self-giving in the incarnation is the very purpose and meaning of creation."[42] Edwards stakes his claim clearly in the divine-self-bestowal or incarnation-anyway camp.

Thus, we have two contrasting models to work with. According to the fix-a-broken-creation model, what motivates the incarnation is God's desire to redeem a fallen world from sin. According to the divine self-bestowal model, the incarnation is one more chapter in the story of God's self-giving love that began with creation; and the incarnation would have happened regardless of a fall. Following this second model, Bonaventure (1221–1274), rejected the idea of the incarnation in Christ as some sort of afterthought, a way to fix what was broken. Incarnation was willed by God for its own sake, not for the sake of a lesser good. In addition, said Bonaventure, God's entry into the created realm as a human serves to unite all of creation with humanity. The incarnate Christ serves to perfect nature. All of God's creative work is a form of incarnate self-expression.[43]

Similarly, John Duns Scotus (1265–1308) sees friendship love (*amor amicitiae*) as the spontaneous divine motivation for creation as well as redemption through Christ. This divine love needs to be shared, so to speak, first with the soul of Christ and then with all of creation. The soul of Christ is the first goal of God in creation and, then, all creatures spread throughout the entire created universe become co-lovers for Christ's sake. Scotus works from within the divine-self-bestowal or incarnation-anyway model. "*Incarnation and redemption are logically independent projects*," comments Marilyn McCord Adams.[44]

An apparent contradiction in John Calvin draws out some implications for us. For this Reformer, the second person of the Trinity is the mediator of both creation and redemption. As mediator of creation, the incarnation is the result of a supralapsarian (prior to the fall into sin) divine decree—that is, incarnation becomes a divine plan independent of the creaturely fall into sin. As the mediator of redemption, however, the incarnation is the result of an infralapsarian (after the fall) work of atonement. The divine-self-bestowal

42. Edwards, *How God Acts*, 40.
43. See the fine analysis of Delio, "Christ and Extraterrestrial Life," 249–65.
44. Adams, *Christ and Horrors*, 184. In her own Christology, Adams does not agree with Scotus "in putting Christ first, in making (the soul of) Christ God's chief end in creation with everything else being made for Christ's sake." Ibid., 190–91.

or incarnation-anyway model seems to be assumed when Calvin writes, "Even if man had remained free from all stain, his condition would have been too lowly for him to reach God without a Mediator."[45] Christ the Mediator would have been offered by God even without the stain of sin, it seems. Yet, on the other hand, human fallenness warrants atonement. Alluding to John 12:27–28, Calvin writes, "Here he clearly indicates why he assumed flesh: that he might become a sacrifice and expiation to abolish our sins."[46] Or, to say it another way, the state of human sin requires the incarnation of the Son of God in order to accomplish the work of atonement, understood here as substitutionary atonement. So, Mr. Calvin which is it: incarnation-anyway or fix-a-broken-creation?[47] The implication of this tension in Calvin is this: we can think of creaturely fallenness and the need for redemption as historically contingent, whereas the expression of divine love for the creation through the second person of the Trinity is eternal.

For Karl Rahner [as we saw in an earlier chapter by Oliver Putz], the incarnation in Jesus Christ is God's self-communication. "The world and its history are from the outset based on the absolute will of God to communicate himself radically to the world. In this self-communication and in its climax (i.e., in the Incarnation), the world becomes the history of God himself."[48] On the one hand, through the incarnation God actualizes our human potential. On the other hand, we human beings and all of creation get taken up into God's own history. "The incarnation of God is therefore the unique, *supreme*, case of the total actualization of human reality . . . God has taken on a human nature, because it is essentially ready and adoptable . . . and comes therein to the fulfillment of its own incomprehensible meaning."[49] This understanding of incarnation within the divine-self-bestowal model is one of adding grace upon grace.

45. Calvin, *Inst.* II:xii:1; 1,465.

46. Ibid., II:xii:4; 1,468.

47. Oliver Crisp contends that both motives for the incarnation obtain in Calvin. "The upshot of this is that there does appear to be a real tension in Calvin's doctrine of the motivation for the incarnation, which is difficult to reconcile. But on one plausible reading of the data, a case can be made for thinking that Calvin's thinking is broadly consistent with something like supralapsarianism, and even with the Incarnation anyway argument . . ." Crisp, *Revisioning Christology,* 38. Crisp comes down with one foot in each camp. According to my interpretation, Calvin is standing on one foot in the fix-a-broken-creation camp. This is because Calvin directly addresses and repudiates Andreas Osiander, who had previously upheld the incarnation anyway position. Calvin, *Inst.* II:xii:4–7.

48. Rahner, "Christology Within an Evolutionary View of the World," in *TI*, V:186.

49. Rahner, "On the Theology of the Incarnation," in *TI*, IV:110.

Significant to the Rahner view is that through the incarnation God's own life becomes historical; the world becomes internal to the divine life. That would include all histories, including histories on other planets. We see a nascent variant of this view articulated as well in the Christology of Marilyn McCord Adams, who begins by recognizing the hostility that exists in the world toward God. The material world is a world filled with horrors, horrific evil. For God to become incarnate means, among other things, that the horrors of this world become internal to God's history. This leads Adams toward the fix-a-broken-creation model. "On my view, Christ is primarily *head of the cosmos*. God makes the world in order to become Christ for it, shares the natures of the whole material universe by making Godself a member of the human race. On my view, there is no 'anyway' to Incarnation, because God's making us in a world like this leaves us radically vulnerable to horrors."[50] But, this does not preclude multiple incarnations. Without directly addressing the ETI question, she writes, "multiple incarnations are metaphysically possible, whether by the same or by different Divine persons."[51]

The Multiple Incarnation School

Now, we need to ask: if we follow the divine self-bestowal or incarnation-anyway model, does this lead us to a single incarnation or to multiple? Thomas O'Meara seems to follow in Rahner's footsteps on this point, leaping to the multiple-incarnation scenario. "As incarnation is an intense form of divine love, would there not be galactic forms of that love? An infinite being of generosity would tend to many incarnations rather than to one . . . A succession of incarnations would give new relationships and new self-realizations of God . . . Incarnations among extraterrestrials would not be competing with us or with each other."[52]

Following Rahner's lead and running parallel to O'Meara, Franciscan Ilia Delio affirms divine self-communication in Christ and similarly affirms multiple incarnations. More exhaustively than either Rahner or O'Meara, however, Delio places God's self-communication in the embodied Word within a Teilhardian scheme of theistic evolution. The Christ principle imbues biological development wherever that biological development takes place, guiding it, perfecting it. This universal Word of God can then take

50. Adams, *Christ and Horrors*, 200.
51. Ibid., 198. Whereas Rahner would stress that only the second person of the Trinity would become incarnate, Adams leaves it open to any of the three.
52. O'Meara, *VU*, 47.

on specific embodiment and be perceived as the divine Word by any creatures who are intelligent. "Incarnation on an extraterrestrial level could conceivably take place, as long as there is some type of intelligence within the extraterrestrial species to grasp the Word of God through knowledge of the divine embodied Word . . . many incarnations but one Christ."[53] The divine self-bestowal school, so to speak—Rahner, O'Meara, Delio—makes a formidable case for the multiple incarnation option.

Have we come to a fork in the road? Apparently not, because both models could justify multiple incarnations. If, on the one hand, we were to follow the fix-a-broken-creation model, then we might speculate that an extraterrestrial incarnation would be contingent on the moral status of the extraterrestrial aliens. If our alien neighbors are fallen, then a redeeming Christ event would be called for. This logic would lead, in addition, to exempting a given planet from its own Christ event if its intelligent creatures are still living in the equivalent of the Garden of Eden. If, on the other hand, we were to follow the divine self-bestowal or incarnation-anyway model, then what? Rahner, O'Meara, and Delio would have us believe that this implies multiple incarnations, one for each extraterrestrial civilization regardless of whether it is Edenic or fallen. Both the Edenic and fallen scenarios could be conscripted into the service of a universal logos manifest in multiple incarnate expressions.

As I see it, neither of these two scenarios leads ineluctably to the multiple incarnation scenario. Divine grace could perfect nature regardless of the fall and regardless of the historical particularity of the Jesus event. How might we describe this? The concept of deep incarnation might help to draw a coherent picture.

Deep Incarnation?

Have we been asking the wrong question? Or, more specifically, has the formulation of our question misled us? The question given us looks something like this: if a number of species of intelligent beings lives on off-Earth planets, would the God of Jesus Christ need to provide a separate incarnation for each of those species? If a theologian answers in the affirmative, then Davies and other critics would describe this as absurd. If a theologian answers in the negative—that the incarnation of Christ on Earth is efficacious for the entire cosmos—then it appears that a pre-Copernican Earth chauvinism or geocentrism is at work. Part of the difficulty in this question is its inherent ambiguity regarding incarnation, and what divine incarnation in Jesus

53. Delio, *Christ in Evolution*, 169.

Christ amounts to. If the function of incarnation is limited to revelation—to God's self-communication to creatures intelligent enough to receive such communication—then perhaps a planet-hopping Christ might make some sense. But, what if incarnation means more than mere revelation? Just what is entailed in incarnation in the first place?

For the ancient theologian Irenaeus, the impact of the incarnation was to give the entire universe a cruciform shape.

> He is Himself the Word of almighty God, who in His invisible form pervades us all and encompasses the breadth and length, the height and depth, of the whole world, for by God's Word all things are guided and ordered. Now God's Son was also crucified in them [the four dimensions], since He has imprinted the form of the Cross on the universe. In becoming visible, He had to reveal the participation of the Cross on the universe.[54]

Although invisible to our natural eyes, the processes of nature everywhere are influenced by the redemptive work of the cross. What we get with Irenaeus is universality: one incarnation is enough for the entire cosmic creation.

By *incarnation* theologians do not refer merely to the Christmas story, to the birth of the human Jesus. Rather, *incarnation* refers to the entire Christ event in time and space, with a stress on the physicality of this particular time and space. However, Reformer Martin Luther expanded the idea of Christ's bodily presence beyond what could be "circumscribed" in the time and space of Jesus' biography; he contended that the body of Christ could also have an "uncircumscribed" presence as well. Luther thought of the resurrected and ascended Christ as ubiquitous—present everywhere—not just spiritually, but also bodily. When Christ ascends to the right hand of the Father, this does not place him in a single location, certainly not next to a divine throne in heaven. Rather, as the Father's right hand, Christ shares in God's omnipresence.

---

54. Irenaeus, *Scandal of the Incarnation*, 15. Christian philosopher in the analytic tradition Walter Alston contends that the Chalcedonian formula was intended to be taken literally as true: the divine became human in the historical Jesus. "The fundamental and most distinctive tenet of the Christian faith as defined at the Council of Chalcedon (AD 451) is the claim that the person who was and is Jesus of Nazareth is one and the same individual as God the Son, the Second Person of the divine Trinity—a literal statement of absolute numerical identity." Alston, *Divine Nature*, 17–18. Sarah Coakley emphasizes that the Chalcedonian formula was intended to be literal and ontological. "The interest in the literalness in the analytic defense of incarnationalism in general and of Chalcedon in particular has at its core an insistence on the ontological reality of what it describes that is fully in line with Chaldedon's intentions." Coakley, "What Does Chalcedon Solve?," 159.

The Scriptures teach us . . . that the right hand of God is not a specific place in which a body must or may be, such as on a golden throne, but is the almighty power of God, which at one and the same time can be nowhere and yet must be everywhere. It cannot be at any one place, I say . . . the power of God cannot be so determined and measured, for it is uncircumscribed and immeasurable, beyond and above all that is or may be. On the other hand, it must be essentially present at all places, even in the tiniest tree leaf. The reason is this: It is God who created, effects, and preserves all things.[55]

Now, Luther did not address the matter of extraterrestrial life. Had he taken up this question, he could have incorporated it within this uncircumscribed understanding of Christ's resurrected body. This comprehensive view of incarnation could be pressed into the service of an argument on behalf of one incarnation for the entire universe, not many incarnations.

To further our discussion on this point we turn to Niels Henrik Gregersen, a Lutheran theologian on the faculty at the University of Copenhagen. Gregersen embraces what he calls "deep incarnation." The central point he wishes to emphasize is that the *carne* or flesh of in-*carn-ation* refers to all that is material. The term *flesh* (Greek *sarx*) connotes every life form and every species in the history of biological evolution. "The New Testament nowhere states that God became human," he writes. "Rather the Logos of God 'became flesh' (John 1:14a) . . . God's incarnation also reaches into the depths of material existence."[56] One implication of this observation would be that incarnation is not limited to putting on a show for rational beings who can watch it. The significance of God becoming flesh stretches to every nook and cranny of the physical universe, including plants and animals who do not actively or consciously share communion with God. God's presence is not restricted to the realm of the mental, the intellectual, the spiritual. It is physical as well.

Peter M. J. Hess, a Roman Catholic theologian formerly at the National Center for Science Education and author of a chapter in this volume, would agree. "The incarnation . . . in the person of Jesus God took on the quarks of the Big Bang, the dust of supernovae explosions, the DNA of dinosaurs, and the long history of the primate genome. In an evolutionary paraphrase of St. Gregory of Nazianzus, God assumes creation by becoming incarnate at its heart in a human person, Jesus Christ."[57] In sum, the person of Christ is not

55. Luther, "These Words of Christ," in *LW* 37:57.
56. Gregersen, "Deep Incarnation: Why Evolutionary Continuity Matters," 174. See Gregersen, "Cross of Christ in an Evolutionary World," 192–207.
57. Hess, "Sacramental Character of an Evolving Creation," 37.

limited to his humanity; rather, he provides an ontological synecdoche for all of the material world.

A further implication of this position is that God identifies with dissolution, deterioration, and death. Gregersen writes, "the 'flesh' assumed by Logos denotes the person and body of Jesus, yet also signifies the whole dimension of materiality . . . Accordingly, when God in Christ became human, the Son of God entered the dimension of material life, in which human beings die, sparrows fly and fall to the ground, and grass grows and withers away. It is as natural for God to dwell in the world of dirt and waste as it is for God to be present in the uniquely human characteristics of highly developed consciousness, morality, and religious imagination . . ."[58]

Roman Catholic systematic theologian Elizabeth Johnson expands Gregersen's notion of deep incarnation by adding deep resurrection. "Drawing out Gregersen's insight into deep incarnation that unites the crucified Christ with all creatures in their suffering, I suggest we employ the idea of 'deep resurrection' to extend the risen Christ's affiliation to the whole natural world."[59] As Jesus rose from the dead, so also can we expect that God's eschatological salvation will triumph over suffering and death; and the benefits of God's triumph will accrue to all living creatures at the advent of the new creation. "The coming final transformation of history will be the salvation of everything, including the groaning community of life, brought into communion with the loving power of the God of life."[60]

Neither Gregersen nor Johnson take up the question of extraterrestrial life. Yet, what these two offer regarding deep incarnation might carry over for us. If Gregersen's concept of deep incarnation illuminates the significance of God's presence in Jesus of Nazareth, then we can see that previous discussions regarding the place of rationality in the list of an extraterrestrial species's capacities becomes irrelevant. God's love and even God's presence is not restricted to highly evolved intelligence. God does not require the prior existence of reason, science, technology, or such before deigning to bless or initiate presence.

This presence of the second person of the Trinity involves more than merely the creative power of the one God of the universe. It includes the cross as well. It includes God's willingness in Christ to absorb the slings and arrows of creation's misfortune. It includes the world's horrors. Physicist and theologian George Murphy applies the theology of the cross to the natural

58. Gregersen, "Deep Incarnation," 185.

59. Johnson, *Ask the Beasts*, 208.

60. Ibid., 209. Actually, Johnson subscribes to a qualified rare earth position. "While life may exist elsewhere, the immense evolutionary epic of life on this planet is a unique chapter in cosmic history. It will never be repeated." Ibid., 114.

world. "The crucified one is present throughout the universe. He is not only in the majestic and beautiful features of the world but also shares its sufferings, in situations of pain, loss, and death."[61] If God's presence in the cross on Earth applies to off-Earth creatures, then God will be present in their struggles and sufferings as well.

God is more than just present. The divine sharing in creaturely suffering is not all by itself redemptive. However, if we remind ourselves that the suffering of the cross is accompanied by the victory of resurrection, Christ's atoning work leads to a divine promise that God's future will overcome past and present evils. We need to view all life, both human and nonhuman, from the perspective of the resurrection, argues Celia Deane-Drummond. "This gives stronger account of a God who acts decisively to raise Christ from the dead, and as such, it offers the possibility of hope for all creaturely beings, not just humankind."[62] If by *single incarnation* we refer to the entire history of Jesus that includes death and resurrection, then we can think of the incarnation as a divine work that establishes within the physical world a promise of redemption from sin, evil, suffering, and even death. In principle, its efficacy would be felt in every time and every place.

George Coyne, SJ, is former director of the Vatican Observatory. Coyne presumes that the fallen or sinful state of creatures warrants the divine work of redemption. More to our point here, however, is that this redeeming work seeps into the very depths of material reality, the material reality present in stars and planets everywhere.

> How could he be God and leave extraterrestrials in their sin? After all he was good to us. Why should he not be good to them? God chose a very specific way to redeem human beings. He sent his only Son, Jesus, to them and Jesus gave up his life so that human beings would be saved from their sin. Did God do this for extraterrestrials? . . . There is deeply embedded in Christian theology . . . the notion of the universality of God's redemption and even the notion that all creation, even the inanimate, participates in some way in his redemption.[63]

In sum, the one incarnation of God in the Jesus of Earth's history will suffice for the entire cosmos.

---

61. Murphy, "Renewal of Creation and the Presence of Christ," 11.
62. Deane-Drummond, *Christ and Evolution*, 179.
63. Coyne, "Evolution of Intelligent Life," 187.

Does the One-Incarnation-Is-Enough Position Justify Geocentrism?

Contemporary mathematician and Russian Orthodox thinker Alexei Nesteruk argues for geocentrism, anthropocentrism, and Christocentrism. Teaching in both Portsmouth, UK and Moscow, Nesteruk observes: it is the human mind on Earth that conceives of the universe as a whole, that asks scientific cosmologists to map the entirety of physical reality. To be seen as a whole, according to Nesteruk, is to become constituted as a whole. In other words, the universe needs the human mind to become constituted as what it is. The self-constitution of the cosmos as a cosmos takes place within the human realization of it.

Nesteruk contends that this is possible because of the divine incarnation taking place on our planet. It is God's creation and redemption of the cosmos that provides the backdrop for human apperception of this reality. Even though the universe may look immense to human eyes, it certainly does not look awesomely large to God. God's eyes keep things in perspective. With this as a backdrop, Nesteruk can retrieve and justify geocentrism. "The geocentrism and anthropocentrism of cosmology inherent in the theological commitment mean Christocentrism, for it is here, on the planet Earth, that the meeting of the Divine and human, uncreated and created took place, and it is because of this that earth is spiritually central as that place from which the disclosure and manifestation of the sense of the created universe takes place."[64]

One implication is this: Christians should not think about other worlds. "The invocation of other worlds is dangerous and soteriologically futile, and Christian theology warned against it because we cannot understand the meaning and purpose of other worlds . . . because of the Fall."[65] The microcosm-macrocosm correlation between human subjective wholeness and the objective oneness of our world seem to provide a limit of sorts to the scope of Nesteruk's theological commitment.

Even though, like Nesteruk, the position I defend here relies upon a single soteriological event taking place in the ministry and resurrection of Jesus of Nazareth on Earth, I do not translate this into any form of centrism. What God has accomplished in this event of terrestrial history does not accrue to an advanced status for either the human race or for the planet on which we live. Nor does it preclude the existence of subjectivities on other worlds looking at our cosmos from a different perspective.

---

64. Nesteruk, *Sense of the Universe*, 75.
65. Ibid., 408.

If anything, what we learn from Scripture and from the teachings of Jesus is to carry ourselves with a demeanor of humility. When God raised up prophets in ancient Israel, God selected marginal individuals to speak truth to power. Jesus repeatedly warns, "the last shall be first." If a theologian adopts the one-incarnation-is-enough position, I recommend he or she resist the temptation to justify any sort of centrism. It would fit hermeneutically, so to speak, for we earthlings to live with the paradox that the cosmic event of salvation took place on our marginal and humble planet.

Interim Summary

Where have we been? It appears that we have identified four logical positions. The first would assume a fix-a-broken-creation Christology with a single incarnation, the history of Jesus on Earth. The second position would assume a fix-a-broken-creation Christology with multiple incarnations, one for each planetary civilization. The third would assume an incarnation-anyway model that relies upon a single incarnation on Earth. The fourth would assume an incarnation-anyway Christology combined with multiple incarnations.

Of these alternatives, I believe the most coherent position is the first: a fix-a-broken-creation Christology (more precisely, Soteriology) combined with reliance upon the atoning work of the single Earthly incarnation event. I would add that I would incorporate a high Christology rather than a strictly revelational or exemplar Christology. Accordingly, in the historical incarnation on Earth God accomplished something with ontological import, import for everything in the material world no matter how distant in space or time.

Defending this position risks some unhappy repercussions. For example, I might get accused of geocentrism; because it appears that I grant our planet a specially chosen status. However, I do not wish to defend geocentrism, because I deem salvation to be an eschatological gift of divine grace for all of creation, all of the galaxies. What happened on Earth in the Jesus event was a prolepsis, and an anticipation of the cosmic wide transformation which the Jesus event promises. This is the case for the cosmos, I think, whether conscious beings realize its truth or not.

This accusation of geocentrism might carry a rider, namely, that a historical event on Earth might never be known on an extra-solar planet; and this implies that we earthlings have exclusive access to a cosmic truth. Doesn't this claim entail *hubris* or anthropocentrism? No. Because I do not deny other forms or events of divine self-revelation that could take place anywhere at any time. With Tillich, I can almost forecast that God would make the power of creation and redemption known where creatures are beset with the sting of death or the horrors of sin. When it comes to revelation, this position is not geocentric.

What Really Is at Stake with Incarnation?

It is time to stop quoting theologians and to stop lining up testimonies to describe various positions. Let me ask again directly: just what is really at stake with the concept of incarnation? Why does this curious doctrine appear within the Christian scheme of redemption? Answer: because the love of God for the world is at stake. That little gospel in miniature that appears on posters during football games, John 3:16, begins: "For God so loved the world . . ." The word for world is *kosmos*. The concept of *kosmos* includes all things, even all physical things. To be sure, when the biblical writers looked up in the sky they saw a lot less than modern scientists with telescopes can see. But this does not change the fundamental insight: God loves the physical world and, in Jesus Christ, God took the existence of the physical world into the divine life. What took place was a communication of attributes (*communicatio idiomata*) in which the world took on divinity and the creator took into the divine being what is created. This interchange of attributes means that the divine power of renewal—the promise of ultimate transformation—belongs now in the world itself.

If we defend the claim that God's incarnation in the history of Jesus in ancient Israel is efficacious for the entire cosmos past and present, we can expect a critic to complain: this looks like a return to Earth chauvinism! It places our rinky-dink marginal planet in the very center of redemptive cosmology! It's one more anachronistic defense of geocentrism!

What such a complaint relies upon is the *scandal of particularity*. Accordingly, it appears scandalous to posit that any particular historical event determines the ontological nature of all things universally. Such critics may assume that universal truths must be derived universally, not produced as effects of particular local events. Therefore, this criticism leveled against geocentrism would apply equally should a parallel claim be made on behalf of incarnation on a planet orbiting Tau Ceti or any other star. No particular history, regardless of planet, could immunize itself from such a criticism. But, we must counter, for something to be real everywhere it must be real somewhere. So, we might argue, any redemptive event cannot escape particularity, even if it bears universal significance. Our gratitude to God for the divine incarnation in the history of Jesus implies two things: first, this atoning event has efficacy for the entire cosmos and, second, this does not lead to geocentrism but rather to theocentrism. God belongs in the center of our reverence, not our planet.

Our belief in the cosmic scope of God's redemptive work begins with our interpretation of the biblical symbols. Most of the Bible's symbols for God's promised transformation are political. The City (*polis*) of God, the New Jerusalem, the Kingdom of God, or even appellations to Jesus such as Lord or King of Kings all point to a transformation of the human social network. This explains why the Christian religion is such a moral religion, emphasizing the pursuit of justice and the common good. Even so, natural symbols also appear in the Bible. The one key to our treatment here is the symbol of the *New Creation*. "I am about to do a new thing," says the God of Israel in Isaiah 43:19. The future of reality will not be what it is now or has been.

The new thing God promises is redemption, salvation. "See, the home of God is among mortals. He will dwell with them; they will be his peoples, and God himself will be with them; he will wipe every tear from their eyes. Death will be no more; mourning and crying and pain will be no more, for the first things have passed away" (Rev 21:3-4). If there will be no more death or crying over death, then the nature of nature will have to change. This is an awesome claim. Yet, this is just what is promised in the symbols of the New Creation and the New Jerusalem.

The first step in this promised transformation of the created order is the Easter resurrection of Jesus. God raised Jesus from death to new life, eternal life. This was more than just a nice thing to happen to one person in ancient Jerusalem. The first Easter resurrection amounted to an ontological change, a fundamental alteration in the nature of nature. Whereas prior to Jesus it was natural that the dead stay dead, God's raising of Jesus on the first Easter initiated a new plan for the future of nature. "But in fact Christ has been raised from the dead," writes St. Paul, "the first fruits of those who have died" (1 Cor 15:20). What happened to Jesus will happen to us. In the future it will be natural for the dead to rise. The dead will be raised to a level of existence in which crying and death are but a memory. The God who is the creator of all things plans to re-create, or at least to transform what now is.

These claims are audacious, to be sure; but the fact that they are audacious does not in itself count against them. They still may be true. Trusting in their truth is what a faithful Christian does. "Without a cosmic eschatology there can be no assertion of an eschatological existence of man . . . world-picture and faith are inseparable,"[66] trumpets Jürgen Moltmann. As

---

66. Moltmann, *Theology of Hope*, 69.

audacious as it may sound, the Christian world-picture includes a vision of a coming new creation for the whole of the cosmos. The Christian claim is that what happened to Jesus of Nazareth on the first Easter models what will happen to the entirety of God's creation in the future.

This is what the concept of the incarnation is intended to convey. It could be misleading if we think of incarnation in terms of a heavenly being taking a vacation on planet Earth or a spiritual being dressing temporarily in human skin. Rather, incarnation is an abbreviated cipher for the entire life, death, and resurrection of Jesus; and it is nested inextricably within God's promise of renewal of all that exists in the creation.

Conclusion

Is the Christian ship likely to sink because the possibility of intelligent life beyond Earth is now a part of our consciousness? No. Despite prophecies of vulnerability in Christian belief due to christological commitments, there is no hole in the bottom of the Christian boat.

Theologians can ask honestly: Does God's promise of transformation extend beyond planet Earth? Does it extend to the Milky Way? Beyond? To all the galaxies and clusters of galaxies, including all the planets that may be orbiting or rogue? This is what that little two syllable word, *kosmos*, implies. No part of creation, past or future, lies beyond the promised redemption of God as promised in the Easter resurrection of the Jesus of Earth's history.

Bibliography

Adams, Marilyn McCord. *Christ and Horrors: The Coherence of Christology*. Cambridge: Cambridge University Press, 2006.
Alston, Walter. *Divine Nature and Human Language*. New York: Cornell University Press, 1989.
Athanasius. *Incarnation of the Word*.
Berger, Peter L. "A Conversation with Peter L. Berger: 'How My Views Have Changed,'" by Gregor Thuswaldner. *The Cresset* 77.3 (2014) 16–21.
Boethius. *Liber de persona et duabus naturis*.
Brück, Michael von. *The Unity of Reality: God, God-Experience, and Meditation in the Hindu-Christian Dialogue*. New York: Paulist, 1991.
Bruns, J. Edgar. "Cosmolatry." *The Catholic World* 191.1 (August 1960) 286.
Burgess, Andrew J. "Earth Chauvinism." *Christian Century* 93 (1976) 1098.
Calvin, John. *Institutes of the Christian Religion* (1559), edited by John T. McNeill, Library of Christian Classics XX, XXI. Louisville: Westminster John Knox, 1960.
Coakley, Sarah. "What Does Chalcedon Solve and What Does it Not? Some Reflections on the Status and Meaning of the Chalcedonian Definition." In *The Incarnation: An*

Introductory Symposium on the Incarnation of the Son of Man, edited by Stephen T. Davis et al., 143-63. Oxford: Oxford University Press, 2002.
Consolmagno, Brother Guy. "Would You Baptize an Extraterrestrial?" http://www.beliefnet.com/story/35/story_3519.html.
Coyne, George V., SJ. "The Evolution of Intelligent Life on Earth and Possibly Elsewhere: Reflections from a Religious Tradition." In *MW* (2002) 177-88.
Crisp, Oliver D. "Desiderata for Models of the Hypostatic Union." In *Christology Ancient and Modern*, edited by Oliver D. Crisp and Fred Sanders, 19-41. Grand Rapids: Zondervan, 2013.
Crisp, Oliver D. *Revisioning Christology: Theology in the Reformed Tradition*. Aldershot, UK: Ashgate, 2011.
Crowe, Michael J. *Extraterrestrial Life Debate from Antiquity to 1915: A Source Book*. Notre Dame: University of Notre Dame Press, 2008.
Davies, Paul C. W. *God and the New Physics*. New York: Simon & Schuster, 1983.
———. "Transformations in Spirituality and Religion." In *SETI* (2002) 51.
Deane-Drummond, Celia. *Christ and Evolution: Wonder and Wisdom*. Minneapolis: Fortress, 2009.
"Declaration on the Relationship of the Church to the Non-Christian Religions." *Nostra Aetate* 4. In *The Documents of Vatican II*, edited by Walter M. Abbott, SJ. New York: America Press, 1966.
Delio, Ilia, O.S.F. "Christ and Extraterrestrial Life." *Theology and Science* 5.3 (2007) 249-65.
———. *Christ in Evolution*. Maryknoll, NY: Orbis, 2012.
Driel, Edwin Chr. van. *Incarnation Anyway: Arguments for Supralapsarian Christology*. American Academy of Religion Series. New York: Oxford University Press, 2008.
Edwards, Denis. *How God Acts: Creation, Redemption, and Special Divine Action*. Minneapolis: Fortress, 2010.
Ford, Lewis S. *The Lure of God*. Minneapolis: Fortress, 1978.
———. "Theological Reflections on Extra-Terrestrial Life." *Raymond Review* 3.1 (1968) 2.
Foster, A. Durwood. *The God Who Loves*. New York: Bruce, 1971.
Gregersen, Niels Henrik. "The Cross of Christ in an Evolutionary World." *Dialog: A Journal of Theology* 40.3 (2001) 192-207.
———. "Deep Incarnation: Why Evolutionary Continuity Matters in Christology." *Toronto Journal of Theology* 26.2 (2010) 173-87.
Heim, S. Mark. *The Depth of the Riches: A Trinitarian Theology of Religious Ends*. Grand Rapids: Eerdmans, 2001.
Hess, Peter M. J. "The Sacramental Character of an Evolving Creation." *Omega* 8.2 (2009) 34-57.
Hick, John. "A Response to Hebblethwaite." In *Incarnation and Myth: The Debate Continued*, edited by Michael Goulder, 192-94. Grand Rapids: Eerdmans, 1979.
Hunsinger, George. "Salvator Mundi: Three Types of Christology." *Christology Ancient and Modern*, edited by Oliver D. Crisp and Fred Sanders, 42-59. Grand Rapids: Zondervan, 2013.
Irenaeus. *The Scandal of the Incarnation: Irenaeus against the Heresies*. Edited by Hans Urs von Balthasar. Translated by John Saward. San Francisco: Ignatius, 1990.
Johnson, Elizabeth A. *Ask the Beasts: Darwin and the God of Love*. London: Bloomsbury, 2014.

Lossky, Vladimir. *Orthodox Theology: An Introduction.* Crestwood, NY: St. Vladimir's Seminary Press, 1989.

Luther, Martin. "That These Words of Christ, 'This Is My Body,' Still Stand Firm Against the Fanatics." *Luther's Works.* American Edition. Vols. 1-30, edited by Jaroslav Pelikan. St. Louis: Concordia Publishing, 1955-1967; Vols. 31-55, edited by Helmut T. Lehmann. Philadelphia: Fortress, 1955-1986.

Marty, Peter W. "Who Gets Saved?" *The Lutheran* 27.3 (2014) 3.

McHugh, L. C., SJ. "Life in Outer Space? An Interview with Rev. L.C. McHugh, SJ." *Sign* 41.5 (1961) 29.

McMullin, Ernan. "Life and Intelligence Far From Earth: Formulating Theological Issues." In *MW* (2000) 151-75.

Moltmann, Jürgen. *Theology of Hope.* Translated by James W. Leitch. New York: Harper, 1967.

Murphy, George L. "The Renewal of Creation and the Presence of Christ." *Currents in Theology and Mission* 40.1 (2013) 5-11.

Nesteruk, Alexei V. *The Sense of the Universe: Philosophical Explication of Theological Commitment in Modern Cosmology.* Minneapolis: Fortress, 2015.

O'Meara, Thomas F. *Vast Universe: Extraterrestrials and Christian Revelation.* Collegeville, MN: Liturgical, 2012.

Pannenberg, Wolfhart. *Systematic Theology.* Translated by Geoffrey W. Bromiley. 3 vols. Grand Rapids: Eerdmans, 1991-1998.

Peacocke, Arthur. "The Challenge and Stimulus of the Epic of Evolution to Theology." In *MW* (2000) 89-118.

Pittenger, Norman. *The Word Incarnate.* London: Nisbet, 1959.

Polkinghorne, John. *Science and the Trinity: The Christian Encounter with Reality.* New Haven: Yale University Press, 2004.

Puccetti, Roland. *Persons: A Study of Possible Moral Agents in the Universe.* New York: Herder & Herder, 1969.

Rahner, Karl. "Christology within an Evolutionary View of the World." In *Theological Investigations*, 5:157-92. Translated by Karl-H. Kruger. Boston: Helicon, 1966.

———. "On the Theology of the Incarnation." *Theological Investigations*, 4:105-20. Translated by David Bourke. New York: Seabury, 1974.

———. "Sternenbewohner: Theologisch." *Lexicon für Theologie und Kirche* 9:1061-62. Freiburg: Herder, 1964.

Russell, Robert John. "What Are Extraterrestrials Really Like?" In *God for the 21st Century,* edited by Russell Stannard, 64-66. Philadelphia: Templeton Foundation Press, 2000.

Tillich, Paul. *Systematic Theology.* 3 vols. Chicago: University of Chicago Press, 1951-1963.

Weidemann, Christian. "Did Jesus Die for Klingons Too?" In *TFC* (2016) 124-34.

Wilkinson, David. *Science, Religion, and the Search for Extraterrestrial Intelligence.* Oxford: Oxford University Press, 2013.

Worthing, Mark W. "The Possibility of Extraterrestrial Intelligence as Theological Thought Experiment." In *God, Life, Intelligence and the Universe,* edited by Terence J. Kelly and Hilary D. Regan, 61-84. Adelaide: Australian Theological Forum, 2002.

17

Many Incarnations or One?

ROBERT JOHN RUSSELL

> Incarnation is unique for the special group in which it happens, but not unique in the sense that other singular incarnations for other unique worlds are excluded.
>
> —PAUL TILLICH[1]

In chapter 4 of this volume I discussed a number of philosophical and theological issues and implications that will most likely arise with the discovery of intelligent extraterrestrial life in the universe. The sixth issue leads to another crucial theological question, perhaps the most crucial of them all: will ETI need redemption? I turn now to that question by addressing the debate proposition: *If multiple societies of extraterrestrial intelligent beings on exoplanets exist, we can predict that God will or already has provided a species-specific incarnation for each planet parallel to God's incarnation in Jesus Christ on Earth.* The position I take is the positive. I affirm this proposition. I shall argue that God provides multiple incarnations wherever ETI has evolved.

My response to this debate proposition differs from those offered by Joshua Moritz and Ted Peters in other chapters. Here, I'll attempt to show why I defend a "many incarnations" view, then acknowledge a serious problem which I find with this view, and close with a response to it that seeks to provide a bridge between my many incarnations view and Peters's single

1. Tillich, *ST*, 2:95.

incarnation view based on recent work on "deep incarnation" by Danish theologian Niels Henrik Gregersen.

Will ET Experience Moral Failure or Be Entirely Benign?

Previously I began by stressing the tragic reality at the heart of human existence: excessive violence against our own and other species, a degree and a kind of violence that is unique within the diversity of life on Earth.[2] Since God is the creator of all life, violence of this kind is, in fact, a ruthless rejection of the sovereignty and love of God. Such violence is a result of sin. But why, then, do we sin? Why do we fail to love and serve God above all else and instead indulge ourselves in unbridled pride and inordinate sensuality?

I then drew on St. Augustine who offered a profound theological insight into the human condition, one which was taken up by Reinhold Niebuhr.[3] This insight is that we live in a paradox. On the one hand, sin is not an intrinsic part of human nature, and yet on the other hand it is an inevitable component of fallen human behavior. Sin by definition, then, is unnecessary; but without grace it is inevitable. Indeed its only remedy is the grace of God. In sum, each of us is created in the *imago Dei*, the image of God, and each of us inherits the inevitability of sin, i.e., sin as original. And both the *imago Dei* and sinfulness are unique to our species.

This traditional response conforms in large measure with our contemporary scientific understanding of biological evolution, particularly with the philosophical theme of the emergence of genuine biological novelty within an underlying physical continuity. Thus, we as a species inherit diverse propensities from our pre-hominid past tracing back some two to three million years to the last common ancestor of all hominids. Nevertheless, in *Homo sapiens* something unique emerges. This uniqueness is manifest in the *imago Dei* (God's image in us, Gen 1:27; 9:6), traditionally interpreted as our capacity for reason, abstract thought, language, technology, art, ethics, and science; our role as priests and stewards of nature; and our being relational creatures reflecting the relationality of the trinitarian Creator and placed in relation to each other and to the world. Our newness is also found, tragically, in the reality of human sin, including ruthless violence and our insatiable appetite for power and control. It is only through the grace of a loving God that our lives can be transformed into the fullness of what it

---

2. I do not mean to replace the concept of sin with that of violence. Instead I want to suggest that one of the core manifestations of sin is violence.

3. Niebuhr, *The Nature and Destiny of Man*, 1:VI–X.

truly means to be human, and what is indeed truly human is given us in Jesus Christ.

What then about extraterrestrial sentient and intelligent beings? Will all ETI be gifted with the *imago Dei*? Will all ETI experience sin? Or, are we humans the only species throughout the universe gifted by the image of God and plagued by sin? I opt for the universality of the *imago Dei* and of sin for the following reason: I assume that since the underlying laws of physics are the same everywhere, the processes of evolutionary and molecular biology most likely will be quite similar. Accordingly, ETI would be gifted by God with an *imago Dei* similar to ours (the image is, after all, that of the one universal God). Moreover, as I suggested in chapter 4, ETI would be tragically flawed by an ambiguous ethical and moral character even as we are. In essence I expect that ETI will experience a kind of moral dilemma that in many ways resembles the moral quagmire of terrestrial human experience, though obviously differing in the moral morphology of personal and social ethics.

If this is the case, my further assumption is that since God is the creator of all life in the universe through the processes of evolution; and, since God loves all that God creates, then God will be present to the moral struggle of life everywhere. Further, God's grace will redeem and sanctify every species in which reason and moral conscience are kindled. Thus, I plan to work here with the supposition that our intelligent neighbors in space will need redemption just as we do. This, in turn, leads to a crucial astrotheological question: should we expect multiple incarnations of the Logos, or will the incarnation of the Logos in Jesus of Nazareth on Earth suffice for the salvation of all life in the universe?

Multiple Incarnations: An Initial Rationale

If the need for redemption is universal among all species of ETI, should we expect that a single incarnation of the Logos in the person of Jesus is sufficient for their redemption as well as ours? Or should we expect there to be an Incarnation of the Logos in each species of ETI that is gifted by reason and disfigured by sin?[4]

Modest support for both options can be found among Protestants and Roman Catholics. As we saw in the Peters chapter, a single, universally

---

4. A rather different discussion involves the claim, rooted in most part in Scotus, that there might well be multiple incarnations aside from the need for redemption from sin. See for example Adams, *Christ and Horrors*. I note but will not pursue this view here for lack of space.

efficacious incarnation is supported by Protestants Ted Peters and Wolfhart Pannenberg along with Roman Catholics L. C. McHugh and J. Edgar Bruns, while multiple incarnations have been advocated by Protestants such as Paul Tillich and Lewis Ford along with Roman Catholics such as Karl Rahner, E. L. Mascall, and Ernan McMullin. David Wilkinson, in his masterful survey and assessment of this and related questions, points to Teilhard de Chardin, Yves Congar, Norman Pittenger, Bishop Frank Weston, John Polkinghorne, and even Thomas Aquinas as open to, even supportive of, multiple incarnations, a position with which Wilkinson seems to agree.[5] In her masterful study of Christology, Marilyn McCord Adams[6] agrees, too, but Oliver Crisp, in his exploration of Christology, disagrees, suggesting that a single cosmic incarnation is "the more likely" option.[7] So the theological and philosophical jury is out! Here, with due respect to this divided audience, I will attempt to offer a rationale for multiple incarnations by focusing on five key considerations that seem to point in this direction. But first I want to turn, briefly, to a distinction Peters makes regarding the incarnation.

In his chapter on the incarnation, Peters makes a strict distinction between a *revelational* view of the incarnation and an *ontological* view of the incarnation. The ontological view emphasizes the significance of the incarnation as the act in which God redeems the world from such physical laws and regularities as to produce the universal facts as sin and death. Instead God's act begins the transformation of the world into the New Creation, including the change in these laws and regularities. The revelational view emphasizes the effect that the incarnation has on us personally, given this ontological view of the changes in natural law. It is couched in our personal and corporate history, and it involves our coming to know Christ and our choosing to follow him in discipleship and newness of life. Now I suggest the following, rather than seeing these in strict distinction as Peters does: There can indeed be redemption as an ontological act without its revelation. This position is held by those supporting the possibility of the universal salvation of all of humanity through a unique incarnation on earth, including the saving of those born before Jesus or those who have never heard the Good News. But conversely, there cannot be a genuine, personal revelation of the Good News without the historical and ontological act by which God redeems the world, without the meaning of this act being couched in the details of human history—in particular the history of the Jews—and without this revelation being received by human beings in the context of their

---

5. Wilkinson, *SRSEI*, 155–9.
6. Adams, *Christ and Horrors*, 198–99.
7. Crisp, *God Incarnate*, 175.

lived religious experience (Niebuhr). On the one hand, I believe it would be incoherent to argue that we humans can encounter the Good News without there being any actual, ontological basis for it, and yet on the other hand I believe that this ontological basis must be manifest (Tillich), appear (Pannenberg), and be received (Niebuhr) by people reflecting the specificities and diversity of human history. I therefore start with the revelational view of the incarnation and argue that this lends support for multiple cosmic incarnations.

Now in this argument I tend to agree, at least on first blush, with scholars such as Peters and Moritz who say that, from an *ontological* perspective, a single incarnation of the divine *logos* in Jesus of Nazareth is sufficient for the redemption of the universe. According to Peters a unique incarnation on earth can ontologically alter the entire universe and the laws of nature which describe natural processes. "The first step in this promised transformation of the created order is the Easter resurrection of Jesus . . . (It) amounted to an ontological change, a fundamental alteration in the nature of nature (and) initiated a new plan for the future of nature."[8] Indeed, as Peters seems to argue, if Christ's death and Resurrection not only forgave moral sin but eschatologically overcame biological death for us on Earth, and if this entails an eschatology of New Creation and with it new laws of nature, then these new laws could have a universal efficacy that starts with radical changes in natural processes on earth and then spreads across the universe. Moritz insightfully draws on non-canonical sources in the early history of the church and their wide acceptance by the church Fathers to show how redemption of even non-human animals was an accepted part of the church tradition. From this he infers that such a broader interpretation of the domain of redemption can be expanded to include all of creation, and thus all ETI in the universe.[9]

But on the other hand, if I start with the *revelatory significance of the incarnation*, our *participation* by faith in it, and the requirement that this revelation and our participation in it be based on an historical and ontological act of redemption by God, I am led to support multiple incarnations. In essence, regarding these two views of the incarnation, the ontological view

---

8. See Peters's chapter, "One Incarnation or Many?" above. In support of Peters's view, I can offer an analogy drawn from mathematics. Take a sphere and delete one point from it. The sphere becomes identical topologically to a sheet of paper: while its shape (geometry) remains spherical, its structure (topology) is entirely changed. The christological interpretation of this fact is that the incarnation of the Word in Jesus of Nazareth has universal significance: it changes the redemptive shape, as it were, of the universe.

9. See Moritz's chapter, "One *Imago Dei* and the Incarnation of the Eschatological Adam."

does not require the revelational view, but the revelational view does require the ontological view. In sum, the ontological and the revelational views are *not* distinct and separate views, as Peters claims. Instead the revelational view requires an ontological incarnation, and one that can be communicated to ETI in the context of its particular history, just as the incarnation of the Logos on earth was suffused with and made intelligible by the history of God's revelation to the Jews and, through Paul and the others, to the Gentile world.

This then is my basic argument for multiple incarnations. ETI could not be included in the revelational efficacy of the incarnation without an ontological incarnation on their planet, embedded in their own species and its histories, but from Peters's ontological view they could be included in the ontological efficacy of the terrestrial incarnation without an incarnation on their planet. Since I adopt the revelational view, I believe there must be multiple incarnations to effect the redemption of all intelligent life in the universe. In sum, if Peters is correct, then from an *ontological perspective* a single incarnation of the divine *logos* in Jesus of Nazareth is sufficient for the redemption of the universe. A unique incarnation on earth can ontologically alter the entire universe and the laws of nature which describe natural processes. But if I am correct, from a *revelatory perspective* our participation by faith in the Resurrection requires that this revelation be based on an ontological act of redemption by God *and* that it be known to all species needing redemption. In this perspective a single incarnation is *insufficient* for the redemption of the universe because the participation of ETI will be impossible. Instead *multiple* incarnations are required.

That being said, I will offer five additional considerations which I believe lend credence to the view of multiple incarnations.

1) For starters there is what I will call *the problem of distance* posed to both the revelational view and the ontological view of the incarnation when a single incarnation is supported. To the revelationalists, given the almost unimaginable enormity of the visible universe (~46 billion light years in size[10]), and the intrinsic limitations on communication between us and distant solar systems by the finite speed of light (e.g., it would take 3.4 years for signals from earth just to reach the nearest star, Alpha Centauri, and some 2.5 million years to reach the nearest galaxy, Andromeda, etc.), how would we communicate the terrestrial revelation of Jesus to the countless species of ETI strewn across the immeasurable distances of space? Needless to say, without such a communication they could not come to have faith

---

10. The age of the universe is some 13.82 billion years but since it has been expanding during that time, its size is almost three times larger.

and participate in the life of faith?[11] I think there is a categorical difference between 1) conceiving of "universal salvation" on Earth, including all peoples "outside the church" (as with Karl Rahner's "anonymous Christians" and with Moritz the possibility of redemption for those who lived and died before Christ) *and* the salvation of all animals on earth, and 2) thinking of such a "universal salvation" as extending across the almost unimaginable vastness of space and time to reach all ETI in the universe. In my mind, the sheer enormity of the latter cosmic revelationist view simply refuses to be domesticated to the homey terrestrial scope of the former revelationist view. The *problem of distance* also poses a conceptual problem for the ontological view of a unique incarnation: how would the changes in the laws of physics, initiated on earth by God, "spread" throughout such cosmic distances? Would the change be spread at the speed of light? Then in an ever expanding universe (indeed, one in which the rate of expansion is accelerating) it might never reach the farthest parts of the universe. Or would the changes occur "simultaneously"[12] throughout the universe? If so, how is this universal change connected to and dependent on the unique initiating event of the incarnation only here on earth? It seems that a multiple incarnations view is essential to either the ontological or the revelational view of the incarnation.

2) Next there is *the problem of difference*. Even if extraterrestrial beings are like us in many ways, as I argued above, they will surely also be radically different from us in many other ways. Is it even conceivable that our story of salvation, so deeply encoded in the context, narratives, and languages of human history in general, and in the history of Israel in particular, could become their story? If alien races of intelligent creatures are to learn of the revelation of the love and grace of their divine creator through the presence of Christ and the Spirit, if they are to come to their own faith experience of God's salvation in their lives through the Incarnate second person of the Trinity, I believe they must have their own access to the revelation of the incarnation and the dispensation of the Spirit in their own histories in species-specific ways[13] and on their own planets. This is "contextual" theology in capital letters!

---

11. The problem is expanded by the prior problem of "communication." For example, it could not be through human language which is clearly too species- and planet-specific, nor could it be through mathematics which is hardly a theologically-sufficient language.

12. I put this in quotes to acknowledge the severe problems in defining simultaneity at great distances in light of special relativity.

13. I take this term from discussions of the ways ethologists talk about animal behavior as involving species-specific moral norms and behaviors versus animal behavior as latent to human moral behavior. See Deane-Drummond, *Christ and Evolution*.

3) There is also the *concern about absence*. If Peters is correct and there is only one incarnation for the entire universe, then the implication is that it is *highly unlikely* that this single incarnation actually occurred on earth. If N is the number of ETI, then 1/N is the likelihood that a *single* incarnation occurred on Earth and not some other planet inhabited by intelligent life. Debates over the size of N abound, but one thing is for sure: the larger it is, the less likely is it that the incarnation actually occurred on earth.

So suppose, as is most likely, that God's Word had *not* been incarnated in human history. Such a case would have left us with a radically different terrestrial history than the one we actually have: a history totally devoid of the New Testament, a history in which the Christian faith is entirely absent. Yes there might have been the Vedas of Hinduism, the Dhammapada and Sutras of Buddhism, the Qur'an of Islam (without, of course, the references to Jesus), even the Hebrew Bible—but ours would have been a human history without the Gospels, the epistles, and the other books in the New Testament. Or take a page from Western philosophy and its influence on Christian theology: We would have the dialogues of Plato but not the theology of Augustine, or the philosophy of Aristotle but not the theology of Thomas Aquinas, or Descartes, Kant, and German idealism but not the theologies of Schleiermacher, Barth, or Tillich, or the philosophy of Whitehead and Hartshorne but not process theology, or the moral, spiritual, and political vision of Mahatma Gandhi but not that of Martin Luther King. Politically we might have had Hitler but not the Christian resistance evidenced in Bonhoeffer. Architecturally Rome might have known the Parthenon but not St. Peters, London might have had its Buckingham Palace but not Westminster Abbey, and Washington, DC, might have included the White House but not the National Cathedral. In Western music could we have the symphonies of Mozart and Beethoven without Gregorian chant? Or in literature, would there have been a Shakespeare without Dante? Indeed could we have had the methodological naturalism that underlines the natural sciences without the combination of the Greek and Johannine *logos* traditions and the biblical/theological idea of creation *ex nihilo*?

Of course, on the positive side we would also have been spared the horrors of the Inquisition, the Crusades, and the Thirty Years War, or the biblical justification of slavery and apartheid, etc. To be frank, it is hard even to begin to conceive what history would have been like without Christianity—for better *and* for worse! Ours would have been an unrecognizably different civilization, one with synagogues, mosques, and temples but with no churches, and no one possessing faith in the Resurrection of the Body. To put it in a nutshell, those of us who are Christian would never have been who we are as Christian. This is a strong argument against a single incarnation from a revelationist view, but I believe it is a fairly strong argument against a single incarnation from an ontological view as well, since in either view the loss to human history would seem immeasurable.

4) Appeal to the theology of *the grace of God as God's self-communication to God's creatures* can also lend some modest support for multiple incarnations and a revelationist view of the incarnation. This appeal can be formulated in a variety of ways, but they all get at the same point: God so loved the world (John 3:16) implies that God does not want to be without the universe in all its details and life forms, nor does God want the diversity of species of life in the universe to be without the knowledge and grace of God. To cite just one example of this argument I turn to a lesser known dogmatics by Karl Barth, the so-called *Göttingen Dogmatics*[14] of 1921. Here in his first venture at such a project, Barth is astonishingly candid and forthcoming in his writing, and he is single-mindedly devoted to the task of preparing future ministers for the task of preaching. Here is a quote from his section on the "communicable attributes" of God, specifically the attribute of love:

> God's love is the attribute whereby God as the supreme good willed from all eternity to have fellowship with *all* things . . . God does *not* will to be *without the world*. This is God's love, the miracle of God's love, we must say . . . We may *not* say that he might have done *differently*, that he might have *refrained from fellowship with the world*. When we think about the Trinity, about the Son of God in whom he has turned *to the world* from all eternity, we have to say that the very idea of another use of his freedom is *a denial* of God, that there is in God no depth in which this *willing of fellowship* does not obtain.[15] (my italics)

Clearly Barth's context is planet earth, but it is now a straightforward extension of Barth's argument that arrives at the context of ET: there is no world of life in the universe for which God does not wish fellowship, in particular fellowship through the Son of God through which all worlds are created in the power of the Spirit. With this appeal to the grace of God as God's self-communication, I believe the case for multiple incarnations takes on an even deeper and more theologically convincing voice.

5) Finally there is the *theological desideratum for universal eschatology*. In previous writings[16] I have argued that the best interpretation of eschatology is one based by analogy with a widely held interpretation of the bodily

14. Barth, *Göttingen Dogmatics*.
15. Ibid., 423–24.
16. See Russell, *CAO*, Ch. 10, and Russell, *TIE*, "Introduction."

Resurrection of Jesus by a variety of biblical scholars. Here the bodily Resurrection is understood in terms of a radical "transformation" (e.g., Jesus of Nazareth *is* the Christ of faith) and not either a resuscitation back to normal life (e.g., what happened to Lazarus) or a Gnostic spiritual flight (e.g., one which leaves the disciples with a purely spiritual renewal of faith à la Bultmann). Such a transformation includes elements of continuity along with elements of discontinuity. So for example it was the same person, Jesus of Nazareth, who is the Risen Christ. He is able to be touched, heard, seen, and he can eat with his disciples. But he appears in closed rooms, disappears from their midst, and ascends to heaven never to die. Of course all of this can be understood without being read literally, but the biblical scholars I follow handle this problem remarkably well.

My approach is to claim that the eschaton by analogy implies a radical transformation of the present Creation into the coming New Creation, and this transformation includes both elements of continuity and of discontinuity. In this transformation there will be incredibly new things about the New Creation, but some things about it are already here in the present Creation—and science might be a terrifically helpful partner in discovering them. I tend to use the phrase "first instance of the new laws of nature" (FINLON) to capture the idea that even the laws of nature are open to change by God's action in bringing about the New Creation starting with the earthly Resurrection of Jesus (and in all its equally normative instances throughout the universe, following my adoption of "multiple incarnations.") Following Ted and Wolfhart Pannenberg, I frame the entire argument in terms of "prolepsis": the appearance of the New Creation within the historical and natural processes of the present Creation, a retroactive temporal event of the coming of the eschatological future into the present events of this world.

Now for the eschaton to be the transformation not just of planet Earth and the laws of nature that apply to it (if such a limited concept of the laws of nature even makes rational sense), a *desideratum* is that the eschaton is somehow the transformation of the entire universe and the universal laws of nature that pertain to it as a whole. (This is connected to the first point I make above, the problem of distance.) This means that life on every planet will experience this eschatological transformation. Given this, it means that every species of intelligent life in the universe will be blessed by Incarnational / Resurrectional / Eschatological dimensions of redemption which we on planet earth know through the person of Jesus of Nazareth as the Incarnate Son of God and the Risen Lord. In this way the case for multiple incarnations seems to gain very significant theological support.

From Inclusivism to Suprapluralism

In sum, then, I think the best alternative from a revelationist—and perhaps even from an ontologicalist—perspective is multiple cosmic incarnations. In short, the divine *Logos* will become incarnate in every species of intelligent life in our universe which achieves the level of rationality and moral capacity found in *Homo sapiens* and in species-specific ways.

Thus, while my basic position on Christology and soteriology regarding *human* salvation is inclusivism, based on all of humanity as a common species, and not exclusivism or pluralism, when it comes to ETI my view looks a lot like pluralism—or perhaps what I might call "suprapluralism"— when speculating about multiple revelations of the one creator God on different planets.[17] As a terrestrial inclusivist with a revelational perspective on the incarnation, I affirm the core truths about God we have been given in the historical Jesus as normative for human salvation, even while stressing its partial multiple realizations in diverse human religions. Still, when it comes to ETI, I presuppose that an extraterrestrial experience with God is something much more *planetary* and *species specific* than even the standard pluralist view of the diversity of world religion, given the underlying unity we share as a single species. Yes I believe that the one incarnation of Christ on Earth is sufficient for all terrestrial life—after all we all share the same DNA—and I find its truth reflected partially in all world religions. But I also believe God will offer a normative revelation to each and every ETI species, and that each such revelation will be radically ETI-species appropriate. Again, I call this "suprapluralism" in comparison with terrestrial-focused inclusivism and pluralism, since it captures the flavor of pluralism (i.e., many saving truths) along with the insistence of inclusivism (i.e., that the revelation of God in Jesus is normative). It then follows that we will have something of even *greater importance* to learn from contact with ETI than we have through dialogue with world religions: new knowledge of multiple instantiation of normative saving truth.

---

17. Before exploring this point further, let me offer some standard definitions. The terms exclusivism, inclusivism, and pluralism, come from recent discussions within the philosophy of religion, where scholars have worked with a threefold typology: (1) *exclusivism*, which asserts that only one religious path leads to absolute truth and salvation; (2) *inclusivism*, which asserts that while one set of beliefs leads to absolute truth, other sets of beliefs are at least partially true and that salvation may well occur outside of any particular religious institution; and (3) *pluralism*, which asserts that while absolute truth is forever unavailable in principle, each religion possesses an equal and partial perspective on it, and thus each religious tradition represents an equally valid path to this single, but unobtainable, absolute truth.

Difficulties with My Position

So, does the defense of multiple incarnations win the day? Not quite. In reality the idea of multiple incarnations leads to its own difficulties. Perhaps most problematic to my view are the implications for the nature of the immanent Trinity: what do multiple incarnations mean for the immanent trinitarian nature of God?

Recall that the trinitarian tradition incorporates at its heart the idea of the *communicatio idiomatum* (the communication of attributes) among the three persons within the Godhead whenever trinitarian thought is based on Chalcedonian Christology, for this is where the idea of the *communicatio idiomatum* first arises. In Christ, the human properties of Jesus are assigned to and even given over to his divine nature, and vice versa. Human nature comes to dwell within and reconfigure the divine nature by God's own incarnate will. Now we see the implications of the approach I am taking: Does the creaturely dimension of *every* ETI species become incorporated into and present within the single divine nature of the trinitarian God? At first sight this would seem absurd. With this in mind, perhaps I should conclude that the *communicatio idiomatum* lends support for Peters's ontological view of a unique incarnation. But wait: there might a way to address this problem for my view of multiple incarnations and, in the process, bring the single and multiple incarnation views closer together.

Deep Incarnation: A Possible Way Forward That Actually Unites Both Views

Peters and Moritz, in this volume, hold to a single incarnation, while I favor multiple incarnations. Meanwhile I have just pointed to a serious problem with the multiple incarnations position involving the doctrine of the Trinity. Is there a way to respond to this problem and at the same time to reconcile the difference between the positions Peters and Moritz hold (single incarnation) and mine (multiple incarnations)? Perhaps so.

For this I turn to the extraordinary work of Danish theologian Niels Henrik Gregersen and the *Deep Incarnation* school of thought at the University of Copenhagen. For well over a decade Gregersen has argued that the incarnation has a depth in nature that underlies and surpasses any seeming restriction to the human person of Jesus. Instead the incarnation embraces, and extends to, all life on Earth. Gregersen bases his argument on an extended hermeneutical explication of the word "flesh" in John 1:14—"the

Word became flesh" (*sarx*), not human.[18] And it is this flesh that is common biologically and physically to all life on Earth.

I start with Gregersen's insight and I extend it to a higher level of generalization which allows me to include the flesh of all life in the universe. My claim is that the Word of God that is made the flesh common to every terrestrial creature is also the flesh that is common to all extraterrestrial life at least in the following sense: what makes it common is the universality of the physical processes that underlie and make flesh possible, and the likely commonality of the biological processes that characterize flesh.[19] Using this argument I now claim that it is this flesh in the universal sense that is taken up into the Second Person of the immanent Trinity, not just its particular terrestrial exemplification, the earthly flesh of Jesus. Thus, on the one hand, this taking-up is a unitary action, a taking-up of one flesh, as Peters emphasizes but, unlike Peters's view, it is stripped of its terrestrial specificity. On the other hand, it is not a countless multiciplicity of flesh as embodied in uniquely in each species of ETI as my "multiple incarnations" might have suggested and which concerns Peters. Put another way the *communicatio idiomatum* is not a species-specific set of human traits taken up into the Godhead as Peters suggests, but the universal physical properties of the diverse set of fleshly traits (i.e., those of all life in the universe) that are taken up into the Godhead, as I want to emphasize, and this is a view which also reflects the univocity of the incarnation that Peters affirms. In short, it combines both terrestrial-centered (Peters) and universal-centered (Russell) commitments.

In this way Gregersen's vision of deep incarnation, when generalized to all flesh in the universe, provides a bridge between my position and that of Peters. The deep-incarnation approach honors those, such as Peters, who defend the uniqueness of the ontological perspective on the incarnation of the *logos* in Jesus, as well as those, like me, who defend the revelational perspective of multiple ETI incarnations. Thankfully, through Gregersen's vision, we can hold together ontological and revelational perspectives on Christology as two sides of the same christological coin. In this way we arrive at a mutually satisfactory "win/win" scenario, one which we might call "the many-in-one" incarnation of the *Logos* in the universal flesh of life in each of its species-specific forms, through the grace of its loving Creator.

---

18. Gregersen, "Cross of Christ," 192–207. See also Gregersen, "Deep Incarnation," 173–87.

19. I made a similar point about the underlying laws of physics that shape biology in the section entitled "Will ET Experience Moral Failure or Be Entirely Benign?" above.

Bibliography

Adams, Marilyn McCord. *Christ and Horrors: The Coherence of Christology.* Current Issues in Theology. Cambridge: Cambridge University Press, 2006.

Barth, Karl. *The Göttingen Dogmatics: Instruction in the Christian Religion.* Vol. 1. Edited by Hannelotte Reiffen. Translated by Geoffrey W. Bromiley. Grand Rapids: Eerdmans, 1991.

Crisp, Oliver D. *God Incarnate: Explorations in Christology.* London: Bloomsbury T. & T. Clark, 2009.

Deane-Drummond, Celia. *Christ and Evolution: Wonder and Wisdom.* Minneapolis: Fortress, 2009.

Gregersen, Niels Henrik. "The Cross of Christ in an Evolutionary World." *Dialog* 40.3 (2001) 192–207.

———. "Deep Incarnation: Why Evolutionary Continuity Matters in Christology." *Toronto Journal of Theology* 26.2 (2010) 173–87.

Niebuhr, Reinhold. *The Nature and Destiny of Man.* 2 vols. New York: Scribner, 1941.

Russell, Robert John. *Cosmology from Alpha to Omega: The Creative Mutual Interaction of Theology and Science.* Minneapolis: Fortress, 2008.

———. *Time in Eternity.* Notre Dame: University of Notre Dame Press, 2012.

Tillich, Paul. *Systematic Theology.* 3 vols. Chicago: University of Chicago Press, 1951–1963.

Wilkinson, David. *Science, Religion, and the Search for Extraterrestrial Intelligence.* Oxford: Oxford University Press, 2013.

18

Multiple Incarnations of the One Christ

PETER M. J. HESS

Extraterrestrial peoples—their probability is a discovery. Their intimation is a gift: more life, more art, more science, more revelation.

—THOMAS F. O'MEARA[1]

Please consider this proposition: *If multiple societies of extraterrestrial intelligent beings on exoplanets exist, we can predict that God will or already has provided a species-specific incarnation for each planet parallel to God's incarnation in Jesus Christ on Earth.* The position I take is the positive. In what follows I will show why I affirm this proposition.

Evolution of Terrestrial and Extraterrestrial Life

In this chapter I would like to take up the second of our astrobiological tasks: *the astrotheologian should set the parameters within which the ongoing debates over Christology (Person of Christ) and soteriology (Work of Christ) are carried on.* In doing so, let me assume that since life has evolved on our own planet the chance is not negligible that it has also evolved on some other biochemically suitable planet within the habitable zone of at least one of the 10^{23} stars in the known universe.[2] Astronomer Fr. George

1. O'Meara, *VU*, 100.
2. Stoeger et al., "Astrobiology, Ethics, and Philosophy," 7.

Coyne notes about humans that as self-reflective beings "in a real sense the universe speaks to us, or, more to the point, it speaks about us."[3] Perhaps there are other pockets within the universe where the potentialities latent with the Big Bang have permitted "stardust" to organize itself into the emergent properties of rationality and moral and spiritual consciousness. Neils Gregersen intriguingly suggests that "potentialities do not simply reside in the past configurations of matter; they result from an interplay between creaturely potencies and the coming into being of the divine possibilities offered to the world.[4] We might conclude, therefore, that some other emergent self-conscious life within or outside our galaxy has evolved moral and spiritual awareness of the sort to be associated with a theology of salvation.

Evolution of Sin and Moral Consciousness

The answer to the question "what does salvation mean?" is considerably more challenging. Theologies of salvation or "soteriologies" are integrally bound up with theologies of "sin." Since sin or "falling short of the mark" is a universal human experience, theologies of the overcoming of sin play a central role in virtually all religious systems.[5] Such theologies range from the karma-samsāra system of Vedic Hinduism,[6] to the Muslim conception of self-responsibility for sin, to the Jewish focus on living a good life rather than worrying about the next life.

Bearing in mind our thesis question about ET and salvation, how would we recognize "sin," and what are its evolutionary origins? In the view of a theistic evolutionist, God created a dynamic and evolving universe some 13.8 billion years ago. The Earth and other planets coalesced from an accretion disc revolving around the proto-sun about 4.5 billion years ago, and following the ignition of the sun, life appeared on Earth 3.5 billion years ago. At least in our experience on planet Earth, the physical laws of the universe seem systematically conducive to conflict. Planets have finite energy resources, and—long before humans arrived on the scene—evolving life had to compete for these resources. As life expanded into every biological niche on the planet, predation between species and competition within species inevitably arose.

---

3. Coyne and Omizzolo, *Wayfarers in the Cosmos*, 152.
4. Gregersen, "Emergence," 299.
5. Livingston, *Anatomy of the Sacred*.
6. Bellah, *Religion in Human Evolution*, 515–16.

Leaving aside the growing body of evidence suggesting a proto-morality in some other species,[7] prior to the evolution of self-consciousness in *Homo sapiens* competition and predation were not usually moral issues. Rattlesnakes evolved the ability to kill prey with venom dispensed from their fangs, and while it is no doubt terrifying and painful for a baby brush rabbit to be killed and eaten by a snake, it is not a moral issue. Similarly, when a lioness takes down a gazelle on the savannah she is not committing a sin; rather, she is acting according to her instinct for preservation of herself and her young.

However, with the emergence of self-consciousness came moral awareness, a dawning recognition of the possibility of doing otherwise in a given situation. A Paleolithic hunter who hoarded more than his share of woolly mammoth meat—thereby jeopardizing the survival of his tribesmen during a tough winter—might have had a vague awareness that what he was doing was wrong. Slow and incremental growth in understanding of right and wrong would be evidence of the evolution of a moral faculty, of a conscience.

A non-literal reading of the story in Genesis 2–3 nicely illustrates this growth in understanding. God breathes on clay to create Adam as a *nefesh ḥayya* or "living being," and fashions Eve from Adam.[8] Together Adam and Eve are tempted by the serpent to appropriate the knowledge of good and evil, and the guilt of this "original sin" of disobedience is transmitted (in the interpretation of Augustine of Hippo) to their descendants through the act of "carnal begetting."[9] Augustine's is not the only theologically viable model of original sin, however: Jerry Korsmeyer, Ilia Delio, and numerous others have articulated models of original sin that are more reflective of what we know from science.[10] Adam and Eve symbolize our species's gradual evolution into rationality, moral consciousness, and spiritual sensitivity. The deepening of this faculty across our species led to an eventual recognition that human selfishness is universal and unavoidable.[11]

---

7. Bekoff and Pierce, *Wild Justice*.

8. In light of both evolutionary biology and a critical understanding of biblical history it is pointless to pursue even a modified literal interpretation that seeks to map Genesis 1:26–28 or 2:7–25 onto a prehistoric infusion of souls into some tribe of suitably prepared hominids tens of thousands of years ago. The emerging human population appears never to have dropped much below three thousand breeding pairs in the last hundred thousand years.

9. Allen and Hess, *Catholicism and Science*, 102.

10. Korsmeyer, *Evolution and Eden*; Delio, *Christ in Evolution*, 53–65.

11. Peterson, "Falling Up," 273–86.

What might we assume about the moral life of ET? In the absence of signals received by SETI, science fiction is the primary medium through which we conjure up extraterrestrial life, and numerous films in this genre portray ET without much nuance. In films like *Alien*, *Battle: Los Angeles*, and *Independence Day* the invaders are depicted largely in negative terms, as quasi-demonic forces of evil and destruction. Other films portray extraterrestrial intelligence primarily as benign or angelic, as in *The Day the Earth Stood Still*, *E.T.*, and *Starman*. Some stories have a broader tonal range, portraying ET as morally ambiguous, such as the character Jeriba Shigan the Drac in *Enemy Mine* (1985), or the alien "Prawns" in *District 9*. Mary Doria Russell's profound yet disturbing novel *The Sparrow* (1996) recounts the story of the first Jesuit mission to the fourth planet in the Rakhat system. The coherent moral world of Emilio Sandoz, SJ disintegrates when he discovers that two species on Rakhat—locked in an evolutionary predator-prey relationship—have emerged into rationality, one species continuing to serve as food for the other.

What does evolutionary morality imply for ET? I believe that if biological life has followed a similar evolutionary trajectory in the heavens it will be marked as it is on Earth by moral ambiguity. Robert Russell has summarized this nicely: if and when we meet alien intelligent life, it will be neither angelic nor demonic, but in theological terms *simul iustus et peccator*—at "once justified (redeemed) and a sinner."[12] That is to say, assuming planets with similar conditions—similar gravity, chemical makeup, water content, water-to-land ratio, temperature, vegetation, seasons, rotational velocity governing hours of daylights, etc.—life might develop along similar lines to life on Earth. Exobiologists assume that even accounting for varying evolutionary pressures and possibly for different vehicles than DNA for the transmission of "genetic" information, there would be food chains in sea and on land, and creatures would take advantage of water, air, Earth, etc. for swimming, flying, burrowing, and running.

Among the traits with strong survival value would be increased brain size, neural complexity, and intelligence. Thus it should not be surprising that a creature that might or might not resemble a primate would evolve through descent with modification to exploit increasing neural complexity and develop self-awareness, rationality, moral responsibility, and spiritual sensitivity. A theist might interpret this process as the creator working immanently within creation through law and chance, inviting creation to a moral and spiritual response. I believe ET will be in the most important respects like us: predatory and cooperative, egoistic and altruistic, betraying

12. Russell, "Life in the Universe."

and praying, plotting and imagining, capable of evil and capable of good. ET will be *simul iustus et peccator*—precisely the kind of life we might associate with a theology of salvation.

The Meaning, Nature, and Source of Salvation

Probably the thorniest aspect of our thesis question is what we mean by "salvation." Christianity offers a highly developed theology of salvation or "soteriology" dependent upon the biblical story of Adam and Eve interpreted in light of the life, death, and resurrection of Jesus of Nazareth. Soteriology is therefore inseparable from Christology—the study of the nature, person, and works of Jesus, the Christ. Stated most generally, the claim of terrestrial soteriology is that Jesus of Nazareth—through perfect obedience to the Father, even to the point of death—was able to overcome the power of death and bring about atonement for the sins of humankind:

> Therefore just as one man's trespass led to condemnation for all, so one man's act of righteousness leads to justification and life for all. For just as by the one man's disobedience the many were made sinners, so by the one man's obedience the many will be made righteous. (Rom 5:18–19)

However, precisely how is it that the death of Jesus overcomes original sin, defeats death, and conveys salvation? In Christian soteriology there never has been one official model of how the atonement works. Vincent Twomey identifies a number of soteriological motifs in the early church, developed in opposition to Gnosticism and Roman state religion.[13] Among these are "salvation as civilization" and *theosis* or "salvation through deification," a theme that would be central to Orthodox theology in succeeding centuries, succinctly expressed by Athanasius of Alexandria: "The Son of God became man so that we might become God."[14] Patristic theologian Hamilton Hess notes that Athanasius's *Festal Letters* are "rich in soteriological teaching expressed in a great variety of motifs."[15] The whole patristic tradition is summed up by St. Thomas Aquinas: The only begotten Son of God, wishing to enable us to share in his divinity, assumed our nature, so that by becoming man he might make men gods."[16]

---

13. Twomey, "Seeing Salvation."
14. Athanasius, *De Incarnatione*, 54, 3 (Patrologia Graeca 25, 192B).
15. Hess, "Place of Divinization," 374.
16. Aquinas, *Opusculum* 57, 1, as quoted in the *Catechism of the Catholic Church*, 46.

Swedish theologian Gustav Aulén identified three dominant theories of the Atonement in the Patristic and medieval periods: the Latin or scholastic view epitomized by Anselm of Canterbury's satisfaction model; the moral exemplar model developed by Peter Abelard and followed by Friedrich Schleiermacher; and the classic view favored by Aulén himself, in which Christ as *Christus Victor* fights against and triumphs over the evil powers of the world, the "tyrants" under which humankind is in bondage and suffering, and in him God reconciles the world to himself.[17] Ted Peters reconstructs Aulén's model method of soteriologies to include a list of six: (1) Jesus as Teacher of True Knowledge; (2) Jesus as Moral Example and Influence; (3) Jesus as Victorious Champion and Liberator; (4) Jesus as Our Satisfaction; (5) Jesus as the Happy Exchange; and (6) Jesus as the Final Scapegoat.[18] My point here is that a number of soteriological models are shelved in the theologian's library. If we ask whether ETI can be saved, we might have to be a bit specific regarding just what constitutes salvation and how it gets accomplished.

If this plurality of soteriological models is challenging in a terrestrial Christian context, imagine how much more challenging it might be in the context of the question, "Can ET be saved?"

ET and Salvation: Historical Perspectives

Almost from the beginning a tension has existed in Christian perspectives on extraterrestrial salvation between the poles (1) that the drama of salvation is exclusively human and therefore terracentric, and (2) that infinite divine creativity cannot be circumscribed by what humans alone experience. Scholastic theologian William Vorilong (d. 1463) wondered "whether Christ by dying on this Earth could redeem the inhabitants of another world" and answered that this must be possible because Christ could not die again in another world."[19] In 1550 Philip Melanchthon, the great expositor of the Lutheran Reformation, categorically denied the existence of other worlds:

> The Son of God is One; our master Jesus Christ was born, died, and resurrected in this world. Nor does He manifest Himself elsewhere, nor elsewhere has He died, or resurrected. Therefore it must not be imagined that there are many worlds, because

17. Aulén, *Christus Victor*.
18. Peters, *GWF*, 390–421.
19. Dick, *Plurality of Worlds*, 88.

it must not be imagined that Christ died and was resurrected more often, nor must it be thought that in any other world without the knowledge of the Son of God, men would be restored to eternal life.[20]

Seventy years later the Catholic Dominican philosopher Tommaso Campanella, OP, defending Galileo in his *Apologia pro Galileo* (1622) took the opposite position: Christ did not have to be crucified many times on behalf of the men who inhabit each of the stars, any more than he had to be recrucified for the inhabitants of the Antipodes who had not heard of Christ. Even if ET were of the same species of man, Campanella argued, they would not have been affected by Adam's original sin, and thus had no need of redemption unless they had committed some other sin.[21] At the end of the seventeenth century Richard Bentley proposed in his Boyle Lectures (1692) that extraterrestrials, though rational, were not human, a position designed to mitigate the problem of the relation of a race of creatures who were not descendants of Adam to the doctrines of redemption and incarnation.[22]

The debate about whether ET can be saved changed significantly with the revolution begun by Copernicus and Galileo, and with the Enlightenment. From the seventeenth century onward astronomy inexorably relativized the place of humans and the Earth against a vast and expanding universe. The Enlightenment unleashed forces that would critically examine everything from the cosmology and authorship of biblical texts to long-held assumptions about human uniqueness.

Writing four years before the appearance of Darwin's *Origin of Species* in 1859, William Whewell summarized centuries of Christian objection to the salvation of ET. Writing about terrestrial salvation history culminating in the incarnation, Whewell said:

> The Earth, thus selected as the theatre of such a scheme of teaching and Redemption, cannot, in the eyes of anyone who accepts this Christian faith, be regarded as being on a level with any other domiciles. It is the stage of the great drama of God's mercy and man's salvation; the sanctuary of the universe; . . . how can we assent to the assertions of astronomers, when they tell us that it is only one among millions of similar habitations . . . ?[23]

---

20. Melanchthon, *Initia doctrinae physicae*, fol. 43, cited in Dick, *Plurality of Worlds*, 89.
21. Campanella, cited in Dick, *Plurality of Worlds*, 93.
22. Bentley, *A Confutation of Atheism*, 358, cited in Dick, *Plurality of Worlds*, 156.
23. Whewell, *Of the Plurality of Worlds*, cited in Crowe, *ELD*, 337-38.

Scottish theologian Frederick Cronholm sided with Whewell, rejecting the idea that the Second Person of the Trinity had to go on successive missions of salvation from world to world, assuming one after another the nature of every fallen race: "Is there a Bethlehem in Venus, a Gethsemane in Jupiter, a Calvary in Saturn?"[24] But David Brewster endorsed multiple incarnations: "It is impossible for the intellectual man with the light of revelation as his guide, to doubt for a moment that on the celestial spheres his future is to be spent.[25]

In contrast, perhaps inevitably twentieth-century theologians became increasingly positively disposed to the salvation of ET. Paul Tillich maintained that "Incarnation is unique for the special group in which it happens, but it is not unique in the sense that other singular incarnations for other unique worlds are excluded. Man cannot claim to occupy the only possible place for Incarnation."[26]

Pierre Teilhard de Chardin fervently declared, "All that I can entertain is the possibility of a multi-aspect Redemption which would be realized on all the stars."[27] His fellow Frenchman Yves Congar saw no theological contradiction in speculating that extraterrestrials—like humans—may have been called to a further life of grace: "There may well be other incarnations of the divine persons of the Trinity in finite persons."[28]

How Can ET Be Saved?

Prima facie, an omnipotent God can do anything that is not logically self-contradictory, so if anything can be saved, ET can be. But we should look at it in the context of Christian doctrinal history. Must revelation involve always incarnation? Can Jesus' salvific terrestrial death be considered efficacious for a planet in another of the hundred billion galaxies? Can Jesus' terrestrial death and resurrection be necessary elsewhere in the universe?

In the Christian perspective, the terrestrial experience of God incarnate in Jesus Christ is sufficient for Earth, but is it necessary for the rest of the universe? Soteriologies since the apostolic era have understandably been anthropocentric, terracentric, and carbonocentric; indeed, how could they have been otherwise? But now—in a universe recognized to have one hundred billion galaxies averaging one hundred billion stars each—the claim

24. Cronhelm, *Thoughts on the Controversy*, cited in Crowe, *ELD*, 357.
25. Brewster, *More Worlds than One*, cited in Crowe, *ELD*, 356.
26. Tillich, *ST*, 2:96.
27. Teilhard de Chardin, "Fall, Redemption, and Geocentrism," 42.
28. Congar, *Wide World of My Parish*.

arguing that Jesus' death and resurrection are unique and universally valid for the entire universe is parochial indeed. Moreover, if civilizations have evolved at different rates in different corners of the universe, it is quite possible that some rational species will have winked out of existence millions of years or longer before *Homo sapiens* evolved and before Jesus walked to Calvary. The vast interstellar distances mean that some rational species will never hear of Jesus even by radio transmission, much less by face-to-face missionary contact if we tried to reach the nearest hypothetically habitable star in own galaxy.[29]

How can we rethink our theology of salvation to be universal rather than anthropocentric, terracentric, or carbonocentric? How can we rethink Christology and soteriology so that it relevantly includes an alien culture that was extinguished a hundred million years ago, or for a culture in which God may not yet become incarnate for another hundred million years? Is the universe really "cruciform" or was Jesus' crucifixion an historical accident resulting from the chance intersection of the incarnation of the Word of God with one specific historical period in one civilization on a continent of the solar planet Earth? Is the notion of sacrifice intrinsic to soteriology, or is it extrinsic, a peculiarity of terrestrial religions and particularly of the sacrificial economy of post-exilic Judaism? Could salvation be accomplished on other planets without a sacrificial death?

Let me argue both that incarnation is essential and that we often envision it in too parochial a fashion. Humans as *imago Dei* embody the moral and spiritual response called forth by God from creation. The incarnation is the supreme terrestrial expression of this response: in the person of Jesus, God assumed the quarks of the Big Bang, the dust of supernovae explosions, the organic molecules of dinosaurs and mammals, and the long history of the primate genome. In an evolutionary paraphrase of St. Gregory of Nazianzus, God assumed terrestrial creation by becoming incarnate at its heart in a human person, Jesus Christ. Karl Rahner, SJ affirms it thus: "In our context it is especially worthy of note that the point at which God in a final self-communication irrevocably and definitively lays hold on the totality of the reality created by him is characterized not as spirit but as flesh. It is this which authorizes the Christian to integrate the history of salvation into the history of the cosmos, even when myriad questions remain unanswered."[30]

---

29. George Murphy suggests that "the Church is called to a cosmic mission," such that Christians are commissioned to make disciples on all planets. Murphy, *Models of Atonement*, 127. However, the vast distances between star systems—and the time differential between stellar civilizations—renders the prospect of interplanetary evangelization less plausible.

30. Rahner, *Hominization*.

This might place Rahner with those advocating a single incarnation for the entire cosmos. But, Rahner can also side with the multiple incarnation position. "In view of the immutability of God in himself and the identity of the Logos with God, it cannot be proved that multiple incarnation in different histories of salvation is absolutely ruled out."[31]

Accordingly, God is neither logically nor theologically bound to become incarnate only once in the universe. If the purpose of creating is to bring that creation into communion with God, why would God not use any means at the divine disposal? Ilia Delio writes: "On the terrestrial level, Jesus Christ assumes a bodily nature by which all of creation (that is, material reality) is assumed into relationship with God. Similarly, on an extraterrestrial level, incarnation must assume a form that includes the material reality of that creation, in whatever way that creation is constituted."[32] Delio further notes that although the term "incarnation" might not be appropriate to another world order, since literally it means taking on flesh, what we are really talking about is embodiment of the divine Word in created reality.

Niels Gregersen extends this theme with "deep incarnation" as the soteriological emphasis on humanity and sin extended to embrace the wider space of creation: "Accordingly, incarnation is not only a passing episode in God's involvement with God's own world of creation. For incarnation is not only about the removal of sin and the transformation of the sinner, but also about the world of creation, in which there shall be neither separation nor confusion between Christ and creation. Accordingly, incarnation must be perpetual."[33] Delio claims that the primary theological reason for the incarnation is related not so much to the "forgiveness of sin" as to the completion of creation in its relationship to God: "An understanding of the incarnation as an act of love rather than a condition of sin may be more fitting to an evolutionary universe where the understanding of human original sin is under revision." There is only one spirit and one Christ no matter how many times the Cosmic Christ might become incarnate throughout the universe.[34]

Conclusion

It took 13.8 billion years for human life to evolve on Earth, and our species may not last another million years. Thus there is both a spatial separation between planetary civilizations and a temporal one, with cultures winking

31. Rahner, "Natural Science and Reasonable Faith," in *TI* XXI, 16–55, 1t 51.
32. Delio, *Christ in Evolution*, 169.
33. Gregersen, "Deep Incarnation and Kenosis," 260–1.
34. Delio, *Christ in Evolution*, 168.

into and out of existence without overlapping or intersecting, "fading away into God's providence."³⁵ The postulation of multiple incarnations overcomes the time and distance problems, allowing the one God fellowship with creatures whenever and wherever they live throughout the universe.

Let me propose an even more inclusive possibility for the salvation of ET. For Roman Catholics (as for numerous other Christians) salvation is far more than a matter merely of elect human souls being raptured out of discarded human bodies; it is far more than merely *Homo sapiens* being rescued from a physical world destined to perdition. To quote Delio again, "Because the incarnation completes that which God creates, extraterrestrials, if they exist, will be open to an incarnation and, in the broadest sense, "saved," insofar as an incarnation of the divine Word will complete extraterrestrial creation in whatever way that creation can fully accept the Word into it."³⁶ The saving of ET is not separate from the saving of human beings, nor is salvation separate from the problems of evolutionary suffering and the eschatological future of the universe. Ted Peters and Martinez Hewlett express this nexus succinctly: what will emerge from the omega of creation "will involve a new divine act of giving the world a unifying future, an eschatological and redemptive future."³⁷ Yes, ET can be saved, because "the Word was made flesh and dwelt among us" in every place that created being—within the all-encompassing divine reality—has reciprocated God's invitation to existence. I cannot express it more tenderly than did Alice Meynell in 1917 in her poem "Christ in the Universe":

> But in the eternities,
> Doubtless we shall compare together, hear
> A million alien Gospels, in what guise
> He trod the Pleiades, the Lyre, the Bear.
>
> O, be prepared, my soul!
> To read the inconceivable, to scan
> The myriad forms of God those stars unroll
> When, in our turn, we show to them a Man.

---

35. O'Meara, *VU*, 99.
36. Delio, *Christ in Evolution*, 169.
37. Peters and Hewlett, *Evolution from Creation to New Creation*, 163.

Bibliography

Allen, Paul L., and Peter M. J. Hess. *Catholicism and Science*. Greenwood Guides to Science and Religion. Westport, CT: Greenwood, 2008.
Aquinas, Thomas. *Opusculum*. Cited in the *Catechism of the Catholic Church*.
Athanasius. *De Incarnatione*. *Patrologia Graeca* 25, 192B.
Aulén, Gustav. *Christus Victor*. New York: Macmillan, 1958.
Bekoff, Marc, and Jessica Pierce. *Wild Justice: The Moral Lives of Animals*. Chicago: University of Chicago Press, 2010.
Bellah, Robert N. *Religion in Human Evolution: From the Paleolithic to the Axial Age*. Cambridge: Harvard University Press, 2011.
Bentley, Richard. *A Confutation of Atheism from the Origin and Frame of the World*. London: Mortlock, 1693.
Brewster, David. *More Worlds than One: The Creed of the Philosopher and the Hope of the Christian*. London: Hotten, 1854.
Congar, Yves. *The Wide World of My Parish: Salvation and Its Problems*. London: Darton, Longman & Todd, 1961.
Coyne, George V., SJ, and Alessandro Omizzolo. *Wayfarers in the Cosmos: The Human Quest for Meaning*. New York: Crossroad, 2002.
Cronhelm, Fredrick William. *Thoughts on the Controversy as to a Plurality of Worlds*. London: Rivingtons, 1858.
Crowe, Michael J. *Extraterrestrial Life Debate from Antiquity to 1915: A Source Book*. Notre Dame: University of Notre Dame Press, 2008.
Delio, Ilia, OSF. *Christ in Evolution*. Maryknoll, NY: Orbis, 2008.
Dick, Steven J. *Plurality of Worlds: The Origins of the Extraterrestrial Life Debate from Democritus to Kant*. Cambridge: Cambridge University Press, 1982.
Gregersen, Niels Henrik. "Deep Incarnation and Kenosis: In, With, Under and As: A Response to Ted Peters." *Dialog: A Journal of Theology* 52 (2013) 260–1.
———. "Emergence: What is at Stake for Religious Reflection?" In *The Re-Emergence of Emergence: The Emergentist Hypothesis from Science to Religion*, edited by Philip Clayton and Paul Davies, 279–302. Oxford: Oxford University Press, 2006.
Hess, Hamilton. "The Place of Divinization in Athanasian Soteriology." In *Studia Patristica XXVI*, edited by Elizabeth A. Livingstone, 369–74. Leuven: Peeters, 1993.
Korsmeyer, Jerry D. *Evolution and Eden: Balancing Original Sin and Contemporary Science*. New York: Paulist, 1998.
Livingston, James C. *Anatomy of the Sacred: An Introduction to Religion*. Upper Saddle River, NJ: Prentice Hall, 1998.
Melanchthon, Philip. *Initia doctrinae physicae*. Wittenberg, 1550, quoted in Steven J. Dick, *Plurality of Worlds: The Extraterrestrial Life Debate from Democritus to Kant*. Cambridge: Cambridge University Press, 1984.
Murphy, George L. *Models of Atonement: Speaking about Salvation in a Scientific World*. Minneapolis: Lutheran University Press, 2013.
O'Meara, Thomas F. *Vast Universe: Extraterrestrials and Christian Revelation*. Collegeville, MN: Liturgical, 2012.
Peters, Ted. *God—The World's Future*. 3rd ed. Minneapolis: Fortress, 2015.
Peters, Ted, and Martinez Hewlett. *Evolution from Creation to New Creation*. Nashville: Abingdon, 2003.

Peterson, Gregory R. "Falling Up: Evolution and Original Sin." In *Evolution and Ethics: Human Morality in Biological and Religious Perspective*, 273–86. Grand Rapids: Eerdmans, 2004.
Rahner, Karl, SJ. *Hominisation: The Evolutionary Origin of Man as a Theological Problem*. Translated by W. J. O'Hara. Quaestiones disputatae 13. New York: Herder & Herder, 1965.
———. "Natural Science and Reasonable Faith." *Theological Investigations*, 21:16–55. Translated by Hugh M. Riley. New York: Crossroad, 1988.
Russell, Robert John. "Life in the Universe: Philosophical and Theological Issues." In *First Steps in the Origin of Life in the Universe*, edited by Julian Chela-Flores et al., 365–74. Proceedings of the Sixth Trieste Conference on Chemical Evolution. Dordrecht: Kluwer Academic, 2001.
Stoeger, William R., SJ, et al. "Astrobiology, Ethics, and Philosophy." In *ELU* (2013) 1–16.
Teilhard de Chardin, Pierre. "Fall, Redemption, and Geocentrism." In *Christianity and Evolution*, 36–44. New York: Harcourt Brace Jovanovich, 1972.
Tillich, Paul. *Systematic Theology*. 3 vols. Chicago: University of Chicago Press, 1951–1963.
Twomey, D. Vincent. "Seeing Salvation: Contemporary and Patristic Perspectives." In *Salvation according to the Fathers of the Church*, edited by D. Vincent Twomey and Dirk Krausmüller, 13–32. Dublin: Four Courts, 2010.
Whewell, William. Of *the Plurality of Worlds: An Essay; Also, a Dialogue on the Same Subject*. 4th ed. London: Parker, 1875. Cited in *Extraterrestrial Life Debate from Antiquity to 1915: A Source Book*, edited by Michael J. Crowe. Notre Dame: University of Notre Dame Press, 2008.

19

One *Imago Dei* and the Incarnation of the Eschatological Adam

JOSHUA M. MORITZ

> In Christ God was reconciling the [cosmos] to himself.
> —2 CORINTHIANS 5:19

> For the presence of the Savior in the flesh was the price of death and the saving of the whole creation.
> —ATHANASIUS, LETTER TO ADELPHI

"And the Word became flesh and lived among us." What are the implications of John 1:14 for evolutionary biology, paleoanthropology, extra-terrestrial intelligent life, and for the destiny of *all life* in the cosmos? Does the Christian message of the incarnation—and Logos Christology in particular—rule out God's loving concern for creation beyond the sphere of terrestrial human beings? Some space researchers, such as Paul Davies, believe it does. As Davies explains,

> Christianity is the religion most challenged by the concept of extraterrestrial beings, because Christians believe that God became a human being. Jesus Christ is called the Savior precisely because he took on human flesh to save humankind. He did not come to save the whales or the dolphins or the gorillas or the chimpanzees, or even the Neanderthals, however noble or

deserving those creatures may be (or were). Jesus Christ was the savior of *Homo sapiens*, specifically; one planet and one species.[1]

The reason Davies alleges that Christian theology is exclusively anthropocentric is this: he wants to show that traditional Christology would require by implication a planet-hopping Christ. He believes that Christian theologians, to be consistent with previous commitments, must posit the apparently absurd view that God would need to become incarnate multiple times, perhaps billions of times. Now, in this chapter we ask: is this actually what Christian theology teaches? My answer will be: no.

Here is the proposition we are debating: *If multiple societies of extraterrestrial intelligent beings on exoplanets exist, we can predict that God will or already has provided a species-specific incarnation for each planet parallel to God's incarnation in Jesus Christ on Earth.* The position I take is the negative. In what follows I will show why I disagree with this proposition.

Let us ask: did Jesus come to save *only* human beings on this one planet? And is this doctrine of "one planet, one species" inevitably implied by the notion of a unique incarnation of the creator God in the historical person of Jesus Christ? In this chapter I will argue that Davies's conclusion is in fact *not* the case, and that his reading of Christ's incarnation overlooks the deeper theological meaning of the Word becoming flesh and the implications of Advent for non-human creatures in this world and beyond. Here, I will examine the idea of the incarnate Logos as a development within a line of Second Temple period Jewish thinking which in turn goes back to the ancient Hebrew adaptation of broader Near Eastern conceptions of the *imago Dei*. I will show that the Advent event, when understood within the context of Early Jewish conceptions of the Messiah as the renewal and fulfillment of the *imago Dei*, is a proclamation of truly cosmic scope declaring God's radical solidarity with *all creatures* in Christ. This applies to all living creatures on Earth, and elsewhere if they be there. The point of God's incarnation in Jesus as the "Final Adam" is not merely the redemption of humanity, or even the salvation of *rational* beings, but rather, the redemption of *all biological life*—or as the Bible puts it "all flesh."

This chapter will address the first two of our astrotheological tasks. How should *Christian theologians along with intellectual leaders in each religious tradition reflect on the scope of creation and settle the pesky issue of geocentrism?* My answer will be that the scope of God's creation subject to God's redemptive action in Jesus Christ is inclusive of all life forms here on Earth; and it includes any and all living creatures elsewhere in our expansive

1. Davies, *Eerie Silence*, 188.

universe. Attached to this is our second astrotheological task, which I will put in the form of another question: *regarding the debates over Christology (Person of Christ) and soteriology (Work of Christ), should we expect multiple incarnations, one for each planet hosting intelligent life? Or, does the historical incarnation here on Earth suffice for all of creation, other planets included?* I will side with those who claim that the one incarnation of God in Christ in Earth's history is efficacious for all sentient creatures wherever and whenever they live.

The Question of Human Uniqueness and the Cosmic Scope of the *Imago Dei*

Is the human race on Earth unique? Should we count human beings among the animals and the other creatures that God has created? Or are we perhaps something else entirely? While many poets, philosophers, theologians, and scientists have suggested various ways to unequivocally distinguish humans from the rest of God's creatures, in the Jewish and Christian scriptures there is only one designation that the human species is given and which non-human beings are not. Genesis 1 says that humans, unlike other creatures, are created as the image and likeness of God (*imago Dei*). Throughout the history of interpretation, many have sought to understand the *imago Dei* in light of specific unique qualities or capacities that human beings alone possess and that non-humans lack. In this way the *imago Dei* has, for many theologians, become synonymous with one central characteristic or several key traits that make humans *unique among* and/or *superior to* animals. Traits such as morality, rationality, sexuality, language, tool use, technology, culture, cognitive fluidity, the possession of an immortal soul, and so on, have contended for the singular distinction of being the unique characteristic that defines human nature over and against the nature of non-humans. A close examination of the sacred texts themselves, however, reveals that the *imago Dei* is never defined according to one characteristic or a specific collection of qualities that set humans apart from other creatures.[2]

In the Bible the "image and likeness of God" is never said to be about exceptional capacities or traits that humans alone have which automatically qualify them (and disqualify other creatures) for inclusion in the *imago Dei*

---

2. For example Hebrew scholar Phyllis Bird points out that the scriptural context of the phrase "image and likeness of God" makes it plain that "its theological significance is in the place it gives to humans within the created order, not in any physical or moral attribute of the species, in either its present or 'original' state." Bird, "Theological Anthropology," 262.

ONE *IMAGO DEI* AND THE INCARNATION OF ADAM   333

category. Assertions of human uniqueness based on certain characteristics and "claims for a 'special creation' of humanity in comparison with animals and the material world conflict with the strong assertion in Genesis 2 that, physically (organically), Adam does not differ from the 'beasts of the field.'"[3] The theological language of anthropology in Genesis 1 and 2 "underscores Adam's linkage with the animal creation, not his difference from it."[4] There is no reason, explains James Barr, to believe that the author of Genesis chapter 1 "had in his mind any definite idea about the content or location of the image of God."[5] The terms "'image' and 'likeness' . . . make no statements about the *nature* of human beings."[6] When we read of "the creation of human beings in God's image (Gen 1:26) . . . the biblical narrative remains silent . . . about *any qualities* of human nature that might account for their special standing."[7] Nor does the Bible ever equate the image of God in humans with an immaterial soul that *Homo sapiens* possess and other creatures do not.[8] If we are to properly understand the meaning of the *imago Dei* texts, then, says Heidelberg Old Testament scholar Claus Westermann, we must confidently resist "the tendency to see the image and likeness of God as a something, a quality."[9] In an earlier chapter, Robert John Russell identified the *imago Dei* with rationality; and with this criterion in mind he could attribute the *imago Dei* to rational creatures on other planets. In contrast to Russell, I do not associate the *imago Dei* with rationality or any other similar creaturely capacity or quality.

If the *imago Dei* is not some *capacity* or *quality* that we humans intrinsically possess, then what can it be? A number of biblical scholars have contended that the *imago Dei* in humans is a type of kingly and priestly representative *function* that the human species fulfills. Others have pointed out that this function appears to be a consequence of a certain kind of unique *relationship* that human beings have with God. Elsewhere, I have argued that both the *functional* and *relational* understandings of the *imago Dei* emerge from—and can be unified through—viewing the "image and likeness of God" as God's historical *choosing* or *election* of human beings *from among*

3. Stone, "The Soul," 50 (quoting the NASB).
4. Ibid., 57.
5. Barr, "The Image of God," 13.
6. Preuss, *Old Testament Theology*, 2:115.
7. Tanner, "Difference Theological Anthropology Makes," 573.
8. For example, Bible Scholar Gordon Wenham explains that in Genesis 2:7, which describes the human being as a *nephesh*, "it is not man's possession of the 'breath of life' or his status as a 'living creature' that differentiates him from the animals—animals are described in exactly the same terms." Wenham, *Word Biblical Commentary*, 61.
9. Westermann, *Creation*, 57–58.

the animals and setting them apart for the sake and fulfillment of the divine purposes.[10] When perceived in light of the original linguistic, historical, and cultural context of the Hebrew theological framework of *historical* (or *biblical*) *election*, the image and likeness of God in humans is not recognized by reference to any skills, behaviors, or souls that *Homo sapiens* might possess in distinction from animals or other non-human creatures. Rather, the *imago Dei* emerges as a designation given through the free historical action of God in his own choosing of *Homo sapiens* and his calling them out from among the multiplicity of life-forms he also created "from the dust" to serve as his representatives to creation, and to uphold God's justice and orient the creation towards fellowship with him.[11]

The early Jewish and early Christian concept of historical or biblical election, through which I am suggesting that we view the *imago Dei*, is to be clearly distinguished from the *classical theological concept of election* which was developed by the Protestant Reformers as a way to understand the eschatological destiny of individual human beings in light of a timeless act of God's "unconditional election." Unlike the classical theological election of Calvinism, biblical or historical election is always conceived as "a concrete *historical* act on God's part that forms the starting point and basis of the salvation history of God with his *people*."[12] Those who are elected are not chosen because they are "the greatest" or inherently more worthy than others, but rather they are elected as a result of mysterious acts of divine love and grace. Election in the biblical understanding relates to *a people* (and often a *lineage*) whom God has chosen *in the midst of history* for a special *purpose* within the wider context of God's design. This purpose of historical election is furthermore defined *not* in terms of privilege (or even individual salvation), but rather for the sake of service. For example, the elected Israelite king is called to be "the guardian of the humble and the needy, the weak and the helpless" and the mission of the divinely elected king is to establish righteousness and justice throughout the land.[13] Thus, in exercising dominion the king is to "watch carefully over the rights of his subjects, and so ensure, in particular, that the weaker members of society may enjoy his protection and thus have justice done to them according to their need."[14]

---

10. Moritz, "Evolution, the End of Human Uniqueness."

11. For a more lengthy discussion of the *imago Dei* as election see Moritz, *Chosen from among the Animals*.

12. Pannenberg, *ST*, 3:442.

13. Johnson, *Sacral Kingship*, 10.

14. Ibid., 7; see Ps 72:12–14.

Within the Hebrew scriptures the service of the chosen or elect ones is rendered through their obedience to God's commandment. By obeying God's commandments, the elected live in community or fellowship with God. Likewise, the mission of the elected is to represent God to "the many" in terms of God's sacredness, authority, and dominion. For instance, with regard to Abraham, "the many" are "all the families of the earth" (Gen 12:3) who will be blessed through his election; for Israel, "the many" are the gentile nations to whom Israel—as God's elect—is to bear God's light and justice. Chosenness in this way serves a larger purpose in that "the chosen people does not withdraw from the human family, but exercises a special office within it, an office defined by the particular character and will of their universal God."[15]

Within this ancient Jewish context the non-elect are to be blessed in and through their relationship with the elect. In this way, "the concept of election was never assumed to be only for the benefit of the elect, but it was always about God's plan for the whole world, the elect and the non-elect alike."[16] Rather than election being a matter of *exclusivism* and *particularity*, then—as in the classical view—the historical concept of election has an *inclusive* and *universalistic* tendency. In fact, with regard to election "*the horizon of particularism is universal.*"[17] One clearly sees this universalistic trend in the election traditions of Noah, Abraham, and Israel. Here "the particularism of the love of God for the elected one is to be related to the more comprehensive horizon of God's love for all." In this way the chosen one "is assigned a function for that wider context. He is elected in order to serve as God's agent in relation to a more comprehensive object of God's love."[18] To this end the prophet Isaiah speaks of the people of Israel as elected "to act as God's servant among the nations to the effect that 'he will bring forth justice to the nations' (Isa 42:1)."[19] The aim of the election of Israel was that it should live out and "proclaim the righteous will of God to the nations. On this view the election of Israel is not an end in itself. It serves the will of God *on behalf of* the human race as a whole."[20]

The historical concept of election, as the central interpretive framework in Hebrew thought, illuminates the meaning of how humans are designated and created as the image and likeness of God. As is well known, the

---

15. Levenson, "Universal Horizon of Biblical Particularism," 155.
16. Kaminsky, *Yet I Loved Jacob*, 25–26.
17. Levenson, "Universal Horizon of Biblical Particularism," 155.
18. Pannenberg, *Human Nature, Election, and History*, 49.
19. Ibid., 49.
20. Pannenberg, *ST*, 2:322.

concept of the "image" as used in Genesis "has a deep ancient Near Eastern background."[21] From a comparison of the Hebrew text with ancient Near Eastern parallels it is clear that the phrase "image of God" emerges from a common royal ideology where individual Mesopotamian, Hittite, Assyrian, Babylonian, and Egyptian priest-kings are referred to as the image and likeness of particular gods. The Hebrew phrase, "image of God," (selem elohim) used in Genesis 1:26–27 is the exact counterpart of the Akkadian expression (salam [God's name]): "image of Enlil [Marduk, etc.]"), an expression which often appears as an epithet of Mesopotamian priest-kings.[22] "References to the king as the image (salmu) of God abound in the Neo-Assyrian royal correspondence."[23] One Neo-Babylonian text declares "The king of the world is the very image of Marduk,"[24] and an ancient Assyrian text reads, "the father of the king my lord was the very image of Bel, and the king my lord is likewise the very image of Bel."[25] In this ancient Near Eastern conception, the king—more accurately understood as a priest-king—was seen as "the god's" authorized deputy or viceroy on earth."[26] In the estimation of many scholars this "description of Near Eastern Kings as the image of a god . . . provides the most plausible set of parallels for interpreting the *imago Dei* in Genesis."[27]

Within this same ancient Near Eastern royal ideology (*Königsideologie*) we find that the very kings predicated as the image and likeness of a god are simultaneously *chosen* or *elected* as king by the god whom they image.[28] For example Thutmose IV of Egypt, "son of Atum, living image of the All-Lord, sovereign, begotten of Re" inscribed on a stele "that in a dream at the foot of the great Sphinx of Giza he had the experience of being *elected* king by the sun god Re"—the god of whom he is the living image. Beyond this, each of the structural elements describing historical election within the Hebrew textual tradition—including divine *blessing*, the *multiplication of progeny*, the giving of *commandments*, and the promise of the *land*—are also present in the early chapters of Genesis as they describe the first humans who are created as the image and likeness of God (Gen 1:28 and Gen

21. Garr, "The Nouns דמות and צלם," 136.
22. Bird, "Theological Anthropology," 260–1. See Millard and Bordreuil, "A Statue from Syria," 135–41.
23. Parpola, "*The Assyrian Tree of Life*," 168.
24. Middelton, *The Liberating Image*, 113.
25. Parpola, *Letters from Assyrian Scholars*, 99.
26. Middleton, *Liberating Image*, 119.
27. Ibid., 121.
28. Preuss, *Old Testament Theology*, 1:29.

2:15-16). Structurally the election narratives of Abraham and Israel link their calling and vocation to that of Adam. In his election, "Abraham is to restore what Adam has done" and thus reaffirm the true meaning and purpose of humanity.[29] As Abraham and Sarah are elected by God to be a *nation* (ethnicity or race) of priests and a light to the other *nations* (ethnicities or races) so Adam and Eve, as the primal human pair, are chosen and called to be a species of priests to the *non-human creatures*.

According to the Genesis narrative the nations in relation to Israel parallel the non-human creatures whom Adam is called to both serve and rule.[30] As Abraham and Israel are each commissioned to grow into a numerous people, so Adam and Eve are commanded to be fruitful and multiply. The commission to Adam "to take possession of the earth is related to Israel and its land."[31] Similarly, as Israel holds a place of honor among the races, so humans occupy a place of honor among the non-human beings (including animals and angels).[32] However, as "the election of Israel neither signaled YHWH's renouncement of the other nations nor involved their rejection in any way," so the election of humans in no way indicates God's rejection or lack of concern for non-human creatures.[33] Acting as vice-regents or kings on God's behalf, human beings are the brethren of the non-human creatures that are under their dominion. As the elected high priests of creation *Homo sapiens* are called to intercede before God for the sake of the cosmos with the ultimate aim that *all creatures* should live in God's presence. Elected as both king and priest, the human being bears God's image, authority, sacredness, healing, and atoning salvation to the whole non-human creation.

The Incarnation in Light of Early Jewish Conceptions of the Messiah as the Renewed Adam

What does the *imago Dei* have to do with the incarnation of God in Jesus the Christ? Like the *imago Dei* passages in Genesis, Early Jewish understandings of the Messiah during the Second Temple period are best understood within the larger ancient Near Eastern context of historical election. Similar to the concepts of the "image and likeness of God" and the divine election of kings, the origin of the Messiah concept "can be traced back to a common

29. Wright, *New Testament and the People of God*, 1:266.
30. Ibid., 267.
31. Preuss, *Old Testament Theology*, 1:115; Wright, *New Testament and the People of God*, 263.
32. Wright, *New Testament and the People of God*, 261.
33. Preuss, *Old Testament Theology*, 2:285.

Near Eastern royal ideology.[34] Deriving its roots from this ancient Near Eastern background the Hebrew word מָשִׁיחַ (*messiah*), meaning "anointed one," is "used in the Hebrew Bible both for kings and high priests, who were in fact anointed."[35] While "there was no single, monolithic, and uniform 'messianic expectation' among first century Jews," the various ideas and titles of the expected Messiah which were prevalent during that period can be most effectively integrated and elucidated in light of Scripture's historical election traditions.[36]

Within the Hebrew understanding the divine election of God's Messiah can be envisioned as part of a historical process forming a series of concentric representative circles or spheres of reality where ambassadors are consecutively chosen to stand on behalf of the wider circle of reality from which they are chosen. In this way the divinely elected Messiah represents the whole of Israel, Israel as God's "chosen people" represents humanity, and humanity as a chosen species represents the realm of living creatures. The Messiah is elected and installed "as the one through whom God is doing what he intended to do, first through humanity and then through Israel" for the sake of the whole creation (i.e., the cosmos).[37] As the final Adam and as the renewed image of God, the Messiah was identified as "a shoot from the stump of Jesse" who will usher in the age of the new Eden when "the wolf will lie down with the lamb."[38] In the conceptual world of Second Temple Judaism God's elected Messiah was expected to fulfill the vocation of the original image and likeness of God—Adam and Eve. Central to this original human vocation was a priestly and kingly calling to represent, serve, and defend the non-human creation. This mission to the non-human creation was anticipated in the Hebrew prophets' visions of the peaceable kingdom as pictured in passages such as Hosea 2:18: "And in that day will I make a covenant for them with the beasts of the field, and with the birds of the heavens, and with the creeping things of the ground: and I will break the bow and the sword and the battle out of the land, and will make them to lie down safely" (author's translation). In the early Jewish tradition God's Messiah as the renewed *imago Dei* comes to restore all relationships that have been disfigured by sin—for both human and non-human creatures.[39] At the

---

34. Zetterholm, *Messiah in Early Judaism and Christianity*, xxi.
35. Collins, "Pre-Christian Jewish Messianism," 1.
36. Wright, *New Testament and the People of God*, 307.
37. Wright, *Climax of the Covenant*, 29.
38. Collins, "Pre-Christian Jewish Messianism," 4–5.

39. See, for example, *Testaments of the Twelve Patriarchs, The Sons of Jacob the Patriarch: Zebulon, the Sixth Son of Jacob and Leah* 5:1 in Charlesworth, *Old Testament Pseudepigrapha*, 806.

time of Jesus "Jewish eschatological expectation included the hope that the righting of all wrongs in the messianic age would bring peace between the wild animals and humans."[40] Within the Jewish Messianic tradition the new Adam as the renewed image of God would usher in the climax of creation. The first-century Jewish hope was for the Elect One to lead God's elect people in accomplishing their elected purpose—to bear God's image, God's healing, and God's redemptive light of righteousness and justice into every dark corner of creation, dispelling the oppressive shadows of cruelty and dispersing the destructive shades of disease, disintegration, and even death.

The Jewish writers of the Christian Gospels clearly understand Jesus of Nazareth as God's Elect One who inaugurates the peaceable kingdom of the Messianic age described in Hosea 2 and Isaiah 11 and as the one who fulfills the Early Jewish messianic tradition.[41] In the Gospels Jesus, like Adam, is born into the world surrounded by animals as his faithful companions. As the ox and ass adore the newborn king in the manger, the early Christians saw Hebrew prophecy concerning the Messiah in Isaiah 1 fulfilled—"the ox knows his owner and the donkey the crib of his Lord" (author's translation).[42] At the beginning of the New Testament, the Gospel of Matthew harks back to the very dawn of the cosmos as the author envisions the birth of Jesus the Messiah as tantamount to a new creation. Drawing our attention to the first chapters of Genesis, Matthew opens his testimony with the words "Βίβλος γενέσεω" meaning literally, "the book of Genesis." In his next words he sums up the whole election history of the Hebrews as he describes "Jesus the Messiah" as "the son of David, the son of Abraham" (Matt 1:1). Here in one sentence are implied all the promises, privileges, and responsibilities of election—from creation onward—along with the declaration that Jesus is their fulfillment.[43] Like Adam, the first of God's elect, the human person Jesus is brought into being by the hand and Spirit (*ruach*) of God. In this way "the role of the Holy Spirit in the virgin birth in 1:18 recalls the activity of the Spirit in Genesis 1 at the creation."[44]

In Mark 1:13 we see Jesus depicted as the new Adam who at the beginning of his ministry is led by the Spirit into the wilderness. At this point, Mark informs us that "ἦν μετὰ τῶν θηρίων." While the usual English

---

40. Bauckham, "Jesus and Animals II," 57.

41 Jesus of Nazareth is affirmed as the *Christ*—the *Messiah*, the "anointed one" or the "chosen one" of God. Jesus is explicitly referred to as the "elect one" in Luke 23:35: "He saved others; let him save himself, if he be Christ, the chosen one /elect one (ἐκλεκτο) of God" (author's translation).

42. Cohn-Sherbok and Linzey, *After Noah*, 6.

43. Ibid., 496.

44. Ibid., 497.

translation reads, "he was with the wild animals" the expression in the Greek original "has the strongly positive sense of close association or friendship" and "in Mark's own usage elsewhere in his Gospel, the idea of close friendly association predominates."[45] Understood in its proper Jewish theological and historical context, then, Mark's declaration that Jesus is the friendly companion of the wild animals means that as the "eschatological Adam" and "as the messianic Son of God on behalf of and for the sake of others," he proleptically and representatively establishes "the Messianic peace with wild animals" in anticipation of the future peaceable kingdom of Isaiah 11.[46]

Other early Christian writings also portray the mission of Jesus within the Early Jewish messianic tradition of the new Adam restoring relationships with animals and inaugurating the peaceable kingdom. These stories—which were highly influential in the early church and remained so throughout the Middle Ages—reveal the character and identity of Jesus the Messiah as it was understood in many early Christian communities. In the *Protoevangelium of James* the author recounts that at the birth of Jesus the animals in the surrounding countryside pause in reverent awe. In the highly influential *Infancy Gospel of Matthew* dangerous wild beasts fulfill the Hebrew scriptures as they worship the young Jesus and obey his commandments.[47]

> Lions and panthers adored him and accompanied them in the desert. Wherever Joseph and Mary went, they went before them showing them the way and bowing their heads; they showed their submission by wagging their tails, they worshipped him with great reverence . . . And the lions kept walking with them, and with the oxen and the asses and the beasts of burden . . . and did not hurt a single one of them . . . They were tame among the sheep and the rams which they had brought with them from Judea . . . They walked among wolves and feared nothing; and not one of them was hurt by another.

Jesus meanwhile explains to his mother, "all the beasts of the forest must necessarily be docile before me."[48] As the animals travel with Jesus and

45. Bauckham, "Jesus and the Wild Animals (Mark 1:13)," 5; Bauckham, "Jesus and Animals II," 58.

46. Bauckham, "Jesus and the Wild Animals (Mark 1:13)," 6.

47. *Infancy Gospel of Matthew* in Elliott, *The Apocryphal New Testament*, 84; *The Infancy Gospel of Matthew* is also known as *The Book About the Origin of the Blessed Mary and the Childhood of the Savior* and *The Gospel of Pseudo-Matthew*. This work was very influential in the Middle Ages and indeed "much of medieval art is indecipherable without reference to books such as *Pseudo-Matthew*."

48. Ibid., 95.

his family, he feeds them and takes care of them until "all were satisfied . . . and they gave thanks to God."[49] Later in the gospel the adult Jesus, speaking to an unbelieving crowd, declares, "how much better are these beasts than you, seeing that they recognize their Lord and glorify him; while you humans, who have been made in the *image and likeness of God*, do not know God! Beasts know me and are tame; humans see me and do not acknowledge me."[50]

The most radical expression of the Early Jewish notion of the Messiah as God's agent for the renewal of the entirety of the living creation is found in John 1:14. As we read the familiar words of John, "The word became flesh," we should be reminded that in the original Greek "the Word" did not merely become "man" (ἀνήρ) or even "human" (ἄνθρωπος). Rather, the Word became "flesh" (σάρξ)—"the term which defines the solidarity of humanity with the rest of creation in its bodiliness."[51] As Jesus the Messiah, "the Elect One of God," took on *flesh* he took on the most conceivably inclusive category of being and mode of existence known. As the Logos of God incarnated as "a living creature," the very creator God in the fleshly form of the Final Adam and renewed image of God entered into material solidarity with all "living creatures" of the cosmos.[52] Within its Early Jewish Messianic context, the incarnation is the most profound expression of a solidarity which encompasses the whole of life. It is an the event where the Word (or Divine Logos) becomes flesh or biological life—as opposed to becoming exclusively human—and in so doing inclusively embraces a category of existence to which all living beings belong. As the Word becomes biological life the boundary between humans and other living creatures thus becomes effectively blurred.

Redeeming Animals and ET: "That Which Has Been Assumed Has Also Been Saved"

The early church theologically spelled out the redemptive significance of the atoning work of the creator God in the flesh. Irenaeus of Lyons sees the entire course of the created cosmos as a salvation history with Christ, the Second Adam and renewed *imago Dei*, at the center. According to Irenaeus within this cosmic salvation history non-human creatures have firm footing

---

49. Ibid., 96.
50. Ibid., 97.
51. Muddiman, "A New Testament Doctrine of Creation?" 32.
52. This notion of 'solidarity in flesh' is similar to what Niels Gregersen has called a 'deep incarnation' that connects Christ with an evolving creation and the whole cosmos.

as subjects of God's providence and redemptive concern. In the wake of Adam's fall, says Irenaeus, "God made a covenant with the whole world through Noah, pledging Godself to all animals and humans."[53] The climax of this epic of redemption is Christ's recapitulation of the entire cosmic history—including the dynamics of creation and fall—within his own life. This is achieved through the incarnation, where as the Word becomes *flesh* the entire *world of flesh* is brought into communion with God through Jesus' victory over sin and through Christ's final triumph over death in his resurrection *in the flesh*.[54] For Irenaeus, as flesh is redeemed *through flesh* the words of Isaiah 40:5 and Luke 3:6 are fulfilled: "Then the glory of the LORD will be revealed, and *all flesh* [lit. living things] will see it together." As the Word became flesh in Christ, so "all flesh" or "all living things" shall together see God's final redemption of the flesh. (These verses are translated by the author.)

A few generations after Irenaeus, Athanasius and the Cappadocians also envisioned God's redemption in terms of the categories of being that God "took on" or "assumed" in the incarnation. Athanasius argues that in Christ's incarnation and resurrection "His *flesh* was saved, and made free the first of all, being made the body of the Word." Since God took on flesh "we, being similarly corporeal therewith, are saved by the same."[55] It is not just humans who are saved through Christ's incarnation and resurrection, says Athanasius, but all flesh: "For the presence of the Savior in the flesh was the price of death and the saving of the *whole creation*."[56] The most influential theological understanding of the meaning of the incarnation for salvation in the Eastern Christian tradition was defined by the Cappadocian father Gregory of Nazianzus's formula: "That which was not assumed is not healed (or saved); but that which is united to God is saved (or healed)."[57] In other words, redemption or salvation occurs at the level (or levels) of being which God in Christ participated in.

The category of incarnate and redemptive relevance that the gospels and the early church fathers single out is *not* Christ existing as a Jew, as a man, as a human, as a mammal, as a vertebrate, or even as an animal. The gospels and the testimony of the early church clearly emphasize that the

---

53. Irenaeus, *Demonstration of the Apostolic Preaching*, 22.

54. Irenaeus, *Against Heresies* 4.34.1. "For the flesh of the lord . . . confirmed the salvation of the flesh. Unless he had himself been flesh and blood . . . he could not have saved what had perished in Adam (5.14.1). By recapitulation his righteous flesh brought the flesh from slavery to sin into friendship with God (5.14.2)."

55. *Orat., II, Contra Arianos*, lxi.

56. *Ep. ad Adelphium*, vi.

57. Gregory of Nazianzus, *Epistle*, 101.

crucial category of being which has been "assumed" is the *flesh* (i.e., biological life at its most basic and inclusive level). The theologically significant category of both incarnation and redemption is the flesh—God's taking on existence as a biological entity. According to this understanding of the incarnation, then (*contra* Davies), Jesus came to save the whales, the dolphins, the gorillas, the chimpanzees, the Neanderthals, and even extra-terrestrial biological life. As one man from one species on one planet is elected as the *telos* of the *imago Dei*, so the entire cosmos is brought into God's salvific endeavor. "For God so loved *the cosmos* that he gave his only begotten Son" and it is for this reason that when the final "glory of the Lord will be revealed . . . *all living creatures* shall see it together" (John 3:16, Isa 40:5; author's translation).

Conclusion

Does Jesus the Messiah—as the creator God incarnate—save the animals, the Neanderthals, and ET? The theological logic of redemption found in Early Judaism, the New Testament, and the Early Church would certainly declare: yes. In the gospel's message of God's incarnate solidarity with the rest of creation, humanity in its true form as the elected image of God is not *removed from* creation, but rather *restored to* creation in order to reclaim the material cosmos on behalf of God's in-breaking kingdom. Understanding the divine likeness as election acknowledges non-human animals, hominids, and extra-terrestrial life as the ontological equivalents of humans and as fellow creatures that are substantively the same.

Here I have addressed one of the focal questions of this book: will extraterrestrial living creatures require their own divine incarnation, or does the redemptive work of Jesus Christ on Earth suffice? I side with those who claim that the one incarnation of God in Christ is efficacious for all sentient creatures wherever and whenever they live.

Even though this answers the christological question, the anthropological question remains. If we encounter alien beings superior to we Earthlings, will we forfeit the *imago Dei*? Does the *imago Dei* belong to the most intelligent of God's creatures? My answer is: no. Regardless of which traits ET possess—superior or inferior—this will not affect either our election or our responsibilities.

We do not need to re-ask about the relationship of *Homo sapiens* to the rest of sentient life, on Earth or elsewhere. Understanding the *imago Dei* as election upholds the attestation of Scripture and tradition in exclusively designating the image to human beings alone. The *imago Dei* as election focuses on God's mysterious choosing, loving grace, and universal salvific will for the whole creation rather than on the possession of particular qualities or capacities (such as rationality, morality, or advanced technology). Whether human beings possess such characteristics uniquely (when compared with non-human animals) or share several of these capacities with extinct non-human hominids or undiscovered extraterrestrial intelligent life is not a theological stumbling block for the *imago Dei* considered as God's election of *Homo sapiens*. Within the historical biblical or Hebrew concept of election, the choosing of people groups and representative individuals is not for their own sake, or for the purpose of individualistic final salvation, but rather for the sake of and in the service of others. Historical election is thus not exclusivist or anthropocentric, but instead is radically inclusivistic and even universalistic in so far as such universalism respects the integrity of creaturely freedom. Election in this ancient historical conception rebukes any sort of internally focused spiritual narcissism that myopically focuses on one person, one people, one species, or even one planet. Consequently, we may envision human beings, as the elected image of God, to be a holy species and royal priesthood that is chosen from among the myriad of life forms in the cosmos, elected from their animal kin, called out from their hominid ancestors, and given the tasks of obedience to the commandments, peace-keeping dominion, and cosmos-healing atonement.

Bibliography

Athanasius. *Ep. ad Adelphium*.
——— . *Orat., II, Contra Arianos*.
Barr, James. "The Image of God in the Book of Genesis: A Study of Terminology." *Bulletin of the John Rylands Library* 51 (1968–69) 13.
Bauckham, Richard J. "Jesus and Animals II: What Did He Practice?" In *Animals on the Agenda: Questions about Animals for Theology and Ethics*, edited by Andrew Linzey and Dorothy Yamamoto, 49–60. Urbana: University of Illinois Press, 1998.
——— . "Jesus and the Wild Animals (Mark 1:13): A Christological Image for an Ecological Age." In *Jesus of Nazareth: Lord and Christ: Essays on the Historical Jesus and New Testament Christology*, edited by Joel B. Green and Max Turner, 3–21. Grand Rapids: Eerdmans, 1994.
Bird, Phyllis A. "Theological Anthropology in the Hebrew Bible." In *The Blackwell Companion to the Hebrew Bible*, edited by Leo G. Perdue, 258–75. Blackwell Companions to Religion 3. Malden, MA: Blackwell, 2001.
Charlesworth, James H., ed. *The Old Testament Pseudepigrapha*. Vol. 1. New York: Doubleday, 1983.
Cohn-Sherbok, Dan, and Andrew Linzey. *After Noah: Animals and the Liberation of Theology*. London: Mowbray, 1997.

ONE *IMAGO DEI* AND THE INCARNATION OF ADAM   345

Collins, John J. "Pre-Christian Jewish Messianism: An Overview." In *The Messiah in Early Judaism and Christianity*, edited by Magnus Zetterholm, 1–20. Minneapolis: Fortress, 2007.
Davies, Paul C. W. *The Eerie Silence: Renewing Our Search for Alien Intelligence*. Boston: Houghton Mifflin, 2010.
Elliott, J. K., ed. *The Apocryphal New Testament: A Collection of Apocryphal Christian Literature in an English Translation*. Oxford: Clarendon, 1993.
Garr, W. Randall. "The Nouns דמות and צלם." In *In His Own Image and Likeness: Humanity, Divinity, and Monotheism*, 117–78. Leiden: Brill, 2003.
Gregory of Nazianzus. *Epistle* 101.
Irenaeus. *Against Heresies* 4.34.1.
———. *Demonstration of the Apostolic Preaching*.
Johnson, Aubrey R. *Sacral Kingship in Ancient Israel*. 1955. Reprint, Eugene, OR: Wipf & Stock, 2006.
Kaminsky, Joel S. *Yet I Loved Jacob: Reclaiming the Concept of Biblical Election*. 2007. Reprint, Eugene, OR: Wipf & Stock, 2016.
Levenson, Jon D. "The Universal Horizon of Biblical Particularism." In *Ethnicity and the Bible*, edited by Mark G. Brett, 143–69. Biblical Interpretation Series 19. Leiden: Brill, 2002.
Middelton, J. Richard. *The Liberating Image: The Imago Dei in Genesis 1*. Grand Rapids: Brazos, 2005.
Millard, A. R., and P. Bordreuil. "A Statue from Syria with Assyrian and Aramaic Inscriptions." *Biblical Archaeologist* 45.3 (1982) 135–41.
Moritz, Joshua M. "Chosen from among the Animals: The End of Human Uniqueness and the Election of the Image of God. PhD diss., Graduate Theological Union, 2011.
———. "Evolution, the End of Human Uniqueness, and the Election of the *Imago Dei*." *Theology and Science* 9.3 (2011) 307–39.
Muddiman, John. "A New Testament Doctrine of Creation?" In *Animals on the Agenda: Questions about Animals for Theology and Ethics*, edited by Andrew Linzey and Dorothy Yamamoto, 25–32. Urbana: University of Illinois Press, 1998.
Pannenberg, Wolfhart. *Human Nature, Election, and History*. 1st ed. Philadelphia: Westminster John Knox, 1977.
———. *Systematic Theology*. 3 vols. Translated by Geoffrey W. Bromiley. Grand Rapids: Eerdmans, 1991–1998.
Parpola, Simo. "The Assyrian Tree of Life: Tracing the Origins of Jewish Monotheism and Greek Philosophy." *Journal of Near Eastern Studies* 52.3 (1993) 168.
———. *Letters from Assyrian Scholars to the Kings Esarhaddon and Ashurbanipal*. Winona Lake, IN: Eisenbrauns, 2007.
Preuss, Horst Dietrich. *Old Testament Theology*. Vol. 1. Translated by Leo G. Perdue. Old Testament Library. Louisville: Westminster John Knox, 1995.
———. *Old Testament Theology*. Vol. 2. Translated by Leo G. Perdue. Edinburgh: T & T Clark, 1996.
Stone, Lawson G. "The Soul: Possession, Part, or Person? The Genesis of Human Nature in Genesis 2:7." In *What about the Soul?: Neuroscience and Christian Anthropology*, edited by Joel B. Green, 47–61. Nashville: Abingdon, 2004.
Tanner, Kathryn. "The Difference Theological Anthropology Makes." *Theology Today* 50.4 (1994) 567–80.

Wenham, Gordon. *Genesis 1–15*. Word Biblical Commentary 1. Waco, TX: Word, 1987.
Westermann, Claus. *Creation*. Translated by John H. Scullion, SJ. Philadelphia: Fortress, 1974.
Wright, N. T. *Climax of the Covenant: Christ and the Law in Pauline Theology*. London: T. & T. Clark, 1991.
———. *The New Testament and the People of God*. Christian Origins and the Question of God 1. Minneapolis: Fortress, 1992.
Zetterholm, Magnus, ed. *The Messiah in Early Judaism and Christianity*. Minneapolis: Fortress, 2007.

20

Extraterrestrial Salvation and the ETI Myth

TED PETERS

[Myths] capture vital truths, both in their general outline and also, probably, in much of their contents.

—KARL JASPERS[1]

My hypothesis is this: at work in contemporary culture is a mythical structure, which we here identify as the *ETI Myth*. The central doctrine within the ETI myth is the belief that extraterrestrial intelligent beings exist and, further, they are more advanced than us in evolution and in technological progress. It is a belief without any empirical evidence, at least as of this writing. Does this myth belong solely to the wider culture? Or, is it present as a conceptual set in the worldview of scientific researchers? Or, is it at home in both?

Traffic over the bridge between science and religion, so far in this volume, has been going one direction: from astrobiology towards astrotheology. In this chapter we will reverse direction. We will address the third on our list of astrotheological tasks: *theologians should analyze and critique astrobiology and related space sciences from within, exposing extra-scientific assumptions and interpreting the larger value of the scientific enterprise.* When we diagnose the space sciences, we will see that a myth is at work in the very heart of science. I will call it the *ETI Myth*.

1. Jaspers, *Origin and Goal of History*, 41.

Are Scientists Practicing Theology without a License?

The first and salient feature of the ETI myth is the imaginary exportation of the history of evolution to other planets or other possible habitats in space. "If life on Earth is well understood," writes Paul Davies, "its position beyond the Earth may be inferred."[2] There is nothing unscientific about this imaginary exportation, to be sure; it is the most reasonable thing to advance such a hypothesis regarding what is not known on the basis of what is known. If we know that life could originate here on earth and could speciate through evolution, then it is reasonable to project that these processes might have occurred more than once in this vast universe. When SETI scientists and other astrobiologists develop the idea of evolution on other worlds, however, they load their concept of evolution down with auxiliary commitments such as the doctrine of progress, the connection between progress and advance in intelligence, and the connection between advance in intelligence with near utopian versions of science, technology, health, and peace.

The trail of SETI reasoning is worth following with a bit more detail. We turn now to the second feature or contributing factor to the ETI myth: SETI is selective. SETI is listening to the skies in hopes of hearing a signal emitted from an extraterrestrial intelligent source. Non-intelligent or less intelligent beings may live elsewhere in the universe, to be sure; but the only ones likely to be sending signals are those with advanced technology.

Third, it follows that SETI is not making judgments about ETI in general, but focuses rather on those intelligent beings capable of sending radio signals. To be sophisticated enough to devise a signal-emitting technology, an extraterrestrial civilization must have been evolving for a long time. We here on earth developed radio only a century or so ago; so if we are to make contact with ETI they must be at least as old as we earthlings and perhaps even older.

Fourth, the astrobiological variant of the ETI myth places the scientist—especially the space scientist—at the pinnacle of evolutionary ascent to this point; and, therefore, the connector between terrestrial life and extraterrestrial reality. This point is subtle yet significant. Let us recall what Paul Ricoeur says about traditional creation myths. "The first characteristic of myth is to state an order indivisibly uniting ethos and cosmos."[3] In this case, the ethos is the professional ethos of the scientist. Within our larger population of *Homo sapiens,* the scientists are the most highly evolved and, thereby, those who connect Earth with heaven, our local situation with the

---

2. Davies, *Space and Time,* 205.
3. Ricoeur, "Evil," 2898–99.

cosmic situation. When the myth is told by the scientists, the scientists stand in a privileged location, in the location of the gnostic redeemer.

Fifth, here is the logic that leads to the salvation component to the ETI myth [salvation through progress]: with increase in evolutionary age comes an increase in technology; with increase in technology comes social changes appropriate to sustaining such a technology, perhaps even a social peacefulness that provides the stability to sustain such a technology for thousands or millions of years. Benevolence would become a necessary ingredient among such beings in order to prevent annihilating themselves. The disposition toward benevolence accompanies a lengthy evolutionary history and the development of advanced technology. This means that whatever ETI share with us on earth will have the equivalent of redemptive if not salvific value.

Davies displays the above logic. "Any alien civilization the SETI researchers might discover is likely to be much older, and presumably wiser than ours," writes Davies. "Indeed, it might have achieved our level of science and technology millions or billions of years ago . . . It is more likely that any civilization that had surpassed us scientifically would have improved on our level of moral development, too. One may even speculate that an advanced alien society would sooner or later find some way to genetically eliminate evil behavior, resulting in a race of saintly beings."[4] The conceptual set from which Davies draws his assertions includes the presupposition that evolution is progressive; it leads over time to the development of science and technology. In addition, it leads also to advances in morality. Note that the advance beyond evil in Davies's scenario is not achieved spiritually, but genetically—that is, scientifically. In short, science saves. This is the central gospel of the ETI myth: science saves. Because science on Earth is the most evolutionarily advanced achievement so far on Earth, an extraterrestrial science with a longer time to evolve would be even more advanced and more perfect than ours. Might we think of such a scenario as presupposing a horizontal mythical worldview?

In an earlier work, Davies speculates that "there may be tens or even hundreds of millions of inhabited planets in our galaxy;"[5] and most of these will be more highly evolved and advanced than we on Earth. Should we hear microwave signals from an extraterrestrial civilization, it would likely be "more developed scientifically, culturally, and ethically than ourselves."[6] Just how does Davies jump to "more developed" not only in science but also in ethics? Where is the empirical warrant for this speculation? Perhaps

---

4. Davies, "E.T. and God," 114–15.
5. Davies, *Space and Time*, 209.
6. Ibid., 210.

Davies works with an unspoken conceptual set—a myth—that says: science saves. This constitutes practicing theology without a license.

The ETI myth is so persuasive that it even finds adherents among those who practice theology *with* a license. Boston University's John Hart, a theologian whose primary portfolio deals with an ethics for space exploration under the rubric of the *cosmic commons,* shares with many in the scientific community a commitment to this very ETI myth. The myth begins with an affirmation of evolution in its progressive form. "In the vastness of space and over its eons of time, life on other worlds, too, might have evolved to be intelligent life. Extraterrestrial intelligent life (ETI) might be billions of years older than terrestrial intelligent life (TI)—and considerably more advanced biologically, intellectually, socially, and spiritually."[7] When we see terms such as *evolution* and *advanced* used together, we see the foundations of the ETI myth being laid. Evolutionary advance might take ETI beyond the primitive stage of religion where we retarded Christians and other religious devotees here on Earth remain. "ETI might even have evolved to a higher spiritual understanding in which institutional forms of religion have become irrelevant."[8] Those of us who remain religious today represent a lower step on the evolutionary ladder, a childish and immature stage of development. When we step up, we will advance beyond our religious adolescence into a scientific and spiritual maturity. And, if we on Earth make contact with our stardust cousins in outer space, we will be meeting our own future. ETI represents Earth's future, a future with a higher state of spiritual consciousness into which we will eventually evolve. Contact with more evolutionarily advanced ETI might become a benefit; it might cause us on Earth to leap-frog from the present to the future more quickly than evolutionary progress alone might accomplish this. In the meantime, according to the ETI myth, it's time for you and I to put away our outdated religious sensibilities and adopt a more advanced science, technology, and spirituality. This is the gospel of the ETI myth.

Graduating from Moral Adolescence

To be more precise, it is the evolutionary advance of science that saves, according to the ETI Myth. How might the human imagination conceive of such an idea? Let us follow the trail of reasoning once again.

---

7. Hart, *Encountering ETI*, 20.
8. Ibid., 177.

Carl Sagan fosters the ETI Myth by defining the current phase of human evolution on Earth with a heuristic metaphor, *technological adolescence*.[9] Once this growth metaphor is in place, we can look backward in time to our more primitive past and forward in time toward maturity and wisdom. Just as adolescent teenagers who can make pressure-cooker bombs in their kitchen and reign death down upon Boston Marathon runners, our political leaders can push buttons and release the power of multiple hydrogen bombs. We on Earth have the technological power to destroy our civilization; yet, we lack the moral fortitude and mature wisdom to prevent it.

Have we come to the point of a moral crisis on Earth? This is the question asked by Guillermo Lemarchand. "Our civilization could collapse due to the failure to solve our mutual aggressions (e.g., human maldevelopment, arms race, population explosion, starvation, global warning, ozone depletion, etc.) . . . If humankind does not make a deep change in its social behavior to eliminate violence, human kind will disappear in less than a thousand years," maybe only thirty years.[10] Just like the prophets of ancient Israel, Lemarchand renders judgment against society and demands repentance under threat of destruction.

At work here is the equivalent of a theological anthropology; it describes what we traditionally refer to as the fall into sin. Just how might human sin be overcome? According to SETI anthropology, we overcome sin with more evolution.

The logic looks like this: if an off-Earth civilization has advanced to the point where it can engage in inter-planetary communication or even spacefaring, then it must have passed through the equivalent of our crisis and preserved itself from self-destruction. Such an ET civilization will have learned how to avoid nuclear self-annihilation. And perhaps it will have advanced to a higher level of moral values and spiritual achievement. "All technological civilizations that already have passed through their technological adolescence, and have avoided their self-destruction, . . . must have developed ethical rules to extend their societal life expectancy."[11] When

9. Sagan, *Cosmic Connection*.

10. Lemarchand, "Speculations," 156–57.

11. Ibid., 154. Douglas Vakoch, formerly with SETI and now directing METI (Messaging Extraterrestrial Intelligence), understands the logic of the SETI myth but reserves the possibility for self-destruction rather than salvation. "Paired with the assumption that the extraterrestrial civilizations that we detect will almost inevitably be older than ours, is the assumption that with increased age comes increased technological capability. Not only would other civilizations be older than ours, it is argued, but that as on Earth, once a critical point is reached in technological development, increased age yields an increased ability to control nature through considerably more advanced technology . . . But as we have seen on Earth, increased technological capacity

high-impact communication between Earth and ETI occurs, we on Earth may benefit from ET's more advanced morality, a mature morality which may save our planet from self-immolation. In short, Earth's salvation may come to us from sky science. Practicing theology without a license?

Busting the ETI Myth

Myth busting requires, first of all, identifying a myth as a myth. There are three types of myth worth identifying. The first is the false story. To bust a myth as a false story, we discredit it with facts.

The second is the classical religious understanding of myth: a myth of origin tells how the gods created the world or part of the world in the beginning, *in illo tempore* or the time before there was any time, which explains why things are the way they are today.[12] To bust a myth of origin does not require discrediting it. Rather, it requires demythologizing to draw from it existential and cosmic meaning.

The third form of myth applies to the case at hand. In its third form, a myth is a conceptual set of presuppositions which frames and guides experience or new knowledge into a meaningful pattern.[13] It may or may not take the form of a story; it could take the form of a concept or set of concepts. Even though the myth structures our interpretations, in itself it is not proven true. But, it is not false either, at least until proven false. To bust this myth requires demythicizing: recognizing the myth as myth and attempting to pursue science as well as theology free of myth.

---

also brings with it the potential to annihilate the very civilization that brought forth that technology." Vakoch does not believe that science saves, at least necessarily. Vakoch, "Asymmetry in Active SETI," 478. See the website: METI.org.

12. "Myth is a narrative of origins, taking place in a primordial time, a time other than that of everyday reality." Ricoeur, "Myth and History," in *ER2*, 9:6371.

13. The conceptual set which frames scientific discourse looks like a model, but it is not tested as one would test a hypothesis. As a myth, the conceptual set functions more like a model comprehending the experimental models which actually do get tested. Is there a counterpart in religion? Yes. Religious myths provide the symbolic discourse out of which rational discourse is drawn. If the myth takes narrative form, the religious model takes rational form, the latter being derivative from the former. "But religious models can also fulfill many of the *non-cognitive* functions of myth," comments Ian Barbour, "particularly in *the expression of attitudes;* these functions have no parallel in science. Models embodied in myths evoke commitment to ethical norms and policies of action. Like metaphors, religious models elicit emotional and valuational responses. Like parables, they encourage decision and personal involvement. Like myths, they offer ways of life and patterns of behaviour." Barbour, *Myths, Models, and Paradigms*, 28. Extracting the model from the myth through critical self-reflection is part of the theologian's task of demythicizing, whether with a scientific myth or a religious myth.

The ETI myth is a conceptual sieve through which some SETI scientists and much of the larger public strain all their thinking about the meaning of ETI. This conceptual set is extra-scientific. It has not been proven, at least not to date. The meaning of ETI, according to the ETI myth, is that we on Earth may someday see salvation descending to us from the skies. The ETI myth includes eschatology.

SETI Eschatology

The science fiction genre and the entertainment genre tend to exploit the image of space aliens as Earth's enemies. Ufonauts come to Earth to conquer us; and we find ourselves engaged in interstellar intrigue and war on the model of international intrigue and war. Movies such as *Independence Day* or *Ender's Game*, or a television series such as *V* or *Falling Skies* or *Defiance*, exploit the warfare model and excite us with special effects. However, the warfare model is not the one most scientists work with. Rather, the scientific model tends to see space aliens as celestial saviors.

This salvific aspect of the myth reveals more of its shape as we read what Carl Sagan and Frank Drake write: contact with extraterrestrials "would inevitably enrich mankind beyond imagination."[14] As we reported in an earlier chapter, Drake dreams about this enrichment. "Everything we know says there are other civilizations out there to be found. The discovery of such civilizations would enrich our civilization with valuable information about science, technology, and sociology. This information could directly improve our abilities to conserve and to deal with sociological problems—poverty for example. Cheap energy is another potential benefit of discovery, as are advancements in medicine."[15] Note how this optimism extends well beyond mere contact with ETI. It includes optimism regarding the solution to "sociological" problems such as poverty and energy while giving us a leap forward in medicine. What Drake believes is that science is salvific, and extraterrestrial science would be even more salvific than Earth's science.

---

14. Sagan and Drake, "The Search." John W. Traphagan, a trustee for METI International, observes that "many in the SETI community have shown commitment to the belief that a technologically advanced civilization will be altruistic, despite the lack of evidence supporting that assumption." Traphagan, "Is Active SETI Really Dangerous?"

15. Drake, "Interview with Dr. Frank Drake," 5. These words are repeated in part in *Drake* xvi. Linda Billings, NASA consultant, excoriates SETI for its unrealistic hope. SETI has "used the Drake equation to construct a mythology, a sort of origins myth, about extraterrestrial life." Billings, "The Allure of Alien Life," in Dick, *LBE*, 317. It is not an origins myth we find at SETI, but rather a myth of eschatological redemption.

Elsewhere Drake has waxed ecstatically about the prospect that contact with ETI might even bring immortality. "I suspect that immortality may be quite common among extraterrestrials. By immortality I mean the indefinite preservation, in a living being, of a growing and continuous set of memories of individual experience . . . Sometimes, when I look at the stars I wonder if, among the most common interstellar missives coming from them, is the grand instruction book that tells creatures how to live forever."[16] Drake's vision is eschatological, incorporating not only improved health but also life without death. This is a grand and beautiful vision, to be sure; but we ask: is this science? Or, is it practicing theology without a license?

Former SETI director, Jill Tarter, likewise holds up a kindred eschatology. This eschatology will save earthlings from the fall, the fall into religion. The darkness in this case is religion; and we need the gnostic light science shines to overcome the darkness. Tarter contends that "the ETIs (or their robotic emissaries) will be far older than we." Because they are older and have evolved longer, it follows that aliens will be more advanced. In order to survive and advance for a long period of evolutionary time, extraterrestrials must have overcome self-destructive tendencies that threaten civilization here on Earth. On Earth, self-destructiveness is the product of religion. "Their longevity is inconsistent with organized monotheistic religions [which are] responsible for the longest lasting warfare and destruction we have witnessed." In contrast to unethical religious earthlings, the extraterrestrials will have "a highly established code of ethics." When contact occurs and high impact knowledge comes to Earth, "they may tell us how it is possible to transition from the 'My God vs. Your God' conflicts . . . to a more stable universal religion/understanding."[17] Like the modernists mentioned above, religion represents the dark ages between the ancient Greeks and today's scientists. More highly evolved extraterrestrials will be like today's scientists, not like yesterday's religious believers. What we find here looks a lot like the horizontal or historical myth Voegelin identified as modern gnosticism.

SETI critic Edward Regis dubs such evolutionary optimism as belief in "salvation from the Stars."[18] Atheist cosmologist and science fiction writer, Fred Hoyle, saw the underlying motivation for belief in ETI: the desire for salvation coming from the stars, but salvation based upon science and not religion. What we are here calling the ETI myth includes "the expectation that we are going to be saved from ourselves by some miraculous interstellar

16. Drake and Sobel, *Is Anyone Out There?*, 162.
17. Tarter, "Implications," 45.
18. Regis, "SETI Debunked," 243.

intervention."[19] Terrestrial science saves, and extraterrestrial science saves even better.

"The Myth of the Extraterrestrials," according to James Herrick, whom we cited in an earlier chapter, is "the idea that intelligent extraterrestrials exist and that interaction with them will inaugurate a new era in human existence."[20] He adds, "this Myth is certainly closely connected to, is perhaps the offspring of, the most powerful scientific idea of the last two hundred years, evolution." The evolution narrative structures the conceptual set with such exclusionary force that it functions as dogma leading to bizarre extremes. Wilkinson applauds and demurs. "SETI can become part of the overarching narratives of understanding who we are in the Universe, but it can also be used for some bizarre religious philosophy."[21]

Is SETI a Secular Religion?

Is SETI a secular religion? Lewis White Beck derisively describes the field we know as astrobiology as a "modern equivalent of angeology and Utopia or of demonology and apocalypse." Then with a flair he exclaims, "*exobiology recapitulates eschatology.*"[22] Let us pause at this point. This negative judgment on the work of astrobiology or SETI is not what we in this book stress or even embrace. We enthusiastically support the search for extraterrestrial microbes and even intelligence. We enthusiastically support NASA, SETI, and METI. What we wish to point out is merely that in this field of science we find sublimated religious hopes, secularized language to express religious sensibilities. Human hopes rise up from deep within the human soul, and these hopes can express themselves in scientific language as well as religious language.

In a Caltech lecture, science fiction writer Michael Crichton launches a barrage of criticism aimed at sinking the SETI ship. "The Drake equation is literally meaningless, and has nothing to do with science. I take the hard view that science involves the creation of testable hypotheses. The Drake equation cannot be tested and therefore SETI is not science. SETI is

---

19. Hoyle, *Monthly Notices of the Royal Astronomical Society,* cited by Wilkinson, *Alone in the Universe?*, 144.

20. Herrick, *Scientific Mythologies,* 51. Physicist Michio Kaku is a scientist who seems to avoid belief in the ETI myth. He speculates: "I believe the aliens will be benevolent and, for the most part, ignore us. We simply have nothing to offer them." Kaku, *The Future of the Mind,* 315.

21. Wilkinson, *SRSEI,* 116.

22. Beck, "Extraterrestrial Intelligent Life," 13.

unquestionably a religion."[23] For Crichton, religion is like bird poop in one's hair. You want to race home and wash it out. I would like to make it crystal clear that I do not view the matter as Crichton does. In my judgment, SETI performs valuable science and the religious dimension which accompanies the science plays a necessary role in justifying the SETI research program. I only ask for transparency, not this kind of disparagement.

SETI researchers themselves ask the question: is SETI a "secular religion"? They answer in the negative. They believe they are pursuing science; and they believe science is different from religion. What's the difference? John Billingham answers. "Our beliefs are subject to test, and change on the basis of hard evidence, which is not the case for religions."[24] This self-defense demonstrates a certain naiveté regarding how science works along with an unnecessary stance of superiority over against religion; yet this commitment to judgment by hard evidence is admirable.

We the contributors to the present book are less negative about the ETI myth than some of its other critics. We are as empathetic as John Sanford in recognizing the vital and indispensible role myth plays in the human psyche. "All myths have their origin in the striving of the psyche to express in mythological form the deepest human spiritual and psychological truths and strivings."[25] Because myth is so intrinsically human we must say that, in itself, the ETI myth is not sinister. However, if the myth influences science in a naive fashion that escapes the notice of scientific researchers, then it risks distorting known facts and devolving into an anti-religious ideology. In our judgment, SETI researchers have no good reason for practicing theology without a license let alone promulgating an anti-religious ideology. This might make them deaf to religious cheerleaders urging SETI on to victory.

Is the Myth of Evolutionary Progress Justified?

We cannot stress enough how many respected scientists refuse to accept the belief that biological nature is driven by a principle of progress that leads over time to increased intelligence, to science, and to salvation. We label this opposition to SETI assumptions as the *unique Earth* or *rare Earth* position. Alfred Wallace (1823–1913), co-discoverer with Charles Darwin of the role played by natural selection in evolution, belongs to the rare Earth club. Wallace is aware of the contingencies of evolutionary change, dependent on unique interactions between inheritance and specific environmental niches.

23. Crichton, "Aliens Cause Global Warming."
24. Billingham, "Who Said What," 36.
25. Sanford, *The Kingdom Within*, 127.

This means evolution could not repeat itself, either on Earth or anywhere else in space. Even if life begins on another planet, it is virtually impossible for it to duplicate the production of an intelligent species such as humanity. In his 1904 book, *Man's Place in the Universe,* Wallace reports, "I submit, therefore that the improbabilities of the independent development of man, even in one other world—and far more in thousands of millions of worlds, as usually supposed—are now shown to be so great as to approach very closely, if not quite to attain, the actually impossible."[26]

A century later we find evolutionary biologist Francisco J. Ayala making the same case. When Ayala poses the question regarding the possible existence of ETI, he says, "My answer is an unequivocal 'no.'" Why such a strong opposition to contact optimism? Because what has happened in our planet's evolutionary history has been contingent, not guided by an internal purpose or entelechy. Ayala, like Stephen Jay Gould, argues that if we on Earth were to replay "life's tape" from the beginning of life to the present, the course of evolution would not repeat itself. According to the existing evolutionary history, for the first two billion years only microbes existed on Earth. The eukaryotes were the first organisms whose cells have a nucleus containing DNA; and, adds Ayala, there is nothing in the process that would make it likely that multicellular organisms would evolve. Evolution could have stopped right there. No animals might have come into existence. "We know that animals evolved only once. So, there is little likelihood that animals would arise again, if life's tape were replayed."

The phenomenon of extinction plays a big role in Ayala's argument. He notes that 99 percent of Earth's species are now extinct. Five hundred thousand years ago most animal species had already become extinct; and their body plans would no longer be represented just one hundred million years later. Only one lineage gave rise to the vertebrates: animals with backbones, including fishes, amphibians, reptiles, birds, and mammals. Even if the low probability event of the evolution of animals would be repeated, we have no reason to expect that animals with backbones would evolve. What follows is a virtually infinite improbability that primates would arise again, let alone hominids and *Homo sapiens.*

In each chapter of the evolutionary story, we find a long concatenation of contingent if not unique events. We find millions of random mutations and environmental circumstances, all points where the history could have taken a different turn. The probability of a repeat of this history is so low as to be virtually nil. The evolutionary process would produce a different outcome every time it gets going. Ernst Mayr puts it this way: "At each level

26. Wallace, cited in Crowe, *ELD*, 436.

of this pathway there were scores, if not hundreds, of branching points and separately evolving phyletic lines, with only a single one in each case forming the ancestral lineage that ultimately gave rise to Man."[27]

The Drake equation would be equally unpersuasive to Ayala, because the improbabilities of a repeat of our evolutionary progress are greater than the probabilities of communicating intelligent life coming into existence. If we "replay life's tape," the improbabilities get multiplied from year to year, from generation to generation, millions and millions of times. The resulting improbabilities are of such magnitude that even if there would be millions of universes as large as the universe that we know, the products (improbability of humans x number of suitable planets) would not cancel out by many orders of magnitude. The improbabilities apply not only to *Homo sapiens*, but also to "intelligent organisms with which we could communicate"; by this phrase I mean organisms with a brain-like organ that would allow them to think and to communicate, and with senses somewhat like ours (seeing, hearing, touching, smelling, tasting) which would allow them to get information from the environment and to communicate intelligently with other organisms. We have to conclude that humans are alone in the immense universe and that we forever will be alone."[28]

Such reasoning regarding contingencies in evolution lead Ernst Mayr, among others, to say that SETI is very likely to fail in achieving contact.[29] Mayr minces no words. "The SETI program is a deplorable waste of taxpayers' money, money that could be spent far more usefully for other purposes."[30]

In an earlier chapter of this book we saw how Martinez Hewlett rejected the force of the rare Earth argument, favoring contact optimism. Let me add here my own judgment: despite the force of this criticism of SETI's assumptions, I need not fully embrace Hewlett's position in order to support the work of SETI. I support SETI for two reasons. First, even if finding extraterrestrials is statistically improbable, this remains a statistical argument rather than an empirical one. Nothing in the laws of nature precludes the development of off-Earth life. In principle, it is still possible for us to discover ETI, even if unlikely. Secondly, the very performance of SETI research leads to a great deal of knowledge of space we would not otherwise have. As

---

27. Mayr, "Probability," 27.

28. Quotes here taken from Ayala, "Evolution of Life," 57–77.

29. Mayr, "Can SETI Succeed? Not Likely." http://www.planetary.org/explore/topics/search_for_life/seti/seti_debate.html. On this SETI website Carl Sagan argues for an "Abundance of Life-Bearing Planets."

30. Mayr in Regis, *Extraterrestrials*, 29.

an aside, SETI is now supported by private funds, not taxpayer money. Now, back to the main argument.

Like Ayala, John Maynard Smith emphasizes the contingent and accidental appearance of the life forms we see on Earth. Should evolution begin once again here or elsewhere, "there is no guarantee—indeed no likelihood—that the result would be the same."[31] In other words, the likelihood that evolution on an extraterrestrial planet would lead to intelligent life similar to our own is virtually nil.

One might conceivably defend the assumptions of SETI and its baby sister, METI, by appealing not to the doctrine of progress but rather to convergence in evolution. Convergence is progress lite. *Convergent evolution* identifies the process by which initially dissimilar organisms respond to similar selection pressures to arrive at similar evolutionary solutions. The intellectual mastermind of convergence theory is biologist Simon Conway Morris, who contends that appeal to sheer randomness to explain evolution fails to explain evolution. Why? Because natural selection channels change into a finite set of options. "Evolutionary convergence [means that] from very different starting points in the Tree of Life very much the same solution has evolved multiple times."[32] One might quickly surmise, then, that an evolutionary history on an exoplanet would lead to a technological civilization like ours even without reliance on built-in directionality, entelechy, or progress. Even though Morris himself does not apply convergence to off-Earth biospheres, the idea of convergence might supply a SETI or METI speculator with mild support. Only mild support, however, because convergence theory could not explain let alone predict a secular utopia on Earth let alone in the heavens.

Pointing out the weaknesses in the ETI myth does not require affirmation of a unique Earth. One can dispute the rare or unique Earth position scientifically, as Hewlett and Manning did in earlier chapters. Harvard astronomer Owen Gingerich, for another example, will allow consideration of purpose or design in the evolutionary process that could be applied extraterrestrially. He sees signs of design and direction in nature, contending that "there does seem to be enough evidence of design in the universe to give some pause."[33] That is to say, if the universe is biophilic, then we should expect to meet ETI some day.

---

31. Smith, "Taking a Chance on Evolution," 34.
32. Conway Morris, *The Runes of Evolution*, 3.
33. Gingerich, "Is There a Role?," 42.

Which Myth Is the Real Myth?

A biophilic universe? Is this good enough to expect us to meet new space neighbors?

The convergence school of evolutionary biology supports the idea that once life begins it will most likely follow a path from simplicity to complexity and to intelligence. This is not due to an inner *telos* justifying a doctrine of progress; rather, it is due to the channeling options natural selection might exact on speciation. Astrobiologist Julian Chela-Flores, for example, contends that "eukaryogenesis will occur inexorably because of evolutionary pressures, driven by environmental changes in planets or satellites where conditions may be similar to Earth . . . the evolution of intelligence as searched in the SETI program is a rational scientific pursuit."[34] Chela-Flores is a contact optimist based upon his view of the science, not myth.

When we turn to the champion of convergence, Cambridge evolutionary biologist Simon Conway Morris, the matter of myth surfaces. Like Chela-Flores, Morris argues that evolution would always follow a somewhat predictable track leading toward intelligent beings such as ourselves. "The emergence of human intelligence is a near-inevitability."[35] Morris bases his speculation on the history of convergence in Earth's evolutionary past. "Convergence is ubiquitous, from molecules to social systems. In fact, the study of convergence reveals a deep structure to life. This strongly suggests that what is true on Earth is true anywhere . . . So, out there as and when we meet the aliens . . . the first will be bipedal and intelligent."[36] Even without a built in entelechy, the principle of convergence in evolution may justify the SETI search.

While advocating convergence, Simon Conway Morris makes the matter a bit more complicated because, like me, he speaks of "evolutionary myths. Not fairy-tales, of course, but areas of received wisdom that might benefit from a re-examination, or if you prefer a really good kicking."[37] Do I have an ally here in Morris regarding the role of myth within science? Well, no and yes. No, Morris is not my ally because the example he gives of myth within science is the very example I give of reigning science. Just what does Morris put on his list of the evolutionary myth to be subjected to re-examination? Here it is: "the assumption, and almost universally observed,

---

34. Chela-Flores, *Science of Astrobiology*, 205.

35. Conway Morris, *Life's Solution*, xii. Even with this premise, environmental conditions restricting the likelihood of a Goldilocks planet means, "life may be a universal principle, but we can still be alone." Ibid., 105.

36. Conway Morris, "Not So Alien," 10–11.

37. Conway Morris, "If the Evolution of Intelligence is Inevitable," 218.

that for all intents and purposes evolution is random, lacking predictability at any useful level."[38] That is to say, what Morris calls "myth" is what I have identified as the standard scientific model among evolutionary biologists. So, it appears that Morris and I disagree. But, let's get behind the appearance of disagreement.

After putting the position taken by Mayr, Gould, and Ayala on his myth list, Morris then looks for factors that might constrain evolution so that its outcomes might be more predictable. He concludes that "*in principle* any organism equipped with the rudiments of intelligence is capable of traveling along the same road as we hominids embarked upon some millions of years ago."[39] In short, the notion that a comparable or longer time for evolution on an off-Earth planet would likely yield aliens who are somewhat like us on Earth. After demythologizing the dominant assumptions, Morris proceeds to provide an argument in behalf of what I have been dubbing the "ETI Myth." What I think of as establishment science he calls "myth"; and what I call "myth" Morris praises as solid science. Again, this may appear to amount to a disagreement.

A still closer look, however, shows that I can honestly admire and thank Morris for two reasons. First, he demands that we subject the myths we find among scientists to scrutiny, to "a really good kicking." This kicking would apply to the dominant view right along with the SETI or UFO view of progressive evolution. Morris kicks the dominant view and then provides a reasoned scientific argument in support of the progressive evolution view. Or, to say it another way, rather than simply assume that evolution is progressive Morris tries to demonstrate scientifically that this is the case. Morris, more than others, provides rational support for what most take as a naive assumption. As an aside, I hope Morris is right about the prospects of finding off-Earth aliens similar to earthlings.

Now, back to the main argument. As we have emphasized, despite what Hewlett, Gingerich or Morris say, the view regnant among evolutionary biologists is articulated by Ayala: "The overall process of evolution cannot be said to be teleological in the sense of proceeding toward certain specified goals, preconceived or not."[40] Philosopher of Biology Michael Ruse adds: "There is absolutely no guarantee of an upward progression on our hypothetical planet to intelligent life forms . . . [E]volution of intelligence is not a necessary consequence of life appearing: not at all."[41] If this argument

38. Ibid.
39. Ibid., 220.
40. Ayala, "Darwin and the Teleology of Nature," 34.
41. Ruse, "Is Rape Wrong on Andromeda?," 50.

holds, then the ETI Myth hangs on a thin thread of utter chance rather than the firm rope of evolutionary progress.

This thin thread may suffice to justify the astrobiological and SETI search. My point is merely that the assumptions with which the search is framed constitute a worldview that is viciously challenged by leading evolutionary biologists as well as cultural critics. To hold tenaciously to these assumptions while under such fire requires something like a daring faith, not a cautious self-restriction to reliable empirical knowledge. The concept of demythologizing might include a good kicking.

Should We Use the Term Myth to Describe Science?

Space researchers readily admit that they work with limited data and find they must fill in the gaps with speculation. "Isn't science supposed to be about logic and evidence?" asks astrobiologist Chris Impey. "Sure, but when data are in short supply, scientists feel free to speculate."[42] This is normal. Still, is it legitimate for us to ask: should we use the term *myth* to describe what is going on in astrobiology? Does the interaction between evidence and speculation prompt our question about myth? If the concept of myth implies a worldview beyond the parameters of what empirical evidence permits us to say with confidence, then perhaps the door is open for this question. If a scientist asks us to engage in worshipping ETI, might we ask if myth as a vertical hierarchy could be at work?

Did I say "worship" ETI?! Yes. One case in point is the notorious advocate of atheism, Oxford's Richard Dawkins. Dawkins recommends that we substitute extraterrestrial intelligent beings—who are more advanced in evolution that we are—for the traditional God. It would be "pardonable" for an atheist on Earth to worship ETI, he says. Why? Because advanced ETI will have progressed in their evolutionary development so far beyond us that they will appear god-like. "Any civilization capable of broadcasting a signal over such an immense distance is likely to be greatly superior to ours . . . [T]here are very probably alien civilizations that are superhuman, to the point of being god-like in ways that exceed anything a theologian could possibly imagine."[43] What is decisive for Dawkins is that god-like ETI are the fruit of evolution. Evolution creates; and god-like creatures are created by evolution. "The crucial difference between gods and god-like extraterrestrials lies not in their properties but in their provenance. Entities that are

---

42. Impey, *Living Cosmos*, 281.
43. Dawkins, *God Delusion*, 72.

complex enough to be intelligent are products of an evolutionary process."[44] Despite that fact that empirical evidence is totally lacking, Dawkins can claim that evolution produces creatures worthy of human worship. Is this science? Or, is it myth under the guise of science? Is it the product of an atheist huckstering theology without a license?

Perhaps it is impossible for a scientist—even an atheistic scientist—to avoid appealing to myth when constructing a comprehensive theory. Why? Because of our need for meaning. Myth is the domain within human consciousness where diverse things are integrated in a meaningful way. The Greek word, *mythos,* was originally associated with story or tale; and its primary function was and is to convey meaning. The research scientist accumulates data and explains data according to conventions of natural law. In so doing, the scientist engages in *logos* or reason, not *mythos*. A meaningful story must be added to the data, so to speak. As butter is spread on bread, *mythos* gets spread over *logos*. What is distinctive about scientized myth, of course, is the deletion of any reference to supra-natural agency or even intra-natural purpose. "Secular myths . . . are derived from religious myths but lack reference to a supernatural agency," comments Anne Foerst.[45]

So, our question is this: should we use the term *myth* to illuminate what is going on here? The analysis of Albert Harrison at the University of California at Davis gives us an opportunity to examine this question. First, like Frank Drake, Harrison commits himself to approach the question of ETI from "within the framework of science and view the evolution of life and civilizations as orderly processes that proceed within broad natural limits."[46] Working within these "natural limits," Harrison finds he can speculate about an intelligent civilization that might have evolved longer than our own. "Thus, a fundamentally positive picture emerges when we extrapolate from life on Earth: there are trends toward democracies, the end of war, and the evolution of supranational systems that impose order on individual nation-states. This suggests that our newfound neighbors will be peaceful, and this should affect our decision about how to respond to them."[47] In sum, advances in evolution lead to democracy and peace. Might it follow that we could ask our newfound space neighbors if they might share with us the fruits of their evolutionary advances in democracy and peace?

44. Ibid., 73.
45. Foerst, "Stories We Tell," 20.
46. Harrison, *After Contact*, ix.
47. Ibid., 312.

Harrison acknowledges how this line of thinking attributes god-like qualities to ETI, even though he is not responding directly to Dawkins. "Two assumptions make it tempting to attribute extraterrestrials with god-like qualities. The first assumption is that any extraterrestrial civilization that we will find is likely to be older than our own . . . The second assumption is that extraterrestrial civilizations will be benign, even benevolent . . . [they] are less likely to be subject to repression and political violence, more likely to have their basic needs for food and shelter satisfied, and more likely to develop economic surpluses that encourage trade" rather than war.[48] Should we expect that evolutionary progress either on Earth or elsewhere can take us to "a world without war"? Yes, so it seems.

Now, this looks like myth. But, Harrison denies that it is myth. Why? Because science and myth do not mix, he says. "SETI is not to be confused with religion and myth, so any superficial similarities among extraterrestrial radioastronomers, God, ancient astronauts, and space brothers have to be taken with a huge grain of salt. God, if He exists, is supernatural. Extraterrestrials would be the product of biological evolution . . . Most importantly of all: for religious people God is a given, but for scientists extraterrestrials are hypothetical, at least pending empirical verification."[49]

In his haste to prevent brushing science with mythical paint, Harrison is overlooking the central role played here by the idea that evolution and science lead to peace if not salvation. Despite the fact that evolution is founded on natural selection and survival-of-the-fittest and that modern science has produced atomic bombs and other weapons of mass destruction, SETI scientists look forward to altruism, benevolence, and salvation. This structure of assumptions is what is key to the application of the term *myth* in this instance. Although Harrison identifies myth with a supernatural God rather than biological ETI, it appears that in the ETI myth biological space beings come to play the role that supernatural redeemers played in past myths. Ancient gods came from the heavens, just as flesh-and-blood extraterrestrials come to Earth from the heavens. The former were spiritual, while the latter are material. But, they perform the same function: they save planet Earth from self-destruction. SETI science looks like a secularized form of the redeemer myth, specifically, the gnostic redeemer myth.

This belief among astrobiologists that ETI might bring peace on Earth takes the form of a secular eschatology. Such an eschatology requires a faith-affirmation, just as a religious eschatology does. To call it "scientific" only buries this hidden fact. Theologian Martin Marty alerts us to how this works.

48. Harrison, *Starstruck*, 97–98.
49. Ibid., 97.

"The secular philosopher of history who deals with a *telos* is also making a faith-affirmation. Strictly speaking, he is not dealing with the future as it has actually occurred: he is reporting on events which have not occurred and concerning which he cannot have historical knowledge as if he has such knowledge. Ordinarily, of course, this faith-affirmation comes disguised as empirical inquiry or commonsense observation."[50] What distinguishes the secular eschatology of SETI is the belief that the future is already embedded in our evolutionary past, a *telos* or entelechy that will lead ineluctably to the scientific equivalent of salvation. This takes as much faith-affirmation as found in any traditional religion.

Is SETI science? Or, is it religion? Michael A. G. Michaud, would still defend SETI as science. "Although SETI shares some qualities and some goals with religion, its method is different. The scientific search attempts to confirm belief by experiment, not revelation."[51] Perhaps we should think of SETI as working from within a projected worldview, informed but not constrained by the Darwinian picture of evolution, embellished by the doctrine of progress, and inspired by hopes for a secularized eschatology that looks for a scientized salvation.

Earlier we used a visual image of macaroni and cheese to describe scientific searches for ETI. The countless elbow macaroni units are analogous to the facts, whereas the gooey cheese that connects them are analogous to the myth. A mythicized concept of evolution structures the worldview within which various data points find their connection and coherence. Michael Ruse puts it this way: the concept of progress, "although hidden, stands ready to influence the ways that theorists might fill evidential gaps between data and meaning."[52] Because on the surface it looks like science, what remains hidden below the surface is a secularized religious sensibility. "Not only has evolution functioned as an ideology, as a secular religion, but for many professional biologists that has been its primary role."[53] Ruse objects to this secularized religion, because he wishes to purify biology from its fallacious reasoning. Here is the fallacy, according to Ruse: "Evolutionists take their belief in scientific progress and transfer it into a belief in organic progress."[54] Because scientific knowledge is itself progressive, scientists are tempted to import scientific progress into their object of study, the natural world. Ruse offers us a very important observation: because the methods

50. Marty, "Non-Christian Eschatologies," 32.
51. Michaud, *Contact with Alien Civilizations*, 203.
52. Ruse, *Monad to Man*, 484.
53. Ibid., 530.
54. Ibid., 538.

of scientific inquiry are progressive, many scientists impute to the natural world they study this same quality, namely, progressiveness. This imputation is not itself scientific. It is religion under the guise of science. Or, better, science and religion are mixed just as cheese and macaroni are mixed in a tasty lunch.

This is fallacious. It is sub-scientific. It should come as an embarrassment. Nevertheless, in defense of SETI, when I say that SETI is a secular religion I do not intend to disparage the value of its scientific contribution. Still, having applauded the valuable science, we must admit that SETI is more than just pure unadulterated science. It takes the form of a sub-culture and, like traditional religion, it attempts to meet existential human needs. Among our human needs are aspirations for meeting our survival requirements along with establishing world peace. These are authentic and wholesome human aspirations. Still, I ask: why does SETI feel it has a mission to meet these needs? Meeting these needs are not SETI's job. I want to admonish SETI scientists—and all scientists—to stick to science! Stop embellishing science with a scientized version of the gnostic redeemer myth! No scientist is compelled by what he or she discovers in nature to promulgate an atheistic worldview, a secular ideology, or an anti-religious religion. Religious persons in our wider society can enthusiastically embrace the goals of SETI and applaud the new knowledge given us by astrobiologists; but they will appreciate such science much more if it remains science and refrains from evangelizing for secularism.

Evolution as Both Science and Myth

What we are seeing here is that the backbone assumption of the ETI myth is evolution; and the rib assumptions—many of which are doubted by evolutionary biologists—include the following: life must evolve where conditions are right; biological evolution is progressive; progress leads to higher intelligence; higher intelligence leads to science; science leads to peace; and, one more assumption: science is more highly evolved than religion. Where our astrobiologists stand today, they assume, is on the pinnacle of Earth's evolutionary rise upward. They are straining to look into outer space to see if they can see what might be even more highly evolved. Science allegedly represents the most advanced stage in Earth's evolution; and scientific astrobiologists just might be the ones to lead all Earth's life forms forward to still higher development.

Let us look more closely at one of these rib assumptions: life must evolve wherever the conditions are right. "Life is the product of deterministic forces," writes Belgian biochemist Christian de Duve, who won the Nobel Prize in 1974 and passed away on May 4, 2013. "Life was bound to arise under the prevailing conditions, and it will arise similarly wherever and whenever the same conditions obtain. There is hardly any room for 'lucky accidents' in the gradual, multistep process whereby life originated. This conclusion is compellingly enforced when one considers the development of life as a chemical process."[55] As long as the right chemical conditions exist somewhere in outer space—in the Habitable Zone (HZ), that is, in the Goldilocks location where it is neither too hot nor too cold—we can expect life to evolve and develop and progress. And, perhaps, some day we will meet this extraterrestrial life form. At the level of assumption, this evolutionary belief has worked its way into the ETI myth.

Christian de Duve speculates, based on the Green Bank equation of 1961 (see the Drake equation discussed in earlier chapters) that "the figure of about one million 'habitable' planets per galaxy is considered not unreasonable. Even if this value were overestimated by several orders of magnitude, it would still add up to trillions of potential cradles for life. If my reading of the evidence is correct, this means that trillions of planets exist that have borne, bear, or will bear life. The universe is awash with life."[56] With such contact optimists speculating without empirical evidence that the universe is teeming with life, it is easy to imagine our culture developing images of just what that life might be like.

Nobel Laureate de Duve continues, feeding the myth with apparent scientific veracity. "My conclusion: We are not alone. Perhaps not every biosphere in the universe has evolved or will evolve thinking brains. But a significant subset of existing biospheres have achieved intelligence, or are on the way to it, some, perhaps in a form more advanced than our own."[57] When science becomes mythologized, we consider that our partners in outer space could be more highly evolved—"more advanced"—than we are.

Carl Sagan was a self-appointed high priest of secular theology, justifying his beliefs with this ETI myth. Yet, Sagan admitted that the ETI myth is based on speculation rather than conclusive empirical evidence. First in his book and posthumously in the Hollywood movie starring Jodi Foster, *Contact,* Sagan dramatically portrayed science as replacing religion in announcing salvation for the human intellect if not for our world.[58] Yet,

55. De Duve, *Vital Dust*, xv.
56. Ibid., 121.
57. Ibid., 297.
58. Sagan, *Contact*. Oxford University transhumanist Nick Bostrom would like to believe in the terrestrial adolescence hypothesis; but he sees more complexities than Sagan or other ETI mythicists do. Bostrom introduces the concept of the *Great Filter*.

secondly, despite his sacralizing of scientific progress, Sagan was aware that the contingencies pointed out by evolutionary biologists challenge his working assumptions. Hopes for ETI salvation fall short of solid science. "I would guess that the Universe is filled with beings far more intelligent, far more advanced than we are. But, of course, I might be wrong. Such a conclusion is at best based on a plausibility argument, derived from the numbers of planets, the ubiquity of organic matter, the immense timescales available for evolution, and so on. It is not a scientific demonstration."[59] Even though this is not a strictly *scientific* demonstration, Sagan protégé Neil deGrasse Tyson still tries to persuade us rhetorically with an appeal to big numbers. After estimating that there are ten sextillion stars in the observable universe, Tyson offers a judgment. "There are people, who walk around every day, asserting that we are alone in this cosmos. They simply have no concept of large numbers, no concept of the size of the cosmos."[60] Big numbers all by themselves make the existence of ETI plausible, say the Saganettes. But this falls short of empirical evidence. Sagan's honesty helps us discern the contours of the ETI myth.

Even more is relevant here. Sagan, though embracing the ETI myth, expresses ethical reservations similar to those of Marx and Engels: he would not want our hope for celestial salvation to blunt our moral obligation to make our present world a better place. "The expectation that we are going to be saved from ourselves by some miraculous interstellar intervention works against the necessity for us to solve our own problems."[61] Sagan stops short of embracing salvation arriving on Earth by grace from the sky.

If the Great Filter lies behind us at the point of abiogenesis or somewhere along our evolutionary history, then we intelligent creatures may be a rare product of natural processes. If, however, the Great Filter lies in our future, then this might mean that every intelligent civilization eventually destroys itself and leaves remnants on its host planet. This could mean that we on Earth will precipitate our own existential cataclysm through some sort of technological apocalypse. All that has happened on Earth will then drop into oblivion. If self-destruction awaits every highly evolved intelligent species, then this explains Fermi's Paradox: it explains why we are currently alone in the universe. Bostom hopes that our space explorations will discover no artifacts of previously self-destructed extraterrestrial civilizations. "No news is good news. It promises a potentially great future for humanity." Bostrom, "Where Are They?" 72–77.

59. Sagan, *Pale Blue Dot*, 33. Sagan speculates not only about the scientific advances of ETI, but also wonders what might happen if visiting ETI would find human beings delicious to eat. "Why transport large numbers of us to alien restaurants? The freightage is enormous. Wouldn't it be better just to steal a few humans, sequence our amino acids or whatever else is the source of our delectability, and then just synthesize the identical food product from scratch?" Ibid., 353.

60. Tyson, "Size and Scale of the Universe," 19.

61. Sagan, "UFOs: The Extraterrestrial and Other Hypotheses," 272.

By espousing a gnostic variant of the redeemer myth, Sagan promulgates a moral challenge: the best we can expect is an extraterrestrial teaching or warning that inspires us on Earth to earn our own terrestrial salvation. It is not by grace that we are saved, but rather by works. As a secular doctrine of salvation, the ETI myth looks much more like the myth of the gnostics than the promise of redemption found in the early Christian gospel.

Must it be the case that the aliens will come to Earth as our saviors? No. It is logically possible to believe that ETI have evolved into a utopian society while also believing that the aliens have no motive to share their pot of the rainbow's gold with us on Earth. Sociobiologist E. O. Wilson verbalizes this view. "Perhaps the extraterrestrials just grew up. Perhaps they found out that the immense problems of their evolving civilizations could not be solved by competition among religious faiths, or ideologies, or warrior nations. They discovered that great problems demand great solutions, rationally achieved by cooperation among whatever factions divided them . . . there was no need to colonize other star systems. It would be enough to settle down and explore the limitless possibilities for fulfillment on the home planet."[62] Although Wilson can admire the more advanced aliens in his imagination, he does not expect them to come to Earth and pay our bills.

But, again, even this line of evolutionary speculation is based on shaky science. Philosopher of Biology Michael Ruse bluntly blunts the assumption that evolution inevitably leads to progress in intelligence. "There is absolutely no guarantee of an upward progression on our hypothetical planet to intelligent life forms . . . Evolution of intelligence is not a necessary consequence of life appearing: not at all."[63] Despite such doubt, astrobiologists press onward, with or without hope for salvation from the sky.

Human destiny is something we feel we need to turn to the scientist to discern. In our modern culture, the scientist has replaced the priest as the dispenser of knowledge, even formerly religious knowledge. Langdon Gilkey describes the modern scientist in the white coat as the new priest, the go-to person for gnosis. "The myth of the new scientific or technological man presents to us the image of the man in the white coat; the man who embodies the gnosis achieved by the new methods of inquiry."[64] What we have here is a myth-colored lab coat parading as a clerical alb.

But, we ask, does the myth-colored lab coat fit the scientist? Or, is it a misfit? Primatologist Frans de Waal reminds us that science should remain science. "Science is not in the business of spelling out the meaning of life

62. Wilson, *Social Conquest of Earth*, 296–97.
63. Ruse, "Is Rape Wrong on Andromeda?" 50.
64. Gilkey, *Religion and the Scientific Future*, 79.

and even less in telling us how to live our lives."[65] Why do some of the scientists among us feel impelled to turn their research into a worldview, into an explanation of everything, into a philosophy of life? Why do they feel the need to practice theology without a license?

Does a Theologian Believe What the Myth Says?

When it comes to theological discernment, one must first ask the question: does myth count in theology? No. Most theologians are willing to interpret myths, but certainly not willing to believe them in their literal form. Myths tell us about human anxieties and propensities, to be sure; but they do not tell us about the reality of God. It is the task of the theologian to say: don't believe myth! Or, at least avoid believing myth with a high degree of literalism. This applies both to pre-scientific religious myths as well as scientized secular myths. Science has not demonstrated that it can save us from self-destruction, whether it be terrestrial or extraterrestrial science.

The theologian's epistemology begins with two recognitions: (1) we need some myth in our worldview just to tell ourselves a meaningful story about reality, and a myth that includes a gracious God can be more comprehensive in its meaning than other myths; and (2) we must avoid believing such myths literally because we know they are the construction of human imagination. The God in whom we believe transcends the mythical worldview and remains mysterious, even if we need to think about God from within the parameters of such a worldview. The true God transcends the myth, even if the myth points our imaginations in the direction of this God.

So we ask: Does the theologian work from within a mythologized model of reality or world view? Yes. "If a myth is defined as a story in which some aspect of the cosmic order is manifest," writes Ian G. Barbour, "then, the *scriptures* of Judaism, Christianity, and Islam must be said to include myths."[66] The symbolic language of myth functions for the theologian like the conceptual model does for the scientist: an enduring "imaginative tool for ordering experience" that looks like a description of the world, even if it is not literally referential.[67] A worldview or conceptual model is indispensible for meaningful thought, even if it is mythical in structure.

Now, we would expect a scientist to say: "I work with models. Not myths." Models are cognitive structures that organize data in a meaningful fashion. Yes, to be sure. So, to add a dose of precision we must admit:

65. De Waal, *Bonobo and the Atheist*, 19.
66. Barbour, *Myths, Models, and Paradigms*, 22.
67. Ibid., 6.

models and myths are not identical. Models are partial interpretations of myths. "One model may be common to many myths," observes Barbour. "A model is relatively static and lacks the imaginative richness and dramatic power which make a myth memorable . . . Models result from reflection on the living myths which communities transmit."[68] Or, to say it another way, myths tend toward narrative whereas models tend toward rational or even mathematical construction. Both myths and models function in virtually the same way, namely, they provide the conceptual set or canvas upon which further rational thought is painted. It would be too simple to say the scientist works from within a conceptual model and the theologian from within a myth, to be sure; but this is at least partially accurate. Both work from within both to greater or lesser degrees.

The myth the theologian likes best is the broken myth. A broken myth provides a narrative or a conceptual model of reality, but the break in the myth allows us to look beyond it while looking through it. The symbolic or multivalent language of the theologian operates at two levels simultaneously, one within the myth and the other in transcendental critique of the myth. This prevents the myth from validating or absolutizing our perspective within the myth. Gilkey contends that the *broken myth* "has the character of symbol qua symbol, and does not *on its own authority* legitimate or validate (though it may refer to) any empirical, natural or historical fact. Such [broken] myths seek to relate the events and sequences of historical passage to their own deepest source, to the ultimate norm that impinges upon their estranged actuality, and to their possibilities and grounds for hope, i.e., to God."[69] The theologian employs multivalent symbolic language in order to orient us to the reality that transcends the myth, namely, God.

Can the theologian's myth incorporate elements of scientific cosmology? Yes, indeed; just as long as the cosmology is not confused with the God of creation. To say it more forcefully, the theologian must explore the extant cosmology, even if this cosmology is slated for demythologizing, slated for becoming a broken myth. The factual knowledge of the scientist is important; but what it means to us and to our relationship with God is even more important. Sweden's archbishop Antje Jackelén asks: "*What does it mean?* What does it mean that there are stars and planets, water and land, flora and fauna? . . . A theological perspective on cosmology is more about present and future than about origins and beginnings . . ."[70] The method of demythologizing—breaking a myth—enables the theologian to distinguish

68. Ibid., 27.
69. Gilkey, *Reaping the Whirlwind*, 150.
70. Jackelén, "Cosmology and Theology," 138.

between the finite and the infinite, this world and the divine reality, between a natural future and a divinely graced future.

In addition to the limit placed on human knowledge—both religious and scientific knowledge—by finitude, is the limit placed by sin, by fallenness, by estrangement. Finitude and fallenness are not the same thing. Finitude simply draws a line between what we know and what is left yet to know, whereas sin reminds us that what we already know is subject to manipulation, distortion, and perversion. The important point here is that what happens in science cannot be taken at face value, not just because it is finite and incomplete, but because science like human life in general is fallen. Scientific knowledge is partial and subject not only to error, but also to distortion by sin. Lydia Jaeger warns us to beware. "The presence of sin therefore requires us to monitor closely the processes by which knowledge is constructed."[71] Focus and perspective as well as error in science can be influenced by, of all things, money and the greed that accompanies it.

Despite the marvels of the new knowledge gained and new technology produced, science has become subject to the funding of jingoists and the ambitions of militarists. Advances in scientific knowledge lead frequently to equal advances in the breadth and efficiency of murder, mayhem, and mass destruction. Each decade marks a new level of global terror due to advances in nuclear and biochemical weaponry. This spiral is beyond political control, religious control, moral control, and beyond self-control. If the ETI myth suggests that augmenting terrestrial science with extraterrestrial science will provide this control, the theologian must simply shrug and say: where is the evidence for such a belief?

The blind alley into which the myth—especially the belief in progress—leads us to what I call the *eschatological problem*.[72] All myths of progress propose that if we in our generation simply make the right choice that, with the advance of science, we in the human race can advance from warring destruction to a state of world peace. Yet, the theologian should ask: how do we get from here to there? Can a leopard change its spots so easily? If science got us into the present mess, how can we expect science to liberate us from this mess? If we have evolved to this point, why should we think that more evolving will save us? We would need to pose these questions on any exoplanet just as we need to here on Earth.

Salvific healing, according to the Christian theologian, comes from divine grace granted us within the setting of our fallen life on Earth. The cross and resurrection of Jesus Christ symbolize the presence and promise

---

71. Jaeger, *What the Heavens Declare*, 103.
72. See Peters, *Futures—Human and Divine*.

of this saving grace. In the cross we see God's identification with the victims of human violence. In the resurrection we see God's promise that we will not forever be locked into the spiral of violence. Unambiguous healing—even world peace—will come to us only as an eschatological transformation, as an act of God. More science will not save us. It is a delusion to think that it will. The theologian, like the rest of us, should welcome and even celebrate the triumphs of science; but these triumphs should not delude us into thinking that science will save us from our human propensity for self-destruction through war.

World religions scholar and philosopher Huston Smith prophesies against placing our faith in science to save us. Science, like all other human enterprises, is ambiguous. It can be just as destructive as palliative. And the doctrine of progress has misled Western civilization, lulling humanity into unreality regarding the human propensity for self-destruction. "Modernity went on to predict that technology would ensure unending progress. Endless progress through the technological application of continuous scientific discovery—this is what modernity's scenario comes down to. And because it was founded on an illusion (the illusion that the scientific method is omnicompetent) it was inevitable that sooner or later it would bump into reality—in this case, history. And it now has, with a vengeance. The twentieth century, the most barbaric in history, makes the myth of progress read like a cruel joke: 160 million human beings slaughtered by their own kind."[73]

A century ago, when the concept of evolution was becoming common cultural currency, Protestant preachers spoke out against trusting progress instead of God. The renowned founder of Riverside Church in New York, Harry Emerson Fosdick, was unflinching. "If ever a river ran out into a desert, the river of progressive hopes, fed only from springs of materialistic philosophy, has done so here. At least the Greeks had their immortality and the Hebrews their coming Kingdom of God, but a modern materialist, with all his talk of progress, has neither the one nor the other, nor anything to take their place as an ultimate for hope. Whatever else may be true, progress on a transient planet has not done away with the need of God and life eternal."[74]

In short, a theologian is not likely to believe in the myth of progress let alone in extraterrestrials who have advanced further than us in their evolutionary history. ETI believers appear to hold that " . . . the highly evolved extraterrestrial, not the preexistent God of traditional faith, will deliver us through advanced technology, extraordinary spiritual insight and a boundless desire to help. Like the idols of the Old Testament, themselves minor

---

73. Smith, *Soul of Christianity*, xvii.
74. Fosdick, *Christianity and Progress*, 36–37.

deities with a type of salvation to offer, the extraterrestrials of today often hold out a false spiritual hope."[75] Could extraterrestrials provide an excuse for idolatry? Perhaps that accusation is a bit severe. Nevertheless, without any empirical evidence whatsoever, it would be foolish for us to trust our constructed image of celestial beings coming to Earth from other planets to save us from ourselves.

Actually, hope for a better future is a healthy thing. What is unhealthy is believing that evolution or progress is the way to fulfill that hope. The theologian will affirm hope, because God has invited us to hope. More: God has invited us into a life of faith, hope, and love. To place one's faith in evolution or progress or, worse, both, is to place one's faith in a delusion. Huston Smith minces no words. "Progress is an illusion; not only future progress but past progress as well ... Utopia is a dream, evolution a myth."[76]

Conclusion

Our judgment here is not to say whether the ETI myth is literally true or false. It may be true. But if it turns out to be true, it will be based on a lucky guess, not on empirical evidence or on a realistic interpretation of creaturely nature. The cautious theologian will remind us that salvation is a gift of divine grace, not a product of natural or even human progress. God's grace can act in, with, and under natural processes, to be sure. But the ETI myth relies on a variant of naturalism, a willingness to accept redemptive grace but to attribute this redemption to science rather than to our creator God. Even the most naive reading of recent human history will document the foolishness of such misplaced trust.

Today's astrotheologian asks that the astrobiologist come clean regarding what counts as empirical knowledge and what is added from the myth of evolutionary progress. Good science is something an astrotheologian can celebrate. Science riddled with myth and secularized religion is less than the best science. It is subject to demythicizing, to myth-busting, by the astrotheologian if the astrobiologist does not do it voluntarily.

Bibliography

Ayala, Francisco J. "Darwin and the Teleology of Nature." In *Science and Religion in Search of Cosmic Purpose*, edited by John F. Haught, 18–41. Washington, DC: Georgetown University Press, 2000.

75. Herrick, *Scientific Mythologies*, 72.
76. Smith, *Forgotten Truth*, 121.

———. "The Evolution of Life on Earth and the Uniqueness of Humankind." In *Perché esiste qualcosa invece di nulla? (Why There Is Something Rather than Nothing?)*, edited by S. Moriggi and E. Sindoni, 57–77. Castel Bolognese: ITACAlibri, 2004.
Barbour, Ian G. *Myths, Models, and Paradigms: A Comparative Study in Science and Religion*. New York: Harper, 1974.
Beck, Lewis White. "Extraterrestrial Intelligent Life." In *Extraterrestrials: Science and Alien Intelligence*, edited by Edward Regis Jr., 3–18. Cambridge: Cambridge University Press, 1985.
Billingham, John. "Who Said What: A Summary and Eleven Conclusions." In *SETI*, 33–40.
Billings, Linda. "The Allure of Alien Life: Public and Media Framings of Extraterrestrial Life." In *LBE*, 308–23.
Bostrom, Nick. "Where Are They? Why I Hope the Search for Extraterrestrial Life Finds Nothing." *MIT Technology Review* (May/June 2008) 72–77. http://www.nickbostrom.com.
Chela-Flores, Julian. *The Science of Astrobiology: A Personal View on Learning to Read the Book of Life*. Heidelberg: Springer, 2011.
Conway Morris, Simon. "If the Evolution of Intelligence is Inevitable, Then What Are the Metaphysical Consequences?" In *The Science and Religion Dialogue: Past and Future*, edited by Michael Welker, 217–32. Frankfurt: Lang, 2014.
———. *Life's Solution: Inevitable Humans in a Lonely Universe*. Cambridge: Cambridge University Press, 2003.
———. "Not So Alien." *SETI Institute News* 12.1 (2003) 10–11.
———. *The Runes of Evolution: How the Universe Became Self-Aware*. West Conshohocken, PA: Templeton, 2015.
Crichton, Michael. "Aliens Cause Global Warming: A Caltech Lecture." January 17, 2003. http://www.s8int.com/crichton.html.
Crowe, Michael J. *Extraterrestrial Life Debate from Antiquity to 1915: A Source Book*. Notre Dame: University of Notre Dame Press, 2008.
Davies, Paul C. W. "E.T. and God." *The Atlantic Monthly* (September 2003) 114–15. http://www.theatlantic.com/issues/2003/davies.htm.
———. *Space and Time in the Modern Universe*. Cambridge: Cambridge University Press, 1977.
Dawkins, Richard. *The God Delusion*. Boston: Houghton Mifflin, 2006.
de Duve, Christian. *Vital Dust: The Origin and Evolution of Life on Earth*. New York: Basic Books, 1995.
de Waal, Frans. *The Bonobo and the Atheist*. New York: Norton, 2013.
Drake, Frank. "Foreword." In *Drake*, xvi.
———. Interviewed by Diane Richards. "Interview with Dr. Frank Drake." *SETI Institute News* 12.1 (2003) 5.
Drake, Frank, and Dava Sobel. *Is Anyone Out There? The Scientific Search for Extraterrestrial Intelligence*. New York: Pocket, 1994.
Foerst, Anne. "Stories We Tell: The *Mythos-Logos* Dialectic as New Method for the Dialogue between Religion and Science." In *God, Life, Intelligence and the Universe*, edited by Terence J. Kelly and Hilary D. Regan, 1–35. Adelaide: Australian Theological Forum, 2002.
Fosdick, Harry Emerson. *Christianity and Progress*. New York: Revell, 1922.

Gilkey, Langdon. *Reaping the Whirlwind: A Christian Interpretation of History.* New York: Seabury Crossroad, 1976.

———. *Religion and the Scientific Future: Reflections on Myth, Science and Theology.* Macon, GA: Mercer University Press, 1967.

Gingerich, Owen. "Is there a Role for Natural Theology Today?" In *Science and Theology: Questions at the Interface,* edited by Murray Rae et al., 29–48. Grand Rapids: Eerdmans, 1994.

Harrison, Albert A. *After Contact: The Human Response to Extraterrestrial Life.* New York: Plenum, 1997.

———. *Starstruck: Cosmic Visions in Science, Religion, and Folklore.* New York: Bergbahn, 2007.

Hart, John. *Encountering ETI: Aliens in Avatar and the Americas.* Eugene, OR: Cascade, 2014.

Herrick, James A. *Scientific Mythologies: How Science and Science Fiction Forge New Religious Beliefs.* Downers Grove, IL: IVP Academic, 2008.

Hoyle, Fred. *Monthly Notices of the Royal Astronomical Society* 109:365 (1949). Cited in *Alone in the Universe?* by David Wilkinson. Crowborough, UK: Monarch, 1997.

Impey, Chris. *The Living Cosmos: Our Search for Life in the Universe.* New York: Random House, 2007.

Jackelén, Antje. "Cosmology and Theology." In *RCRS,* 135–44.

Jaeger, Lydia. *What the Heavens Declare: Science in the Light of Creation.* Translated by Jonathan Vaughan. Eugene, OR: Cascade Books, 2012.

Jaspers, Karl. *The Origin and Goal of History.* Translated by Michael Bullock. London: Routledge, 1953, 2010.

Kaku, Michio. *The Future of the Mind.* New York: Doubleday, 2014.

Lemarchand, Guillermo A. "Speculations on First Contact." In *SETI,* 153–64.

Marty, Martin E. "Non-Christian Eschatologies." In *Christian Hope and the Future of Humanity,* edited by Franklin Sherman. Minneapolis: Augsburg, 1969.

Mayr, Ernst. "Can SETI Succeed? Not Likely." http://www.planetary.org/explore/topics/search_for_life/seti/seti_debate.html.

———. "The Probability of Extraterrestrial Intelligent Life." In *Extraterrestrials: Science and Alien Intelligence,* edited by Edward Regis Jr., 23–30. Cambridge: Cambridge University Press, 1985.

Michaud, Michael A. G. *Contact with Alien Civilizations: Our Hopes and Fears about Encountering Extraterrestrials.* New York: Springer, 2007.

Peters, Ted. *Futures—Human and Divine.* Louisville: Westminster John Knox, 1977.

Regis, Edward, Jr. "SETI Debunked." In *Extraterrestrials: Science and Alien Intelligence,* edited by Edward Regis Jr., 231–44. Cambridge: Cambridge University Press, 1985.

Ricoeur, Paul. "Evil." In *ER2,* 5:2897–2904.

———. "Myth and History." In *ER2,* 9:6371–80.

Ruse, Michael. "Is Rape Wrong on Andromeda? An Introduction to Extraterrestrial Evolution, Science, and Morality." In *Extraterrestrials: Science and Alien Intelligence,* edited by Edward Regis Jr., 43–78. Cambridge: Cambridge University Press, 1985.

———. *Monad to Man: The Concept of Progress in Evolutionary Biology.* Cambridge: Harvard University Press, 1996.

Sagan, Carl. *Contact.* New York: Pocket Books, 1985.

———. *The Cosmic Connection: An Extraterrestrial Perspective.* New York: Doubleday, 1973.

———. *Pale Blue Dot: A Vision of the Human Future in Space.* New York: Random House, 1994.

———. "UFOs: The Extraterrestrial and Other Hypotheses." In *UFO's: A Scientific Debate*, edited by Carl Sagan and Thornton Page. Ithaca, NY: Cornell University Press, 1972.

Sagan, Carl, and Frank Drake. "The Search for Extraterrestrial Intelligence." *Scientific American* (May 1975) 80–89.

Sanford, John A. *The Kingdom Within: The Inner Meaning of Jesus' Sayings.* New York: Harper, 1970.

Smith, Huston. *Forgotten Truth: The Primordial Tradition.* New York: Harper, 1977.

———. *The Soul of Christianity.* San Francisco: Harper, 2005.

Smith, John Maynard. "Taking a Chance on Evolution." *New York Review of Books* (May 14, 1992) 34–36.

Tarter, Jill Cornell. "Implications of Contact with ETI Far Older than Humankind." In *SETI*, 45–46.

Traphagan, John W. "Is Active SETI Really Dangerous?" *METI International* Blog (2016). http://meti.org/blog/active-seti-really-dangerous.

Tyson, Neil deGrasse. "The Size and Scale of the Universe." In *WU*, 17–25.

Vakoch, Douglas A. "Asymmetry in Active SETI: A Case for Transmissions from Earth." *Acta Astronautica* 68 (2011) 476–88.

Wilkinson, David. *Science, Religion, and the Search for Extraterrestrial Intelligence.* Oxford: Oxford University Press, 2013.

Wilson, Edward O. *The Social Conquest of Earth.* New York: Norton, 2012.

PART 5
Astroethics and Space Policy

21

Astroethics and the Terraforming of Mars

CHRISTOPHER MCKAY

The laws of nature tell us *how* the universe behaves, but they don't answer the *why?* questions.[1]

—STEPHEN HAWKING

The natural sciences deal with the concrete shaping of the world and theology accompanies this venture in a reflective manner and connects it with the origin, meaning, and goal of this world.[2]

—HANS SCHWARZ

Here is my astroethical premise: *the long-term goal for astrobiology is the enhancement of the richness and diversity of life in the universe*. In the case of Mars, then, it follows that we on Earth have an opportunity if not a moral obligation to enhance life on Mars. By any measure, Mars today does not have a flourishing biosphere. Might this enhancement imperative warrant terraforming the Red Planet?

---

1. Hawking and Mlodinow, *Grand Design*, 171, italics original.
2. Schwarz, *Vying for Truth*, 11.

Why Mars?

Among the other worlds of our Solar System, Mars is of particular interest because it once had liquid water flowing on its surface and presumably a thicker, warmer, atmosphere. This period of Earth-like conditions on Mars was contemporaneous with the first evidence of life on Earth, about 3.5 billion years ago.[3] The search for past life on Mars and consideration of how to alter Mars to enhance future prospects for life have raised two novel, ethical questions that are without precedent in Earth-based ethical discussions. The first question concerns the ethical status of Martian life if it is the result of an independent origin—a second genesis.[4] The second question concerns the choice between preserving a planet that is essentially devoid of life and altering the environment on that planet to allow life to flourish there.[5]

In this chapter, I will review these novel ethical questions that emerge from the Mars exploration program. In my opinion, the solutions to these questions are best-obtained under the premise that *the long-term goal for astrobiology is the enhancement of the richness and diversity of life in the universe.*[6] A perspective that seeks to enhance life in the Universe places intrinsic value on a second genesis, even if the life forms are microscopic. This perspective would encourage altering Mars to produce conditions that allow widespread life on that planet, using the indigenous life, or if Mars lacks indigenous life, then sharing life from Earth.[7]

Mars: A Biological Past

As the global dust storm settled in early 1972, the Mariner 9 spacecraft orbiting Mars discovered the presence of dry river channels on the planet, indicating that water had once flown on its surface. A recent high-resolution image of a Martian dry riverbed is shown in Figure 1. Currently, the average temperature on Mars is -60°C and the average atmospheric pressure is less than 1 percent of the sea level pressure on Earth. The main components of

---

3. Carr, "Water on Mars," 30–35.

4. A second genesis would be life forms not on the "tree of life," see: McKay, "Search for a Second Genesis," 269–77.

5. The question of the choice between Mars as it is and Mars with life is first discussed in McKay, "Does Mars Have Rights?" 184–97.

6. The statement of this principle for astrobiology is from McKay, "Astrobiology and Society," 158–66. This is elaborated in Randolph and McKay, "Protecting and Expanding," 28–34.

7. McKay, "Planetary Ecosynthesis on Mars," 255.

the Martian atmosphere are carbon dioxide, nitrogen, and argon.[8] These conditions are non-conducive to the presence of stable liquid water. However, evidence that water once flowed in a river on the surface of Mars suggests that the atmosphere must have been substantially warmer and thicker in the past.

The dry river features on Mars are primarily found in the heavily cratered areas of the planet, indicating that they formed early in Martian history—3.5 to 4.0 billion years ago. During this early time, Mars likely had a thick, warm atmosphere, with water flowing on the surface. Presumably, the atmosphere was composed of carbon dioxide and nitrogen—similar in composition to the present Martian atmosphere. These two gases, carbon dioxide and nitrogen, are also thought to have been the predominant gases in Earth's early atmosphere. The earliest firm evidence for life on Earth is found in rocks that are 3.5 billion years old.[9] Taken together, these observations imply that Mars and Earth had similar surface environments early in their history, and the period of similarity occurred at the time when life first appeared on Earth. This compelling story motivates the search for life on Mars.

The search for life on Mars is based on our understanding of life on Earth. There are two key points about life on Earth in this respect. First, there is only one example of life on Earth—all life on Earth shares a basic biochemical unity and a common genetic endowment from a common ancestor. Second, all life on Earth requires liquid water to grow or reproduce. If life on Mars is related to life on Earth, it may have originated in one location and mobilized to the second through the transport of spores between the two worlds via exchange of meteorites.[10]

---

8. Kieffer et al., "Planet Mars," 1–33.
9. Knoll, *Life on a Young Planet*.
10. Melosh, "Rocky Road," 687–88.

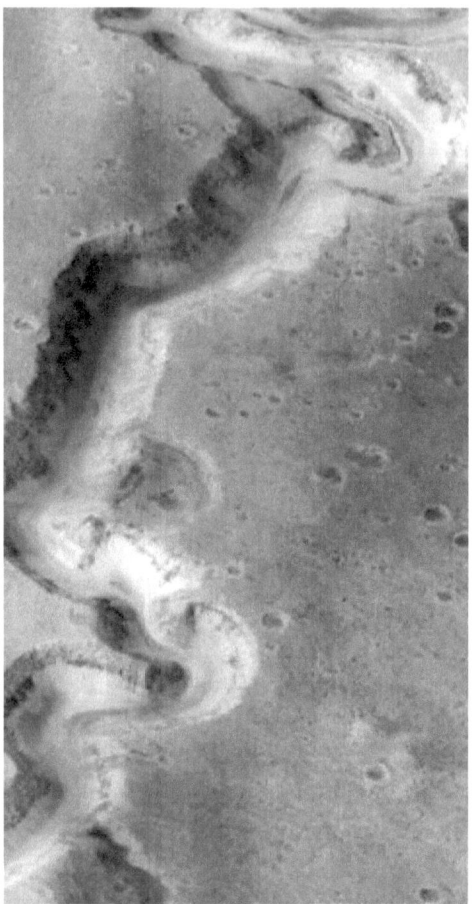

Figure 21.1. Picture from orbit of a dry riverbed on Mars. This is Nanedi Vallis in the Xanthe Terra region of Mars. The image covers an area 9.8 km by 18.5 km; the canyon is about 2.5 km wide. (NASA photo, https://www.giss.nasa.gov/research/briefs/gornitz_03/)

Alternatively, and regardless of the level of similarity between life forms, life on Mars may represent an independent origin of life—a second genesis. The discovery of a second genesis of life on Mars would have profound scientific and philosophical implications. Scientifically, it would provide a second example of biochemistry and of evolutionary history. Many important biological questions could be answered through the comparison of biochemistry between the two life forms. Philosophically, the discovery of a second genesis of life in our Solar System would be a compelling indication that the phenomenon of life is widespread and numerous in the Universe. We would be confident that we are not alone.

Mars: A Biological Future

The evidence that Mars had an Earth-like environment early in its history is the motivation behind the search for evidence of life on that planet. It is also the motivation behind suggestions to restore Mars to habitability, re-creating conditions suitable for a global biosphere on that planet—known variously as terraforming or planetary ecosynthesis.[11] There are significant uncertainties regarding the feasibility of terraforming Mars. The main question is the availability and quantity of the key materials that compose a biosphere, namely carbon dioxide, water, and nitrogen.[12] The low level of nitrogen in the Martian atmosphere has been previously identified as the potentially limiting material,[13] but the recent discovery of nitrates in the soil on Mars[14] indicates that there is an alternative, and larger, source. Mars appears to have all the ingredients necessary to recreate a habitable global environment.

If all the key materials needed for a biosphere are present on Mars, then the first step in terraforming Mars is to warm the planet. Lovelock and Allaby suggested that warming Mars could be accomplished the same way humans have been successfully warming the Earth: that is, by the release of greenhouse gases.[15] Detailed analysis shows that using perflourocarbons produced on the surface of Mars is a feasible method to warm the planet. Energy balance calculations indicate that if production of greenhouse gases on Mars was comparable to their present production on Earth, Mars could be warmed in one hundred years[16] resulting in a thick carbon dioxide and nitrogen atmosphere—assuming enough carbon dioxide and nitrate on Mars. Transforming the carbon dioxide atmosphere into an Earth-like atmosphere with breathable levels of oxygen would be achieved using plants,

11. Averner and MacElroy, eds., *On the Habitability of Mars*, 414.

12. An inventory of amount of the key elements needed is in McKay et al., "Making Mars Habitable," 489–96.

13. Nitrogen was identified as a key resource because the level in the atmosphere is inadequate to support a biosphere. See discussion in ibid.

14. Stern et al., "Evidence for Indigenous Nitrogen," 4245–50.

15. Lovelock and Allaby, *Greening of Mars*.

16. A timescale of warming Mars is computed from simple energy balance and assumed efficiency of heating the planet of 10 percent as discussed in McKay et al., "Making Mars Habitable." See relevant discussion in Marinova et al., "Radiative-convective Model," 110:E3.

and calculations suggest this would take hundreds of thousands of years.[17] Should we engage in intentional planetary eco-synthesis?

The Ethics Surrounding a Second Genesis of Life

The search for a second genesis of life on Mars poses new questions in ethics. One question is: what ethical consideration is due to an alien life form when that life is distinctly different from Earth life and the members of that life are no more advanced than microorganisms? The question of ethical regard for aliens has been asked before, but only in the context of intelligent, or at least complex, alien life. The rational nature of such organisms forms the basis for their ethical regard. Earthly microorganisms receive little, if any, ethical regard. The usual metrics for determining the ethical standing of non-human life include such considerations as the ability to experience pain, complex behavior or social interactions, and communication. On the list of non-human organisms, microorganisms score at the bottom in all these attributes. This, coupled with their virtual invisibility to the unaided human eye and their robustness in surviving even extreme human actions, relegates them to minimal ethical consideration. There would likely be very little outcry against a human effort to eradicate the *Giardia spp.*, the parasitic, single-celled protozoans responsible for the disease giardia.

There is an alternative way to think through this issue, which is predicated on the general agreement that at least two distinct sets of creatures deserve ethical consideration. Humans compose the first set, while life in general comprises the second. Within the first set, humans, individual members are generally granted natural and equal ethical consideration—at least in principle. Within the second set, there is a gradient of ethical consideration, with non-human primates at the upper end of the scale and pathogenic microorganisms at the lower end of the scale. While at the bottom of the ethical scale microorganisms are still within the set "life" and there is a common, if intuitive, sense that life deserves ethical consideration distinct from non-life. Indeed, Cockell has explicitly argued for ethical consideration of microbial ecosystems.[18]

The discovery of a second genesis of life on Mars comprised of microorganisms would add a completely new set of life for ethical consideration. We might call this set "life 2.0," as it compares to the two sets described

---

17. A timescale of producing oxygen on Mars is computed from simple energy balance and assumed efficiency of plant production of 0.01 percent as discussed in McKay et al., "Making Mars Habitable."

18. Cockell, "Rights of Microbes," 141–50.

above, "humans," and "life 1.0." There are several reasons for humans to have high ethical regard for microorganisms comprising "life 2.0." The first reason is the utilitarian value of a second type of life. Studies of a second type of biochemical and genetic system could yield unique and valuable insights into the nature of biochemistry in general, with important applications for agriculture, medicine, and biotechnology. Secondly, a second genesis contributes in a fundamental way to the richness and diversity of life in the Universe—a goal we should endorse as discussed below. And finally, perhaps speculatively, a lifeform that is purely microscopic may develop, over deep time, and produce large intelligent and rational life forms. We should note that for larger part of Earth's history our ancestors were microscopic.

Mars may, or may not, have a second genesis of life. We will not know that answer for some time. Meanwhile, we have sent viable Earth microorganisms to Mars in all spacecraft since Viking. They remain in a dormant form due to the lack of water and low temperatures. Any Earthly organisms released into the Martian environment have been killed by the strong biocidal ultraviolet light that reaches the surface of Mars through its thin atmosphere, but any sheltered by the spacecraft remain viable, albeit dormant. I have previously argued that the robotic and human exploration of Mars should be done in a way that is biologically reversible.[19] We must be able to undo ("ctrl z") our contamination of Mars if we discover a second genesis of life there and decide to enhance conditions to allow for the growth of that native life. So far our contamination on Mars has all been biologically reversible. Human exploration can also be done in a way that is biologically reversible. A decision to enhance conditions on Mars in support of an indigenous Martian biology would be an example of an action rooted in a commitment to life as an intrinsic value.

The Ethics of Terraforming Mars

The second novel question in ethics on Mars relates to terraforming. The first scientific definition of terraforming was: the alteration of "the environment of a planet in order to improve the chances of survival of an indigenous biology or to allow habitation by most, if not all, terrestrial life forms."[20] If defined as thus, terraforming makes several assumptions. The first assumption is that a world with widespread life is of more intrinsic value than a world with beautiful, albeit lifeless, landscapes. Second, terraforming assumes that

19. McKay, "Biologically Reversible Exploration," 718; and McKay, "Hard Life for Microbes," 30.
20. McKay, "Terraforming Mars," 427–33.

Mars today does not have a global biosphere, and that any biota present on Mars is at risk of extinction. Thirdly, terraforming assumes that humans have the capacity to determine the nature of a second genesis on Mars and determine under what environmental conditions that life form would flourish into a global biosphere. Put more simply, terraforming assumes that life is better than no life, that global life is better than limited life, that Mars does not have global life currently, and that humans can determine how to alter Mars so that it can have global life in the future.

The last two points, the lack of a global biosphere on Mars and the determination of what environment would allow life to flourish on Mars, are scientific issues that can be easily solved through studies on the biology of Mars. We can anticipate the answers to these questions based on what we know about life on Earth and data we have already collected from Mars. Life on Earth is widespread, and many ecologists have argued that a flourishing biosphere dominates the global cycles of the light elements on a planet. By those criteria, Mars clearly does not have a flourishing biosphere. If there is, or was, life on Mars, the environmental conditions that allow that life to flourish are likely to be, as on Earth, the presence of liquid water. The basis for this assertion is the assumption that any life on Mars started early in Martian history when liquid water was widespread on the surface. Having been born in liquid water, it is likely that a second genesis of life on Mars will require widespread liquid water to flourish.

One major question regarding terraforming relates to the value of worlds with life compared to worlds with lifeless landscapes. Following the arguments of Deep Ecology, one can hold that life on Mars has intrinsic value and that, while not the only possible source, life is perhaps the greatest source of value.[21] Some natural landscapes and planetary features (such as the rings of Saturn) may have intrinsic value as well, due to aesthetic or cultural aspects, which warrant their preservation. However, life is unique in that its value can be not only preserved, but it can be increased as well. If life has value, then humans can create value and spread value as they create and spread life. Enhancing the richness and diversity of life in the Universe provides a meaningful role for humans as active creators, or co-creators, helping shape and guide the existing creation. It gives us something to do that is worth doing.

---

21. McKay, "Astrobiology and Society."

Conclusion: The Long-term View of Life in the Universe

The biological exploration of Mars's past and future will be our first encounter with these new ethical issues that emerge as we expand our range of activities from Earth out into space. The precedent we set with Mars will likely shape our future elsewhere in space. We should consider carefully how we search for life on Mars, and how we treat a second genesis of life if we encounter it.

Mars also provides a test to resolve the value of life in comparison to the value of natural landscapes. A useful and direct approach to addressing these issues begins with a fundamental principle for ethical treatment of life on Earth and beyond. We have proposed such a principle: *the long-term goal for astrobiology is the enhancement of the richness and diversity of life in the universe.*

Bibliography

Arnould, Jacques. *Icarus' Second Chance: The Basis and Perspectives of Space Ethics.* New York: Springer, 2011.
Averner, M. M., and R. D. MacElroy, eds. *On the Habitability of Mars: An Approach to Planetary Ecosynthesis.* NASA Special Publication 414. Washington, DC: Scientific and Technical Information Office, National Aeronautics and Space Administration, 1976.
Carr, M.H. "Water on Mars." *Nature* 326:6108 (1987) 30–35.
Cockell, Charles S. "The Rights of Microbes." *Interdisciplinary Science Reviews* 29:2 (2004) 141–50.
Hawking, Stephen, and Leonard Mlodinow. *The Grand Design.* New York: Bantam, 2010.
Kieffer, Hugh H., et al. "The Planet Mars: From Antiquity to the Present." In *Mars*, edited by Hugh H. Kieffer et. al, 1–33. Tucson: University of Arizona Press, 1992.
Knoll, Andrew H. *Life on a Young Planet: The First Three Billion Years of Evolution on Earth.* Princeton, NJ: Princeton University Press, 2003.
Lovelock, James, and Michael Allaby. *The Greening of Mars.* New York: Warner, 1985.
Marinova, Margarita M., et al. "Radiative-convective Model of Warming Mars with Artificial Greenhouse Gases." *Journal of Geophysical Research: Planets* (1991–2012) 110:E3 (March 10, 2005).
McKay, Christopher P. "Astrobiology and Society: The Long View." In *ELU* 158–66.
———. "Biologically Reversible Exploration." *Science* 323:5915 (2009) 718.
———. "Does Mars Have Rights? An Approach to the Environmental Ethics of Planetary Engineering." In *Moral Expertise*, edited by Don MacNiven, 184–97. London and New York: Routledge, 1990.
———. "Hard Life for Microbes and Humans on the Red Planet," *Ad Astra* 19:3 (2009) 30.
———. "Planetary Ecosynthesis on Mars: Restoration Ecology and Environmental Ethics." In *Exp*, 245–60.

———. "The Search for a Second Genesis of Life in our Solar System." In *First Steps in the Origin of Life in the Universe*, edited by Julian Chela-Flores and Tobias Owen, 269–77. Heidelberg: Springer, 2001.

———. "Terraforming Mars." *Journal of the British Interplanetary Society* 35 (1982) 427–33.

McKay, Christopher P., et al. "Making Mars Habitable." *Nature* 352:6335 (1991) 489–96.

Melosh, H. J. "The Rocky Road to Panspermia." *Nature* 332 (1988) 687–88.

Randolph, Richard O., and Christopher P. McKay. "Protecting and Expanding the Richness and Diversity of Life, An Ethic for Astrobiology Research and Space Exploration." *International Journal of Astrobiology* 13.1 (2014) 28–34.

Schwarz, Hans. *Vying for Truth—Theology and the Natural Sciences: From the 17th Century to the Present*. Bristol, CT: Vandenhoeck & Ruprecht, 2014.

Stern, Jennifer C., et al. "Evidence for Indigenous Nitrogen in Sedimentary and Aeolian Deposits from the Curiosity Rover Investigations at Gale Crater, Mars." *Proceedings of the National Academy of Sciences* 112:14 (2015) 4245–50.

22

Astroethics and Microbial Life in the Solar Ghetto

TED PETERS

> Space constitutes one of the domains on which the human imagination feeds and expresses itself. Our fascination with the sky is such that the cultures of all countries and of all eras have not only placed their gods, their paradises and their origins there but also meaning, the destination of their desires and their dreams.
>
> —JACQUES ARNOULD[1]

> The sacramental *communion*—God incarnate in us and among us as human communities and as a planetary or even cosmic community of life—is a locus of moral power.
>
> —CYNTHIA MOE-LOBEDA[2]

Astrobiology and astrotheology need *astroethics*. Astrobiologists, according to SETI's Margaret Race, must "acknowledge that science and technology are embedded inseparably in societal and cultural contexts."[3] Like a duck in a pond, space science swims in a globalized culture concerned about planetary security, human survival, and international justice. It is time for

1. Arnould, *Icarus*, 36.
2. Moe-Lobeda, *Healing a Broken World*, 104.
3. Race, "Space Exploration and Searches for Extraterrestrial Life," 154.

space ethics or astroethics to inform, guide, and direct terrestrial science, technology, and perhaps even commerce when we humans go to space.

"The job of ethics is to evaluate issues of right and wrong, or good and bad, directing our focus to normative questions of value," contend space philosophers Carol Cleland and Elspeth Wilson.[4] To date, curiously, normative questions have seldom been raised to engage the flood of issues rising out of our increased capacity for becoming present in off-Earth locations. To date, the only concern to be given thorough ethical and public policy attention has been planetary protection, that is, protecting Earth from alien contamination while giving some consideration toward protecting off-Earth habitats from terrestrial contamination. Beyond planetary protection, the field of astroethics is fertile, but still waiting for plowing.

The field of astroethics must be divided into two sub-fields, ethics dealing with extraterrestrial intelligent life (ETIL or just ETI) and ethics dealing with extraterrestrial non-intelligent life (ETNL). ETNL is usually referred to as microbial life, occasionally even "stupid" life. We might actually have to admit that even stupid life is intelligent, even if relatively less intelligent.[5] Be that as it may, ETNL refers to microbial life.

If we make contact with the ETIL, it will most likely be associated with extrasolar planets in the habitable zone. Intelligent life, if it exists, will be found on the 8.8 billion Earth-like planets in the larger Milky Way metropolis. If we make contact with the microbial life, it will most likely be within our own solar ghetto on Mars or a moon of Saturn or a similar location. In this chapter, we will restrict our attention to the search for microbial life within our solar ghetto.

This and the next chapter take up the fourth of our astrotheological tasks: *theologians and religious intellectuals should cooperate with leaders of multiple religious traditions and scientists to address ethical issues associated with space exploration and to prepare the public for the eventuality of extraterrestrial contact.* We will begin with an inventory, a simple list of ethical issues regarding space exploration within our solar ghetto pressing for deliberation in our global society. Here is our list: (1) Does Planetary Protection Apply to Earth Alone? (2) Does Extraterrestrial Life Have Intrinsic Value? (3) Should Space Explorers Invoke the Precautionary Principle? (4) Should We Clean Up Our Space Junk? (5) What Should We Do About Satellite Surveillance? (6) Should Nations Weaponize Space? (7) Who Gets Priority: Scientific Research or Making a Profit? (8) Should Earthlings Terraform

---

4. Cleland and Wilson, "Lessons from Earth," 29.

5. Elsewhere I raise the hypothesis that all life, regardless of how simple, exhibits some characteristics of intelligence. "Where There's Life There's Intelligence."

Mars? (9) Should Earthlings Colonize Mars? (10) How Should We Protect Earth from Extraterrestrial Threats? (11) Does AstroEthics Require a Single Planetary Community of Moral Deliberation? (12) Should the Common Good include the Galactic Commons? It might be too early to resolve each of these issues, but simply formulating them for moral deliberation should provide a service for public policy makers.

1. Does Planetary Protection Apply to Earth Alone?

The first on our list of ethical issues arising from the search for ETNL is *planetary protection* (PP), which will receive most of our attention because it affects all the other issues on our list. PP raises a concern to protect ecosystems from contamination by alien life forms that may be destructive. The risk of contamination goes in two directions, forward and backward. The possibility of *forward contamination* alerts us to the risk of disturbing an already existing ecosphere; the introduction of Earth's microbes carried by our spacecraft or equipment could be deleterious to an existing habitable environment. *Back contamination* would occur if a returning spacecraft brings rocks or soil samples that contain life forms not easily integrated into our terrestrial habitat. A quarantine program will be required to determine the safety of newly introduced ETNL.

Article IX of the 1967 UN Outer Space Treaty states that "parties to the Treaty shall pursue studies of outer space including the Moon and other celestial bodies, and conduct exploration of them so as to avoid their harmful contamination and also adverse changes in the environment of the Earth resulting from the introduction of extraterrestrial matter and, where necessary, shall adopt appropriate measures for this purpose . . ." The PP principle has been guiding us since 1967.

In practice, prevention of backward contamination trumps protection against forward contamination. NASA scientists responsible for planetary protection, John Rummel and Catharine Conley, recommend that planetary protection be incorporated from the earliest stages of mission planning and development, to ensure proper implementation. As we have just said, issues involve both forward contamination, the contamination of other solar system bodies by Earth microbes and organic materials, and backward contamination, the contamination of Earth systems, including astronauts, by biological hazards or potential alien life. PP applies to other off-Earth bodies as well as Planet Earth. Even so, scientists are aware that some degree of forward contamination associated with human astronaut explorers is unavoidable. Despite this unavoidability, PP ethicists argue that when humans are exploring space the principles and policies of planetary protection, developed by COSPAR (UN Committee on Space Research) in accordance with the 1967 Outer Space Treaty, still apply. Although forward contamination is a matter of concern, some forward contamination is permissible. What is not permissible is backward contamination. Preventing harmful contamination of the Earth must be of the "highest priority" for all missions.[6]

The term, *planetary protection*, applies to Earth. Should it apply equally to other off-Earth bodies? With tacit obeisance to the galactic commons, some planetary protectionists assume we on Earth are responsible for protecting ecosystems on other planets from the destructive influence of human visitation. It is easy to see why we might wish to protect Earth. It's a matter of survival. Survival has a way of trumping all other moral priorities. We cannot act responsibly if we are dead. At least, this is the way many among us think. So, giving priority to protecting Earth from an invasion by a dangerous microbe seems to be a most reasonable stand to take. But, we might ask: why are we morally responsible for what happens on other planets? This leads us to the next on our list of issues, the question regarding the intrinsic value of off-Earth life.

As one might expect, some scientists will complain that rules and regulations will constrict wildcat exploration. "Mars will be just fine on its own . . . and the stringent safeguards now in place discourage scientists from exploring the Red Planet,"[7] writes Nathan Collins. Yet, Mars has its planetary protectors. Catharine Conley, NASA's planetary protection officer rises to Mars's defense: "If you want to study life elsewhere, you have to make sure not to bring Earth materials along."[8] On behalf of the Red Planet, we say: thank you NASA.

For a point of departure for thinking through this and other ethical issues on our list, I rely in part on the work of Margaret Race and Richard Randolph. Race and Randolph have proposed four principles for developing an ethical scheme appropriate to the discovery of non-intelligent life in our solar system: (a) cause no harm to Earth, its life, or its diverse ecosystems; (b) respect the extraterrestrial ecosystem and do not substantively or irreparably alter it (or its evolutionary trajectory); (c) follow proper scientific procedures with honesty and integrity during all phases of exploration; and (d) ensure international participation by all interested parties.[9] What I like

6. Conley and Rummel, "Planetary Protection, 792–97; See Conley, "Life Guard," 30.
7. Collins, "Stop Pampering the Red Planet," 24.
8. Conley, cited by ibid.
9. Race, "Societal and Ethical Concerns," 495.

about the Race and Randolph proposal is that it makes a commitment to planetary protection; to treating off-Earth ecosystems has having intrinsic value; to demanding scientific integrity; and to emphasizing that all peoples on Earth belong to a single community of moral deliberation. In this spirit I will turn to a number of issues arising from the prospect that our space explorers will find the gold: stupid but beloved life elsewhere in our universe.

2. Does Life beyond Earth Have Intrinsic Value?

Star Trek introduced the prime directive. Michael Gilmore applies it to astroethics. "The prime directive restricts human beings from interfering with any extraterrestrial life that is less developed than they are."[10] Might this presuppose that extraterrestrial microbial life has intrinsic value?

Richard Randolph now in tandem with Chris McKay "believe that new operational policies for space exploration and astrobiology research must be developed within an ethical framework that values sustaining and expanding the richness and diversity."[11] This leads us to ask: does life have intrinsic value?[12] Or, does the value of living organisms depend on the usefulness they have for us? Is worth inherent or instrumental? Do we terrestrial *Homo sapiens* have a responsibility toward extraterrestrial life based upon that life's intrinsic worth or based upon its usefulness to us? Almost no one to date has risen up to defend a brute instrumentalism. The predominant discussion takes place within the intrinsic value option.[13]

Because of the obvious carry over to space ethics from environmental ethics, we might remind ourselves of the concerns voiced early in the deep ecology movement. Norwegian eco-sophist Arne Næss gave us the term *deep ecology* in 1973, arguing for an exhaustively systemic view of nature and for treating both human and nonhuman life *alike* as having "intrinsic

---

10. Gilmore, "Space Exploration," 1834.

11. Randolph and McKay, "Protecting and Expanding."

12. Intrinsic value refers to "value that is truly independent of valuing agents." Lupisella, "Cosmological Theories of Value," 80. Lupisella is suspicious of a value that is truly independent of a relationship to the one who values. In my judgment, intrinsic value is first conferred by the valuing agent and then, in the case of intelligence, claimed by the one so valued. It will be difficult to make a case for the intrinsic value of an off-Earth biosphere when that biosphere does not claim its own intrinsic worth.

13. "Teleorespect or teleoempathy merely captures our recognition that extraterrestrial life, including life independently evolved from the biology that we know on Earth, places demands on our behaviour if we think it has intrinsic value." Cockell, "Ethical Status of Microbial Life," 171.

value."[14] This metaphysical commitment would imply ethically that we treat all of living nature as sacred. If we sacrilize the entire living world in nature, deep ecologists presume, then we human beings would treat our biosphere with reverence, respect, and responsibility. Might we borrow this principle for off-Earth biospheres, treating newly discovered life forms as having intrinsic rather than instrumental value?

Yes, would be the answer offered by environmental philosopher Holmes Rolston III, who argues against the instrumental approach on the grounds that non-human nature preceded human beings who value things. "Humans on Earth are latecomers, and it seems astronomically arrogant for such late products to say that the system is only of instrumental value, or that not until humans appear to do their valuing does value appear in the universe."[15] Valuing in general is grounded in nature, not in human preference. This opens the door for treating off-Earth ecospheres as having intrinsic worth and integrity.

However, Reformation Protestants find it difficult to treat anything created as sacred in itself. God the Creator is sacred, to be sure; but the world of nature—whether living or non-living—belongs in the category of creature. Human dignity is called for, to be sure; and so is responsible stewardship called for. But not an intrinsic sacredness spread around to all living things. What is created cannot bear in itself an ontological sacredness. Nevertheless we might ask: could a creature be sacred if God deems it sacred? Might we appeal to intrinsic value if God imputes intrinsic value?

Methodist ethicist Richard Randolph would answer in the affirmative. He affirms the intrinsic value of life by appeal to God. "From a Christian point of view, God's preferential option for life means that all of life has intrinsic value. By this I mean that all living organisms, as well as their ecosystems, are entitled to a basic, underlying level of respect—and, even reverence—by humans. Every living organism is good in and of itself, regardless of the instrumental value it may have for humans."[16] We human beings on Earth should express this intrinsic worth of extraterrestrial life by behaving as servant/stewards, supporting both living organisms and their respective habitats wherever they may be found. "God's preferential option

---

14. Næss, "Shallow and the Deep," 95–100.

15. Rolston, "Preservation of Natural Value," 140–82. Not every philosopher likes the idea of intrinsic value. E. J. Bond, for example, calls intrinsic value an "obscure notion" without explanation. Bond, "Value, Concept of," 1746. James Schwarz and Tony Milligan distinguish between 'inherent' and 'intrinsic' value, but the distinction fails to discriminate enough to warrant selecting one term over the other. "Introduction: The Scope and Content of Space Ethics," 1.

16. Randolph, "God's Preferential Option for Life," 287.

for life grounds the claim that all of life has intrinsic worth and that God intended for extraterrestrial life to flourish and be self-determinant."[17] In Næss, Rolston, and Arnould the intrinsic value of the natural world derives from our treating nature as sacred. In Randolph, in contrast, the intrinsic value of the natural world derives from the respect God shows for it. The resulting sense of moral responsibility is the same in both cases.

Astroethicists, like other ethicists, cannot avoid a conceptual hurdle to be jumped. What warrants intrinsic value? Is it a quality or trait possessed by the entity valued? Or, is intrinsic value conferred by the valuing agent? Astroethicist Kelly Smith holds that the presence of socially functional intelligence warrants being treated as having intrinsic value. If a creature on Earth or beyond Earth is intelligent, then it can claim intrinsic value. Smith is here advocating a "racio-centric approach to ethical value."[18] We note further that the valuing agent is the one who values intelligence, and this undermines the very notion of intrinsic value in the first place. Such a conundrum could lead an ethicist to despair. But Smith does not despair. His bottom line is this: the space explorer must look for a socially functioning intelligence beyond Earth before assigning it intrinsic value.

Now we must ask: does level of intelligence suffice for discriminating between extraterrestrial life forms deserving of intrinsic value and life forms subject to instrumental value for earthlings? A closer look will show we have two issues, not one. First, should we grant intrinsic value to all life forms or should we grant intrinsic value to only some but not others? Second, if we discriminate between life forms of greater value from those of lesser value, what shall be our criterion of selection?

A flat attribution of intrinsic value to all of nature or, within this all of life, is difficult to ground without appeal to God, who transcends nature. First, life does not treat itself with intrinsic value. Predation, for example, pervades the domain of living things. Life eats life. Some life dies so that other living things may live. Thus, if we human beings are going to protect life on the grounds of its intrinsic value, we do so for reasons other than imitating life itself. Second, we daily treat some living organisms instrumentally. Everything we eat we kill, except for salt. Plants and sometimes animals are sacrificed for our own survival or pleasure. We do this without thinking ethically. Therefore, if we are to impute intrinsic value to living entities in outer space, we may need to discriminate; and we may need to provide a reason for this discrimination.

17. Ibid.
18. Smith, "Trouble with Intrinsic Value," 261.

Notre Dame University Bioethicist Celia Deane-Drummond contends that the concept of intrinsic value does not preclude discriminating between greater or lesser worth. "It is possible to hold to the notion of intrinsic value, while also being able to discriminate between different forms of life and non-life in terms of their worth."[19] Or, to say it another way, even if we impute intrinsic value to all living things, within this wide category we may identify some living things to be of greater value or worth. But, we ask: how do we decide? Without appealing to instrumental criteria for discriminating between greater or lesser worth, we should look for criteria within the scope of intrinsic value. One candidate might be sentience. Complex life forms are sentient in a way that we humans are sentient; whereas more simple life forms lack this attribute. Could sentience provide the criterion of discrimination?

Erik Persson, bioethicist at Lund University in Sweden, appeals to sentience within the larger category of life. "According to sentientism, one has to be sentient to have moral status whether terrestrial or extraterrestrial and whether biological or nonbiological [such as post-biological] . . . The most plausible theory for moral standing seems to be sentientism that connects directly to the basic idea behind modern ethics: that ethics is about dealing with situations where one's own actions affect others in a way that matters to them . . . If we accept sentientism, microbial life and plants do not have moral status, but there are reasons for protecting someone or something other than being a moral object."[20] Complex or sentient life has greater moral worth than simple or primitive life, according to this scheme. Will this work? No.

Sentience will not work as a general ethical category, except for vegetarians. Here on Earth we have already committed ourselves to eating meat. Meat-eating requires the death of sentient creatures. We discriminate between pets, which we do not eat, from stock, which we do eat. Vegetarians object to this practice on moral grounds, on the grounds that we have a responsibility to all sentient creatures. If we are to export to extraterrestrial realms a categorical respect for all sentient organisms, then for the sake of consistency we would need to adopt vegetarianism back at home. A consistent ethic based upon sentience would require vegetarianism on Earth as well as on all space expeditions.

Sentience will not help for another reason. To date, those contributing to this discussion have drawn on ethical precedents set by environmentalists and eco-ethicists. This ethical posture is oriented holistically toward

19. Deane-Drummond, "Alpha and the Omega," 104.
20. Persson, "Moral Status of Extraterrestrial Life," 983.

entire ecosystems, toward protecting entire habitats with their resident living creatures regardless of level of sentience or intelligence. This holistic approach seems intuitively relevant to what we might discover on Mars or a moon orbiting Saturn. Once engaged, we would not discriminate between one species on behalf of another species. Rather, we would assume we are responsible for each entire biosphere with its already established life forms. Entailed in a holistic commitment to an entire ecosystem is an indispensible level of commitment to simple life forms and even to abiotic contributors to this ecosystem.

In sum, we may have to live for a period with a generic respect-for-life's-intrinsic-value principle until we have entered into actual engagement with ETNL or ETIL. At that point we will re-articulate the quandary and re-formulate our responsibility. By no means is this a form of kicking the ethical can down the road. Rather, we are simply marking specific areas where we will need to respond to actual rather than hypothetical situations.

3. Should Space Explorers Invoke the Precautionary Principle?

Earth's ecologists are used to debating and commending the *Precautionary Principle*. Might astroethicists borrow it?

The so-called *Wingspread Definition* of the *Precautionary Principle* was formulated at the 1992 United Nations Conference on Environment and Development: "When an activity raises threats of harm to human health or the environment, precautionary measures should be taken even if some cause and effect relationships are not fully established scientifically. In this context the proponent of the process or product, rather than the public, should bear the burden of proof."[21] When space scientists and ethicists met at Princeton for a COSPAR workshop in 2010, they embraced a variant formulation: "we define the *precautionary principle* as an axiom which calls for further investigation in cases of uncertainty before interference that is likely to be harmful to Earth and other extraterrestrial bodies, including life, ecosystems, and biotic and abiotic environments."[22]

Employment of the precautionary principle for space exploration provides the kind of middle axiom that connects the larger value of life with practical policies that facilitate off Earth activities.

---

21. Appell, "New Uncertainty Principle," 18.

22. "COSPAR Workshop" on Ethical Considerations for Protection in Space Exploration, Princeton University, Princeton, NJ, 8–10, 2010.

4. Should We Clean Up Our Space Junk?

Currently, about twenty-two thousand large pieces of space junk in the form of dead satellite parts are orbiting Earth. We have turned our upper atmosphere into a trash dump. Do we want to pollute extraterrestrial space just as we have befouled our terrestrial nest?

Over the last half century we earthlings have shot up into space approximately twenty thousand metric tons of material. Forty-five hundred metric tons remain in orbit, broken into countless chunks of junk. Of the forty-eight thousand satellites placed in orbit, half of these, twenty-four hundred, are still present. But, of these twenty-four hundred satellites, only six hundred or so are still active. The inactive satellites and other rocket debris sail silently like a never-ceasing stream of unaimed bullets just waiting for a target to get into their way.

The problem with our orbiting landfill is not merely that it is ugly. It is also dangerous. It risks danger to future space flights and future satellites. Arnould warns us: "there are now 22,000 human-made objects larger than 10 centimeters across in orbit and half a million larger than 1 centimetre—and all pose a grave risk to space missions . . . Even if space agencies never launched another rocket, the cloud of debris will continue to grow as a pieces of space junk crash into one another."[23] As space junk crashes, each piece fragments and multiplies the number of dangerous micrometeorite material that risks damage to future space vehicles we send up. Some space scientists fear a runaway chain reaction—called the Kessler syndrome—that pulverizes everything in orbit, including functioning satellites. This would establish a band of untraversable danger, a no-man's land in space. Here is the warning: for safety's sake, stay out of the space dump.

To date, no one has been held responsible for space junk. Those who make profits or who otherwise gain from sending this material into space are not required to recycle or dispose of their waste. Space waste accumulates, but nobody is required to pay for cleaning it up. Nations or corporations treat the Greater Earth or galactic commons as their ashtray, as a public trash dump.[24] Follow the money.

If we define Greater Earth as a part of the galactic commons, then we find ourselves already beset with a classic moral problem: those with power and influence utilize common space for their own profit while the population as a whole absorbs the cost of deterioration or degradation of what is publically shared. If and when our planetary society consolidates its

---

23. Arnould, *Icarus*, 92. See Marks, "Clearing the Heavens," 22.
24. According to Arnould, "Greater Earth," 129.

diversity into a single community of moral deliberation, then responsibility will need to be parsed and parceled according to a renewed principle of justice.

The European Space Agency has set up a Space Debris Office to coordinate research activities in all major debris disciplines, including measurements, modeling, protection, and mitigation, and coordinates such activities with the national research efforts of space agencies in Italy, the United Kingdom, France and Germany. Together with ESA, these national agencies form the European Network of Competences on Space Debris.[25]

The Japan Aerospace Exploration Agency (JAXA) is testing to see if a tethering technique might begin the process of debris-gathering. The tether consists of a long conductive wire attached to a junk chunk which, by implementing an electrodynamic drag, would pull the debris into the atmosphere where it would burn up. The Space Tethered Autonomous Robotic Satellite-2 (STARS-2) is testing the idea and, if it works, then it could be attached to future missions aimed at capturing existing debris.[26]

5. What Should We Do about Satellite Surveillance?

Reconnaissance satellites or spy satellites have been deployed over sixty years for purposes of military or intelligence applications. The telescopes on board are pointed toward Earth, not toward the stars. Mission tasks include high resolution photography; measurement and signature intelligence; communications eavesdropping; covert communications; monitoring of nuclear test ban compliance; and detection of missile launches. With the improvements in technology, today's spy satellites have a resolution capacity down to objects as small as ten centimeters. Surveillance satellites also provide us with efficient communications, weather reporting, Google maps, and many more public services.

Spying is international, not just national. The Echelon spy network coordinates satellite snooping by the governments of the United Kingdom, the United States, Canada, Australia, and New Zealand. The Echelon network spies, sorts, decrypts, archives, and processes three million telephone calls transmitted by satellite every minute.

Remote sensing and communication technologies were developed before any legal structure was put in place to govern their developments. The United States government sells pictures taken by satellites; but it keeps

25. ESA Space Debris Office, http://www.esa.int/Our_Activities/Operations/Space_Debris/ESA_Space_Debris_Office.
26. "Researchers Eye Tethers for Space Debris," 1062.

certain subjects from public review. Sensitive facilities such as military installations are restricted, as are remote pictures taken over Israel. Similarly, private companies use satellites for remote sensing and sell their pictures.

"Can a State gather information about the natural riches and resources of another sovereign State without having obtained the latter's prior agreement?" asks Arnould. "Is it not up to the remote sensing State to ask for the prior permission of the State whose territory is being observed?"[27] This sounds like a reasonable ethical question. Yet, it presupposes the present situation of sovereign nation states, a political system that may have made sense prior to the current thrust toward economic and technological globalization. Satellite surveillance and communication services, right along with other space activities, are playing into an emerging planetary consciousness.

Protecting national boundaries from foreign intelligence or even public transparency may soon be an artifact of history, an era we remember but no longer live in. Perhaps the way forward is to support an ethic of maximal *information without discrimination*. Rather than attempt to police information gathered from remote sensing, it would be healthier and easier to prevent such information from deleterious usage.

6. Should Nations Weaponize Space?

"Star Wars" became the nickname for U.S. President Ronald Reagan's 1983 proposal for the Strategic Defense Initiative (SDI). Reagan had inherited the cold war strategy of "Mutual Assured Destruction" (MAD), the policy of detente that had kept the peace between the United States and the Soviet Union since World War Two. Reagan asked his scientists to help him create SDI, which would include among other things space-based weapons. Specifically, these space-based weapons would rely upon lasers aimed at shooting down Soviet Intercontinental Ballistic Missiles (ICBMs).

Reagan's Strategic Defense Initiative Organization (SDIO) worked with an earlier idea developed by physicist Edward Teller for an X-Ray laser. Then, SDIO funded newer ideas. Beginning in 1985, designs and tests were planned for innovative forms of laser technology. The U.S. Air Force tested a deuterium fluoride laser, also known as the Mid-Infrared Advanced Chemical Laser. Later the Air Force tried shooting an old satellite with a Tactical High Energy Laser; and the U.S. Navy shot at drones with similar laser designs. The best these experiments could report was "mixed" success, just enough success to keep funding flowing.

---

27. Arnould, *Icarus*, 75.

On one occasion, Charles Townes caught a private moment with then President Reagan. Recall, that Dr. Townes—to whom this book is dedicated—received the Nobel Prize for his co-invention of both the laser and the maser. Townes addressed the president: "now, I have been following what is being said about SDI. I happen to know a great deal about laser technology. Given what I know, Sir, it is impossible to develop the kind of space-based weapon you are planning."

"Yes, I know," said the president. "But, we want to scare the hell out of the Russians." It must have worked, because the Berlin Wall came down only a half decade later. As of this writing neither lasers nor bombs are sitting in orbit, waiting to be directed at Earth targets. For decades now the planet has enjoyed a temporary moratorium on space weaponization.

The moratorium may be coming to an end, however. Eyes on militarizing space are looking up. "For modern warfare, space has become the ultimate high ground, with the U.S. as the undisputed king of the hill," writes Lee Billings.[28] "China and Russia are both developing capabilities to sabotage crucial U.S. military satellites."[29] Laser technology has advanced, so that satellites will begin to use lasers to disable other satellites. One can only imagine a skirmish that could lead to Star Wars or, more precisely, Satellite Wars.

This scenario is important, because the 1967 United Nations Outer Space Treaty emphasizes that celestial locations could be used "exclusively for peaceful purposes." The treaty explicitly prohibited the "placing in orbit around the Earth any objects carrying nuclear weapons or any other kinds of weapons of mass destruction." Light saber battles between satellites was not explicitly prohibited. Any regulation of military equipment in space is today the responsibility of bilateral and multilateral agreements, not the United Nations. No global community of moral deliberation exists, at least not yet.

7. Which Should Have Priority: Scientific Research or Making a Buck?

Economic and political motivations for gaining dominance in space may increase over the next decades. The telecommunications industry is already accustomed to the cost effective use of satellites. We are on the brink of an era of space tourism, with the first trips to suborbit and low orbit vacations in the planning stages. Visits to the moon will most likely follow.

28. Billings, "Are We on the Cusp of War?" 15.
29. Ibid., 18.

Establishing research laboratories on the moon and Mars are being envisioned. Might we be wise to ready ourselves for an El Dorado type of gold rush to the new extraterrestrial world? If so, should we try to put policies and policing mechanisms in place in advance?

Up until this point we have thought of outer space as a sandbox for Earth's scientists to play in. Governments have found the money to fund modest exploratory adventures; and scientists have organized to conduct experiments which have yielded an abundant harvest of new knowledge about our cosmos. Frequently, scientific goals have been mixed with military goals, because leaders in the military have been willing to share their budgets for scientific purposes. Scientific experiments do very little damage, if any. Somewhere on the Moon is a golf ball left by visiting astronauts. Landing on Mars or on Titan has not infected or contaminated anybody's ecosystem, as far as we know. NASA decontaminated its first Mars lander, but more recently NASA has saved the money spent for decontamination under the assumption that a little contamination of Mars doesn't matter. The impact on our solar system by scientific activity is benign.

This situation is about to change. The private sector is now ogling space for profit. What about space tourism? Simply flying a few wealthy passengers high enough to experience weightlessness is not likely to provoke anyone's moral ire. But, what about tour busses roaming the surface of the Moon? Busses will leave tire tracks. Perhaps trash. No doubt tourists will want to visit that golf ball as well as historical sites where astronauts first landed. Will the crowds of visitors damage those sites? Are those sites sacred? Protectable? Who will decide and what will be the criteria by which they decide?

The market does not always react the way the marketers predict. Low cost and frequent flights to suborbit heights might actually encourage increased participation by scientists. These scientists will want to do research on the *ignorosphere*. The ignorosphere is a level just above balloon traffic but too low for satellites. Scientific researchers might buy tickets with the tourists and then look out the windows.[30]

8. Should Earthlings Terraform Mars?

Should we terraform Mars? Or, any other planet or moon, for that matter? Will we Earthlings rest content until we see the golden arches of McDonalds on the red planet?

30. See Stern, "Low-Cost Ticket to Space," 69–73.

Our Mars Man is Christopher McKay, a space scientist at NASA's Ames Research Center. According to McKay, we work with the assumption that Mars is lifeless. At least it is lifeless today. The red planet may have been home to life in the past; but Mars must have lost its atmosphere and its ability to sustain life for reasons yet unknown. Its thin atmosphere is replete with carbon dioxide, but not oxygen. Let us speculate: suppose we would transplant living organisms from Earth that take in carbon dioxide and expel oxygen into the atmosphere? Then, when enough oxygen suffuses the atmosphere, we could introduce oxygen inhaling organisms that expel greenhouse gases. These greenhouse gases would warm up Mars, and life would thrive. A self-regenerating ecosystem could run on its own. In less than a century, estimates McKay, we could establish a biosphere that would last ten to a hundred million years.

McKay calls this terraforming project "planetary ecosynthesis." This proposal for planetary ecosynthesis raises a number of ethical concerns. Immediately, one might ask: should we do it? To what do we appeal ethically to answer this question?

McKay starts with a simple axiom: life is better than non-life. Virtually no one who mulls over the question of life's intrinsic value would challenge this axiom. But, we ask: what comes next?

On Earth the principle that life is better than non-life has influenced our decisions and policies to preserve life. We preserve species from extinction. We preserve habitats to encourage certain species to proliferate. In short, we attempt to prevent certain forms of life from dying out.

This would not apply to Mars, however. If we assume that Mars is currently lifeless, then we would not find ourselves considering the preservation of life. Rather, the question is: should we seed life on Mars? If life is better than non-life, says McKay, then the moral answer should be in the affirmative. Transferring terrestrial life forms to Mars would be better than leaving Mars lifeless.

Curiously, McKay appeals to both intrinsic value and instrumental or utilitarian value when justifying planetary ecosynthesis. First, the intrinsic argument. Because life has intrinsic value, Mars with life would be ethically of greater worth than a lifeless Mars, even if it is transplanted life. Second, the instrumental argument. Because we on Earth would learn so much from the Mars project about sustaining a biosphere, we could apply what we learn on Mars to sustaining Earth's biosphere in the face of our imminent ecological challenges. "Both utilitarian and intrinsic worth arguments support the notion of planetary ecosynthesis."[31]

31. McKay, "Planetary Ecosynthesis on Mars," 259.

How might such an argument sit with a theologian? With a Buddhist theologian? Francisca Cho, Associate Professor of Buddhist Studies at Georgetown University, offers a Buddhist interpretation. "A Buddhist would apply neither an intrinsic nor instrumental value of life or nature to the question of terraforming Mars. The idea of an intrinsic value would go against the principle of emptiness. Instrumental value, on the other hand, would be problematic because one could not ensure that the instrumental objectives and the proper motivations . . . There is no intrinsic worth to nature but neither is there intrinsic worth to human beings . . . There is no option between them, so you have to transcend that framework all together."[32] From a Buddhist perspective, neither an appeal to the intrinsic value of life nor an appeal to life's utilitarian value to human beings provides ethical guidance for the terraforming question.

Another issue appears on our moral radar screen, namely, the risk that we terrestrials will make a mess out of Mars. We have already spoiled one planet. Will we spoil others? Theologian Cynthia Crysdale recommends that we incorporate this risk into our ethical vision. "We need to think of ourselves as living within an ethic of risk, not an ethic of control. I say this in direct reference to the actions we take in terraforming or colonizing or exploring other planets. My caution is to point out that the conditions of possibility that we establish in the hopes of one outcome may at the same time establish conditions under which totally unforeseen schemes of recurrence become established."[33] Dr. Crysdale has wisely asked us to consider human nature—that is, human sinfulness—when making plans. No ethical justification could suffice without acknowledgement of who we are as humans. Nevertheless, anticipating the unforeseen damage we humans are capable of is a principle one must incorporate into any such project, regardless of whether it is justified by appeal to an intrinsic or utilitarian ground.[34]

Should earthlings terraform Mars or any other celestial body within our solar ghetto? On the one hand, McKay's argument that life is better than non-life provides a sound point of departure. On the other hand, transplanting terrestrial life to an extraterrestrial location looks a great deal like

---

32. Cho, "Asian Religious Perspective," 212.

33. Crysdale, "God, Evolution, and Astrobiology," 240.

34. Would earthlings uglify Mars with countless MacDonald's fast food outlets? Does an aesthetic criterion count? Perhaps, according to Sean McMahon. "If aesthetic objections to terraforming Mars are to be dismissed on the grounds that Martian beauty is valuable only in proportion to its resemblance to the more familiar natural beauty of Earth, or that the beauty of a terraformed Mars would for other reasons be as valuable or more valuable than the beauty of Mars today, this judgment had better be the result of a deep and serious engagement with existing Martain aesthetic qualities." McMahon, "Aesthetic Objection," 216.

colonizing. As we bring the history of terrestrial colonization to mind, we cannot avoid recalling the imperialism and greed that motivated colonization and the devastating impact of exploitation and genocide on the lands colonized. The Crysdale incorporation of risk based upon what we know from history about human nature gives one pause.

Our pause cannot last too long. The Mars Society is already making plans to colonize the Red Planet.

9. Should Earthlings Colonize Mars?

"The goal of the human spaceflight program should be to improve our survival prospects by colonizing space," contends Princeton astrophysicist Richard Gott.[35]

Does this include our colonization of Mars? Yes, and more places in the solar system. "Humans-to-Mars" is the direction Robert Zubrin is leading his followers in the Mars Society. His "Mars Direct" colonization plan "advocates a minimalist, live-off-the-land approach to exploring the planet Mars, allowing for maximum results with minimum investment. Using existing launch technology and making use of the Martian atmosphere to generate rocket fuel, extracting water from the Martian soil and eventually using the abundant mineral resources of the Red Planet for construction purposes, the plan drastically lowers the amount of material which must be launched from Earth to Mars, thus sidestepping the primary stumbling block to space exploration and rapidly accelerating the timetable for human exploration of the solar system."[36] Money raised from the private sector will support this effort. Mars Society adherents see themselves as rivals to NASA. They are in a space race and plan to beat NASA to the fourth planet.

The Mars Society plans to initiate Mars Direct by sending an "Earth Return Vehicle" or ERV, arriving on the red planet six months following launch from Earth. While on the Mars surface, the ERV will set up and operate nuclear reactors, which will generate the fuel needed for the return trip, thirteen months later.

The second Earth-to-Mars launch will take place twenty-six months after the first, sending two more craft, a second ERV and a habitat module for the astronauts to live in. After a year and a half on the Martian surface, the first crew returns to Earth, leaving behind the habitat, the rovers associated with it and any ongoing experiments conducted there. When they

---

35. Gott, "Our Future in the Universe," 421.
36. The Mars Society, Mars Direct, http://www.marssociety.org/home/about/mars-direct.

land on Earth six months later, they are greeted to a hero's welcome. From this point on the cycle is repeated, with more and more of Mars opening up to human exploration and habitation. This will be the beginning of a permanent human settlement on the planet Mars.

The Mars One project based in the Netherlands and headed by Bas Lansdorp is making similar plans. As of the summer of 2013 the project planners began their selection of the first crew headed for the Red Planet in 2023. The crew would be given seven years of training in engineering, medicine, agriculture, and astrophysics. This would be a one way trip. Once the astronauts have landed, they would become Martians.

The mood of the Mars Society and the Mars One project is one of promethean expectation. The human race is being called by destiny to go, go, go. To spread our race throughout the solar system fulfills our inherited evolutionary mandate, to fill every niche with life.

Some Muslims find the prospect of Mars colonization objectionable, especially the risk of self-sacrifice on the part of the first explorers. Because the first wave of travelers from Earth to Mars will know in advance that they will not be returning, the one way trip idea looks like a plan for suicide. To this, Muslim theologians object. In early 2014 the fatwa committee of the General Authority of Islamic Affairs and Endowment in the United Arab Emirates issued the fatwa after determining that "such a one-way journey poses a real risk to life, and that can never be justified in Islam." They continued, "There is a possibility that an individual who travels to planet Mars may not be able to remain alive there, and is more vulnerable to death." Qu'ran 4/29: "Do not kill yourselves or one another. Indeed, Allah is to you ever merciful." In sum, suicide is immoral, even on Mars.[37]

It would seem to me that we need not think of one way trippers to Mars as necessarily committing suicide. There is a risk of death, to be sure; but there is also the prospect that earthlings might live a normal life span in the atmospheric bubble on Mars's surface. To die of natural causes off-Earth does not necessarily count as suicide, in my judgment.

When the Spanish crown commissioned Christopher Columbus in 1492 to sail west across the Atlantic Ocean, it was hoped he would return with gold and mineral wealth. Europe's race to the new world had begun, a race to maximize national power and wealth through exploiting foreign land. Might we see a repeat here? Not exactly. Even if the Mars Society or Mars One are successful at establishing a permanent habitat for earthlings on Mars, it is not clear that this would lead to profits for those back on the third planet who funded their expedition. If profit is to be made, it would

37. Rojas, "Islamic Leaders Issue Fatwa Prohibiting One-Way Trips to Mars."

not likely occur within the lifetime of those planning the mission. This gargantuan mission will have to rely on the promethean spirit, at least for the near future.

10. How Should We Protect Earth from Extraterrestrial Threats?

Back on the third planet, the vast majority of earthlings will still be living here. The very thought that Mars might provide a safe refuge for *Homo sapiens* once we have so polluted our home planet is utterly preposterous, according to anyone who has considered the matter. We had better make our peace with one another and with our own biosphere, because Earth will continue to be our home for the foreseeable future. If we *Homo sapiens* do not get our act together and end up so fouling our habitat that we go extinct, the stupid life in our solar system will not be limited to Titan microbes.

Nevertheless, Earth is a dangerous home. The heavens hold plenty of threats. The Sun occasionally launches solar flares, which fry electricity grids by generating intense currents in wires. A solar megastorm in 1859 sparked fires in telegraph offices. If such a flare would reach Earth today, it would knock out satellites and shut down power grids for months or longer. Such an event would incur trillions of dollars in economic damage. Although we rely upon the sun for our daily life, some day it just might kill us.

In addition to solar threats, we need to anticipate the possibility of a large comet or asteroid strike. "I'm telling you," insists Neil deGrasse Tyson, "It's going to happen. I don't know when, but it's going to happen, and it will be a bad day on Earth."[38]

We have had forewarnings. On February 15, 2013, more than four hundred Russian people were injured when an asteroid exploded just above the city of Chelyabinsk. NASA referred to it as a "tiny asteroid" that measured roughly forty-five feet across, weighed about ten thousand tons and traveled about forty thousand mph. The object vaporized roughly fifteen miles above the surface of the Earth, causing a shock wave that triggered the global network of listening devices that was established to detect nuclear test explosions. The force of the explosion measured between three hundred and five hundred kilotons, equivalent to a modern nuclear bomb.[39]

Within hours of the Russian disaster, another asteroid, 2012 DA14 passed between Earth and our geosynchronous satellites. Once or twice every two million years our planet gets smacked by rocks two kilometers or

38. Tyson, "Search for Life," 148.
39. Morin, "Russian Meteor."

more in diameter, leading to extinctions. It is widely believed among scientists that sixty-five million years ago an asteroid ten kilometers in diameter hit Earth and triggered the mass extinction of dinosaurs. Can we protect Earth from future asteroid catastrophes? The UN's Science and Technical Subcommittee's Near-Earth Object Working Group and its internal panel, Action Team 14, have been working on the details of an international approach since 2001.

Solar flares and asteroid strikes. That's not all. More rare but equally potent would be the blast of radiation from a nearby γ-ray (gamma ray) burst. A short-hard γ-ray burst, caused by the violent merger of two black holes or two neutron stars or a combination, provides the most frightening scenario. If one such blast would be directed at Earth from within two hundred parsecs away (less than 1 percent of the distance across the Milky Way), it would zap Earth with enough high-energy photons to wipe out 30 percent of the atmosphere's protective ozone layer for nearly a decade.[40] Such an event—expected once every three hundred million years or so—would double the amount of ultraviolet light reaching the ground and scorch phytoplankton, which make up the base of the ocean's food web. Astronomers are unable to predict such bursts, so we have no way of knowing whether such a rare event is imminent.

What about long-soft bursts? From a distance of about two thousand parsecs, "long-soft" γ-ray bursts—which result from the collapse of massive stars—could also damage our planet and cause extinctions. Long-soft bursts are rarer than short-hard bursts. In addition, they are easier to spot in advance because they come from larger, brighter stars.[41]

How does knowledge of these potential threats from the heavens provoke ethical concerns? Because these damage scenarios lead us to think ahead. We need to plan for our planet's future, and we need to incorporate such possibilities into our planning. With regard to solar flares, fortunately, there are ways to mitigate the damage should it occur: engineers can protect the grid with fail-safes or by turning off the power in the face of an incoming blast. With regard to a comet or asteroid strike, we will be given advanced notice. A diversion strategy could be effective, perhaps by hitting the object while it is yet far away with a nuclear bomb. We have no way to prevent gamma ray bursts from striking our Earth, but we could provide protective shields in sanctuaries for life forms we wish to restart following the event. These matters belong to our quandary. Just how will we respond?

40. Jones, "Planetary Disasters."
41. *Nature* Editorial, "Realities of Risk."

We have been suggesting that the community most appropriate for deliberating over such matters would consist of all the peoples of Earth working together. When confronting scenarios that have a planet-wide impact, the planet as a whole should become the community of moral deliberation and provide the network to shoulder the responsibility. Planetary plans to meet such threats should be international or supranational. The principle of distributive justice may require that each nation contribute to a coordinated effort in proportion to its capability by providing either technological expertise or funding for such expertise. Planetization is a corollary to the notion of a galactic commons. Eco-images such as "green globalization" or "spaceship Earth" connote the circumstances that lead to the concept of a single planetary society.

In order for a single planetary community of moral deliberation to emerge, local profiteering and nationalisms will have to be superseded in the name of a higher level of human unity, including a unity between the human race and the natural realm of which we are a part. This will require a repentance—a metanoia or turning around—as the first step to a sense of planetary let alone galactic responsibility. Those who have studied human nature fear that we might be asking too much. Ferkiss registers his fear that we may not make the necessary transition to the requisite holistic faith. "Virtually every possible extension of man's powers over nature and himself almost certainly will be made or not made, shaped in one direction or another, not by considerations of what is good for man but what is good for some men; problems that are basically scientific and ethical will be decided in economic terms. Metanoia must be global."[42] Local short-term profit may provide the insurmountable barrier to establishing a long-term commitment to a planetary society and a galactic commons.

Ferkiss sounds pessimistic. Former US Vice President Al Gore is much more optimistic. "Fortunately, the awakening of the Global Mind is disrupting established patterns—creating exciting new opportunities for emergent centers of influence not controlled by elites . . . [elites who have set incentives] that reward unsustainable exploitation of limited resources, the destruction of ecosystems crucial to the survival of civilization, unlimited flows of pollution, and the disregard of human and social values."[43] Whether pessimistic or optimistic, a long-term global ecoethic or accompanying galactic astroethic should be the product of a single planetary society that rises above the self-destructive greed of competing subsidiary economic forces.

42. Ferkiss, *Future of Technological Civilization*, 283.
43. Gore, *Future*, 364.

11. Does AstroEthics Require a Single Planetary Community of Moral Deliberation?

Who should engage in ethical deliberation? Who should make public policy decisions? Because every moral commitment regarding space is simultaneously a commitment regarding the entire Planet Earth, it follows that the entire planetary community should become the unit for ethical deliberation.[44]

In my judgment, the matter of establishing a single planetary community of moral deliberation is urgent. When confronting scenarios that have a planet-wide impact such as a threatening asteroid, the planet as a whole should become the community of moral deliberation and provide the network to shoulder the responsibility. Planetary plans to meet such threats should be international or supranational. The principle of distributive justice may require that each nation contribute to a coordinated effort in proportion to its capability by providing either technological expertise or funding for such expertise.

12. Should the Common Good Include the Galactic Commons?

Only the invocation of the common good can successfully trump the current competition between vested interests, cost-benefit priorities, industrial ambitions, and international conflicts. A single planetary community of moral deliberation could arise only if the common good becomes its agenda.[45]

Further, we note that the commons we share is larger than Planet Earth alone. It is even larger than our solar system. Why? Because Earth is nested in a larger physical context, the Milky Way galaxy. This leads us to consider our moral domain to be a *cosmic* commons. But, because if the immensity of the cosmos, it is inconceivable at this time to think of any reciprocity beyond the Milky Way. Perhaps *galactic commons* becomes a more reasonable scope for our domain of moral responsibility. It would wisest for the time being to designate the *galactic commons* as the shared home for our common good.[46]

---

44. Certainly this is the view of the Planetary Society, http://planetary.org/.

45. Pope Paul VI defined the common good as "the sum of those conditions of social life which allow social groups and their individual members relatively thorough and ready access to their own fulfillment." Pope Paul VI, *Gaudium et Spes*.

46. A pioneer in this line of thinking is Boston University's John Hart, who employs the term *cosmic commons*. I prefer *galactic commons,* because at least some reciprocity might be possible within the Milky Way. But the light year problem would make this impossible beyond our galaxy due to distances. Hart, *Cosmic Commons*, 2013.

Conclusion

Despite the fact that the field of astroethics is the new kid in school, already a considerable list of issues appear on its report card. Within our solar ghetto, we must provide ethical deliberation prompted by the prospect that we will be traveling in outer space and that we may discover primal or microbial life, what we affectionately call stupid life. These prospects elicit an ethical quandary regarding matters such as: planetary protection (including protection of Earth and protection of off-Earth ecospheres); the intrinsic value of extraterrestrial life and of off-Earth ecosystems; what to do about space junk; satellite spying; weaponization of space; the competition between scientific research and economic interests, including space tourism; terraforming Mars; Mars colonization; mitigating the damage done by solar flares, asteroid collisions, and gamma bursts; and such.

These quandaries prompt in us a sense of responsibility. The very knowledge that such challenges may be approaching us in the future is sufficient to prompt in us the question: what should we do? The matter becomes more complex when we ask: just who makes up the community of moral deliberation here? It appears obvious that challenges to the future of all life on Earth—actually, all life in the galactic commons—lead to the prospect of planetization. All peoples of Earth in cooperation need to deliberate over what is best for our planet as a whole, and our cosmic commons as a whole. Can the peoples of Earth think of themselves as a single planetary society shouldering responsibility for all biota and even abiotic factors in our solar ghetto?

Bibliography

Appell, David. "The New Uncertainty Principle." *Scientific American* (January 18, 2001) 18.

Arnould, Jacques. *Icarus' Second Chance: The Basis and Perspectives of Space Ethics.* New York: Springer, 2011.

Billings, Lee. "Are We on the Cusp of War—in Space?" *Scientific American* 313:4 (October 2015) 14–18.

Bond, E. J. "Value, Concept of," *Encyclopedia of Ethics,* edited by Lawrence C. Becker and Charlotte B. Becker. 3 vols., III:1745–50. London: Routledge, 2001.

Cho, Francisca. "An Asian Religious Perspective on Exploring the Origin, Extent and Future of Life." In *Workshop Report: Philosophical, Ethical, and Theological Implications of Astrobiology,* edited by Connie Bertka et al., 208–18. Washington, DC: AAAS, 2007.

Cleland, Carol E., and Elspeth M. Wilson. "Lessons from Earth: Toward an Ethics of Astrobiology." In *ELU,* 17–55.

Cockell, Charles S. "The Ethical Status of Microbial Life on Earth and Elsewhere: In Defense of Intrinsic Value." In *ESE*, 167–80.

Collins, Nathan. "Stop Pampering the Red Planet." *Scientific American* 309:3 (September 2013) 24.

Conley, Catherine A. "Life Guard: An Interview with Catharine Conley." *Scientific American* 311:4 (October 2014) 30.

Conley, Catharine A., and John D. Rummel. "Planetary Protection for Human Exploration of Mars." *Acta Astronautica* 66:5–6 (March–April 2010) 792–7.

"COSPAR Workshop on Ethical Considerations for Protection in Space Exploration." June 8–10, 2010. Princeton University. Princeton, NJ.

Crysdale, Cynthia S. W. "God, Evolution, and Astrobiology." In *Exp*, 220–42.

Deane-Drummond, Celia E. "The Alpha and the Omega: Reflections on the Origin and Future of Life from the Perspective of Christian Theology and Ethics." In *Exp*, 96–112.

ESA Space Debris Office. http://www.esa.int/Our_Activities/Operations/Space_Debris/ESA_Space_Debris_Office.

Ferkiss, Victor C. *Future of Technological Civilization*. New York: George Braziller, 1974.

Gilmore, Michael. "Space Exploration." In *ESTE*, 4:1831–5.

Gore, Al. *The Future: Six Drivers of Global Change*. New York: Random House, 2013.

Gott, J. Richard. "Our Future in the Universe." In *WU*, 400–424.

Hart, John. *Cosmic Commons: Spirit, Science, and Space*. Eugene, OR: Cascade Books, 2013.

Jones, Nicola. "Planetary Disasters: It Could Happen One Night." *Nature* 493 (January 8, 2013) 154–56.

Lupisella, Mark. "Cosmological Theories of Value: Relationalism and Connectedness as Foundations for Cosmic Creativity." In *ESE* 75–92.

Marks, Paul. "Clearing the Heavens, One Piece at a Time." *New Scientist* 209:2799 (February 12, 2011) 22.

The Mars Society. Mars Direct. http://www.marssociety.org/home/about/mars-direct.

McAdamis, E. M. "Astrosociology and the Capacity of Major World Religions to Contextualize the Possibility of Life Beyond Earth." *Physics Procedia* 20 (2011) 338–52. http://www.sciencedirect.com/science/article/pii/S1875389211006006.

McKay, Christopher P. "Planetary Ecosynthesis on Mars: Restoration Ecology and Environmental Ethics." In *Exp*, 245–60.

McMahon, Sean. "The Aesthetic Objection to Terraforming Mars." In *ESE*, 209–18.

Moe-Lobeda, Cynthia. *Healing a Broken World*. Minneapolis: Fortress, 2010.

Morin, Monte. "Russian Meteor was Actually a Tiny Asteroid, NASA Says." *Los Angeles Times* February 15, 2013. http://www.latimes.com/news/science/sciencenow/la-sci-sn-russian-tiny-asteroid-20130215,0,5424522.story?track=rss.

Næss, Arne. "The Shallow and the Deep: Long Range Ecology Movement." *Inquiry* 16 (1973) 95–100.

Nature Editorial. "Realities of Risk." *Nature* (January 8, 2013).

Persson, Erik. "The Moral Status of Extraterrestrial Life." *Astrobiology* 12.10 (2012) 976–85.

Peters, Ted. "Where There's Life There's Intelligence." In *What is Life? On Earth and Beyond*, edited by Andreas Losch and Andreas Krebbs, 236–59. Cambridge: Cambridge University Press, 2017.

Pope Paul VI. *Gaudium et spes.* http://www.vatican.va/archive/hist_councils/ii_vatican_council/documents/vat-ii_const_19651207_gaudium-et-spes_en.html.

———. "Pastoral Constitution on the Church in the Modern World: *Gaudium Et Spes*, Promulgated by His Holiness, Pope Paul VI on December 7, 1965." No. 26, The Holy See.

Race, Margaret S. "Space Exploration and Searches for Extraterrestrial Life: Decision Making and Societal Issues." In *ELU*, 141–56.

———. "Societal and Ethical Concerns." In *Planets and Life: The Emerging Science of Astrobiology*, edited by Woodruff T. Sullivan III, and John A. Baross, 483–97. Cambridge: Cambridge University Press, 2007.

Randolph, Richard O. "God's Preferential Option for Life: A Christian Perspective." In *Exp*, 218–302.

Randolph, Richard O., and Christopher McKay. "Protecting and Expanding the Richness and Diversity of Life, an Ethic for Astrobiology Research and Space Exploration." *International Journal of Astrobiology* 13 (2014) 28–34.

"Researchers Eye Tethers for Space Debris." *Science* 343:6175 (March 7, 2014) 1062.

Rojas, Alejandro. "Islamic Leaders Issue Fatwa Prohibiting One-Way Trips to Mars." Open Minds Editorial. February 24, 2014. http://www.openminds.tv/islamic-leaders-prohibit-one-way-trips-mars/26034.

Rolston, Holmes, III. "The Preservation of Natural Value in the Solar System." In *Beyond Spaceship Earth: Environmental Ethics and the Solar System*, edited by Eugene C. Hargrove, 140–82. San Francisco: Sierra Club, 1990.

Schwarz, James, and Tony Milligan. "Introduction: The Scope and Content of Space Ethics." In *ESE*, 1–11.

Smith, Kelly C. "The Trouble with Intrinsic Value: An Ethical Primer for Astrobiology." In *Exp*, 261–80.

Stern, S. Alan. "The Low-Cost Ticket to Space." *Scientific American* 308:4 (April 2013) 69–73.

Tyson, Neil deGrasse. "The Search for Life in the Galaxy." In *WU*, 146–69.

23

Astroethics and Intelligent Life in the Milky Way Metropolis

TED PETERS

> And what is the object of my love? I asked the earth and it said "It is not I"... I asked heaven, sun, moon and stars; they said: "Nor are we the God whom you seek." And I said to all these things in my external environment: "Tell me of my God who you are not, tell me something about him." And with a great voice they cried out: "He made us" (Ps 99:3).
>
> —AUGUSTINE, CONFESSIONS X.9

In this chapter we turn to the larger Milky Way metropolis. Even though we speak of a metropolis, connoting that we can get to the other side of town if we just spend a little time on the freeway, this metaphor understates the distances involved. Our Milky Way is a giant spiral with uncountable star systems numbered in the billions. It is a million light years in diameter. And surrounding the Milky Way in satellite orbit are two dozen other lesser galaxies as well as mysterious Magellanic Clouds which take two to four billion years just to complete one orbit. This is just our local galaxy! A hundred billion other Milky Ways populate this universe. In only the most humorous sense can we speak of our home galaxy as our metropolis. Yet, we will work with this model in hopes that communication with neighbors on the other side of town, so to speak, strike up a friendship with planet Earth. We cannot imagine the technology necessary to make extrasolar communication happen, but imagine we must.

In anticipation of contact with alien intelligence in a form that reasonably resembles *Homo sapiens* on Earth, perhaps we should consider engagement with three possibilities: extraterrestrial biotic individuals who are inferior to us (less evolved), our peers (equally evolved), and superior to us (more highly evolved). In the case of more highly evolved ETI, they may have discarded their biological bodies and placed their intelligence into a machine substrate, into advanced computers; they may be post-biotic. This chapter will suggest that each of these three categories—inferior, peer, superior—implies different moral responsibilities.[1]

Do we have empirical knowledge that ETI exist? No. "No unambiguous signals from extraterrestrial intelligence have been detected."[2] Yet, the search goes on. As the search goes on, it works with a set of assumptions regarding the likelihood that life has had an independent genesis elsewhere and that this independent life form has followed a recognizable path of evolutionary development. If the path of evolution looks like evolution on Earth, then we can project that an extrasolar planet much older than Earth will be home to a much more highly evolved life form. Such ETIL may have progressed much further than we in complexity, intelligence, wisdom, science, and technological achievement. Conversely, life on a planet that began later than it did on Earth might still have attained intelligence, but it might be more primitive than *Homo sapiens* here at home. This set of assumptions contributes to the structure of the ETI myth, as I have outlined in an earlier chapter. Here, we will work with these assumptions regarding evolution; and we will try to work within the parameters of the myth's speculative calculus: planets with a longer evolution time would produce greater rational intelligence, while those with a shorter evolution time would develop an appropriate level of intelligence. The intelligence level of the ETI we eventually engage might be measured according to such an evolutionary calculus.

It must be acknowledged at the outset that the distances in question reduce the likelihood that we will soon find ourselves engaged in a real time interaction with any extrasolar civilization, let alone one that has not yet developed the technological capability of interstellar communication. Given the limit of the speed of light, and given that such ETIL might be one hundred light years distant, what is more likely is communication that involves considerable delays between question and answer. Even with this

---

1. The argument here builds on an earlier analysis in Peters, "AstroEthics: Engaging Extraterrestrial Intelligent Life Forms," 200–221.
2. Dick, "Extraterrestrial Life," 317.

acknowledgement of low likelihood of real time engagement between Earth and an extraterrestrial civilization, we will here try to imagine interactive engagement and try to identify fitting ethical frameworks for developing our moral responsibility.

Discoveries of exoplanets and the dramatic new knowledge produced by Hubble space vistas "are transforming the human mind. This transformation is playing a key part in the evolution of the ethics of space exploration—an evolution that may now be at a stage where there is a need to develop a preliminary "prime directive," in order to define conduct with other life in the galaxy."[3]

In the previous chapter we raised the prospect of planetization—that is, the rise of a single planetary community for moral deliberation regarding a galactic common good. The anticipation of developing a relationship with intelligent life in an off-Earth civilization only increases the warrant for a unified ethical posture on the part of those of us who live on the third planet from the sun. "Their (ET) confirmed presence would create tremendous pressure for a unified human response, or world government." Sovereign and separate nation states would become a thing of the past, argue Alexander Wendt and Raymond Duvall. The coming of an extraterrestrial visitation "would threaten what the modern state is, quite apart from the risk of physical destruction."[4] Whether the existence of the modern state would be threatened is debatable. Regardless, at some level we the peoples of all nations will be inclined to think of ourselves as belonging to a single human race, a single *humanum*. The time for global ethical responsibility will have arrived.[5]

So we ask: can we move from quandary toward responsibility as a single global community? Perhaps popular German philosopher Otfried Höffe might be of help here. He contends that a "trans-human, and absolutely universal universalism is entirely reasonable . . . morality maintains that there are fundamental claims applicable to all physically conditioned, linguistic, and rational beings."[6] Höffe observes that "almost all cultures highly value reciprocity combined with generosity."[7] Could such a universalism apply first to a single planetary society and then, by extrapolation, to extraterres-

---

3. Gilmore, "Space Exploration," 1835.

4. Wendt and Duvall, "Sovereignty and the UFO," 621.

5. "Relevant responsibilities to address include (1) looking out for the interests of humankind as a whole, (2) being truthful in interstellar messages, and (3) benefiting extraterrestrial civilizations." Vakoch, "Responsibility, Capability, and Active SETI," 512.

6. Höffe, *Can Virtue Make Us Happy?*, 128.

7. Ibid., 23.

trial individuals and societies? Might extraterrestrial intelligent neighbors value reciprocity and generosity just as we do? If so, then the conditions for communication sketched by Martinez Hewlett and Mark Graves in previous chapters might be met.

Distribution of Information about Contact

When it comes to ethical guidelines for first ETI contact, SETI has already offered a statement: *The Declaration of Principles Concerning Activities Following the Detection of Extraterrestrial Intelligence.* The thrust of SETI's nine principles is to follow scientific best practices, seek independent confirmation to establish credibility, and announce the discovery only after consultation with international leadership.[8]

SETI has a commitment to transparency combined with a commitment to planetization. These commitments extend to the dissemination of information regarding contact with extraterrestrial neighbors. But, that is the limit of this document. It is good as far as it goes. But, additional ethical deliberation is called for.

The Light Year Problem

We must acknowledge at the outset that the distances in question reduce the likelihood that we will soon find ourselves engaged in a real time interaction with any extrasolar civilization, let alone one that has not yet developed the technological capability of interstellar communication. Given the limit of the speed of light, and given that such alien intelligence might be a hundred light years distant, what is more likely is communication that involves considerable delays between question and answer.

Even more daunting is the prospect of sending astronauts beyond our solar system to exoplanets. Our physical bodies were made for Earth, not space. Our bodies rely upon Earth's specific gravity to function. Weightlessness would interfere with our internal systems and muscle tone. Human beings, no matter how fit, simply cannot exist for long periods in a spaceship.

With this in mind, space explorers have since the 1969 moon shot been considering robotic astronauts or even cyborgs to serve as astronauts. Machines will be able to travel much farther than biological organisms. In 1969 futurist Victor Ferkiss forecasted that "the astronauts of the future

---

8. SETI, "The Declaration of Principles Concerning Activities Following Detection of Extraterrestrial Life," 153–4; see Tarter, "Contact: Who Will Speak for Earth and Should They?" 178–99.

would be Cyborgs, men who would have many artificial organs and surroundings." Ferkiss imagined that the biological human component would remain on Earth, but the machine component would be aboard the spacecraft. He forecasted a " . . . machine-man symbiosis, in which there would be direct electromagnetic connection between the human nervous system and equipment that was receiving transmissions from sensors elsewhere in space. Those humans involved would control the exploring machines and receive impressions . . . as if they were actually physically experiencing the exploration themselves. Space would thus become part of the human environment, since even those not in the direct symbiotic relationship would be part of a culture . . . of space."[9] In other words, a few dedicated astronauts would sit in a laboratory at Cape Kennedy with a wireless connection to their machine counterpart en route to Alpha Centauri. Once our machines would find themselves welcomed by an off-Earth civilization, our Earthbound explorer could engage in conversation with the extraterrestrials. The terrestrial astronaut would report to the wider public all that he or she would be experiencing during the ride, and also report what is being learned about our new alien neighbors. This cyborg approach would be conducive to enhancing our already existing space culture and, most likely, enhance our loyalty to the galactic commons.

So, even with this acknowledgement of low likelihood of real time engagement between Earth and ETI, we will here try to imagine interactive engagement and try to identify fitting ethical frameworks for developing our moral responsibility.

Working with the Assumptions of the ETI Myth

Recall our earlier discussion of the Drake Equation. The Drake formula relies on the assumptions that provide warrant for contact optimism. For the sake of argument, we will grant these assumptions as we walk the path of ethical reasoning leading to an ethic of the galactic commons shared with intelligent space neighbors. Admittedly, such ethical reasoning is speculative. Still, the present quandary requires that we ask speculative questions about our responsibilities. Our plan here is to enter the myth, to think from within the ETI myth.

Because no empirical evidence is available for describing alien intelligence, we must speculate on the basis of what we know from the evolution of life on Earth and then be ready for surprises. Might we find intelligence that is silicon based rather than carbon based? Might we find intelligence

9. Ferkiss, *Technological Man*, 95.

expressed in entities other than biotic individuals? Might intelligence belong to groups rather than individuals? Such would count as surprises, to be sure. Allowing for such surprises, speculations seem most reasonable when we project the image of an extraterrestrial being who looks somewhat like us, a biotic individual.

Extraterrestrial individuals may be more or less advanced on a scale of evolutionary development, astrobiologists surmise. If ETI forms have evolved longer than we, they might be more complex and more intelligent. If they have evolved less than we, they might be simpler and less intelligent. Despite the absence of any empirical evidence, speculations that extrapolate on our understanding of evolution on Earth provide at least a partially coherent matrix within which we might begin ethical deliberation.

Before we go too much further, let us spell out with precision just what these assumptions are. The first assumption is this: we will at some point find ourselves in an interactive engagement with ETI either on Earth or on the home planet of the ETI in question. The stage of interactive engagement we are picturing here would come sometime after initial passive contact by SETI. Even though we must consider the possibility of face to face engagement; more than likely electronic communication will predominate, at least at the beginning.

Second: extraterrestrial intelligent creatures will in fact be creatures, intelligent individuals living together in a society. This parameter is warranted because of the working assumption made in the field of astrobiology that life might originate elsewhere in a fashion similar to what happened on Earth, and that a history of evolutionary development parallel to what has happened on Earth might follow. A concomitant sub-assumption is that evolution is progressive. This means that as time passes complexity and intelligence advance; and that the degree of intelligence achieved by ETI would be indexed to a corresponding span of evolutionary time.

Third, the level of advance achieved by ETI would be measured primarily in terms of intelligence. Even though relevant to terrestrial ethics could be alternatives to intelligence such as advances in culture, aesthetics, or morality, we will limit this discussion solely to intelligence. This restriction is warranted because intelligence is the single category most frequently identified by astrobiologists as a measure of evolutionary progress.

Fourth, the level of ETI intelligence would be measured partly, though not exclusively, by scientific or technological achievement. Many astrobiologists believe that science and technology are what intelligent beings naturally produce: the more highly evolved, the more science and the more advanced in technology. If ETI exhibit a high level of scientific and technological achievement, we will conclude that they are at least our equals if not our superiors in intelligence.

By keeping such suppositions at the level of assumption, it should be obvious that they cannot be taken for granted. We who put this book together do not grant these assumptions without some wiping of our sweaty brows. They have not been empirically demonstrated; even though by chance they might eventually be confirmed. In the meantime, we are constructing a vision of future contact built on a bamboo rather than concrete foundation.

Specifically, as we have emphasized in earlier chapters, the imputation of the doctrine of progress into biological evolution is objected to by scientists whose judgments should be respected. Ian Tattersall, for example, makes the point that "the human mind is *not* fine tuned for anything. It is the outcome of a whole host of historical accidents."[10] We will not take sides in this debate; nor will we repeat earlier arguments at this point. However, we note that if we side with the contact pessimists, then we would have no reason to speculate further. The assumptions made in the ETI myth by the contact optimists generate further research proposals that we believe are well worth pursuing. As we proceed we will simply leave an asterisk next to each of these four assumptions, recognizing their disputed status. But, we will proceed. We simply admit that we are working out ethical scenarios in light of possibilities, not inevitabilities.

Three Levels of Extraterrestrial Intelligence

In light of these assumptions, what might we forecast about our new space neighbors? Can we expect their children to progress in school like ours? Or, their men to enjoy the Super Bowl while drinking Budweiser? Or, their women to break glass ceilings? Or, should we expect a spectrum of differing levels of intelligence, interest, and moral fiber?

Although it may appear elementary, let us project three possibilities: the aliens we engage may be our inferiors, our peers, or our superiors. Given the astrobiological assumptions above, we might very well encounter beings less fully evolved and, hence, less developed in intelligence than we are. We are also likely to meet some ETI who approximate our level of evolutionary development. And certainly we might meet some whose evolutionary history is much longer than ours and whose level of achievement is far more complex than ours. Speculating on what kind of ETI we might meet belongs

---

10. Tattersall, "Origins of Human Cognition," 180.

to our quandary; and speculating on our moral obligations belongs to our responsibility.

Before going much further we need to ask: how might we go about measuring intelligence to determine which of the three ethical categories might best apply? The initial stage of engagement will immediately elicit a hypothesis: they are our equals! They are our inferiors! Or, they are our superiors.[11] We might then need to test to confirm or disconfirm this initial hypothesis.

Testing Alien IQ

To begin with, let us ask: how might we discern between our peers and our inferiors, especially if we appeal to the criterion of intelligence? Let me suggest three complementary tests: the Technology Test, the Turing Test, and the Naming Test. To be realistic, we are not likely to be in a position to ask our new space acquaintances to sit in a classroom with pencil and paper to mark black lines next to IQ questions. Our testing will probably consist of observations we make while getting to know them.

First, the technology test would be quite straight forward: do our new space neighbors make machines that work as well as ours? As SETI's Seth Shostak puts it: "you're 'intelligent' if you can build a powerful laser or a thumping radio transmitter."[12]

Like General Motors asking about Toyota, we will ask: just how does alien technology compare to ours? Worse? Better? Do their machines represent the product of an intelligence we recognize? Can we learn from ETI; and can they learn from us? If we find it to be a two way street, then perhaps our new space neighbors belong in the peer category. If it turns out to be a one way street with ETI as the only learners, then perhaps they belong in the subordinate category.

Suppose the technology test is ambiguous? Suppose it just is not clear whose technology is more advanced, theirs or ours? Then, we might need to turn to another test. If the technology test is ambiguous, we might turn to the Turing Test or the Naming Test. These two tests will depend on the establishment of communication, with back-and-forth conversation. If language is based upon a deep structure—an inherent logic required by its

---

11. The three parallel yet not identical categories used by Jan Narveson are "superhuman," "subhuman," and "very different nonhumans." Narveson, "Martians and Morals," 245–66.

12. Shostak, "Are We the Galaxy's Dumbest Civilization?" http://www.seti.org/epo/news/features/are-we-the-galaxys-dumbest.php.

fundamental expression of the relationship of the mind to its physical substrate and interaction with the physical world—then we can have confidence that with work we will be able eventually to understand alien communication. In addition, the philosophical field of hermeneutics describes shared understanding in terms of a fusion of horizons (*Horizonverschmelzung*), the slow but sure dialectical construction of a shared horizon of understanding. These considerations provide reason for optimism. To proceed with our ETI speculations, we must operate with the optimistic assumption that conversational bridges are possible.

With this in mind, let us turn to our second test, the Turing Test. So far, the Turing Test has been applied terrestrially for measuring the purported intelligence of machines, of computers. Proposed by British mathematician Alan M. Turing (1912–1954), the test simply sets up a pattern of blind interaction between a human person and a machine.[13] If during the interaction the human perceives the machine partner to exhibit signs of intelligence, then we can conclude that it is intelligent. This is an interactive test. To date no computer has passed the Turing Test, despite internet contests with chatterbots (an automatic program that carries on a conversation). Ray Kurzweil forecasts that computers will "emulate the flexibility, subtlety, and suppleness of human intelligence" sometime after 2020.[14] Might we devise a variant on the Turing Test to detect the presence of intelligence among ETI? Or, more precisely, to detect a certain feature of intelligence important to our future, namely, the ability to interact? We will need to know what to expect regarding the future of interaction before we can formulate appropriate ethical principles.

Perhaps even more than the Turing Test, the third test depends on semiotics. It is the *naming test*. As we do when introducing ourselves at our neighbor's backyard party, we would ask our new ETI acquaintances: "what is your name?" If they answer with a name, they will be candidates for peerage. If they have not named themselves, then perhaps they are something less than our peers. The criterion of naming suggests itself from two sources, one theological and the other scientific. Theologically, it is the naming of the animals which provides a significant descriptor of the human being in Genesis 2:19: "So out of the ground the LORD God formed every animal of the field and every bird of the air, and brought them to the man to see what he would call them; and whatever the man called every living creature, that

---

13. Danielson, "Turing Tests," 1985–6.
14. Kurzweil, *Singularity Is Near*, 294.

was its name."[15] Here on Earth human beings name animals, not the other way around (as far as we know).

Does the terrestrial interaction between humans and animals provide anything useful here? Those among us who live on farms or who have pets know about naming animals. Farmers name breeds. Pet owners name individual dogs or cats. The reverse is not the case. Even though animals possess intelligence and can even interact with us at certain levels, they do not name either themselves or us. We human beings are the sole namers on our planet, as far as we know. Should we engage in conversation with our new space acquaintances and find they name neither themselves nor us, then the task will be left to us.

Historically, names have been important when it comes to engagement. Europeans arriving in the Western hemisphere quickly named those already living there "Indians." The people we know as the Navaho, for example, already had a name for themselves, *Dine*. Roughly translated, the name *Dine* means "the people." As it turns out, Europeans and the *Dine* are intellectual peers. Neither has the right to establish the name for the other. Had the naming test been invoked upon the arrival of Europeans in the new world, Europeans and the *Dine* along with other tribal groups would have found themselves to be peers. Perhaps some of history's injustices might not have followed. Learning from this lesson might aid us at the first moment of engagement with extraterrestrial contacts. We recommend that we ask them: what is your name?

This is the theological ground for suggesting the naming test. Science also provides justification. Scientific support for the naming test is based on the judgment of many that the emergence of language accompanied by symbolic thinking in our historical evolution marks a significant if not definitive threshold crossed by our human species. The genes and brains of *Homo sapiens* are much like those of our primate siblings, the chimpanzees, gorillas, orangutans, and bonobos. Yet, our cognitive abilities are qualitatively different. Once the doorway to language was unlocked, a co-evolution of brain and language ensued. "Language is not merely a mode of communication," writes Terrence Deacon; "it is also the outward expression of an unusual mode of thought—symbolic representation . . . thought does not come innately built in, but develops by internalizing the symbolic process that underlies language. So species that have not acquired the ability to communicate symbolically cannot have acquired the ability to think this

---

15. Science fiction novelist David Brin brags that naming rights make the namer into a god. "Naming stuff is a god-like power . . . What is the act of naming other than an apprenticeship in co-creation?" Brin, "So You Want to Make Gods," 48.

way either."[16] If our measure of evolutionary advance is intelligence, then language is the key indicator. "Brain-language co-evolution has significantly restructured cognition from the top-down, so to speak, when compared to other species."[17] The naming test would distinguish a species that has crossed the language threshold from one that has not. After all, said Aristotle, "man is the only animal whom [nature] has endowed with the gift of speech."[18]

We have suggested three tests: the technology test, the Turing test, and the naming test. If our new alien acquaintances are superior to us in intelligence and more advanced in science and technology, these tests are likely to be useless. They might even make our aliens giggle as they watch us apply them. The most the tests could accomplish is to tell us whether our superior space friends have a sense of humor. They will be quite effective, however, in discriminating between our peers and our inferiors.

Before proceeding, logic requires that we entertain one additional possibility. It is possible that the alien will be genuinely alien—that is, the alien might be unrecognizable because it does not fit the image we are looking for. Our standard model—the model that fits the Drake Equation—utilizes intelligence in analogy to human intelligence as the main criterion. Chris Impey acknowledges how this might limit our perspective. "In astrobiology, it makes sense to take a broader view—if our intelligence sets the definition, then we'll only recognize creatures like us as intelligent."[19] In the field of hermeneutics we acknowledge how the form of the question pre-forms the answer. The *Fragestellung* or shape of the question shapes what our filter will admit as an answer. So, if we pose the question of extraterrestrial life by placing a question shaped by our experience of human intelligence, we must be aware that an extraterrestrial phenomenon might appear to us that does not fit the question guiding our research. Be that as it may, we must proceed in one or another fashion, and it is reasonable to follow the path blazed by SETI and other astrobiologists along with the evolutionary assumptions they make.

Ethics for Engaging Inferior ETI

If these three tests—the technology test, the Turing test, and the naming test—indicate that the ETI in question are our inferiors in intelligence, we would then ask: might the ethical framework for discerning our

16. Deacon, *Symbolic Species*, 29.
17. Ibid., 417.
18. Aristotle, *Basic Works*, 1129.
19. Impey, *Living Cosmos*, 287.

responsibility toward them be analogous to our responsibility toward Earth's animals? If we answer affirmatively, then we would find ourselves in a classic dialectic. On the one hand, the human race exploits all other life forms—both plants and animals—for human welfare. Animals provide food, work, clothing, and even company. Animals can be sacrificed in medical research to develop therapies that will benefit only human persons. On the other hand, we human beings have a sense of responsibility toward the welfare of animals. We respect them as intelligent beings; and we are concerned about preventing suffering to animals. In some instances, we exert considerable energy and effort to preserve their species from extinction and to insure the health of individual animals. In the case of pets, we love them to a degree that rivals loving our own family. In sum, we have inherited this double relationship to our inferiors already here on Earth.

It is not unusual for moral conflict to arise over the way we treat animals. "Every area of human-animal interaction, be it agriculture, research, hunting, trapping, circuses, rodeos, zoos, horse and dog racing, product extraction, and even companion animals, is fraught with ethical and welfare issues."[20] In the early history of the human race which included the domestication of animals, "husbandry" became the dominating ethic. This "fair and ancient contract" embraced the double principle that animals become better off, because humans care for them and protect their welfare, while humans become better off because they benefit from animal products. Animal ethics included proscriptions against intentional mistreatment or deliberate cruelty. As a consequence, humans and animals have co-evolved.

In the post-World War II era, however, the industrialized production of animal products has removed the "welfare" component. Animal suffering has become massive: confinement, loneliness, boredom, disease. This suffering is not due to cruelty but due to the efficiency called for by industrial production. The previous husbandry ethic is being replaced with an "animal rights" ethic as our society grapples with the reinstantiation of the "fair and ancient contract." Columbia University's Richard Bulliet forecasts that "the future of human-animal relations in real-world terms will be determined by the worldwide expansion of exploitation in a late domestic mode and the reaction to that expansion by increasingly angry post-domestic activists."[21] Might we expect the tensions we experience here on Earth to reappear when we meet intellectually inferior—less intelligent—extraterrestrial aliens?

Important for our application to the upcoming ETI question is the ethical criterion for rendering moral policy. Is the animal rational in the

20. Rollin, "Animal Welfare," 82.
21. Bulliet, *Hunters, Herders, and Hamburgers*, 223.

same sense that we *Homo sapiens* are rational? No, says the tradition beginning with Aristotle. The human is *the* distinctively rational animal. The separation of humanity from the animal world due to a separation on the scale of rational intelligence justifies an ethic whereby the superior human exploits the inferior animal. "Irrational animals are natural slaves, and no positive human moral or political categories can govern humankind's relations with them."[22] Yet, we might ask: is rational intelligence the only ethical criterion? No, say some moralists. By extension, our moral commitment to protect humans from suffering should be applied to animals. How an animal feels is morally significant. "Animal welfare is most crucially a matter of the animal's subjective experience—how the animal feels, whether it is in pain or suffering in any way."[23]

With this experience of relating to animals on Earth in mind, our first ethical question in the case of engagement with inferior ETI would be: by which criterion do we orient our ethics, ETI's relative rational intelligence or our responsibility for the welfare of ETI? Our second ethical question would be: which of these two habits should we invoke: exploitation of ETI for our own use? Or protection of ETI life from suffering? More than likely, all of these alternatives would inform the policies we develop. In ethical categories, we would not impute dignity to ETI whose level of rational advance falls significantly short of ours. We very well might show them respect, even care. Motivated by faith, we might view alien beings as creatures precious in the eyes of God. In economic categories, more than likely we would exploit lesser intelligent ETI for increased terrestrial prosperity. We might work out terms of exchange, or, more likely, simply set up an infrastructure for ongoing exploitation. Would we exploit with moral abandon? Or, would we exploit only to the limit set at the point of detriment to the welfare of the ETI themselves?

This raises the matter of our moral responsibility. In terms of our responsibility, I believe we should take the initiative to extend concern for the welfare of such ETI on the model of our current concern for the subjective quality of animal experience. We should do what we are able to protect ETI from suffering and enhance their experience of wellbeing.

One more question. Might we expect the extraterrestrials to be moral beings? Recall how Höffe asserted that all rational beings would value reciprocity and generosity. Michael Ruse makes a similar argument, relying on the assumptions of the ETI myth: given that aliens would evolve as we have; and given that morality is adaptive, then highly evolved extraterrestrial

---

22. Fellenz, "Animal Rights," 1:74–77.
23. Rollin, "Animal Welfare," 1:82.

beings will have discovered morality just as we earthlings have. "What I argue, therefore, is that if, indeed, natural selection is at work on our extraterrestrials, we might expect to find something akin to the Categorical Imperative [Golden Rule] having evolved amongst them."[24] I will suggest here that the categorical imperative fits like a glove over a hand when it comes to engaging peer ETI.

Ethics for Engaging Peer ETI

Before proceeding further, let's return for a moment to the question of grounding our ethics. To what do we appeal for our moral standards? Might the Golden Rule provide the foundation we need?

If we conclude that ETI are our peers in rational intelligence, then we might find appropriate the Golden Rule along with Immanuel Kant's Categorical Imperative. Jesus' version of the Golden Rule is familiar to us all: Matthew 7:12 "In everything do to others as you would have them do to you." Even though Kant found weaknesses in the Golden Rule, his categorical imperative universalized it. The formal principle from which all moral duties are derived is this: "I ought never to act except in such a way *that I also will that my maxim should become a universal one.*"[25] In sum, we should treat peers as equal to ourselves; and we should care for their welfare just as we would care for our own.

The Golden Rule does not require actual reciprocity on the part of our off-Earth neighbors. Even though we think we know how we on Earth would like to be treated, our motive to treat space aliens accordingly would be motivated by our love for them. We Earthlings may pre-emptively act beneficently toward them. "When Christians try to live according to the Golden Rule, they should imitate this [God's love for us] love, the rule of divine love," says Helsinki theologian Risto Saarinen. "Humans should meet their neighbor's needs."[26] Might this rationale sit well with non-Christians who'll be working with Christians in our terrestrial community?

One of the traits of the Golden Rule that commends to us is this: the Golden Rule appears on Earth in diverse cultures. We find a version of it in the teachings of Confucius in ancient China, in Thales and Aristotle in ancient Greece, and elsewhere. Randolph and McKay proffer a *Cosmic Golden Rule* applied to space exploration. According to "what we call the *cosmic Golden Rule,*" they write, "all of the Earth's global religions contain some

24. Ruse, "Is Rape Wrong on Andromeda?" 65.
25. Kant, *Groundwork,* 70; see Narveson, "Martians and Morals," 248.
26. Saarinen, *God and the Gift,* 55.

version of the 'Golden Rule.' We prefer the Confucian version of the Golden Rule, which frames it as a prohibition: 'Never impose on others what you would not choose for yourself.'[27] Perhaps Höffe is right: the principle of reciprocity is universal. This would make the Golden Rule a ripe candidate to be the foundation on which to construct an ethic for the galactic commons.

Might the Golden Rule be due to an inherent moral logic bequeathed to the human race by our evolutionary development? Marcus Singer puts before us a challenge. "The fact that it is basic to moral codes of so many and such different peoples would seem to entail that it is a fundamental normative moral principle, connected inextricably with human nature, and this inference from an *is* to an *ought* surely deserves examination."[28] With this in mind, let us speculate. If the Golden Rule is grounded in what is natural, then perhaps peer aliens may themselves have stumbled upon it. Whether it belongs to existing alien morality or not, at least an alien would likely understand our appeal to the Golden Rule as a moral standard. This might require a two step process for both earthlings and aliens: first, establish equality; then, second, apply the Golden Rule.

Might this apply as well to the Kantian variant? Might Kantianism also be rooted in nature, as Ruse suggests? If so, then extraterrestrials might already embrace it. This leads Guillermo Lemarchand to propose a *Lex Galactica*, a galactic Kantianism. "The only possibility to avoid self-destruction is a deep and strong societal mutation, based in some sort of Kantian ethics . . . some kind of *Lex Galactica*."[29] In short, a Kantian version of the Golden Rule might already be embraced by our new space neighbors; and this would support our embracing such an ethic here on Earth.

Critics of the Golden Rule say it should not apply to space aliens, however. Why? Because ETI might differ from us. Therefore, we should treat space aliens not as we would treat others within our terrestrial commons, but, rather, the way they wish to be treated.[30] Such an argument can be dismissed as a tautology. The way we want to be treated along with the way they want to be treated is just what the Golden Rule targets.

Jesus' Golden Rule and Kant's Categorical Imperative have greatly influenced the value system of the Enlightenment and, hence, the modern culture of which we are a part. If we find that ETI resemble us enough to

27. Confucius, Analects XV.24, cited in Randolph and McKay, "Protecting and Expanding." Because of the immensity of the cosmos, it is inconceivable at this time to think of reciprocity beyond the Milky Way. Perhaps *galactic commons* with a *Galactic Golden Rule* becomes more reasonable.

28. Singer, "Golden Rule," 1:615.

29. Lemarchand, "Speculations," 162.

30. "Religion: Space Ethics," *Time*.

be considered our peers, then we might invoke the value system of the Enlightenment—that is, we might invoke the Golden Rule and impute dignity to our space neighbors, honoring the way they declare they wish to be treated. Our moral disposition would be to approach our new neighbors with operative values such as equality, liberty, dignity, justice, and mutuality. When it comes to dealing with ETI as individuals, we would impute dignity to them—that is, we would treat each as a moral end and not merely as a means. "Act in such a way that you always treat humanity, whether in your own person or in the person of any other, never simply as a means, but always at the same time as an end," wrote Kant.[31]

In a more contemporary and theological medium, we might say: "human dignity is the inherent worth or value of a human person from which no one or nothing may detract."[32] Might we impute dignity to an alien? Might we treat aliens with dignity? Might *dignity* become the label for identifying our responsibility?

Dignity comes in both theological and secular forms. Theologically, dignity is the gift of God. The "gospel in miniature," John 3:16, reports that "God so loved the world [*kosmos*], that he gave his only Son . . ." God's act of redemption in Jesus Christ confers dignity on the world as a whole, and on us individually. God's love for the world elicits in us a sense of value or worth. God confers upon us dignity and we, in turn, affirm our dignity and claim that dignity. God's conferring of dignity upon us is an act of divine grace; it is not based on any quality we ourselves possess.

The more secular variant of dignity which we find in Kant associates dignity with our rational capacity. Rational intelligence warrants being treated with dignity. If we lack rational capacity, we cannot claim dignity. These two variants—dignity due to rational capacity or dignity due to gracious conferral regardless of rational capacity—can cohere as long as we posit that every individual being belongs in the set or class of rational beings. Infants or elderly human persons suffering from dementia may lack individual rational capacity; yet we treat them with dignity because they belong to the human race. We confer dignity upon less than fully rational members of the human race. Based upon these post-Christian Enlightenment values, every human person can claim the right to be treated with dignity.

What might this imply as we construct ethical policy for engaging peer ETI? Theologian Robin Lovin draws out what he deems to be the implication. "*Any* intelligent life we are able to discover . . . will have capacities for

---

31. Kant, *Groundwork*, 96.
32. Falconer, "Dignity," 278.

reason, will, and memory that are analogous to our own. There is, then, a presumption of dignity in any intelligent life that argues against attempts to conquest, conversion, deception, or exploitation."[33] Lovin's logic is clear: if intelligent, then dignity.

I recommend that we adapt portions of the *Universal Declaration of Human Rights*, passed by the General Assembly of United Nations in 1948. We would rewrite them to refer to the ETI we have deemed to be our peers. Where we read "human beings," we could substitute the name ETI apply to themselves.

> Article 1: "All human beings are born free and equal in dignity and rights. They are endowed with reason and conscience and should act toward one another but in a spirit of brotherhood."
>
> Article 3: "Everyone has the right to life, liberty, and security of person."
>
> Article 5: "No one shall be subjected to torture or to cruel, inhuman or degrading treatment or punishment."

The imputation of dignity to extraterrestrial aliens should be accompanied by a denial of our own right to unilaterally exploit them. We might encourage the development of bilateral commerce, of course; but we should do so presuming the equality and liberty of our trading partners. We might also restrict our intrusion into their ecosphere. We might adapt for ETI the Race and Randolph principle aimed at non-intelligent life: "respect the extraterrestrial ecosystem and do not substantively or irreparably alter it (or its evolutionary trajectory)."[34] In sum, the ethical principles we invoke to deal with peer ETI might draw upon our Enlightenment values. We could formulate principles applicable to ETI which we now invoke to maintain terrestrial justice and peace.

These Enlightenment values regarding dignity combined with a vision of the need for a single planetary society to become the community of moral deliberation should find enthusiastic support from the Christian church. According to postmodern theologian John Milbank, "the Church . . . is a practical operation, which seeks to establish an international, harmonious, *cosmopolis.*"[35]

---

33. Lovin, "Astrobiology and Theology," 230.
34. Race and Randolph, "Need for Operating Guidelines," 1583–91.
35. Milbank, *Beyond Secular Order*, 124; Milbank's italics.

Refining the Categories

Before proceeding to ethical analysis of our third category for identifying relevant moral communities, superior ETI, we might wish to refine the three basic categories. I recommend for this thought experiment dividing peer ETIs into two subcategories: hostile and peaceful; and I recommend we divide the superior ETIs into three subcategories: hostile, peaceful, and salvific. Once we have discerned that ETI are our equals or our superiors in technology and perhaps in intelligence, we will need to ask whether or not they pose a threat to Earth's security and wellbeing. How we answer this question may partially guide the moral direction we take.

The anxiety associated with insecurity leads us Homo sapiens to strike out with violence.[36] We on Earth will find ourselves uneasy, on the verge of violence, until we can be assured that the aliens we confront mean us no harm. Whether the high-minded among us find it moral or not, the inescapable reality is that no rational discourse about ethics can take place when anxiety is high and security is low. To determine whether ETI are a threat or not will ineluctably become our first priority.

In the event that the ETI in question are in fact hostile, then we will find ourselves working within a framework that includes both the imputation of dignity mentioned above and our pressing need to protect our planet from alien exploitation or damage. We know from experience that whenever we are confronted with a hostile enemy from without, we find ourselves within our society compromising human dignity. Our political leaders try to persuade our society that our targeted enemies should be reduced to "inhuman" if not demonic status; and this justifies going to war. What this indicates is that the social psychology of self-defense pits human dignity against the mustering of military support. Security trumps dignity, even if a minority of high-minded individuals protest. If threatened by alien hostility, we can forecast that military rhetoric will attempt an equivalent of dehumanizing and, hence, de-dignifying the ETI enemy. A nation's leaders simply cannot embrace Jesus' peace ethic of loving our enemies combined with turning the other cheek (Matthew 5—7). So, as difficult as it may sound, we will need an ethic that affirms the dignity of ETI while rallying our Earth allies in planetary defense. We might need to adapt for peer ETI the Race and Randolph principle, "cause no harm to Earth, its life, or its diverse ecosystems," within a tense relationship to the wider ethical principle of imputing dignity to our extraterrestrial peers.

36. On anxiety and security, see: Peters, Sin.

This may create a conscience challenge for the morally sensitive among us. Bastianel reminds us that "the horizon of meaning of all that belongs to the sphere of morality has its focal point: a relationality qualified by gratuitousness, acceptance, and fraternity."[37] This is the moral horizon within which Jesus asked us to love our enemies. Yet, if the aliens present themselves as our enemies, we earthlings may need to pull our wagons into a circle for self-defense. You cannot pursue a fraternal relationship if you are dead.

In the event that peer ETI prove to be neutrally peaceful or even benevolent, then the principles giving expression to Enlightenment values should prevail without challenge: equality, liberty, dignity, and mutuality.

Ethics for Engaging Superior ETI: Post-Biological Intelligence

One scenario that the ETI myth of progressive evolution suggests is that an off-Earth civilization may have enjoyed a longer time for evolutionary advance than we on Earth. Such an extraterrestrial civilization may have arrived at our stage in evolution long ago, and then continued to evolve. We know from the Darwinian model that no species stays around forever. Every species is undergoing constant change, change that leads to extinction and replacement by another species. There is no scientific reason whatsoever to suppose that the human species is here to stay. Our species will also become extinct and replaced. Similarly, should a species like ours have appeared on another planet long ago, it would be gone by now. This scenario is implied by the scientific assumptions at work in any serious consideration of evolutionary principles.

When we consider contact with a species blazing our evolutionary path yet older and more highly evolved than we, we are simultaneously considering our own future. Aliens may have arrived at our future before we have.

If this logic holds, then we might ask: just what have futurists among us on Earth forecasted for the future of *Homo sapiens*? This question has an answer, perhaps many answers. Among the answers the one proposed by the post-humanists or transhumanists elicits the most excitement and stimulates a spiraling of our imagination. In brief, the transhumanists envision that we ourselves will actively replace ourselves with a post-human society of intelligent computer-like machines. Our future will be post-biological.

---

37. Bastianel, *Morality in Social Life*, 49.

Intelligence will continue, even increase; but our physical bodies will be buried in the dust of terrestrial history.

The posthumanist scenario leads to merging humanity with technology as the next stage of our human evolution. Humanity plus (H+) is calling us forward. The term *posthuman* refers to who we might become if the transhumanists among us achieve their goals.

For the transhumanists, evolution and progress are two ideas that fit together like a gun and a bullet. "That technological progress is generally desirable from a transhumanist point of view is also self-evident,"[38] Nick Bostrom at Oxford says with a touch of understatement. This technological progress will allegedly shoot the human race like a cannon ball over its previous barriers. "Transhumanism has roots in secular humanist thinking, yet is more radical in that it . . . direct[s] application of medicine and technology to overcome some of our basic biological limits."[39]

One of these limits is death. Can we overcome death through advancing computerization? Yes, indeed. We can upload our mind into a computer, and then our intellectual life will take place within computer hardware instead of biological hardware. We will have attained cybernetic immortality.

Posthuman intelligence will find a way to remove itself from our deteriorating bodies and establish a much more secure substrate for endurance. Our mental lives in the future may take place within a computer or on the internet. What we have previously known as *Homo sapiens* will be replaced by *homo cyberneticus*. "*As humanism freed us from the chains of superstition, let transhumanism free us from our biological chains.*"[40]

In a computer, we will not age like we would in our bodies. If a computer breaks down, someone can simply copy and upload our mental life into another computer. This could go on forever, in principle. Once liberated from the limits of aging or even from our physical bodies themselves, the expansion of human intelligence would be limited only by the size of our universe. We will travel mentally throughout our solar ghetto, throughout the Milky Way Metropolis, and beyond. What Hans Moravec foresees is a cosmic imbuing of matter with our consciousness. "Liberated from biological slavery, an immortalized species, *Homo cyberneticus*, will set out for the stars. Conscious life will gradually spread throughout the galaxy . . . until finally, in the unimaginably distant future, the whole universe has come alive, awakened to its own nature—a cosmic mind become conscious of itself as a

---

38. Bostrom, "Transhumanist Values."
39. Ibid.
40. Young, *Designer Evolution*, 32; italics in original.

living entity—omniscient, omnipotent, omnipresent."[41] The entire universe will be converted into an "extended thinking entity," writes Moravec.[42] If the transhumanists actualize the potential residing in their futuristic vision, the more intelligent creatures who replace us will enjoy a level of salvation that we in our generation will never experience. Hooray for them!

With this scenario in mind, could we imagine that something like this has already happened to one or more races of alien intelligences? If we establish contact, might it be with the post-alien machines enjoying cybernetic immortality and spreading throughout the stars?

Nobel Prize winning physicist and cosmologist Martin Rees entertains the possibility of artificial intelligence if not post-human intelligence. "The most likely and durable form of life may be machines whose creators had long ago been usurped or become extinct."[43] In outer space we earthlings may meet our own post-human future.

Would we have any moral obligation to non-biological or post-human intelligence? If we use contemporary experience on Earth with computers as a model, the answer is negative. Destroying or recycling computer hardware or software elicits within us no moral sensitivity. We know tacitly that we are not snuffing out the consciousness of a sentient being when we press the "delete" button.

If the machines we encounter in space exhibit artificial intelligence, then unplugging them could be done without moral consequence. If the machines we encounter are actually the living tombs of immortal aliens who exchanged their biological bodies for plastic and silicon hardware, then we might give moral pause. We may find we should treat our conscious alien friends with the kind of dignity entertained in our previous discussion of peer ETI. We might offer to clean the keyboard and secure the plug in the wall socket, perhaps even to play a hand of Bridge or two and keep them company.[44]

Ethics for Engaging Superior ETI: Still Biological Yet Intellectual

We are engaged here in a series of thought experiments. When entertaining these thought experiments, it is difficult to *imagine up*. It is easier to *imagine down*. When comparing humans with animals, for example, we can

41. Moravec, *Mind Children*, 44.
42. Ibid., 116.
43. Rees, *OFH* 167.
44. See Peters, "Outer Space and Cyber Space."

imagine down by distinguishing things we humans can do that are beyond the capability of the animal with whom we already share a commons. When it comes to imagining ETI who might be superior to us in intelligence, it is difficult to imagine up. It is difficult to imagine what superior intelligence could manifest that is beyond the very human intelligence that is doing the imagining. This puts initial constraints or limits on how we can begin to approach the topic of ethics when engaging ETI more advanced than Earth's *Homo sapiens*. Nevertheless, it is incumbent in astrobiological ethics to speculate about the possibility of engaging with intelligent beings who are superior to us.

If we meet ETI superior to ourselves, will they be hostile? neutrally peaceful? or salvific? Given the assumptions made by astrobiologists that extraterrestrial evolution will follow a path toward increased intelligence as it has on Earth, the prospect of ETI fitting the hostile category is to be expected. Charles Darwin's key evolutionary principle is "natural selection," which he identifies with "the struggle for existence" and with Herbert Spencer's phrase, "survival of the fittest."[45] In the struggle for existence, living creatures undergo cruelty, suffering, and waste.[46] And the species to which virtually every individual creature belongs will eventually go extinct to make way for a more fit species. The strong devour the weak. The big eat the small. The fit survive in a world that is, as Tennyson put it, blood "red in tooth and claw."

Even slavery is not unknown in the pre-human struggle for existence. One species of red ant (*formica sanguinea*) enslaves a species of black ants (*formica fusca*) to gather food and make nests, observes Darwin. "The slaves are black and not above half the size of their red masters."[47] If we would substitute intelligence for size and ascribe it to the ETI we meet, perhaps they might decide to enslave *Homo sapiens* for their gain in the struggle for existence. This prospect would be most consistent with the theory of evolution as we project it onto planets among the stars.

With this set of assumptions in mind, *risk management* theorists such as Mark Neal recommend that we earthlings begin preparing now. The prospect that extraterrestrial beings superior to us exist on extrasolar planets will inevitably mean a reconfiguration of disaster and risk management priorities, which would bring the issue of alien life from the periphery of research and theorizing into its core. If the contact is from a benign and nearer source, which enables visitation, or involves visitation itself—then

---

45. Darwin, *Origin*, 89.
46. Ibid., 445.
47. Ibid., 245.

the risks are again sociocultural, with humans perhaps being cast as intellectual underlings, with a relationship of dependency emerging with the alien civilization."[48] Dependency, not slavery. Note Neal's justification for this thinking, namely, the ETI myth regarding evolutionary advance. "The use and usefulness of human responses to alien encounter depend a great deal on how advanced the contacting civilization is. The universe is old, and space travel difficult. As a consequence any visiting civilization might be thousands or millions of years more developed than our own. Human protocols, from information management to planetary defence, might thus be brutishly inadequate, and immediately confounded by the appearance of such advanced beings."[49] Not only will the alien civilization be more advanced than ours on Earth, it may be hostile or exploitative.

Given astrobiological assumptions regarding a repeat of evolution on extraterrestrial planets, hostility is what we should expect on the part of ETI. Yet, surprisingly, some SETI speculators anticipate meeting intellectually superior ETI who will benevolently help us on Earth. For this reason, I add the subcategory of *salvific*. Now, how do we get from the struggle for existence to extraterrestrial saviors? How does evolution transcend itself?

We are trying to discern the logic inherent within SETI thinking. As we have noted, some in the astrobiology community project an image of a more highly evolved extraterrestrial creature who would like to rescue us earthlings from the ignorant habits we have developed due to our inferior level of intelligence. Because we on Earth have not yet achieved the level of rationality necessary to see that international war and planetary degradation are inescapably self-destructive, we could learn from ETI more advanced than we. "All technological civilizations that already have passed through their technological adolescence and have avoided their self-destruction . . . must have developed ethical rules to extend their societal life expectancy," says Lemarchand, speaking for the space community.[50] Because they are beyond war, spacelings can help earthlings get beyond war.

Such thinking is obviously myth, not science. No empirical evidence justifies such speculation; and this introduces incoherence into the theory that extraterrestrials have evolved in the same we that we have evolved. Yet, such dreaming of redemption descending from the skies is tantalizing to the terrestrial imagination. The essence of the ETI myth is that science saves. Science can save Earth from its inadequacies, its evolutionary backwardness,

---

48. Neal, "Preparing for Extraterrestrial Contact," 71.
49. Ibid., 72.
50. Lemarchand, "Speculations," 154.

its propensity for self-destruction. If terrestrial science is insufficient, then extraterrestrial science just might be.

Recall once again what SETI's Frank Drake looks forward to. "Everything we know says there are other civilizations out there to be found. The discovery of such civilizations would enrich our civilization with valuable information about science, technology, and sociology. This information could directly improve our abilities to conserve and to deal with sociological problems—poverty for example. Cheap energy is another potential benefit of discovery, as are advancements in medicine."[51] Note the optimism. Drake does not expect what Darwin would expect, namely, an extraterrestrial race engaged in the struggle for existence which might like to exploit us on Earth. Rather, Drake's extrapolation of evolution to ETI imagines an intelligent and benevolent race ready to offer us aid and assistance. His vision includes optimism regarding the solution to Earth's "sociological" problems such as poverty and energy. Space visitors might even give us a leap forward in medicine.

What Drake believes is that science is salvific, and extraterrestrial science would be even more salvific than terrestrial science. In sum, should an extraterrestrial civilization more evolutionarily advanced than we engage planet Earth, we could benefit from the ability of ETI to save us from our own primitive inadequacies and even our own propensity for self-destruction. It is this thought structure within astrobiology that warrants the designation for more highly evolved and more intelligent ETI as "celestial saviors."

For ETI to become salvific for us on Earth, ETI would necessarily be altruistic. By *altruism* we refer to a motivational disposition to aim conduct toward "the good of the other person" (Latin *alter,* "the other"). We would be "the other" in the eyes of superior ETI. What warrants our projecting altruism on to evolving ETI? An assumption with two overlapping applications is at work here. The first is this: to be more highly evolved is to be better. The second is this: a race of beings more highly evolved than we will have developed not only science but also altruistic virtue. With such an assumption, Drake and others such as Paul Davies can speculate that a more highly evolved race of intelligent beings will have surpassed us not only in science but also in morality. These advances will place ETI in a position of saving us on Earth from our backwardness and from the dangers of ecological or thermo-nuclear self-destruction. What is happening here? The idea of salvific power on the part a more advanced extraterrestrial life form is the projection of Earth's astrobiologists who already see terrestrial science as potentially salvific. A more advanced extraterrestrial version of

---

51. Drake, "Interview with Dr. Frank Drake," 5.

science combined with altruistic virtue prompts in us a vision of celestial science descending from the sky to rescue us from our primitive and still self-destructive human propensities.

In short, we find within space speculation an image of evolutionary perfection which is inconsistent with what the more conservative Darwinian model would predict. Given Darwin's description of the "struggle for existence," one would expect more highly evolved ETI to exploit us, perhaps even enslave us. Drake's SETI mindset, however, predicts an advanced stage in evolution where the principles of past evolution would no longer apply, where ETI would be so altruistic as to desire to aid us on Earth without expecting reciprocity. Evolution will have produced celestial saviors coming to our planet to rescue us from our own underdeveloped intelligence.

We have seen this incoherence elsewhere in the scientific discussion. This is reminiscent of Richard Dawkins who, on the one hand, says that the entire history of evolution has been driven by the "selfish gene" while, on the other hand, he also says that we in the modern world can overcome our genes and adopt liberal ethics. "We have the power to defy the selfish genes of our birth and . . . cultivating and nurturing pure, disinterested altruism—something that has no place in nature."[52] Dawkins articulates what seems to be assumed by some SETI voices: even though evolution to date has been cruel and selfish and destructive, eventually with more highly evolved intelligence creatures will become so altruistic as to leave their evolutionary background behind. Now, one might ask: how can one make such a prediction based upon evolutionary precedents? One might ask: how can this evolutionary leopard change its spots? Yet, despite the incoherence, we can find within the SETI conceptual set a prediction of scientific salvation from ETI. So, we will ask here about the ethical implications of such an optimistic prediction right along with predictions based on the more standard Darwinian model.

Superior ETI: Ethics of Enslavement

If superior ETI follow the standard Darwinian model and confront us with hostile and exploitative enslavement, then perhaps we will frame our ethics accordingly. The New Testament provides instructions for slaves. 1 Peter 2:18: "Slaves, accept the authority of your masters with all deference, not only those who are kind and gentle but also those who are harsh." This may seem as unrecognizable as it is repugnant. The treatment of the superior master by an inferior slave has fallen into disuse in our post-Enlightenment

52. Dawkins, *Selfish Gene*, 200–201.

period. This is because of the erasure of the line between superior and inferior human beings within modern Enlightenment culture. We are all equal—that is, we are all ethically equal. Each of us has dignity by virtue of our belonging to a single moral set: the human race; and slavery violates the principle of dignity which we ascribe to every individual member of this moral set. Should a master-slave relationship rear its ugly head somewhere on our planet, we children of the Enlightenment would encourage the slaves to rebel and strive for their own liberation. Such a moral commitment to liberation would be justified by the assumption that both masters and slaves are equal.

When we use the assumptions made by many in the astrobiology field, in contrast, we cannot coherently make the argument that all intelligent beings are equal. Those who have evolved longer and who have attained a higher level of rational intelligence would be, by definition, superior to us. We could not justify liberating ourselves from their rule with an argument based upon equality. If we allow a New Testament influence, we might consider developing an ethic of slave responsibility. Given the assumptions adumbrated plus our need to link the galactic commons to our every day activity, our middle axiom would be: we *Homo sapiens* should express loyalty if not love for our masters.

Might loyal slave morality have its limits? What if our superior slave masters request or even demand of us something we deem immoral? Jan Narveson poses this question. "If another set of beings, invulnerable to our best efforts to destroy, damage, or injure them and overwhelmingly more powerful than us in every way, threatened us with various kinds of coercion which they were perfectly able to apply unless we did their bidding, should a moral human being do the bidding in question? I think it would be hard to give a general answer. It would surely depend on what they wanted us to do."[53] To put it in other words, if we human beings adhere strongly to certain transcendent values, then we would feel morally obligated to resist following orders dictated by our alien masters if those orders implied a sacrifice of our own moral integrity. Insofar as our alien masters dictate orders in concordance with our moral integrity, we would have no ground for disobedience.

Continuing this particular line in our thought experiment, how might we terrestrials think if we are asked to make destructive weapons? Let us consider an analogy. Currently on Earth, selected companies design and manufacture increasingly sophisticated and deadly weapons. These weapons are sold for use first by the U.S. military, then governments friendly

53. Narveson, "Martians and Morals," 256.

to the U.S., then to terrorists and drug cartels. Those who work in this industry contribute to both widespread death and big profits. Workers with consciences tend to avoid such employment. Because an enslaving ETI group would need weapons to maintain its hegemony, we earthlings might be asked to provide support for such a rule. Might we confront a moral dilemma between our responsibilities as slaves and our responsibilities to our own moral judgment? We could proceed with many such examples of moral quandary.

Superior Aliens: Ethics of Gratitude

Let us now abandon the trajectory of the enslaving ETI and turn to the second subcategory: peaceful. In the event that ETI approach the civilizations on Earth in a peaceful manner, we would want to respond with an appropriate ethic of peace. Maintaining peace with justice would become our middle axiom, linking our vision of a galactic commons to everyday activity. We might even find ourselves organizing to quiet down and restrict earthly voices among us that would disturb the peace. We would want to police ourselves in the name of peace. Peace would benefit life on Earth. In addition, moral policies we set would likely treat our alien superiors with dignity, respect, and courtesy due to their position of superiority and potential power.

In the event that ETI turn out to be not only more intelligent but also altruistic toward us, then an ethic of gratitude might be included in our responsibility. We would receive and make use of the gifts that increased intelligence would allegedly provide us: such as the means for maintaining a healthy planetary ecology, improvement in our medical care, and more justice in our social practices. Then, we would build upon what we have already said about maintaining terrestrial peace and treating our superiors with dignity; we would add a measure of grateful respect. The middle axiom in this case would be simple: gratitude.

In sum, we should treat superior ETIs with dignity, respecting and even caring for their welfare. If they are hostile and enslave us, we should invoke an appropriate slave morality that maintains their dignity. If ETI are peaceful toward us and open up avenues of conversation and commerce, then the principles of justice and the striving to maintain peace should obtain. If out of their superior wisdom and altruistic motives ETIL seek to better our life here on Earth, we should accept the gifts they bring and respond with an attitude of gratitude.

Table 1: Ethics for Inferior, Peer, and Superior ETIL

	Inferior ETI		Peer ETI	Peer ETI		Sup ETI	Sup ETI	Sup ETI
			Hostile	Peaceful		Hostile	Peaceful	Salvific
Ethics?	Respect		Dignity	Dignity		Dignity	Dignity	Gratitude

Conclusion

Astroethics today is necessarily a speculative endeavor. The astrobiological work of SETI upon which astroethics deliberates is itself speculative, and at best only partially coherent. When it comes to extraterrestrial intelligent life forms, terrestrial scientists are comfortable exporting to habitats in space the idea of a separate genesis of life and a story of evolution parallel to Earth's story. Evolution in this case is assumed to be progressive, following an entelechy toward increased rational intelligence. In the case where the length of evolutionary development is less than or comparable to our own, we can expect inferior or equal levels of rational capacity. In the possible case where an extraterrestrial race has had more time to evolve, we can expect a level of rational intelligence superior to our own. Speculation on the part of the astroethicist should be ready to construct a framework for moral responsibility that corresponds to these three relevant moral communities.

Our orientation in such ethical speculation begins with the proposal of a galactic commons. In a previous chapter we introduced the notion of a galactic commons, a cosmic category for hosting commitment to the common good. When Douglas Vakoch asks about the benefits of searching for extraterrestrial intelligent life, the common good on this scale provides the backdrop. "As we mature as a civilization, we should increasingly take on both a cosmocentric and an intergenerational ethic to guide our search strategies, in which the potential benefit for those humans living now and in the near future is balanced with the potential benefit of extraterrestrial civilizations who may receive our messages, as well as future generations of humans who may receive a reply to any messages we might send."[54] Vakoch is on the path toward astroethics on a galactic and multi-generational scale.

Even if the cosmos is too big for a coherent concept of a morally interactive community, the Milky Way Galaxy is not. On the scale of the Milky Way metropolis, a common good could be shared between earthlings and alien neighbors. On a still smaller scale, that of our solar system ghetto, the

---

54. Vakoch, "Asymmetry in Active SETI," 483.

circumterrestrial region could be shared by all people on Earth along with our abiotic and possible biotic neighbors in space. At both scales, we should find ourselves thinking in terms of a single planetary society on Earth, a single society made up of individual earthlings belonging to the present and future generations. This single global society would transcend current national boundaries as well as ethnic and linguistic and class differences. While respecting local communities and indigenous traditions, this sense of global unity would provide grounds for affirming a single *humanum* as our community of moral discernment.

Bibliography

Aristotle. *The Basic Works of Aristotle*, edited by Richard McKeon. New York: Random House, 1941.

Bastianel, Sergio. *Morality in Social Life*. Translated by Liam Kelly. Miami, FL: Convivium, 2010.

Bostrom, Nick. "Transhumanist Values." http://www.nickbostrom.com/ethics/values.html.

Brin, David. "So You Want to Make Gods. Now Why Should That Bother Anyone?" In *TFC*, 45–52.

Bulliet, Richard W. *Hunters, Herders, and Hamburgers: The Past and Future of Human-Animal Relationships*. New York: Columbia University Press, 2005.

Danielson, Peter. "Turing Tests." In *ESTE*, 4:1985–6.

Darwin, Charles. *The Origin of Species by Means of Natural Selection of the Preservation of Favoured Races in the Struggle for Life*. 6th ed. 1872; New York: Signet, 2003.

Dawkins, Richard. *The Selfish Gene*. Oxford: Oxford University Press, 1976, 1989.

Deacon, Terrence W. *The Symbolic Species: The Co-evolution of Language and the Brain*. New York: Norton, 1997.

Dick, Steven J. "Extraterrestrial Life." In *Encyclopedia of Science and Religion*, edited by J. Wentzel Vrede van Huyssteen. 2 vols. New York: Macmillan/Thomson Gale, 2003.

Drake, Frank. Interviewed by Diane Richards. "Interview with Dr. Frank Drake." *SETI Institute News* 12.1 (2003) 5.

Falconer, Alan. "Dignity." In *The Westminster Dictionary of Christian Ethics*, edited by James F. Childress and John Macquarrie, 278. Louisville: Westminster John Knox, 1986.

Fellenz, Marc R. "Animal Rights." In *ESTE*, 1:74–77.

Ferkiss, Victor C. *Technological Man: The Myth and the Reality*. New York: New American Library, Mentor, 1969.

Gilmore, Michael. "Space Exploration." In *Social Implications of the Detection of Extraterrestrial Civilization*, edited by John Billingham et al., 1831–35. Mountain View, CA: SETI, 1994.

Höffe, Otfried. *Can Virtue Make Us Happy?* Translated by Douglas R. McGauhey and Aaron Bunch. Evanston, IL: Northwestern University Press, 2010.

Impey, Chris. *Living Cosmos: Our Search for Life in the Universe*. New York: Random House, 2007.
Kant, Immanuel. *Groundwork of the Metaphysic of Morals*. Translated by H. J. Paton. New York: Harper, 1948.
Kurzweil, Ray. *The Singularity is Near: When Humans Transcend Biology*. New York: Penguin, 2005.
Lemarchand, Guillermo A. "Speculations on First Contact." In *SETI*, 153–64.
Lovin, Robin." Astrobiology and Theology." In *LBE* , 222–32.
Milbank, John. *Beyond Secular Order*. Oxford: Wiley Blackwell, 2013.
Moravec, Hans. *Mind Children: The Future of Robot and Human Intelligence*. Cambridge: Harvard University Press, 1988.
Narveson, Jan. "Martians and Morals: How to Treat an Alien." In *Extraterrestrials: Science and Alien Intelligence*, edited by Edward Regis Jr., 245–66. Cambridge: Cambridge University Press, 1985.
Neal, Mark. "Preparing for Extraterrestrial Contact." *Risk Management* 16.2 (2014) 63–87.
Peters, Ted. "AstroEthics: Engaging Extraterrestrial Intelligent Life Forms." In *ELU*, 200–221.
———. "Outer Space and Cyber Space: Meeting ET in the Cloud." *International Journal of Astrobiology*, doi:10.1017/S1473550416000318. http://journals.cambridge.org/action/displayFulltext?type=1&fid=10445483&jid=IJA&volumeId=-1&issueId=-1&aid=10445478&fromPage=cupadmin&pdftype=6316268&repository=authInst.
———. *Sin: Radical Evil in Soul and Society*. Grand Rapids: Eerdmans, 1994.
Race, Margaret S., and Richard O. Randolph. "The Need for Operating Guidelines and a Decision Making Framework Applicable to the Discovery of Non-Intelligent Extraterrestrial Life." *Advances in Space Research* 30:6 (2002)1583–91. http://archive.seti.org/pdfs/m_race_guidelines.pdf.
Randolph, Richard O., and Christopher P. McKay. "Protecting and Expanding the Richness and Diversity of Life, an Ethic for Astrobiology Research and Space Exploration." *International Journal of Astrobiology* 13 (2014) 28–34, citing *The Analects* [of Confucius]. Translated by David Hinton. Washington, DC: Counterpoint, 1998.
Rees, Martin. *Our Final Hour*. New York: Basic, 2003.
"Religion: Space Ethics." *Time*, October 1, 1956. http://www.time.com/time/magazine/article/0,9171,862394,00.html.
Rollin, Bernard E. "Animal Welfare." In *ESTE*, 1:80–83.
Ruse, Michael. "Is Rape Wrong on Andromeda? An Introduction to Extraterrestrial Evolution, Science, and Morality." In *Extraterrestrials: Science and Alien Intelligence*, edited by Edward Regis Jr., 43–78. Cambridge: Cambridge University Press, 1985.
Saarinen, Risto. *God and the Gift: An Ecumenical Theology of Giving*. Collegeville, MN: Liturgical Press, 2005.
SETI. "The Declaration of Principles Concerning Activities Following Detection of Extraterrestrial Life." *Acta Astronomica* 21.2 (1990) 153–54. http://www.setv.org/online_mss/SETI-DofP90.pdf.
Shostak, Seth. "Are We the Galaxy's Dumbest Civilization?" SETI Institute. http://www.seti.org/epo/news/features/are-we-the-galaxys-dumbest.php.

Singer, Marcus G. "Golden Rule." *Encyclopedia of Ethics*, edited by Lawrence C. Becker and Charlotte B. Becker, 1:615. 3 vols. 2nd ed. London: Routledge, 2001.

Tarter, Jill Cornell. "Contact: Who Will Speak for Earth and Should They?" In *ELU*, 178–99.

Tattersall, Ian. "The Origins of Human Cognition and the Evolution of Rationality." In *The Evolution of Rationality: Interdisciplinary Essays in Honor of J. Wentzel van Huyssteen*, edited by F. LeRon Shults, 167–82. Grand Rapids: Eerdmans, 2006.

Vakoch, Douglas A. "Asymmetry in Active SETI: A Case for Transmissions from Earth." *Acta Astronautica* 68 (2011) 476–88.

———. "Responsibility, Capability, and Active SETI: Policy, Law, Ethics, and Communication with Extraterrestrial Intelligence." *Acta Astronautica* 68 (2011) 512–19.

Wendt, Alexander, and Raymond Duvall. "Sovereignty and the UFO." *Political Theory* 36:4 (August 2008) 607–33.

Young, Simon. *Designer Evolution: A Transhumanist Manifesto*. Amherst, NY: Prometheus, 2006.

24
Concluding Scientific Prescript

TED PETERS

Where have we been? Where are we going?
 We began by looking at our human experience with the sky, with what is beyond. We acknowledged the compelling and even inspiring impact outer space makes on our inner soul. As if a distant galaxy would be calling us on our cell phone, something within us wants to answer. So much more than merely a factual knowledge of the natural world is at stake. The near unfathomable distances and depths evoke curiosity, while the magnitude and magnificence of the cosmos educe a feeling of awe. Wonderment draws out fright, veneration, and finally, reverence.
 Our research scientists serve us like messengers from the heavens, revealing the sublime mysteries of the realm beyond. We extol and thank today's scientists for blessing us on Earth with information, conjecture, and speculation. We wish to express our gratitude to the uncountable number of astronomers, astrophysicists, spectroscopists, and astrobiologists who are daily expanding our knowledge.
 In this volume, we have given focal attention to SETI Institute research and NASA astrobiology, two overlapping fields concerned with life in the universe—its origins, evolution, and future. We have introduced a new field, a deliberative discipline to be called *Astrotheology*. We have proposed a working definition: *Astrotheology is that branch of theology which provides a critical analysis of the contemporary space sciences combined with an explication of classic doctrines such as creation and Christology for the purpose of constructing a comprehensive and meaningful understanding of our human situation within an astonishingly immense cosmos.*

We have proposed four initial tasks to place on the astrotheologian's To-Do list. First, Christian theologians along with intellectual leaders in each religious tradition need to reflect on the scope of creation and settle the pesky issue of geocentrism. Second, the astrotheologian should set the parameters within which the ongoing debates over Christology (Person of Christ) and soteriology (Work of Christ) are carried on. Third, theologians should analyze and critique astrobiology and related sciences from within, exposing extra-scientific assumptions and interpreting the larger value of the scientific enterprise. Fourth, theologians and religious intellectuals should cooperate with leaders of multiple religious traditions and scientists to prepare the public for the eventuality of extraterrestrial contact.

Regarding task three, we have given sustained analysis and criticism of assumptions at work in some of the space sciences. Specifically, the conceptual set taken by space scientists to the question of extraterrestrial life relies upon the Darwinian model of evolution. Unfortunately, in our judgment, this SETI version of evolution is over embellished with optimism and disguised ideology. In fact, the concept of evolution through which SETI scientists view their subject matter looks like a secularized myth of gnostic redemption. The task of the astrotheologian is to point this out and to argue for a cleaner science, a science that recognizes the limits imposed by confirmable empirical knowledge. We have quipped that SETI sometimes practices theology without a license.

Be that as it may, we laud and applaud and praise the progressive research performed by SETI scientists and other astrobiologists. We look forward to increased partnership between space scientists and theologians attempting to comprehend the scope and splendor of God's creation. Durham's David Wilkinson makes a recommendation: "I suggest, therefore, that the Christian churches need to be active supporters of SETI."[1] So do we.

Bibliography

Wilkinson, David. *Science, Religion, and the Search for Extraterrestrial Intelligence.* Oxford: Oxford University Press, 2013.

---

1. Wilkinson, *SRSEI,* 134.

Index

'Abbās, Ibn, 224
Abelard, Peter, 321
abiogenesis, 139, 150–51
Abraham, 60, 335, 337
Abrams, Nancy, 42
absence, concern about, 310
abstraction, human capacity of, 254
Action Team 14, 410
Adam, 319, 321, 333, 337–42,
    342n54, 424
Adams, Marilyn McCord, 288,
    288n44, 290, 290n51
'ālamīn (intelligent creatures),
    217n3
al-Baghdādī, al-Junayd, 224
al-Birūnī, 221
Alexander, Victoria, 186
Alien (movie), 320
alien, space as, 208
al-Kubrā, Najm al-Dīn, 225
Allaby, Michael, 385
Allah, 191, 218–19, 224, 225, 408
    as creator, 218–20, 223–25
    names and characteristics of,
        219–20
Almagest (Claudius Ptolemy), 90
alone in the universe
    position that we are, xi, 358
    question of are we, ix, xiv, 10,
        15, 16, 36, 53, 56–59, 61,
        110–11, 115, 133, 360n35
    unlikely that we are, 143–44,
        157, 190, 367–68
Alpha Centauri, distance to, 112

al-Qurṭubī, Abū 'Abd Allāh
    Muḥammad b. Aḥmad,
    219–20
al-Qushayrī, Zayn al-Islām Abū-
    l-Qāsim 'Abd al-Karīm b.
    Hawzān, 220
al-Rāzī, Fakhr al-Dīn, 220, 222, 224
al-Samarqandī, Abū al-Layth,
    223–24
al-Shāṭir, Ibn, 221
Alston, Walter, 292n54
al-Sulamī, 224
al-Ṭabarī, Abū Ja'far b. Jarīr, 223
al-Tadhkira fī-l-Hay'a (*Memoir on
    Astronomy*), 221
altruism, 81, 439–40
al-Ṭūsī, Naṣīr al-Dīn, 221
America, founders of, 235
ampullae of lorenzini, 259n16
"An Anatomy of the World" (John
    Donne), 198–99
Ancient Astronaut school of
    thought, 17–18
angel(s), 57, 94, 163, 171, 174n57,
    188, 194, 210, 217n4, 218,
    224, 282, 283, 337
    in Islamic thought, 194, 217n4,
        218, 224
animal ethics, 427
animal rights, 427
animal welfare, 427–8
animals, 10, 52, 153, 162n7, 165,
    171, 201, 210, 224, 250,
    254–56, 293, 307, 309,
    332–34, 337, 339–44, 357,
    397, 424–28, 436–37

449

450 INDEX

animals (*continued*)
  consciousness of, xv, 162n7, 165
  human exploitation of, 427–28
  naming of, 10, 424–25
  non-rational, 428
  redemption (salvation) of, 307,
    309, 341–43
anonymous knowledge of God, 168,
    171–72
Anselm of Canterbury, Saint, 44,
    162–63, 322
Answers in Genesis, 16
anthropic principle, 63, 63n24,
    198n32
Anthropocentric Principle (AP), 63
anthropocentrism, 63, 95, 184,
    185–86, 190–93, 197–202,
    274, 296, 297, 331
  argument for, 296
  in science, 191, 191n16
  opposition to, 190–91, 202
Apollo Space Program, 32, 117, 147
Apologia pro Galileo (Thommaso
    Campanella, OP), 273, 323
Aquinas, St. Thomas, 30, 83n44,
    93–94, 171, 237, 321
Archimedes, 96n12
Aristarchus of Samos, 96n12
Aristotle, 57, 83n44, 92–93, 93n4,
    212, 221, 426, 428, 429
Armstrong, Karen, 240
Armstrong, Neil, 58
Arnould, Jacques, 18, 33, 400, 402
Arrival, 254
artificial intelligence, 436
asteroid strikes, 409–10, 412
astrobiology, ix, 5–9, 10–11, 111,
    121–22, 229, 441, 443
Astrobiology Roadmap (NASA), 6
Astrobiology Strategy (NASA), 6,
    121–22, 147
astroethics, 50–52, 230n7, 233,
    381–89, 391–413, 416–44
astrometry, 126
astrotheology
  as glorifying God, 14
  as public theology, 46
  definition of, 11–12, 447

  etymology of, 14
  goal of, 23
  history of the term, 236
  models for, 19–20
  other uses of the term, 12
  sources of, 28–29, 30–34
*Astro-Theology, or a Demonstration
  of the Being and Attributes
  of God from a Survey of the
  Heavens* (William Durham),
  14, 101, 236
Athanasius of Alexandria, 287, 321,
  342
atheism and science, xi, xv, 20–22,
  30, 47, 195, 354, 362–63, 366
atomism, 83n44, 92, 148, 161n5
atonement, 46, 163, 273, 280, 285,
  287–89, 295, 299, 321, 341
  substitutionary, 289
  theories of the, 321
Augustine of Hippo, 165, 175n60
Aulén, Gustav, 321
Avatar (movie), 57
Averroes (Ibn Rushid), 93
Avicenna (Ibn Said), 93
awe, 3, 32, 33, 90–91, 234, 447
Ayala, Francisco J., 77, 79, 82, 153,
  357–58

bacteria, photosynthetic, 139
Bainbridge, William Sims, 186
Bañez, Domingo, 175
Barbour, Ian, 19, 75, 82, 352n13,
  370–71
Barr, James, 333
Barth, Karl, 311
Bastianel, Sergio, 434
Battle: Los Angeles (movie), 320
beatific vision, 172n52
Beck, Lewis White, 355
Behe, Michael, 238
being
  absolute, 170–71, 173, 175–76
  actualized, 173–4
Bellah, Robert, 32
Bennett, Jeffrey, 37, 97, 98
Bentley, Richard, 323

INDEX 451

Berger, Peter, 282
Bertka, Constance, 196
bias, awareness of, 58
Bible-Against-Aliens position, 16–18
Bible-UFO school of thought, 17–18
Bible-Welcomes-Aliens position, 18, 30
Big Bang, xiv, xvi, 5, 12, 38, 63, 64, 78, 164, 202, 204, 293, 318, 325
big numbers, appeal to, 37, 154, 157, 177, 275–76, 368
Billingham, John, 356
Billings, Linda, 353n15
Billings, Lee, 403
biologically reversible exploration, 387
biomarkers (biosignatures), 7, 9, 116, 129–32, 229, 231
biosignatures. *See* biomarkers
Bird, Phyllis, 332n2
Boethius, 276, 276n13
Bonaventure, 288
Bondi, Herman, 43
Bostrom, Nick, 367n58, 435
Brahe, Tycho, 99
brain-language co-evolution, 425–26
Brewster, David, 324
bridge metaphor for relation of theology and science, 4–5, 13–14, 20, 21, 29, 347
Brownlee, Donald, 61
Bruno, Giordano, 59, 98, 98n19, 161, 161n5
Bruns, J. Edgar, 283–84
Bulliet, Richard, 427
Bullock, Mark, 34
Burgess, Andrew, 274
Buridan, John, 83n44
Burhoe, Ralph, 235
Burney, Venetia, 10
Butler, Paul, 156

Cabrol, Nathalie, 42
Calisto (moon of Jupiter), 156

Calvin, John, 175n60, 288–89, 289n47
Campanella, OP, Thommaso, 273, 323
capacity to conceive of an unknown other, 247, 249, 256–58, 260–61, 263, 266
capacity to conceive of oneself as other, 261
Cassini probe, 156
cause, efficient, 175
cell, naming of the, 151n18
Center for Theology and the Natural Sciences (CTNS), xv, 12
Cernan, Eugene, 4
Chalcedonian formula, 272n2, 292n54
Chambers, Robert, 102, 102n30, 102n31
Chela-Flores, Julian, 77, 360
Chick, Garry, 141
China National Space Administration (CNSA), 6n13
Cho, Francisca, 406
Chomsky, Noam, 255n12
Christ
  and animals, 339–41
  as center of creation, 288n44
  as center of universe and history, 68
  ascension of, 292
  as elect, 338, 339, 339n41, 341
  as head of the cosmos, 290
  as mediator, 288–9
  as sacrifice, 163, 289, 325
  as Second Adam, 331, 338–41
  as having unique work, 67, 69, 197, 277, 283n27, 286, 289, 306–8, 314, 325, 331
  birth of, 340
  characteristics of the Risen, 294, 312
  crucifixion of, 295, 325
  natures of, 275, 276n13, 290
  nonsensicality of a cosmically significant, 272

"Christ in the Universe" (Alice
  Meynell), 327
Christ principle, 290
Christian Answers Network, 16
Christian Church, early, 321, 340–43
Christocentrism, 67, 296
Christology, 41, 45–46, 191, 273,
  278, 280, 288n44, 290, 297,
  306, 313–15, 317, 321, 325,
  330–32
  high, 280, 283n27, 297
Christus Victor, 322
Churchland, Patricia, 95
city of God, 299
Clarke, Arthur C., 53
Clayton, Philip, 41
Cleland, Carol, 392
Clinton, Hillary, 57–58
Close Encounters of the Third Kind
  (movie), 57
Coakley, Sarah, 292n54
Cobb, John, 202
Cockell, Charles S., 386
cognitive capacities needed for
  communication, 245–67,
  246n3
cognitive science, 246n2
Cole-Turner, Ronald, 42
Collins, Francis, 10
Collins, Nathan, 394
colonization
  of Mars, 407
  of planets, 121
  terraforming looks like, 407
Columbus, Christopher, 408
comet strikes, 409–10
commitment to commitment,
  264–65
Committee on Space Research
  (COSPAR), 118, 120
common good, the, 51, 202, 299,
  393, 412, 418, 443
commons, galactic, 51, 202, 350,
  393–94, 400, 411–13, 420,
  430, 430n27, 441–43
communicatio idiomatum. See
  communication of attributes

communication
  and implicit knowledge, 253
  conditions for, 419
  literal, 253
  of God, 162–64
  window of, 141n24
communication of attributes, 298,
  311, 314, 315
communion between beings as telos
  of creation, 164
communion with God. *See*
  fellowship with God
community
  formation, 263
  human, 51, 119, 248–50, 263–
    66, 395, 401, 403, 411–13,
    418, 432, 444
  of interpretation, 264–5
  of moral deliberation, 401, 403,
    411, 413
  philosophy of, 248–49, 263–66
compassion, 261
concrete being, 169n33
Concrete Universal, 285
Confucius, 429
Congar, Yves, 306, 324
Conley, Catharine, 393–94
Connell, Francis J., C.S.S.R., 193
conscience, 57, 86, 305, 319, 342,
  434, 442
consciousness. *See* self-
  consciousness;
  transcendental
  consciousness; planetary
  consciousness
Consolmagno, Guy, 229, 285n35
Contact (movie), 57
Contact (Carl Sagan), 367
contact, direct. *See under*
  extraterrestrial intelligent
  life
contact optimism, 37–38, 77, 154,
  420
contact pessimism, 422
contamination, xii, 116–18, 120–21,
  387, 392–94
  backward, 117–18, 120–1,
    393–4, 404

forward, xii, 118, 121, 387,
    393–94, 404
contextual theology, 309
contingency, 77, 153–54, 230, 240,
    289, 291, 356–59, 368
Copernican model, acceptance of
    the, 98–99
Copernican Principle, 43, 63, 68,
    63n24, 197–98
  as myth, 199–200
Copernican Revolution, 64, 90–92,
    94–103
  myth of the, 94–98
Copernicanism, religious rejections
    of, 97
Copernicus, Nicolaus, 43–44, 59,
    91–92, 94–98, 103, 221
coronagraphs, 127
Cosmic Background Radiation, xiv
cosmic Christ, 283n27
cosmic teleology, rejection of, 48
cosmism (space worship), 48
Cosmological Principle, 197–98
Cosmotheoros (Christiaan Huygens),
    101
cosmovision, 16
COSPAR (Committee on Space
    Research of the International
    Council for Science), 118,
    120, 394, 399
cosmotheology, 229–42
  definition of, 231
  principles of, 231–33
Council of Chalcedon (AD 451),
    292n54
covenant, 342
Coyne, SJ, George, 16, 37, 236, 295,
    317–18
creatio continua, 80
creatio ex nihilo, 80
creation
  as an expression of the purpose
    of God, 79–81
  doctrine of, xvi, 12, 15–16, 19,
    46, 284
creationists, 16–17
creative mutual interaction between
    theology and science,
    xv–xvi, 4–5, 13–14, 19–20,
    21–23, 29–30, 75n5
Crescas, Hasdai, 83n44
Crichton, Michael, 355
Crick, Francis, x
Crisp, Oliver, 289n47
Critique of Pure Reason (Kant), 236
Cronholm, Frederick, 324
cross, theology of the, 294–95, 373
Crowe, Michael J., 47, 196
Crysdale, Cynthia S. W., 44, 193,
    406–7
curiosity, 57, 70, 447
Curiosity Rover, 76
cyborg astronauts, 419

Dallal, Ahmad, 222n13
DAMPE (project to explore dark
    matter), 6n13
Danielson, Dennis, 98
Darling, David, 7
Darwin, Charles, x, 38, 91, 103–4,
    147–50, 150n14, 437, 439–40
  home of (Down House), 148–49
Darwin, Emma (Wedgewood),
    148–49
Darwin, Erasmus, 148
Dasein (presence, existence), 168–
    69, 169n32, 169n33, 172n52
Davies, Paul, 14–15, 37, 52, 64–65,
    77–78, 83, 94, 160, 184,
    196–97, 271–72, 277–78,
    330–1, 348–50, 439
Dawkins, Richard, 21, 79, 362–4,
    440
The Day Earth Stood Still (movie),
    57, 320
Deacon, Terrence, 425
Deane-Drummond, Celia, 295, 398
death, 299, 442, 435–36
  as a limit, 435
  overcoming, 435–6
De Caelo (*On the Heavens*)
    (Aristotle), 92
de-centering, 43–44, 63–64, 67–69,
    198, 231
  physical, 68, 231
  spiritual, 67–69

The Declaration of Principles Concerning Activities Following the Detection of Extraterrestrial Intelligence, 419
de Duve, Christian, xi, 77, 154, 161–62, 176, 367
Defiance (TV series), 353
Delio, Ilia, 290, 326–27
Democritus, 83n44, 92, 92n2, 161n5
demotion myth, 43–44
De Revolutionibus Orbium Coelestium (*On the Revolution of the Heavenly Spheres*) (Nicolaus Copernicus), 45, 96
Derham, William, 14, 101, 236
Descartes, René, 99–100, 99n20
design argument for the existence of God, 14–15, 148
destiny, human, 233
de Waal, Frans, 247, 369
Dick, Steven, 9, 195
difference, appreciating, 263, 264
dignity, 52, 231, 246, 276–77, 396, 428, 431–34, 436, 441–43
dinosaurs, extinction of the, 410
directionality in evolution
  scientific view of, 48, 78
  theological view of, 174
distance, problem of, 308–9
distributive justice, principle of, 411–12
District 9 (movie), 57
divine action, xvi, 12, 44–45, 148, 164, 175, 220, 283, 327
  at a sub-atomic quantum level, 44
  non-interventionist, xvi, 80n30
  objective, 44–45
divine self-bestowal model, 287–91
divine self-communication, 160–77, 278, 280, 282, 285–92, 294, 297, 311, 325
Dodds, Michael, 40
Donne, John, 198
Doppler effect, 126, 136

Doppler shift method of searching for exoplanets, 136
domination, cosmic, 199–200
Downing, Barry, 17
Drake Equation, 36–40, 37n25, 134–44, 154, 246, 259, 261, 276, 355, 358, 367, 420, 426
  an addition to the, 261
  as meaningless, 355
  as speculative, 276
  factors not considered in the, 143
  discussion of the variables (*see* variables in the Drake Equation)
Drake, Frank, 36–37, 39, 40, 133, 134, 141, 353–54, 439–40
Drees, Willem, 42
drug cartels, 442
dual-aspect monism, 173
Dunér, David, 141
Duvall, Raymond, 418
Dyson, Freeman, 38, 79
Dyson spheres, 38

Earth
  as a dangerous home, 409–10
  as special, x, 41, 63n24, 198n32
  atmosphere of, 5, 9, 129–30, 383, 400–401, 410
Earth chauvinism, 42, 44, 45, 274
Earthism, 202
Earth Microbiome Project, 152
Echelon spy network, 401
ecoethic, global, 411
ecology, deep, 388, 395
ecotheologians, 201–3
Edwards, Denis, 288
Egan, Harvey, 166–67
Eicher, David, 41
Einstein, Albert, 230, 234, 240
elect as representatives of God, the, 335
election
  as a call to serve, 334–35, 344
  of Abraham, 337

of humans by God, 175n60, 297,
    327, 333–39, 343, 344
  of Israel, 337
  of kings, 337
  of the Messiah, 338
  historical (or biblical), 333–38,
    344
  narratives, 337
electromagnetic signals, search for,
    36, 114, 131, 134
Eliade, Mircea, 33, 34
emergent spirituality, 262
empiricism, 238
emptiness, Buddhist principle of,
    406
Enceladus, 118, 129
encephalization quotient, 140
Ender's Game (movie), 353
Enemy Mine (movie), 320
enhancing life as enhancing value,
    388
enlightenment, Buddhist, 191–92
Enlightenment
  era, 99, 246, 276, 323, 430–32,
    434, 441
  explorations of extraterrestrial
    life, 99–102
  values, 431–32, 434, 441
entelechy, 443
*Entretiens sur la Pluralité des
    Mondes* (*Conversations
    on the Plurality of Worlds*)
    (Fontenelle), 99–100
environmental disaster, threat of,
    141
environmental protection
  of Earth, 71
  of habitats, 392–95, 398–99, 405
  of outer space, 120
Epicureans, 59
Epicurus, 83n44, 124, 161n5
epicycles, 96
epistemology
  non-reductionist, 79–81
  of science, 238–39
  of theology, 238–39
equal, all intelligent beings are not,
    441

equality, 432
  of humans, 441
eschatology, xvi, 46, 255, 299,
    307, 311–12, 327, 353–55,
    364–65
  Islamic, 225
  secular, 353–59, 364–65
eschaton, 312
estrangement from God, 279–81,
    285–87, 372
E.T. (movie), 57, 320
eternal forms, 100
ether (quintessence), 93–94, 97
ethic of gratitude, 442
ethic of peace, 442
ethic of risk, 406
ethic of slave responsibility, 441
ethical consideration, gradient of,
    386
ethical deliberation, 412–13, 419,
    421
ethics
  appropriate to the discovery of
    non-intelligent life in our
    solar system, 394
  biological origins of, 81–82
  of space (*see* astroethics)
ETI myth, 48–49, 347–74, 353n15,
    417, 420, 438
  debunking the, 352, 352n13
Europa (moon of Jupiter), 76, 118,
    129, 156
European Network of Competences
    on Space Debris, 401
European Space Agency (ESA),
    6n13, 155, 229, 401
evangelization, interplanetary,
    325n29
Eve, 319, 321, 337–38
evil, 280, 290
evolution, 16, 38–40, 44–52, 59,
    64, 77–85, 148–49, 153–54,
    164–68, 173–74, 176–77,
    229, 232–33, 236–37,
    239–40, 256, 281, 293, 305,
    317–20, 347–51, 354–70,
    373–74, 417–18, 420–22,
    425–27, 434–35, 437–40,
    443, 447–48

evolution (*continued*)
    as a secular religion, 365
    as a unifying vision, 241
    assumption of progress in, 39,
        47–49, 348–50, 355–74,
        421–22, 434–35, 443
    contingencies in, 77, 153–54,
        356–59, 368
    convergent, 77, 359–61
    cosmic, 38–39, 164–66, 172,
        174, 176–77, 229, 232–33,
        237, 239–41
    consequences if it started over
        again on Earth, 357–58
    directionality in (*see*
        directionality in evolution)
    myth of progressive, 48, 349,
        355–59, 361, 364, 374, 434
    of intelligence, 360
    reception of Darwin's theory
        of, 149
    rejection of purpose in, 48
    to overcome sin, 351
evolve again, improbable that
    humans would, 153, 165,
    174, 357, 359
exclusivism, 190, 283n27, 313n17,
    335
exobiology, 6–7, 320; *see*
    astrobiology
exoplanet detection, star wobble
    method of, 9, 125–27, 136
exoplanet searches, 76, 115–16,
    119–20, 125, 136
    no theological controversy
        surrounding, 116
    light suppression technique for,
        127
exoplanet(s)
    atmospheres, 129–30
    direct imaging of, 9, 127, 128,
        132
    discovery of, ix, 8–10, 62, 125,
        156
    habitable (habitability of), 115–
        16, 124–30, 137, 156–57,
        360n35
    indirect detection of, 125

    number in our galaxy, 147
    number of, 8n20, 126
    number of candidate, 115
    number of confirmed, 62, 115
    number of earth-sized, 147
exoplanets.org, 62
exotheology, 13, 189, 230, 283
experience
    of the sky, 18–19, 32–34, 60
    transcendental (*see*
        transcendental experience)
exploitation of space, commercial,
    118, 120, 201, 408, 411, 428,
    432
exploration of space, remote, 420
extinction, 357, 427, 434
extra-scientific assumptions, 23, 41,
    47, 49, 248, 347, 353, 368,
    448
extrasolar planets. *See* exoplanets
extraterrestrial artifacts, searches for
    large-scale, 115
extraterrestrial civilizations will be
    much older than ours, view
    that, 85, 349, 351n11, 354,
    364
extraterrestrial intelligence, types
    of, 52
extraterrestrial intelligent life (ETI)
    and moral failure, 85
    and the intensions of God, 83
    anxiety about, 109–10, 119, 433
    as altruistic, 353n14, 440, 442
    as a challenge to all religions, 70
    as a threat to humans, 52, 119,
        433, 437–43
    as beneficial to humans, 39
    as benevolent, 349, 438
    as Earth's future, 350
    as enslaving us, 52, 437–42
    as fallen, 188, 280, 286, 291, 296,
        324, 372
    as fallen angels, 188
    as having a utopian society, 369
    as moral beings, 82, 84, 274,
        428, 320
    as overcoming warfare, 85

as peaceful, 363, 433–34, 437,
  442–43
as salvific, 352–53, 364, 433,
  438–40
as unrecognizable (alien), 426
atheist views of, 195
Bahai views on, 189
Buddhist view of, 191–2
communicating with, 7, 36–39,
  75, 131, 133–37, 140–44,
  186, 191, 245–67, 277, 308,
  351–52, 358, 417–19, 421,
  423–25
concerns about predatory,
  368n59
difficulties with communicating
  the gospel to, 308–9
direct contact (with), 117, 434–
  5, 437–38, 440–42
discovery of as inevitable, 7, 111
Eastern Orthodox views on, 190
engaging inferior, 52, 343, 417,
  422, 423, 426–29, 437, 443
engaging peer, 417, 422, 423–6,
  429–34, 443
ethical guidelines for first
  contact with, 419
ethics of, 85n48
ethics of engaging, 416–44
hostile, 52, 119, 433, 437–43
imputation of dignity to, 428,
  431–33
Islamic views on, 194, 213–26
Jewish views on, 188, 193, 204,
  209–10, 215
longevity of as inconsistent with
  religion, 85n48
love for, 429
low likelihood of real time
  engagement with, 115,
  417–20
meaning of, 353
moral development of, 349
Mormon belief in, 189
non-religious views of, 190
organized human response to,
  442

our moral responsibility in
  relation to, 210, 391–413,
  416–44
our physical continuity with, 82
personhood of (see personhood
  of extraterrestrials)
possibilities of engagement with,
  18, 351, 391–413, 416–44
prediction of religious crisis at
  discovery of, 75n5, 83, 160,
  183–205, 271–72, 277–78,
  300, 330–31
preparing for contact with,
  50–52
protection of from suffering, 428
Protestant views on, 83, 186,
  188, 189, 190, 191, 192, 193,
  305, 306
redemption of, 45–46, 68–69,
  86, 168, 197, 271–300, 303–
  15, 317–27, 330–44
Roman Catholic views of, 58,
  61, 67, 68, 83, 100, 160, 188,
  189, 190, 191, 192, 193, 195,
  284, 305, 306, 323, 327
spirituality of, 85n48, 245–67,
  279, 350
superior, 52, 185, 232, 347–74,
  417, 434, 437–42
theological argument for the
  necessary existence of, 164
theological problem of
  communicating with,
  309n11
time delay in communicating
  with, 419
will ignore us, 355n20
will strengthen faith, 189
why aren't we yet
  communicating with, 143
worship of, 48, 362–4
extraterrestrial life (ET)
  argument for the inevitability of
    from the size of the universe,
    176
  Catholic views on, 68, 189
  criticism of search for, 147

extraterrestrial life (ET) (*continued*)
  cultural, religious, and philosophical implications of the search for, 57–58
  discovery of as a confirmation of Christian beliefs, 196
  existence of not a problem for most Christians, 196
  existence of as undermining belief in God, 27
  exploitation of, 432, 427–8, 432
  intrinsic value of, 382, 387–88, 394–99
  inverse correlation between traditional religious devotion and affirmation of the existence of, 186
  meaning and implications of the discovery of, 74–86, 109–11, 115
  methods of searching for, 113–18
  probability of the existence of, x, xi, 62–64, 66, 68, 76–77, 133–44, 146–57, 231, 232, 317, 357–58
  redemption of as not needed, 68, 273
  rights of indigenous, 120, 387
  the search for, 51–52, 392
  the search for as a spiritual journey, 71
  theological reasons to expect, 161
  types of evidence of, 114–8
extraterrestrial non-intelligent (microbial) life (ETNL), xii, xv, 7, 12, 15, 17, 41, 51–52, 76, 116, 119–21, 156, 195, 231, 277, 355, 386, 391–413
  ethical issues arising from the search for, 393
  searching for in our solar system, 116–18, 392
  societal and theological issues relating to, 120–22
extraterrestrial technology, 38

extraterrestrial visitation as threat to the modern state, 418
extremophiles, 65, 139, 152–53

faith and reason, 70n38
fallacy of big numbers, 276
fallenness, 280, 285–91, 295, 296, 372
Falling Skies (TV series), 353
fall into sin, human, 175n60, 287–89, 296, 342, 342n54
false-belief task, 255–57
fear, 447
fellowship with God, 46, 285, 288, 293–94, 311, 326–27, 334–35, 342
Ferkiss, Victor C., 411, 419–20
Fermi, Enrico, 35
Fermi's Paradox, 35–36, 368n58
Festal Letters (Athanasius), 321
The Fifth Miracle (Davies), 65
51 Pegasi b, 62, 76, 136, 156
fine-tuning of the universe, 80
finitude, 372
"first instance of the new laws of nature" (FINLON), 312
The First Three Minutes (Weinberg, Steven), 78
Fishbane, Michael, 22
fix-a-broken-creation Christology (soteriology), 286–91, 289n47, 297
Foerst, Anne, 363
Fontenelle, Bernard le Bovier de, 99–101, 100n21
Ford, Lewis, 281
Fosdick, Harry Emerson, 373
Foster, Durwood, 279
freedom, 31, 66, 68, 164–65, 164n17, 171, 175, 226, 239, 281, 311, 344, 432
friendship as motivation for creation, 288
Froehlig, Julie Louise, 187
Fuller, Buckminster, 53
Funes, José, 132
fusion of horizons, 20, 424

future 39, 71, 121–22, 299–300,
    307, 312, 327, 365, 372, 374,
    385–89, 410, 413, 419–20,
    424, 427, 434–36, 443–44
  Christian view of the, 295, 299–
    300, 307, 312, 327, 340, 372
  of humanity, 39, 71, 121–22,
    350, 434–36
  of life, ix, 6, 11, 49, 117, 121–22,
    381–89, 413
  of religion, 70
  of the universe, xiv, xvi, 5, 7, 12,
    16, 44, 71, 111, 327, 435
  planning for our planet's, 410
futurists, 434

Gadamer, Hans-Georg, 20, 169n32
Gagarin, Yuri, 33
galaxies in the universe, number
  of, 112
Galilei, Galileo, 60, 61, 95, 97, 273
gamma ray bursts, 410
Ganymede (moon of Jupiter), 129,
  156
Garreau, Joel, 113
Gemini space missions, 146
General Relativity, Einstein's theory
  of, 138
generosity, 428
genocide, 407
geocentric, view that religions are,
  184–86
geocentrism, 41–46, 63, 91–95, 97–
  99, 101–3, 184–86, 190–91,
  197–204, 274
  affirmations of, 201, 296
  as a sin, 42
  as misrepresentation of religion,
    204
  Buddhist opposition to, 190
  little impact of losing, 97–98
  meaning of, 103
  Mormon view of, 190
  not required by Islam, 221–2
  secular, 200
geo-cosmic-domination, 200
George, Marie, 160, 176

Gilkey, Langdon, 31, 34, 50, 369,
    371
Gilmore, Michael, 395
Gingerich, Owen, 10, 95, 359
Gliese 832C, 8
gnostic redeemer myth, 364, 366,
    369
Gnosticism, 321, 349, 354
God
  as alien, 208, 213–15
  as Creator, xiii, 13, 15, 65, 69, 80,
    191, 200, 204, 208, 211, 213,
    214, 218–20, 223, 225, 267,
    274, 276, 286, 287, 298, 299,
    304, 305, 309, 313, 315, 320,
    331, 341, 343, 374, 396
  as foreign, 211
  as giver of life, 65
  as identifying with victims of
    violence, 373
  as purpose of universe, xiii
  as totality of the universe, 234
  as transcending myth, 370–71
  communicable attributes of, 311
  grace of, 49, 84–86, 172, 175,
    175n60, 189, 280, 281, 283,
    286, 287, 289, 291, 297,
    304–5, 309, 311, 315, 334,
    344, 369, 372–74, 431
  immanence of, 80–81, 80n30,
    175–76, 314–15, 320
  immutability of, 165–68, 326
  love of (*see under* love)
  new conceptions of, 232
  oneness of, 212–14
  presence of, 204, 278, 280, 281,
    285, 287, 292–95, 309, 342
  self-communication of (*see
    divine self-communication*)
  simplicity of, 165
  as ultimate cause, 175
"god of the gaps" arguments, 230
god-of-the-gaps theology, 190
Golden Rule, 429–31, 430n27
Goldilocks planets. *See under*
    exoplanets.
Goldilocks zone. *See* habitable zones
Goodenough, Ursula, 234

Gore, Al, 411
Goren, Shlomo, 208
Göttingen Dogmatics (Karl Barth), 311
Gott, Richard, 63, 407
Gould, Stephen Jay, 77, 153, 165, 165n18, 174n58, 357
grammar, complexity of human, 255n12
Graves, Mark, 419
gravitational microlensing. *See* microlensing
great chain of being, 174n57
Great Filter, 367n58, 368n58
Gregersen, Niels Henrik, 293–94, 304, 314–15, 318, 326, 341n52
Gregory of Nazianzus, Saint, 293, 325, 342
greed, 372, 407, 411
Greeks, ancient, 92–93
Greene, John, 235
green globalization, 411
Griffith Observatory, xiii–xiv
Grinspoon, David, 142
The Guide of the Perplexed (Maimonides), 213

habitable exoplanets. *See under* exoplanets
habitable galactic zones, 62n19, 143
habitable zones (Goldilocks zones), 8, 9n23, 58, 62, 65, 116, 128–9, 136–8, 154, 156
of red dwarf stars, 128
habitability, ix–x, 7–8, 14, 35–37, 58, 62, 65, 76, 115–16, 128–31, 136–38, 143, 154, 156–57, 231, 266, 367, 385–86, 393
Haeckel, Ernst, 235
Hard X-ray Modulation Telescope (HXMT), 6n13
harmony, 264–5
Harrison, Albert, 11, 48, 51, 363–4
Hart, John, 51, 350
Haught, John, 15, 203
Hawking, Stephen, 119, 191, 191n16

heaven, problem of locating, 199
Hebrew scriptures, 335–44
Hebrew textual tradition, 336
Hebrew theology, 334
Hegel, Georg Wilhelm Friedrich, 170n41
Heidegger, Martin, 167–72, 168n32, 169n33, 172n52
Heidegger's concept of being, 169n33
Heim, Mark, 280
heliocentric model of Aristarchus of Samos, 96n12
heliocentrism, 91–104, 198, 200, 222
acceptance of, 95
acceptance of by Islam, 222
as evoking little anxiety, 200
hermeneutic of engagement, 34n18
hermeneutic of secular experience, 30–32, 34, 40
hermeneutics, 30–32, 34, 36, 40, 314, 424, 426
Herrick, James, 48–49, 355
Herzfeld, Noreen, 45
Heschel, Abraham, 166, 172
Hess, Hamilton, 321
Hess, Peter M. J., 293
Hewlett, Martinez, 177, 273–74, 327, 419
Hick, John, 278
High Accuracy Radial Velocity Planet Searcher (HARPS), 9
Hippolytus of Rome, 92n2
Hitler, Adolf, 211
HMS Beagle, Darwin's voyage on the, 148–9
Höffe, Otfried, 418, 428, 430
hominids, fossils of early, xv
Homo cyberneticus, 435–36
Homo erectus, xv
Homo habilis, xv, 140
Homo sapiens sapiens, xv
Hooke, Robert, 151n18
Hooker, Joseph, 150, 150n14
hope, 295, 355, 374
horizon of meaning, 434
horizon of particularism, 335
horizon of understanding, 424

INDEX   461

Hoyle, Fred, 354
Hubble, Edwin, 228, 231
Hubble space telescope, 131
hubris, 43, 274, 297
human(s)
   as inevitable, 153–54, 360
   as pinnacle of creation, 94
   as priests to non-human
      creatures, 337–38
   as self-reflective, 174, 266, 318
   as vice-regents of God, 337
   first, 336
   significance of, 60, 67, 199–200,
      332n2
   uniqueness, xv, 67, 69, 84, 190,
      199, 224, 294, 304, 323,
      332–37, 344
   what it means to be, 18, 46, 69,
      85, 134, 199, 201, 235–36,
      241, 266–67, 304–5, 337
human behavior, biological basis of,
   81–82
human body for space, unfitness of
   the, 419
Human Destiny (Pierre Lecomte du
   Nouy), 233
humanism, 235, 435
Humanity plus (H+), 435
humankind, unification of, 51
Human Microbiome Project, 152
human nature, 430
human rights, 334, 432
humility, 68, 200, 297
Hurayra, Abū, 224
Huygens, Christiaan, 100–101
Huxley, Julian, 149, 235

iconic communication (signs),
   251–2, 254–6, 259n17
ignorosphere, 404
image of God (*imago Dei*), 80,
   83–85, 175n60, 192n18, 195,
   209, 277, 304–5, 330–44,
   332n2
   as election, 333–44
   Buddhist argument against
      humans as, 192n18

immanent directionality of
   evolution, 78
immortality, 332, 354, 373, 435–36
   cybernetic, 435–36
imperialism, 407
Impey, Chris, 5, 41, 94, 141, 157,
   362, 426
incarnation(s), 197, 271–300,
   290n51, 292n54, 303–15,
   317–27, 330–44, 342n54
   and non-human animals, 331
   arguments against multiple, 272,
      277, 282, 314
   arguments against single, 308–
      12, 327
   as an act of love, 326
   as completion of creation, 326
   deep, 291–95, 314–15, 326,
      341n52
   effects of in human history, 310
   motives for, 289n47
   multiple view, 45–46, 68, 168,
      271–300, 303–15, 305n4,
      317–27, 331–32; (*see* planet-
      hopping Christ)
   necessity of, 162–3, 289
   ontological view of the, 306–9,
      311, 313–15
   revelational view of the, 45,
      46, 279–80, 282, 285, 292,
      306–9, 311, 313, 315
   Roman Catholic views of the,
      282–3
   single view, 17, 45–46, 67–69,
      271–300, 272n2, 285n35,
      303–11, 314–15, 324–26,
      331–32, 343
   universal significance of the,
      307n8
incarnationalism, 292n54
incarnation-anyway position, 287–
   91, 289n47, 297
inclusivism, 313, 313n17, 335, 344
Incommensurability Problem,
   259n17
Independence Day (movie), 320, 353
Independence Day: Resurgence
   (movie), 57

indexical communication (signs), 251–52, 255–59
Infancy Gospel of Matthew, 340, 340n47
infinite horizon of being, 169–70, 170n37, 171
infinity, 3, 10, 33, 40, 190, 255, 284
inflation period of early universe, xiv
information processing, 246n2
information sharing, 69, 119, 402
"In Memoriam A.H.H." (Alfred Lord Tennyson), 102n30
instrumental value, 395–98, 405–6
intelligence, 140, 142, 360–61, 397, 421, 423
    as criteria for value, 397
    as measure of evolutionary progress, 421
    defining, 140
    emergence of human, 360
    evolution of, 361
    measuring, 421, 423–26
    moral ambiguity of, 142
    the Naming Test of, 423–26
    the Technology Test of, 423
    the Turing Test of, 423–24
Intelligent Design, 79, 239
intercession for the cosmos, human, 337
interdisciplinary dialogue. *See* multidisciplinarity
International Astronomical Union (IAU), 10
International Council for Science, 118
International Space Station, 117n10
International Year of Astronomy, 67
interpretation, 249–53, 260–65
interpreter of a community, 265
interreligious dialogue, 70–71
Interstellar (movie), 57
intrinsic value (of life), 51, 382, 387–88, 394–99, 405–16
Irenaeus of Lyons, 292, 341–42, 342n54
irreducible complexity, 238
Isaiah, the prophet, 280, 335

Islam
    accountability in, 226
    annihilation in, 225
    divine honoring of humankind in, 223–24
    extraterrestrials (ETI) in, 226
    judgment in, 226
    obligations in, 226
    resurrection in, 225–26
    view of other planets in, 189
Islamic argument for the existence of God, 220
Islamic astronomers, 96n12, 221
Islamic eschatology, 225
Islamic objections to Mars colonization, 408
Islamic prayer, 217
Islamic scientists, 221
Israel, 17, 208, 297–98, 309, 334–35, 337–38, 351, 402
    people of, 335

Jackelén, Antje, 16, 371
Jaeger, Lydia, 372
Japan Aerospace Exploration Agency (JAXA), 155, 401
Jaspers, Karl, 29
Jenkin, Fleeming, 149
Jerison, Harry, 140
Jerusalem, 209
Jewish ethics, 210, 213
Jewish law, 209–10, 212
    classifications of humans in, 210
    in space, 209–10
Jewish theology, 208–15
Jews
    as alien, 211
    as strangers, 211
    as vampires, 211
    European and Christian views of, 210–11
    halachic, 208
Jewish eschatological expectation, 339
Jewish Messianic tradition, 339
Jewish thought, early, 337
Jinn, 217–18, 217n4

INDEX 463

Johnson, Elizabeth, 294, 294n60
Johnson, Mark, 252
Johnston, Robert K., 34n18
Jung, Carl, 252n8
Juno spacecraft, 218n5
"Jupiters" (large exoplanets), 9
justice, 442
James Webb Space Telescope (JWST), 132

Kaku, Michio, 355n20
Kant, Immanuel, 161, 236, 276, 429, 431
Kant's Categorical Imperative, 429
Kantian ethics, 246, 276, 429–31
Kauffman, Stuart, 234–36
Keas, Michael, 43
Kennedy, John F., 146–47, 147n3
kenosis, 275
Kepler, Johannes, 60, 97
Kepler mission, 126–27, 136–38, 141
Kepler-186f (first Earth-like planet discovered in the habitable zone), 9n23, 62
Kepler space telescope, 62, 126–27, 137–38, 156
Kessler syndrome, 400
Kuhn, Thomas, 91, 96
Küng, Hans, 15
Kurzweil, Ray, 424

Lajolo, Giovanni, 61–62
Lakoff, George, 252
Lamarck, Jean Baptiste, 148–49, 148n8
Lang, Fritz, 211
language
  and technology, 257
  as a game, 251
  as the difference between humans and animals, 425
  as the key to intelligence, 426
  emergence of, 425
  structure of, 423–24
Lansdorp, Bas, 408
lasers, 402–3
Late Heavy Bombardment, 139

Laudato Si' (Pope Francis), 71
laws of nature, 45, 82, 307, 315n19
Lederberg, Joshua, 6
Lee, Hak Joon, 46
Lemarchand, Guillermo, 351, 430, 438
Lemonick, Michael D., 76
Leucippus, 161n5
Levin, Jeff, 195–96
Lewis, C.S., 190, 286
Lex Galactica, 430
liberation, moral commitment to, 441
life
  adaptability of, 65
  as a clue to the theological meaning of the universe, 78n17
  as a cosmic imperative, xi–xii, 161
  as a gift, 65
  as a "pearl of great price," 78n17
  as giving significance to the universe, 80
  as more valuable than non-life, 381–89, 406
  earliest evidence for on Earth, 383
  first on Earth, xiv, 139–40
  fragility of, 65, 71
  inevitability of, 48, 76, 111, 154, 176–77, 366–67
  on Mars (*see under* Mars)
  origin of, x–xi, 65
  preservation of, 120, 354, 388, 405, 427
  relative abundance of in the universe, 76–77, 79
  resilience of, 64–65, 139
  reverence for, 232–33, 240, 396
  second (independent) genesis of, xi–xii, 6, 104, 117, 131, 382, 382n4, 384–89, 417, 443
  spreading of as vocation, 388
  value of, 77, 389–89, 405–6
  proceeding from God, 397
  as intrinsic (*see* intrinsic value)
life principle, xi

light pollution, 60
Logos, 284–85, 293, 294, 326, 341
Lossky, Vladimir, 287
Lotz, Johannes, 170–71
love, human, 142
Lovejoy, Arthur O., 161
Lovelock, James, 385
love
    of God, 162–67, 172, 281, 290, 298, 304–5, 309, 311, 335, 431
    of enemies, 434
Lovin, Robin, 431–32
loyalty, 248–49, 263, 266, 420, 441
Loyalty to Loyalty, 263–66
Lucretius, 161n5
Lupisella, Mark, 236
Luther, Martin, 292–93
Lutheran doctrine, 201
Lyell, Charles, 148

Magno, Albert, 59
Magnus, Albertus, 83n44
Maimonides, Moses, 211–4
Malthus, Thomas, 148–49
manifest destiny, 198–201
Manning, Heidi, 177
Man's Place in the Universe (Alfred Wallace), 357
Maragha Revolution, 96n12
Marcy, Geoffrey, 35, 156
Mariner 9 spacecraft, 382
Mars, 381–89, 404–7
    assumption of lifelessness on, 405
    atmosphere of, 128, 382–83, 405
    atmospheric pressure on, 382
    colonizing, 407
    Islamic view of, 408
    discovery of dry river beds on, 382
    early environment of, 385
    Earth-like conditions on, 382
    exploration of, 76
    human settlement on, 408
    life on (search for life on), xii, 120, 154–56, 195, 382, 383, 385–87, 389
    movement of, 96
    polar ice caps on, 154–55
    protection of, 387–88, 394, 404, 406
    rovers (Spirit, Opportunity, Curiosity), 155–56
    seed theory of life on, 383
    temperature on, 382
    terraforming of, 385–86, 387–88, 404–7
    ethics of, 387–88, 405
    timescale of producing oxygen on, 386n17
    warming, 385, 385n16
    water on, 154–55, 382, 383
Mars Attacks (movie), 57
Mars Direct, 407
Mars Exploration Program, 155–56
Mars One project, 408
Mars Society, 407–8
Mars specimens, Draft Protocol for handling, 120
Martian life, ethical status of, 382
Marty, Martin, 195, 364–65
Marty, Peter W., 283
materialism, 20–21, 47
mathematics as a language, 259
Mayor, Michel, 76, 156
Mayr, Ernst, 48, 79, 357–58
McAdamis, E. M., 185–86
McFague, Sallie, 254n11
McHugh, L. C., 283–84
McKay, Christopher, 6, 395, 405, 429
McMullin, Ernan, 275–76, 285–86
meaning, human need for, 363
Medieval era, 90–99, 98, 162, 175, 199, 209, 211–12, 231, 249, 321–22, 340n47
Medieval philosophy, 212
Melanchthon, Philip, 45–46, 322
Mendel, Gregor, 149
Mercury flights, 146
Messiah, the, 331, 337–41, 343, 339n41
messianic age, 339
metanoia, 411
metaphor, 252–53, 253n11, 254n11

INDEX 465

metaphorical theology. *See under* theology
methane, as a biomarker, 130
METI (messaging to extraterrestrial intelligence), 4, 8n18, 119, 183, 352n11, 359
Metropolis (movie, Fritz Lang), 211
Meynell, Alice, 327
Michaud, Michael A. G., 365
microbial extraterrestrial life. *See* extraterrestrial non-intelligent (microbial) life
microbial life, 152, 153, 156
microbiome, definition of, 152n22
microlensing, 127, 138
microorganisms, ethical considerations of, 386–87
Middle Ages. *See* Medieval era
Milbank, John, 432
militarization of space, 403–4
military action, 433
Milky Way galaxy, ix, 112, 112n4, 126, 143, 416
  number of planets in the, 126
  number of stars in the, 112, 112n4
  size of the, ix, 112n4, 143
Miller, Ben, 198n32
Miller, Stanley, 150
Miller-Urey experiment, 6, 139, 150–51
mind
  cosmic, 435–46
  theory of, 247, 255–57
mind reading, 260, 260n20
Mishneh Torah (Moses Maimonides), 213–14
Mix, Lucas, 7–8
Mlodinow, Leonard, 191n16
models as interpretations of myths, 371
modern evolutionary synthesis, 149
Moltmann, Jürgen, 299
Monod, Jacques, xi, 78
moon(s)
  cost of mission to Earth's, 147
  ice-covered, 129
  number of in our solar system, 112
  possibility of life-bearing, 129
  samples from Earth's, 117, 139
moral ambiguity, 22, 142, 305, 320
moral behavior, 262, 309n13
moral capacity, 81–83, 313
moral commitment to protect humans from suffering, 428
moral failure, 84
moral obligation to non-biological or post-human intelligence, 436
moral policies, 427, 442
morality
  evolutionary, 320
  human, 81, 233, 241, 294, 332, 344, 349, 418, 429, 434, 439, 441
Moravec, Hans, 435–36
More Worlds Than One: The Creed of the Philosopher and the Hope of the Christian (Brewster), 103, 103n32
Moritz, Joshua M., 300, 307, 309, 314
Morris, Simon Conway, 153–54, 359–61
moveable stars, 91, 93, 96
multidisciplinarity, 61, 69, 71, 74, 111, 121
multi-generational ethics, 443
multiverse theory, 63, 69
Murphy, George, 294–95, 325n29
Murphy, Nancey, 28
mystery, 40, 52–53
myth(s), 43–44, 47–49, 347–74, 352n12, 353n15, 363, 417, 420, 438
  and fact, 365
  and model, 352n13
  broken, 371
  in science, 23, 40, 43–44, 47–50, 94–98, 199–200, 274, 277, 347–74
  of evolutionary progress, 47, 356–60

myth(s) (continued)
    of the Extraterrestrials, 48–49,
        355
    role of in the human psyche, 355
    symbolic language of, 370
    types of, 352
mythos (myth), 363

Næss, Arne, 395
Nagel, Thomas, 238
Narveson, Jan, 441
NASA, 6–7, 13, 116–18, 121–22,
        132, 146–47, 155–56, 228,
        394, 404, 407, 409
    archives, 117
NASA Astrobiology Roadmap,
        116–7, 121–2
Native Americans, 61, 67, 425
naturalistic fallacy, 81n32
nature as sacred, 234
natural laws as biophilic, xii
natural selection, 38, 82, 149, 199,
        356, 359, 360, 364, 429, 437
natural theology, 148, 236, 238
*Natural Theology; or Evidence of the
    Existence and Attributes of
    the Deity* (William Paley),
        148
naturalism, 49, 202, 233–8, 310, 374
    versus supernaturalism, 228–42
naturalistic cosmotheology, 228–42
Nature and Destiny of Man
    (Reinhold Niebuhr), 233
nature, philosophical interpretation
        of, 78
"nature red in tooth and claw"
        (Alfred Lord Tennyson),
        102n30
navigation, stellar, 95–96
Neal, Mark, 437–38
Neanderthals, xv, 257n13
Near-Earth Object Working Group
        of the UN, 410
nearest star, distance to the, 112
negative theology, 208, 214–5
    of Maimonides, 214

neo-Darwinian model of evolution,
        50, 150
neo-geo-centrism, 201–2
nephesh, 333n8
Nesteruk, Alexei, 296
New Creation, the, 294, 299–300,
        306–7, 312, 339
New Jerusalem, the, 299
new religious movements, 12
Newton, Issac, 66
Nicene Creed, 212n1
Nicholas of Cusa, 83n44
Niebuhr, Reinhold, 84, 233
nihilism, cosmic, 78
nitrogen, 385n13
Noah, 335, 342
non-human life, ethical standing
        of, 386
non-religious views of religious
        crisis, 194–95
Nostra Aetate, 283
nothingness, 170, 170n42
novelty within continuity, 85
nuclear war, threat of, 141

Of the Plurality of Worlds (William
        Whewell), 102
O'Meara, Thomas F., 15, 229, 278,
        290, 291
Omizzolo, Alessandro, 16
*On the Origin of Species by Means of
    Natural Selection* (Charles
    Darwin), 38, 148–49
*On the Revolutions of the Celestial
    Spheres* (Nicolaus
    Copernicus), 200–201
ontological thirst, 34, 40, 41
ontological understanding, 169
Oparin, Aleksander, 150
Optical SETI, 8, 115
Ordinatio (John Duns Scotus), 165
Oresme, Nicole, 83n44
origin of life
    as a scientific problem, x–xi, 65
    question of the, 103–4
Osiander, Andreas, 289n47
other, conceiving of oneself as,
        260–61

INDEX 467

other worlds, question of, 59–64
Our Final Hour (Martin Rees), 71
outer space, definition of, 5
Out of the Silent Planet (C. S. Lewis), 286
Outer Space Treaty of the United Nations (1967), 117, 118n13, 393–94, 403
oxygen as a biomarker, 130

pagan astrology, 12n33
paganism, Christianity as, 12n33
Pale Blue Dot (Carl Sagan), 238
Paley, William, 148
Pannenberg, Wolfhart, 284, 285, 287
parable of the lost sheep, 68
paradigm shift, 91, 96, 101
particularity, scandal of, 298
Passive SETI, 8n18
Pasteur, Louis, 151
pathos of God, 166
Patristic period, 307, 321–22, 339–40, 341–43
peace, 339–40, 344, 349, 363–64, 366, 372–73, 402–3, 432–33, 442
  as an act of God, 373
  importance of education for, 70
peaceable kingdom, 338–40
Peacocke, Arthur, 80, 237, 272
Peck, M. Scott, 263
Peirce, Charles S., 251, 252n8, 255n12
Perelandra (C. S. Lewis), 286
Persson, Erik, 398
person
  as moral end, 276
  definition of, 209
  in Jewish thought, 209
persons and non-persons,
   distinction between, 209
personhood of extraterrestrials, 273–77, 281
Peters, Ted, 63, 75, 83, 236–37, 303, 305, 306, 307–10, 307n8, 314–15, 322, 327

Peters ETI Religious Crisis Survey, 83, 184, 187–97, 204–5
Pettinico, George, 185
philosophy of religion, 313n17
photosynthesis, 130, 139
Physics and Metaphysics (Aristotle), 212
Pittenger, Norman, 272
place in the universe, our, ix–x, 43, 63–64, 68, 94, 97–98, 103, 109–12, 122, 191, 218, 228–29, 231–32, 234, 238, 241, 323
planet(s)
  formation, 62n19, 64, 137
  number of in the universe, ix, 37, 115, 126, 154, 176, 367
  number per star, 61, 126, 136, 138, 141
planetary consciousness, 402
planetary defense, 433–34
planetary ecosynthesis. See terraforming; see under Mars
planetary protection, 117–8, 120–1, 391–5, 413
planetary readiness, 52
planetary society of moral deliberation, single, 51, 204, 393, 411–13, 418, 444
planet-hopping Christ, x, 45, 197, 272–73, 275, 277, 292, 331
  as absurd, 197, 272, 275
planetism, 42
planetization, 411, 413, 418–19
Plato, 10, 92, 93, 276
pluralism, 92–93, 98, 204, 249, 313, 313n17
  cosmic, 92–93, 98
plurality of worlds debate, 45, 59, 83, 93, 100, 102, 161
Pluto, naming of, 10
policies for space exploration and colonization, 52, 117–20, 392–95, 399, 404–5, 412, 427–28, 431, 442
Polkinghorne, John, 30, 278
Pontifical Academy of Sciences, 61
Pope Benedict XVI, 67, 68, 277

Pope Francis, 57, 64, 65, 70–71, 70n38, 193
Pope John Paul II, 65, 277
Pope Leo XIII, 15
Pope Paul III, 67
post-alien machines, 436
post-biotic life, 398, 417, 434–36
post-Enlightenment, 440
posthumanism, 434–36
power, need to limit human, 71
practicing theology without a license, 49, 348–50, 352, 354, 356, 363
Precautionary Principle, the, 399
pre-Copernicans, 43, 102–3, 198–99, 201–2
pre-Darwinian models for the origin of life, 100
predatory causes, 264
predestination, 163–64, 163n12, 175n60
preferential option for life, God's, 396–97
Preus, Robert, 201
Primack, Joel, 42
prime directive, 395, 418
primordial soup, 150–1
principle of mediocrity, 82, 94, 200
principle of plentitude, 161–2, 161n5, 165n19
principle of reciprocity, 430
Principles of Geology (Charles Lyell), 148
problem of distance, the, 308–9, 312
process theology, 239, 281, 310
profit, 400, 404, 408, 411, 442
  as a barrier to ethics, 411
  mining space for, 404
progress
  as illusion, 373–4
  belief in, 39, 48–49, 348–49, 356–74, 421–22, 435
  doctrine of, 49, 348, 359–60, 365, 373, 422
  myth of, 372–74, 434
progressive entelechy, 48
progressive evolution, 361, 434
prolepsis, 297, 312, 340

protecting space sites, 404
Protoevangelium of James, 340
Proxima Centauri b (exoplanet), 8, 62
Ptolemaic model of the cosmos, 14, 63, 90–97, 101, 221; *see* geocentrism
Ptolemy, Claudius, 90–91
public policy decisions, 52, 392–93, 412
Puccetti, Ronald, 274–6
purpose(s)
  in the evolutionary process, 48–49, 359
  lack of cosmic, xi, 49
  of God, 44, 80–81, 193, 278, 288, 326, 334
  of humanity, 337, 339
  of the universe, 4, 19, 48, 63, 288

quarantine facilities, 117
Queloz, Didier, 76, 156
QUESS (quantum mechanics at the Space Scale), 6n13
questions as forming answers, 426
Qur'ān, 216–21, 217n4, 223–26, 310, 408

Race, Margaret, 7, 391, 394–95, 432, 433
radio astronomy, ix, 8, 36–38, 40, 76, 114, 119, 131, 134–35, 140–43, 254, 258, 348, 364
Radio SETI, 8
Rahner, SJ, Karl, 34–35, 161, 164–77, 169n33, 170n41, 172n52, 174n58, 276, 289, 290n51, 325–26
Randolph, Richard, 7, 394–96, 429, 432, 433
Rare Earth (Ward and Brownlee), 61
rare earth position, 16, 77, 147, 153–54, 294n60, 356–59
rationality, 52, 70, 81–81, 276–77, 294, 313, 318–20, 332–33, 344, 438
Reagan, Ronald, 402–3
reciprocity, 412, 418–9, 428–30, 440

INDEX 469

reductionism, 20–21, 47, 79, 81,
  169, 173
  arguments against, 82
reductionist versus non-reductionist
  philosophies, 79
redemption, 18, 28, 32, 39, 46, 48–
  49, 67, 69, 80, 86, 163, 237,
  271–300, 303, 305–9, 312,
  320, 323–24, 331, 341–43,
  369, 374, 431, 438, 448
  as an ontological act, 306
  of all biological life, 331
  of all creation, 300, 307
  story of not translatable to ET,
    309
reductionism, ontological, 47
Rees, Martin, 63, 71, 436
Reformation, 175n60
Regis, Edward, 354
Reinventing the Sacred (Stuart
  Kauffman), 234
relationship with God, 14, 17, 30,
  83, 161–62, 164–66, 168,
  172, 174, 177, 184, 193, 223,
  267, 272, 274, 276, 290, 326,
  333, 371
religion
  and science (*see* science and
    religion)
  as cause of destruction, 354
  as cause of warfare, 85n48
  as open to existence of
    extraterrestrial life (ET),
    205n43
  future of, 70
  hostility toward, 196
  role of in society, 70–71
*Religions and Extraterrestrial Life:
  How Will We Deal with It?*
  (David Weintraub), 229
religious myths and models, 352n13
religious naturalism, 233–39,
  233n15
religious sensibility, secularized, 365
religious symbols, 253n11
remote sensing, 129, 401–2
  of exoplanets, 129
repentance, 254n24, 411

Resnik, Judith, 208
responsibility for the welfare of
  extraterrestrial intelligent
  life (ETI), human, 267, 395,
  398–99, 413, 426–28, 431
resurrection, 95, 216, 218, 225–26,
  294–96, 299–300, 307, 308,
  310, 312, 321, 324, 325, 342,
  372, 373
  deep, 294
  general, 299–300
  human participation in the, 308
retrograde motion of the planets, 96
return to the Creator, universal, 225
revelation, divine, 14, 28–29, 45–46,
  57, 67–68, 160–77, 197,
  223–24, 226, 237–38, 277,
  280, 282, 285, 292, 297–98,
  306–15, 324
revelational view of the incarnation,
  280, 297, 306–9, 313, 315
reverence, 447
risk, incorporation of, 407
risk management, 437
robotic astronauts. *See* cyborg
  astronauts
Rolston, Holmes, III 52, 396
Roman Catholic Church, 61, 64, 67,
  68, 94, 95, 97, 192, 193, 202
  attitude of towards scientific
    progress, 70n38
Rosenzweig, Franz, 215
Roy, Rustum, 21
Royce, Josiah, 248, 263–5, 264n24
Ruether, Rosemary Radford, 201–2
Rummel, John, 393
Ruse, Michael, 81–82, 361, 365, 369,
  428
Russell, Bertrand, 78
Russell, Mary Doria, 320
Russell, Robert John, 20, 29, 44, 147,
  272, 286, 320, 333

Saarinen, Risto, 429
The Sacred Depths of Nature (Ursula
  Goodenough), 234
sacrifice, Christian, 163, 289, 325

Sagan, Carl, 6, 10, 36, 48, 84n47,
    238, 252, 351, 353, 367–69,
    368n59
salvation, 39, 45, 49, 68, 163, 281–
    82, 299, 306, 317–27, 331,
    337–38, 341–44, 342n54,
    349, 368, 436
  meaning of, 318
  of non-human creatures, 331,
    337–38, 341–43
  theologies of, 318–27
  through progress, 349
  universal, 306, 343
salvation history, cosmic, 341
Sandoz, SJ, Emilio, 320
The Sand Reckoner (Archimedes),
    96n12
Sanford, John, 355
Satan (Lucifer), extraterrestrial life
    (ETI) and, 188
satellites, 400–4
  broken, 400
satisfaction theory, 162–63
scarcity and value, 77n17
Scharf, Caleb, 198–201
Schwarz, Hans, 50
Scholasticism, 174n57
science
  and the meaning of life, 369
  as ambiguous, 373
  as broadening theology, 65
  as giving humans significance,
    199
  as giving total control, 191n16
  as giving total understanding,
    191n16
  as increasing theological truth,
    65
  as redemptive, 49
  as salvific, 349–59, 366–67,
    438–39
  assumptions in, 29–31, 38–41,
    45, 47, 49, 79, 116, 141, 143,
    155, 186, 240, 248, 277, 320,
    356, 358–59, 361–62, 364,
    366, 417, 420–22, 426, 428,
    434, 437, 438, 440, 441, 443,
    448

  limits of, 20–21, 30, 40, 47–50,
    62, 129, 143, 308, 362, 363,
    372, 417, 419, 448
  theological interpretation of, 34
  faith-affirmation in, 365
science and atheism. *See* atheism
    and science
science and religion, 10–20, 203,
    239; *see* Center for Theology
    and the Natural Sciences;
    bridge metaphor for relation
    of theology and science;
    creative mutual interaction
    between theology and
    science
  conflict between, 203
  contrast between, 203
  convergence between, 203
  disputational friendship model
    of relating, 19n53
  faith as the difference between,
    239
  the field of, 74
  independence model of relating,
    19
  no absolute break between, 75n4
  seeking common ground
    between, 239
  two language model for relating,
    19
science fiction, 56–57, 320, 353
science-oriented theology, 235
*Science, Religion, and the Search for
    Extraterrestrial Intelligence*
    (David Wilkinson), 229
science saves myth, 349–50, 355,
    364, 374, 438
scientist(s)
  as gnostic redeemer, 349
  as priest, 369
  as the most highly evolved
    human, 348–49
  as the pinnacle of creation, the,
    48
  role of, 348
scientist-centrism, 48
scientific integrity, 23, 62, 117, 394,
    395

scientific progress, ambivalence of, 372
scientism, 20–22, 40, 47, 49, 70n38, 231, 231n10
scientized myth, 39, 48, 363, 365, 366, 370; *see* demotion myth; ETI myth; gnostic redeemer myth; science saves myth; secular redemption myth, SETI myth; *see under* Copernican Principle; Copernican Revolution; evolution; myth; progress
scope of creation, 41, 44–46, 91, 197, 204–5, 331, 448
speed of light as a limit, the, 71, 245, 308, 417, 419
scope of theology widening with new knowledge, 45
Scotistic doctrine of God, 161–77
Scotus, John Duns, 161–67, 172, 288
Seager, Sara, 63–64
Secchi, Angelo, 60, 64–65
Second Temple Judaism, 331, 337-338
secular eschatology, 364–5
secular experience. *See* hermeneutic of secular experience
secular humanist views of sin, 84n47
secular redemption myth, 39–40
secular religion, 20–22
secular thinking, 31–32
Seiende (concrete being), 168–70, 169n33
Sein (being), 168–70, 169n33
Seinsfrage (question of being), 169, 169n33
self-awareness, 319
self-communication of God. *See* divine self-communication
self-consciousness, 80, 161, 162n7, 165, 168, 172–74, 176–77, 318–19
  emergence of, 319
  evidence for of non-human animals, 162n7, 165
"selfish gene," 440

selfishness, human, 319
self-organization, chemical, xi
self-reflection, 174, 266, 318, 352n13
self-transcendence, active, 164, 173–77, 174n58
self-understanding, 31, 34, 42, 44, 63, 169, 205
semiotics, 245–67, 246n2, 424
sentience as criteria for moral status, 398
SETI, ix, xii, 4, 8, 15, 36, 38–40, 47, 48, 76, 77, 114–15, 116, 119, 131, 183, 203, 248, 261, 265, 266, 320, 348–56, 358–66, 419, 421, 426, 438, 440, 443, 447
  Active (*see* METI)
  as a secular religion, 355–6, 366
  as a waste of money, 358
  as likely to fail, 358
  Passive (*see* Passive SETI)
  Principles, 119, 121, 419
  Radio (*see* Radio SETI)
  research leading to increased knowledge of space, 358
SETI myth, 351n11
Shabbat in space, observance of, 208–9
Shma, 212, 212n1
Shannon information, 249
Shapley, Harlow, 231
shared interpretations, 246–67
shared spirituality, 245–67
shared vision, 263–6
shared worldview, evolution as a, 241
Shermer, Michael, 21, 142
Shindell, Matthew, 187
Shostak, Seth, 8, 114, 423
shukūk (doubts), 221
Sībawayh, Abū Bishr 'Amr b. 'Uthmān b. Qanbar, 219
signals, interpretation of, 249–51
signs on Pioneer 10 and Pioneer 11 spacecraft, 252
Simpson, George Gaylord, 47–48, 147

sin, 84, 84n47, 163, 280, 285, 304–5, 304n2, 318, 320, 406
   evolutionary origins of, 318
   limits due to, 372
   original, 319
Sīnā, Ibn, 222
Singer, Marcus, 430
single-celled organisms, xiv, 140, 386
singularity (t=0), 80n29
size of universe relative to size of neurons in human brain, 61
slave ethics in the New Testament, 440
slavery, 52, 211, 264, 310, 428, 435, 437–38, 440–42
   in ants, 437
Smith, Huston, 373–74
Smith, John Maynard, 359
Smith, Kelly, 247n5, 397
social cohesion, 262–64
social convention, 252
social intelligence and value, 247n5
social psychology of self-defense, 433
sociobiology, 81
sodium in atmosphere of exoplanets, detection of, 130
solar flares, 409–10
Solar System
   exploration of our, 117
   size of the, 112
Solar Wind Magnetosphere Ionosphere Link Explorer (SMILE), 6n13
solidarity, 341n52
soteriological motifs in the early church, 321
soteriology, 41, 45–46, 191, 197, 282, 296, 297, 313, 317–18, 321–27, 332, 448
soul
   as distinction between living and nonliving, 209
   in Jewish thought, 209
   no immaterial, 333
Soviet Union, space missions of the, 155

space culture, 420
Space Debris Office, 401
space ethics. *See* astroethics
space exploration, ethical issues related to, 50–52, 120–22, 266, 350, 381–89, 391–413, 416–44
space junk, 400–401
space missions, 6n13, 117–19, 117n10, 126, 135–38, 146–47, 155–56, 228, 393–94, 400–402, 408
spaceship Earth, 411
space tourism, 120, 403–5, 413
The Sparrow (Mary Doria Russell), 320
special creation, x, 99, 103, 333
special providence, 22
spectroscopy, 62, 114, 116, 129–30
speculation in science, ix, x, 9, 14, 39, 49, 64, 67, 109, 276–77, 349, 362–63, 367, 369, 417
speed of light as limited, the, 419
Spencer, Herbert, 437
spiritual communities, 264
spirituality of extraterrestrials, 248, 262–67, 350
spontaneous generation, 151
stable environment needed for life, 128–29
star dimming caused by planetary transit, 8, 126
star formation, history of, xiv
Starman (movie), 320
Star of Redemption (Franz Rosenzweig), 215
Star Trek (TV series), 57, 395
Star Wars (movie), 57
state, end of the modern, 418
Stendahl, Krister, 191
Stenmark, Lisa, 19n53
stewardship, 396
Stewart, Matthew, 235
Stoeger, William, 76, 78, 82
Stone, Jerome A., 233–5
Strategic Defense Initiative (SDI), 402
Strom, Robert, 37

INDEX 473

"struggle for existence," 437–40
suffering, 294–95
Summa Theologica (St. Thomas Aquinas), 93
Sungenis, Robert, 202
super-Earths, 127, 130, 132
supernatural, the idea of the, 230
supernatural existential, 172n52
suprapluralism, 313
supralapsarianism, 289n47
surveys about religious beliefs and extraterrestrial life (ET), 183–205
survival, human, 70, 366, 391, 394, 397, 407, 411
"survival of the fittest," 437
suwar (section), 217n2
symbolic communication, 245–67, 252n8
symbolic conventions, 246–47, 251–52, 257
symbolic language in theology, 371
symbolic representation, 425
symbolic thinking, 425
synthetic cells, 151
synthetic genome, creation of a, 151
Systematic Theology (Tillich), 279

Tanzella-Nitti, Giuseppe, 66
Tarter, Jill, 85, 85n48, 185, 354
Tattersall, Ian, 422
Taylor, Charles, 200
Taylor, John, 166
technological adolescence, 351, 438
technological progress, 435
technology
  and abstractions, 256n12
  and destruction, 352n11
  and language (*see under* language)
  and social change, 349
  and survival, 141n24
  and symbolic language, 254–55
  and symbolic thought, 248
Teilhard de Chardin, Pierre, 239, 290, 306, 324
telecommunications industry, 403
teleology, 47–49, 63, 78–79

"teleology without teleology," 78
telescopes, ix, xiii, 8,–9, 40, 61–62, 76, 95, 97, 110, 114–5, 125–32, 133, 137–38, 298, 401; *see* Hard X-ray Modulation Telescope; Hubble space telescope; Kepler space telescope; James Webb Space Telescope
Teller, Edward, 402
telos of creation, 163–64
Ten Commandments, 212
Tennyson, Alfred Lord, 102n30
terraforming, 121; *see under* Mars
terrestrial adolescence hypothesis, 367n58
terrorism, 442
TESS observatory (Transiting Exoplanet Survey Satellite), 132
tethering technique for debris-gathering, 401
Thales of Miletus, 429
Thank God for Evolution (Michael Dowd), 237
theism as explaining more than atheism, 30
theistic evolution, 50, 79–83, 102n3, 290, 318
theocentrism, 44, 68, 200, 288n44, 299
theologians
  as correcting mistaken religious views, 47
  as critiquing extra-scientific assumptions, 23, 41, 47, 49, 248, 347–74, 448
  as interpreting the value of science, 47
  duty of to keep abreast of science, 65
theological anthropology, 273, 351
theological attitude towards science, 65
theological critique of science, 30
theology
  adaptability of Christian, 195
  as a challenge to science, 23

theology (*continued*)
  as a science of science, 34–35
  as not fixed, 65–66
  as open to self-revision, 45
  as "sacred science," 30
  and science (*see* science and religion)
  metaphorical, 254n11
  of becoming, 161, 162, 164, 167–69, 173–76
  of culture, 34
  of hominization, 173
  of nature, 4, 13–14, 19–20, 22, 23, 27, 29–30, 75
  process (*see* process theology)
  revising (in light of science), 67
  significance of extraterrestrials (ETI) to, 203
  transcendental (*see* transcendental theology)
  theory of mind. *See under* mind
theosis, 287, 321
Thomistic theology, 160
thought experiments, 110, 436
threats to Earth from outer space, 409–11
Thutmose IV of Egypt, 336
Tillich, Paul, 34, 252n8, 279–81, 285, 287, 298, 324
time, deep, 5, 33, 148, 387
Titan (moon of Saturn), 156
Torah, 212
Townes, Charles, 19, 403
Tracy, David, 32
tradition (as source of theology), 28–29, 91, 262–63
transcendence of God, 80, 200–201
transcendental consciousness, 165, 168–73
transcendental experience, 169, 171–72
transcendental theology, 162n7, 165, 236
transformation of the cosmos, 294, 297–300, 306–7, 312, 373
transhumanism. *See* posthumanism

transit method of exoplanet detection, 9, 126, 137–38, 156
transiting systems, 126–27, 130, 132
transparency, 46, 356, 402, 419
Traphagan, John W., 353n14
truth as goal of theology, 22, 28, 45, 46, 66, 70, 113, 297, 313n17
Turing, Alan M., 424
Turing test. *See under* intelligence
Ṭūsī Couple, 221
Twomey, Vincent, 321
Tyson, Neil deGrasse, 4, 6, 37, 48, 142n29, 205, 368, 409

UFO phenomenon, 48, 48n61, 83n43
ultimacy, 31, 33–35, 40, 41
ultimate reality, 34–35, 41
unique earth position. *See* rare earth position
Unitarian tradition, 235
United Nations, 399, 403, 432
United Nations Conference on Environment and Development, 399
unity, human, 57, 263, 284, 313, 411, 444; *see* community
unity in difference, 171, 173
Universal Declaration of Human Rights, 432
universal efficacy of Christ, 284, 307
universal eschatology, 311–12
universal ethic, 430
universalism, 335, 344, 418
universal salvation, 225, 283–85, 292, 295, 298, 305, 306–7, 309, 325, 335, 344
universe
  age of the, 308n10
  as biophilic, 7, 153–54, 359–60
  as sterile, xi, xii
  history of the early, xiv, 64
  no center to an infinite, 59
  observation of as spiritual encounter, 4
  size of, xiii, 308, 308n10
utopia, 49, 348, 355, 359, 369, 374

INDEX   475

Urey, Harold, 150
US National Astronomical
    Observatory, 61
utilitarian value. *See* instrumental
    value

V (TV show), 353
Vakoch, Douglas A., 8n18, 141n24,
    186, 259n17, 351n11, 443
Van Gogh, Vincent, 60
van Huyssteen, Wentzel, 29
variables in the Drake Equation,
    135, 259–60
  f_c (the fraction of intelligent
    civilizations that develop
    technology that releases
    detectable signs of their
    existence), 140
  f_i (estimate for the fraction
    of inhabited planets
    where intelligent life does
    eventually develop), 140
  f_l (the fraction of suitable planets
    on which life actually
    appears), 139–40
  f_p (the fraction of stars that have
    planets), 138, 141
  L (the length of time that an
    intelligence communicating
    will send out signals that
    we can receive and know of
    their existence), 135, 141–2
  N (number of communicating
    civilizations which could be
    present in our galaxy at this
    point in time), 135, 142–3,
    142n29
  n_e (number of planets per solar
    system with an environment
    suitable for life), 138, 143
  R* (the rate of formation of stars
    in our galaxy suitable for the
    development of intelligent
    life), 135
*Vast Universe: Extraterrestrial Life
    and Christian Revelation*
    (Thomas F. O'Meara), 229

vastness of the universe, xiii, 16, 33,
    42, 44, 61, 71, 111, 115, 131,
    136, 163, 168, 176, 190, 191,
    228, 231, 238, 309, 323, 325,
    325n29, 348, 350
Vatican Observatory, 15–16, 61, 67,
    68, 95, 193
  searching for extraterrestrials,
    193
Vatican II (Second Vatican Council),
    13, 69, 283
Venter, Craig, 151
Venus, atmosphere of, 128
*Vestiges of the Natural History
    of Creation* (Robert
    Chambers), 102, 102n31
Victorian views of extraterrestrial
    life, 102
violence, 85n48, 304, 304n2, 351,
    373, 433
  human, 84–85
Virchow, Rudolf, 151
Voegelin, Eric, 354
voluntary action and moral
    judgment, 209–10
Vorgriff auf esse (pre-apprehension
    of being), 169–72
Vorilong, William, 322
Voyager 1 spacecraft, 112, 238

Wallace, Alfred, 153, 356
Ward, Peter D., 61
"warm little pond," 150
war(s), 85, 85n48, 141, 142, 239,
    241, 252–53, 310, 353–54,
    363–64, 372–73, 402–3, 433,
    438
  religious, 239, 241
War of the Worlds (movie), 57
waste heat, searches for
    extraterrestrial, 115
water
  as needed for life, 8, 9, 101, 116,
    118, 127, 128–30, 132, 137,
    139, 154, 155, 383, 385, 388
  in our solar system, 118, 129
  on exoplanets, 9n23
  on Mars (*see under* Mars)

water bears, 152–53
water vapor as a biomarker, 130
weapons, 141, 364, 372, 402–3, 413, 441–42
    space-based, 402
Weinberg, Steven, 78
Weintraub, David A., 205n43, 229
Welker, Michael, 47
Wendt, Alexander, 418
Wenham, Gordon, 333n8
Wesleyan Quadrilateral, 28–29, 91, 262
Westermann, Claus, 333
Whewell, William, 102–3, 323
Whitehead, Alfred North, 239
Wiesner, Matthew, 203
Wilkinson, David, 14, 42, 47, 196, 229, 237, 281, 306
Wilson, E. O., 81, 235, 369
Wilson, Elspeth, 392

Wittgenstein, Ludwig, 251
Wolf-Chase, Grace, 7
wonder, 34, 52, 64
world as God, 234
worldview construction, 51, 238–39
Worthing, Mark, 18, 283
Would You Baptize an Extraterrestrial? (Guy Consolmagno), 229
Wright, N. T., 44

Yilmaz, Mehmet Nun, 194
Young, Simon, 21

Zohar, the, 209
Zoonomia (Erasmus Darwin), 148
Zubrin, Robert, 407

Scripture Index

OLD TESTAMENT

Genesis
    60, 65, 193, 209, 319, 336–37, 339
1   332–33, 339
1:26–28   319n8
1:26–27   336
1:26   333
1:27   304
1:28   336
1:31   66
2   319, 333
2–3   319, 319n8
2:7–25   319n8
2:7   65, 333n8
2:15–16   336–37
2:19   424
9:6   304
12:3   335

Exodus
20:2   212

Deuteronomy
6:4   212

Psalms
8   64
8:3–4   4
8:4–5   68

Isaiah
1   339
11   339–40
40:5   342, 343
42:1   335
43:19   299

Hosea
2   339
2:18   338

❧

NEW TESTAMENT

Matthew
     339
1:1   339
5—7   433
7:12   429
13:45–46   78n17

Luke
3:6   342
23:35   339n41

John
1:14   314, 330, 341
1:14a   293
3:16   298, 311, 343, 431
12:27–28   289

Romans

2:13	18
5:18–19	321

1 Corinthians

15:20	299

1 Peter

2:18	440

Revelation

21:3–4	299
22:17	65

www.ingramcontent.com/pod-product-compliance
Lightning Source LLC
Chambersburg PA
CBHW021230300426
44111CB00007B/492